Indonesia Malaysia Research Consortium Seminar (IMRCS 2018)

Development Research Potential in Renewable Energy

IOP Conference Series: Materials Science and Engineering
Volume 588

Surabaya, Indonesia
21 – 22 November 2018

ISBN: 978-1-5108-9461-7
ISSN: 1757-8981

Printed from e-media with permission by:

Curran Associates, Inc.
57 Morehouse Lane
Red Hook, NY 12571

Some format issues inherent in the e-media version may also appear in this print version.

Institute of Physics
Dirac House, Temple Back
Bristol BS1 6BE UK

Phone: 44 1 17 929 7481
Fax: 44 1 17 920 0979

techtracking@iop.org

Additional copies of this publication are available from:

Curran Associates, Inc.
57 Morehouse Lane
Red Hook, NY 12571 USA
Phone: 845-758-0400
Fax: 845-758-2633
Email: curran@proceedings.com
Web: www.proceedings.com

Table of contents

Volume 588

Indonesia Malaysia Research Consortium Seminar 2018 (IMRCS 2018)

21–22 November 2018, East Java, Indonesia

Accepted papers received: 8 July 2019
Published online: 20 August 2019

Preface

Papers

Energy

Engineering

Material

IOP Conf. Series: Materials Science and Engineering **588** (2019) 011001 doi:10.1088/1757-899X/588/1/011001

PREFACE

IMRCS 2018 stands for Indonesia Malaysia Research Consortium Seminar 2018 that will be held on 21-22 November 2018 in Surabaya, Indonesia.

I'MRC was established in Jakarta on 22nd December 2014 when the delegation from the Universiti Teknologi Malaysia (UTM) held a meeting with the Indonesian Minister of Research, Technology and Higher Education. In that bilateral talks, the UTM delegation proposed a high Indonesia-Malaysia Research Consortium. Subsequently, another meeting was held in Jakarta on 10th July 2015 between the UTM delegation and the Indonesian Agency for the Assessment and Application of Technology (BPPT) to establish Indonesia-Malaysia Research Consortium. To strengthen this strategic relationship, I'M Research Consortium Forum was held at UTM campus in Kuala Lumpur on 18th May 2016. The Forum was attended by more than 70 delegates from more than 30 institutions of higher education from Indonesia, Malaysia, Japan, Taiwan, Thailand, Vietnam and others.

Recently, the Malaysian minister of higher education, Dato' Seri Idris Jusoh, and the Indonesian Minister for Research, Technology and Higher Education, Mohamad Nasir, led a bilateral meeting on 2nd October 2017 in Jakarta to discuss further cooperation within the I'MRC platform. UTM from Malaysia and Institut Teknologi Sepuluh Nopember (ITS) from Indonesia have been asked to act as coordinators for this cooperation.

As the continuation of the initiative and to refine the strategic issues on cooperation in the fields of higher education and research between Indonesian and Malaysian institutions, I'MRC Seminar will be held on 21st -22nd November 2018 in Surabaya, Indonesia. This event will be open not only for institutions from both Indonesia and Malaysia but also for the partners from both countries.

The theme of this international seminar is "**Development Research Potential in Renewable Energy**" involving the development of Advanced vehicles, Biomass energy, Bioenergy, Gasification systems, renewable energy, Hydroelectric power, Hydrogen energy, Ocean thermal energy conversion, Solar energy, Wind energy, ICT for energy, Infrastructure for energy and related interest with these topics

Peer review statement

All papers published in this volume of *IOP Conference Series: Materials Science and Engineering* have been peer reviewed through processes administered by the proceedings Editors. Reviews were conducted by expert referees to the professional and scientific standards expected of a proceedings journal published by IOP Publishing.

IOP Conf. Series: Materials Science and Engineering **588** (2019) 012001 doi:10.1088/1757-899X/588/1/012001

Criteria analysis in the selection method of mud pumping in coal mining project

Aryoseno[1, a)] and Ervina Ahyudanari[2, b)]

[1]Magister Management of Technology, Institut Teknologi Sepuluh Nopember, Jl. Tjokroaminoto 12, Surabaya Indonesia

[2]Department of Civil Engineering, Institut Teknologi Sepuluh Nopember, Jl. Arif Rahman Hakim, Sukolilo, Surabaya, Indonesia.

[a)]aryoseno.1984@gmail.com
[b)]ervina@ce.its.ac.id

Abstract. The process of discharging water from the pit is one of the stages prior to the continued excavation of coal deposits. Due to the increase of coal price, several areas that were not economically valuable became visible to re-mined. In some mine ponds that are considered economically viable, are necessary to pump the mud before carried out the water before further digging. This requirement needs decision-making for the determination of sludge pumping methods. A large number of criterion variables for selecting the best mud pumping method drives this study. The distance of three mine ponds under the company is quite far each other. Some steps were executed to have the best method to cater to the problem. The results indicate that there are six criteria that suitable to be considered in selecting a method for mud pumping. Those are the productivity of pump, man power, mobilization, supporting facilities, flexibilities, and operational cost.

1. Introduction

Coal become one source of energy resources. Most of the power plants in Indonesia requires coal as the source of energy. Coal mining becomes an attractive business for some mining companies. The mining cycle starts from the clearing, top soil, overburden excavation, mining material excavation, and the transportation of minerals. [1]. Pumping, both mud and water are carried out when there are puddles generated from rain that cover the surface of the area to be excavated.

There is a coal mining company in South Kalimantan which is a government contractor based on Law No. 4 of 2009 having the permission of PKP2B (Coal Working Agreement of Coal Mining) of generation 1. After the world energy climate is gradually improving and coal prices are rising again, a company plans to do redeployment (cutback) at locations that were previously not economically valuable. The temporarily abandoned location is currently covered by rainwater and there is mud in its basin as a result of the sedimentation process. To continue the mining sequence in the area that is considered economical, this water and mud must be removed from the mine by pumping. Because muds have different densities or densities with water, special pumps are required to remove the mud material because the specifications of conventional water pumps are not capable of pumping. From the new mine economic analysis, there are several sumps or voids in the three work sites that are still economically

IOP Conf. Series: Materials Science and Engineering **588** (2019) 012001 doi:10.1088/1757-899X/588/1/012001

viable. Therefore, water and mud pumping are required to extend the life of the mine and to keep the company sustain.

From the initial meeting session, the top management of the company decided that the method for the pumping option would only choose one of three alternatives. Based on Article 7 paragraph (2) the letter "o" of Law Number 1 of 2004 on State Governance all asset purchases are state-owned. Therefore, this dives this research to focus on the best method of mud pumping where the main component of the drum pump (drag-flow) will be held through the rental method.

The three alternatives proposed are pumping by installing a mud pump (drag-flow) on the excavator (excavator) which is then transported to the dumping site by means of a haulage dump truck (HD). The second alternative is pumping by installing a mud pump on a pontoon vessel, where pumped mud is flowed through a pipe from the ship directly to the dump. The third alternative is pumping by installing a drag-flow pump on the excavator (excavator) and the tool is raised to the pontoon then where the pumped mud is flowed through the pipe from the ship directly to the dump. The explanation of these three alternatives is briefly shown in Table 1. The selection of these three alternatives is based on several things such as the availability of supporting tools (excavator/haulage dump truck, and sewer pipe) and the method of pumping sludge as mentioned above has been applied to many other companies.

Due to the demand from the top management for pumping mud is not only in job location Y allocations, but it can also be used in location X and Z. Job location, where the three locations are located far apart from each other, there are several variables that can influence the outcome of the decision. The criteria for selecting the mud pumping method are not based solely on operational cost calculation variables. This is caused by the designation of the mud pumping method to be used on three different job sites so that other criteria (non-cost) variables appear that must be considered in the decision-making process.

Because of the many criteria considered, and each of these criteria is considered to have no relationship with each other, the determination of decision making will be appropriate when using the Analytical Hierarchy Process (AHP) method introduced by Saaty [2].

According to Ghazinoory [3], AHP is a method that is well known in solving decision-making problems. In this method, paired comparisons are carried out by decision makers to determine the parameters weight in a problem. This weighting helps decision-makers in choosing the most superior alternatives. In this study involves many criteria variables as a basis for choosing the best alternative pumping method. With AHP estimated, it can facilitate decision makers with well-organized perceptions and assessments that will influence decision making. Saaty [4] in Kursunoglu & Onder [5] also said that the strength of this AHP is to make the basis of elections that were previously general and poorly controlled to be detailed and can be controlled. Likewise, this method can make decision makers make comparisons between each of the criteria and alternatives so that there is a reciprocal ratio between each criterion and the alternatives compared. This is in accordance with this study where the basis for the assessment and weighting of criteria and alternative pumping methods can be controlled and consistent.

The criteria used in selecting this pumping option were obtained from the FGD led by Operation General Manager PT ABC. In relation to the mud pumping application plan, it will be applied in three different locations, in the next session FGD, weights will be obtained for each criterion and alternative, and the final results will be obtained in determining the rankings that make up the decision of the pumping method.

The method of selecting the pumping method is one of the most important factors of mining engineering activity [6]. Dewatering is one of the most important activities in mining operations which goal is to continue the mining sequence, maximize recovery from reserves and provide a safe environment. According to Bitarafan & Ataei [6], no single method is sure to be used for mining. There are two or more methods worth applying, so is the pumping methods, both water, and sludge pumping. Each method has each different problem with each other. Some of the factors highlighted by Bitarafan & Ataei [6] include economic factors such as capital cost, operating cost and productivity factors such as efficiency, tool productivity, and environmental considerations.

For applications in the mining world, there are studies in which to make decisions using multiple applications of multi-criteria decision making (MCDM) such as fuzzy applications in open mine modeling [7], fuzzy applications to determine optimum underground mining methods [8]; non-linear fuzzy in safety evaluation at mine [9]; analytical hierarchy process (AHP) for selecting fan in tunnel [5], selection of mining equipment [10], crusher selection [11]. For more details of the considered variables can be seen in Table 1.

Table 1. The considered variables

No	Variable	Title	MCDM Tools	Author	Year
1	Productivity	Application of analytical hierarchy process to selection of primary crusher	AHP	Rahimdel	2014
		The application of fuzzy analytic hierarchy process (FAHP) approach to the selection of optimum underground mining method for Jajarm Bauxite Mine, Iran	Fuzzy AHP	Naghadehi, et al	2009
		ADecision Support System for Optimal Equipment Selection in Open Pit Mining: Analytical Hierarchy Process	AHP	Bascetin	2013
		Determining proper strategies for Iran's dimensional stone mines: a SWOT–AHP analysis	SWOT - AHP	Tahernejad, et al	2013
		Selection of an appropriate fan for an underground coal mine using the Analytic Hierarchy Process	AHP	Kursunoglu, et al	2014
2	Personel / Manpower	An application of nonlinear fuzzy analytic hierarchy process in safety evaluation of coal mine	AHP	Qiaoxiu, et al	2016
		Determining proper strategies for Iran's dimensional stone mines: a SWOT–AHP analysis	SWOT - AHP	Tahernejad, et al	2013
3	Mobility	A Decision Support System for Optimal Equipment Selection in Open Pit Mining: Analytical Hierarchy Process	AHP	Bascetin	2013
4	Support	A Decision Support System for Optimal Equipment Selection in Open Pit Mining: Analytical Hierarchy Process	AHP	Bascetin	2013
		Determining proper strategies for Iran's dimensional stone mines: a SWOT–AHP analysis	SWOT - AHP	Tahernejad, et al	2013
5	Flexibility	A Decision Support System for Optimal Equipment Selection in Open Pit Mining: Analytical Hierarchy Process	AHP	Bascetin	2013
		Selection of an appropriate fan for an underground coal mine using the Analytic Hierarchy Process	AHP	Kursunoglu, et al	2014
		Selection of material handling equipment system for surface mines by using a combination of fuzzy MCDM models	Fuzzy MCDM	Yazdani-Chamzini, et al	2013
6	Operating Cost	Selection of an appropriate fan for an underground coal mine using the Analytic Hierarchy Process	AHP	Kursunoglu, et al	2014
		Selection of material handling equipment system for surface mines by using a combination of fuzzy MCDM models	Fuzzy MCDM	Yazdani-Chamzini, et al	2013

This research is similar to some previous researchers who intended to find or choose the best alternative compiled by many existing criteria. Reference to previous research is in the scope of the mining project, such as the selection of tools and methods, but the object is different, namely the mud pumping project located in the former void of the mine.

Some previous studies only used a number of criteria in determining the best alternative, but this study used all the criteria in compiling a decision hierarchy to choose the best alternative so that it can enrich the science of dewatering project management in the mining world. For more details, Table 2 presents the summary of variables concerned in this research.

2. Research Methodology

This research was conducted based on the following steps: determination of location, data collection, and stages in selection the criterion of the mud pumping.

The research location is the coal mining project in Kintap District, Tanah Laut Regency, South Kalimantan province, where the location map is shown in Figure 1.

Table 2. Proposed variables

Item	*Variabels*						
	Produktivities	*Maintenance*	*Mobilization*	*Spareparts*	*Man Power*	*Flexibility*	*Operational Cost*
Equipment Support Selection	v	V	v				v
Mining Tools Selections	v					v	v
Material Handling Selections		V		v		v	v
Strategy / Mining Methods	v	V		v	v		v
Best Method Selections for Mud Pumping Projects	v	V	v	v	v	v	v

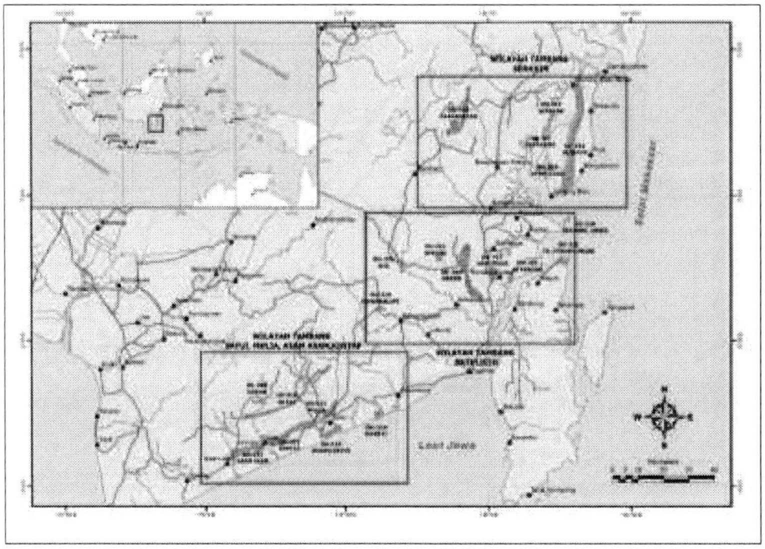

Figure 1. The location of this research

This project has three different areas in three different districts. The northernmost area goes into Kotabaru Regency, the middle area is entered into the location of Tanah Bumbu Regency and the southernmost area is in the location of Tanah Laut Regency. The southernmost area is divided into three different work/job site locations. The research location takes place in the southernmost area where the area is divided into the X-Y-Z work location.

2.1. Data Collection

In conducting the data collection process, there are two stages namely comparative study and interview to expert.

The comparative study is aimed to have some data on work conditions, equipment, human resources/manpower, equipment productivity data (excavation, hauling and pumping equipment) from the other project. These data are used for decision-makers' consideration to assign assessment weightings for each criterion and alternative. To collect this data, it is necessary to conduct a direct survey at the prospective pumping area at the coal mining site at the owner of the project, at X - Y -Z working site. Then, conducted a comparative study at the pumping area at another coal mine company in Batu Kajang, East Kalimantan. This comparative study was conducted because in that area there are three methods that become alternative candidates that have been run regularly. The result of the comparative study is to get an idea on how the three methods are implemented, the productivity and flexibility of each method, the resources needed for both human and supporting equipment such as haulage dump truck, excavator, spare parts, pipe needs, and technical unloading assembly and mobilization.

The second stage in data collection is an interview to the experts. The selection of representative respondents is an important factor in this study. Decisions that involve many criteria require a lot of expert participation [12]. As in the optimal study of equipment selection in open-pit mining by Bascetin [10], which uses respondents who are experienced in tool selection and have the qualifications to fill out questionnaires so the results can be justified. Likewise in the crusher selection study at the Golehogar Mine, Iran by Rahimdel & Ataei [11] who used respondents who truly understood and understood the various advantages and disadvantages of each type of crusher that would be used as an alternative. Respondents here are employees and management as well as dewatering experts from contractors involved in this research, namely a). General Manager Operations, Head Office, Jakarta; b). Head of Mining Engineering Site X, Site Y, Site Z; c). Crushing Plant - Overland Conveyor - Port Subtend site Y; d). Dewatering Expert, Site Y; e). Plant - Electric & Mechanical Expert, Site Y; The 5 (five) prospective respondents were chosen because all were involved in the pumping work (a, b, c,d, e).

2.2. *Forum Group Discussion.*

The main purpose of the study using focus group discussion is to get respondents' attitudes, feelings, beliefs, experiences and reactions which are difficult to obtain using other methods such as observation, one-to-one interview or questionnaire. Focus group discussion is especially useful when there are different forces between participants and decision makers [12].

There are several definitions of focus group discussions (FGDs) including:

First, FGD is an individual group selected by the researcher to discuss and provide opinions on the subject of research [12] interviews involve the number of people interviewed at the same time, and the results depend on the questions and responses between the researcher and the participants. For group discussions focused on interactions within the group based on the topic presented [13].

Second, focus group discussion is one of the techniques used by researchers to extract the data needed by the researcher. The resulting data is accurate and has high validity because all information is the result of the agreement of all participants.

Based on two definitions mentioned, this research conduct FGD to have the criteria of the best method to pump the mud from the coal mining basin. The participants and decision makers have different rules in the projects. The characteristics of the participants are classified based on the experience in dealing with mud pumping. Table 3 presents the characteristics of the participants.

Table 3. Number of participants based on work experience

No	Work Experience (year)	Number of Respondents	Percentage	Job Description
1	0-5	1	14%	Dewatering Engineer Contractor Site Y
2	5-10	1	14%	Senior Engineer Site Y
3	10-15	2	29%	Senior Engineer Site X Engineering Superintendent Site Y
4	>15	3	43%	Project Manager of Contractor Site Y Dewatering Exper – Head Office Contractor Head of Mining Engineer Site Y

For a more succinct, the flow of the research method that shows the stages - the process at that stage - and the output to be used for the next stage is shown in Table 4. as follows:

IOP Conf. Series: Materials Science and Engineering **588** (2019) 012001 doi:10.1088/1757-899X/588/1/012001

Table 4. Summary of Research Methodology

Step	Item	Input	Process	Output
1	Identifications	Field surveys include pond conditions, number of support equipment, terrain conditions and access to conformity, candidates for alternative management proposals	Field survey, interview with dewatering expert	Data of support equipment, comparative study object
2	Survey	Data of support equipment, comparative study object	Comparative Study	Criteria and alternative variable
3	Literature Review	Criteria and alternative variable	Variabel from previous researchers	Syntesis of literature review, empirical data of alternatives proposed
4	Interview & comparative study	Syntesis of literature review, empirical data of alternatives proposed	Discussion	Criteria proposed
5	Focus Group Discussion	Criteria proposed	Discussion	Selected Criteria

3. Results

Table 2 showed the seven proposed criteria for the mud pumping method. Those criteria are productivity, maintenance, mobilization, spare parts, manpower, flexibility, and operational cost. Through the process of research as presented in Table 4, the result comes to six criteria as shown in Table 5.

4. Concluding Remark

The goals of this paper are to get what criteria are needed in determining best method priorities in implementing the mud pumping option. Step of research started with empirical identification, literature study to get proposed criteria, data collecting which is consist of both of local survey condition and comparative study, focus group discussion and get a synthesis of criteria. For the final criteria, there are six variables selected out of seven in the initial criterion because indeed these six variables are important variables in the mud pumping project. Six final criteria that have influence with method selection of pumping are: productivity, support, mobilization, manpower, flexibility, and operational cost

Table 5. The Selected Criteria

No	Main Criteria Variables	Data Source	Description and Operational Definition
1	Spare part	FGD, Comparative Study, Interview	Each alternative proposal has a spare parts type - each where there is a difference in the difficulty of spare parts for each option (Pontoon, Highway Dumptruck, Dredger, PC / Excavator have different spare parts). Pontoon: The floating vessel where the main component of the dredger is located Highway Dumptruck: A giant truck with large capacity Dredger: The main component of the mud pump PC / Excavator: Digging tool with boom / long arm
2	Manpower	FGD, Comparative Study, Interview	Since this pumping will be done by switching at several different sites, there is a need for manpower at every job site to operate this mud pump (not picking up from job site Y). Manpower could be an operator, mechanic, welder, crew sampling, where the ability and experience of each - different job sites will vary.
3	Mob-Demobilization	FGD, Comparative Study, Interview	Mobilization - demobilization is needed when going to move location. Each alternative method/option has different difficulties in unpacking the equipment when it is about to move
4	Productivity	FGD, Comparative Study, Interview	Productivity is the ability to produce production output per unit time. This variable is used in this study because the value of pump productivity in each alternative greatly determines how quickly the mud pool is dry .
5	Flexibility	FGD, Comparative Study, Interview	Each pumping method/option has different work characteristics/principles, so the degree of flexibility in the field will differ also between method
6	Operational Cost	FGD, Comparative Study, Interview	Each method has different operational costs so it will be considered in choosing the pumping method

References

[1] Committee on Technologies for the Mining Industries, *Evolutionary and Revolutionary Technologies for mining*. Washington, DC: NATIONAL ACADEMY PRESS, 2002.

[2] T. L. Saaty, *The Analytic Hierarchy Process*. New York: McGraw Hill. International, 1980.
[3] S. Ghazinoory, "Using AHP and L. P . for choosing the best alternatives base on the gap analysis," *Appl. Math. Comput.*, vol. 184, pp. 316–321, 2007.
[4] T. L. Saaty, "Decision making with the analytic hierarchy process," *Int. J. Serv. Sci.*, vol. 1, no. 1, pp. 83–97, 2008.
[5] N. Kursunoglu and M. Onder, "Selection of an appropriate fan for an underground coal mine using the Analytic Hierarchy Process," *Tunn. Undergr. Sp. Technol.*, vol. 48, pp. 101–109, 2015.
[6] M. R. Bitarafan and M. Ataei, "Mining method selection by multiple criteria decision-making tools," *J. South African Inst. Min. Metall.*, no. October, pp. 493–498, 2004.
[7] A. A. Bazzazi, M. Osanloo, and B. Karimi, "A new fuzzy multi-criteria decision-making model for open pit mines equipment selection," *Asia-Pacific J. Oper. Res.*, vol. 28, no. 3, pp. 279–300, 2011.
[8] M. Z. Naghadehi, R. Mikaeil, and M. Ataei, "Expert Systems with Applications The application of fuzzy analytic hierarchy process (FAHP) approach to the selection of optimum underground mining method for Jajarm Bauxite Mine, Iran," *Expert Syst. Appl.*, vol. 36, no. 4, pp. 8218–8226, 2009.
[9] W. Qiaoxiu, W. Hong, and Q. Zuoqiu, "An application of nonlinear fuzzy analytic hierarchy process in safety evaluation of coal mine An application of nonlinear fuzzy analytic hierarchy process in safety evaluation of coal mine," *Saf. Sci.*, vol. 86, no. July 2018, pp. 78–87, 2016.
[10] A. Bascetin, "A decision support system for optimal equipment selection in open-pit mining : Analytical Hierarchy Process," *Istanbul Univ. Min. Eng. Dep.*, vol. 16, no. 2, pp. 1–11, 2003.
[11] R. Mohammad and A. Mohammad, "International Journal of Mining Science and Technology Application of analytical hierarchy process to selection of primary crusher," *Int. J. Min. Sci. Technol.*, vol. In Press, 2014.
[12] E. J. Sobczyk, J. Kicki, W. Sobczyk, and M. Szuwarzy, "Support of mining investment choice decisions with the use of the multi-criteria method," *Resour. Policy*, vol. 51, no. November 2016, pp. 94–99, 2017.
[13] D. L. Morgan and R. A. Krueger, "When to Use Focus Groups and Why," in *Focus Group as Qualitative Research*, 2nd ed., London: SAGE, 1997, pp. 3–9.

Design of output power control system based on mass flow rate comparison of air and fuel ratio (AFR) on dual fuel generator set by using pid control method

M B Toriki , B L Widjiantoro and A Abdurrakhman

Department of Physic Engineering, Institut Teknologi Sepuluh Nopember, Jl. Arif Rahman Hakim, Sukolilo, Surabaya, Indonesia.

bereltoriki11@gmail.com

Abstract. Generator set is a tool that is used to generate energy or electrical power. The electric power generated by the generator set is used to supply the electrical loads in this thesis is lamp. Power that can be produced by a generator set is ranged between 100-1200 watt. The power generated by generator set is used influenced by a mixture ofair and fuel. Generator set used is dual fuel generator set. From the results of this study, a stable response with overshoot value below 0% and its error is 2%. In addition, the best obtained value of Air Fuel Ratio (AFR) is 15.06, Furthermore, the stability of the power generated by the generator set is also influenced by the flowrate mass of the fuel injected into the combustion chamber. From the simulation results when given a power set point at 1200 watt, obtained value of the air mass flowrate is 0.03754 kg/s, mass flowrate of biogas is 0.002367 kg/s, and gasoline constant mass flowrate is 0.000125 kg/s. Meanwhile, when given a set point of 100 watts of power, a value of 0.009928 kg/s air mass flowrate is injected to the chamber, mass flowrate of biogas is 0.0005341 kg/s and the mass flowrate of gasoline is 0.000125 kg/s. The goal of output power control on generator set is to make generator working to produce output power in accordance to required load.

1. Introduction

The surge of world oil prices, followed by rising prices of fossil fuels. On the other hand, global environmental issues that demand high levels of environmental quality, encourage the various energy experts to develop more environmentally-friendly energy and support the sustainable supply security. Currently fossil energy reserves had been nearly depleted, and the research of new source has been difficult to do, even predicted in 2030 Indonesia will truly become a nett importer of energy, because the balance between production and domestic energy consumption occurred in 2030. Starting this year, the production of domestic energy (fossil and EBT) is no longer able to meet domestic consumption and Indonesia changed its status into a net importer of energy. [1]

This makes the renewable energy development is growing especially Indonesia has huge potential for it. Based on the geographic state, Indonesia is blessed with a high amount of energy resources ranging from hydro, geothermal, biomass, solar, wind, even the ocean. But unfortunately such property has not been widely used as an energy source that can replace fossil energy sources that nearly depleted. One source of renewable energy that are being highly used is biogas, especially for household scale. However, the development of biogas in Indonesia is still relatively slow due to various factors, ranging from people who are still not comfortable with the energy of dirt, to biogas purification problems and the implementation of the generator set. [2]

Based on data from the Directorate General of Livestock and Animal Health, while the number of beef cattle numbers in 2015 reached 15 million. One cow capable of producing dirt as much as 23,6 kg of solid waste a day and 9,1 kg / of liquid waste a day. [3] In 1 kg of cow / buffalo manure can be expected 0,023 to 0,040 m^3 of biogas is Produced [4]. So that the maximum potential value of biogas from manure as high as 14.16 million m^3 /day. Methane amount is 9,912 million m^3 /day and CO_2 of 4,248 million m^3 /day. In other words, within a year Indonesia will contribute 1,55 billion m^3 CO_2 or the equivalent of 2,77 million tons per year and amount of 3,617 billion m^3 methane, equivalent to 6,466 million tons per year. The effect of greenhouse gases of methane gas are equivalent to 21 times of CO_2 greenhouse gas. That means with the contribution of methane gas, will be equivalent to 135,8 million tons of CO_2 per year, Meaning that in addition of this gasses in the atmosfer, the methane gas is equivalent to 135,8 million tonnes worth of CO_2 gasses a year, and if summarized with Indonesia will contribute 138,573 million tonnes worth of CO_2, meaning that if cow waste or manure is not used properly will negatively affect the environment wildlife as the greenhouse gasses is the cause of depletion of ozone gasses in the atmosphere.

Dual fuel generator set is the standard gasoline engine with biogas fuel added in its combustion chamber and the ignition of the engine is carried by the gas spray called pilot fuel. In other word liquid or gaseous fuels can be injected to the engine by making a hole in the intake manifold of the engine itself. Depending on the type of the added fuel , when the fuel type is liquid such as ethanol or methanol, carburetor is added to the system and make the fuel pumped at certain pressure and atomized as the fuel are injected to the air intake of generator engines. As for the fuel gas is not needed anymore because the gas fuel carburetor already has its own pressure.[3]

Dual fuel system benefit is to conserve the use of gasoline as a fuel, production costs can be reduced, as well as generator set modification costs relatively cheaper than converting to the whole gas engine. Furthermore, the application of biogas with a dual fuel system on the generator set can improve the performance and efficiency of the engine [8]. The utilization of biogas generator set would not change the composition of machine tools and only adds to the system equipment such as mixer ventury on the suction channel. While the use of dual fuel intended to reduce the use of gasoline in the combustion process, as it would be the partial substitution of gasoline by biogas.

On generator set using dual fuel system is still a lot to do in development phase, one is controls of the output power of the generator set. Control of output power on generator set needs to be done, so that the generator set can produce a maximum and stable output power, and in accordance with the needs of electrical energy that can be used for everyday needs. In order to make the generator set produce a stable output power for the demand, the mass flow rate of fuel and air in the combustion chamber of engine should be in accordance with the required power demand. If higher power produced by generator output is required, then the mass flow rate of air and fuel (petrol and biogas) that entering the combustion chamber is also getting bigger. Additionally, the mixture of fuel and air must be in accordance with the rules of the combustion reaction. Therefore to obtain optimum combustion, calculation of Air Fuel Ratio (AFR) is done by stoichiometric equation.

2. Research Methodology

On this research, dual fuel generator set is used as research object. The characteristic of engine is spark ignition engine with 4 step. Gasoline and biogas is fuel that is used to get combustion on the engine cylinder. While the use of dual fuel intended to reduce the use of gasoline in the combustion process. On the other hand, to get complete combustion and optimum power, the air is also injected into cylinder chamber. To inject gasoline, biogas, and air into combustion chamber is used ratio control algorithm. Ratio control is a algorithm that working based on value that was determine by using dual fuel Air Fuel Ratio (AFR) calculation.

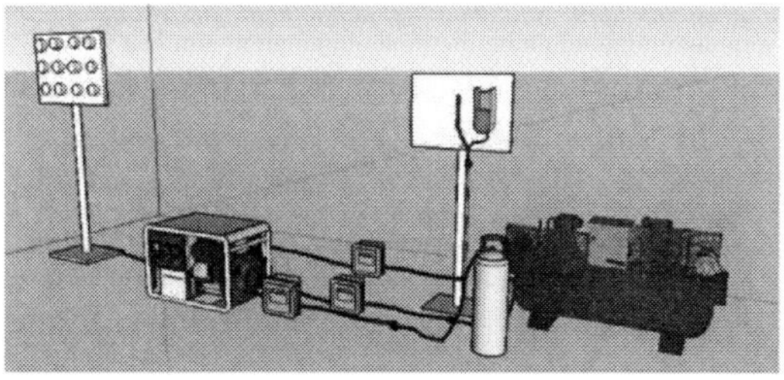

(a)

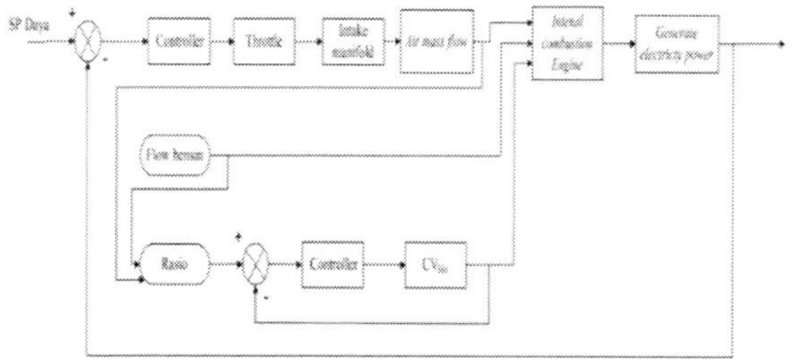

(b)

Figure 1. (a) Experiment Tools, (b) Block Diagram of Power Control Output System Dual Fuel Generator Set

3. Modelling of The System

The calculation of the value of AFR (Air Fuel Ratio) for dual fuel is used to determine the ratio between the mass air flowrate and mass flowrate of fuel (petrol and biogas). To obtain dual fuel AFR value calculation using the following equation.[4]

$$AFR_{sto} = \frac{\left[\alpha\left(x_d+\frac{y_d}{4}\right)+\beta\left(x_g+\frac{y_g}{4}\right)\right]MW_{air}}{\alpha(x_d.MW_c+y_d.MW_H)+\beta(x_g.MW_c+y_g.MW_H)} \quad (1)$$

Where :

MW C = *molecular Weight* atom C
MW H = *molecular Weight* H atom
MWair = *molecular Weight* air
α = Mass of gasoline to the total mass of fuel
β = Mass biogas to the total mass of fuel
Xd = Number of C atoms in the gas compound
Yd = The number of H atoms in the compound of gasoline
xg = The number of carbon atoms in the compound of biogas

Throttle valve is an actuator which regulates the amount of air mass flow into the combustion chamber of the engine. To calculate the number of mass air flow (ideal gas) that passes through the throttle valve can be used follows Bernoulli equation.

$$q_i(t) = C_d[1 - \cos(u)]\frac{\pi}{4}D^2\frac{P_{in}}{\sqrt{RT_{in}}}f(P_{in,}P_{out}) \quad (2)$$

Based on the equation 2 above, the air flow can be adjusted via the throttle valve opening (u) between 0^0 to 90^0 . The non-linear function that connects the input pressure and output pressure is as follows

$$
(P_{in}, P_{out}) = \begin{cases} \sqrt{k\left[\frac{2}{k+1}\right]^{\frac{k+1}{k-1}}} & P_{out} \leq P_{cr} \\ \left[\frac{P_{out}}{P_{in}}\right]^{\frac{1}{k}} \sqrt{\frac{2k}{k-1}\left[1-\left(\frac{P_{out}}{P_{in}}\right)\frac{k-1}{k}\right]} & P_{out} \geq P_{cr} \end{cases} \tag{3}
$$

Mathematical modeling of the subsystems of the intake manifold can be approximated using the ideal gas equation as follows.

$$
\frac{d}{dt}m(t) = \dot{m}_{in}(t) - \dot{m}_{out}(t) \tag{4}
$$

$$
\frac{d}{dt}U(t) = \dot{H}_{in}(t) - \dot{H}_{out}(t) \tag{5}
$$

It is assumed that the fluid flow is modeled using the ideal gas equation as follows

$$
p(t)\,.\,V = m(t)\,.\,R\,.\vartheta(t) \tag{6}
$$

When connected with a caloric value

$$
U(t) = c_v\,.\vartheta(t).m(t) \tag{7}
$$

$$
U(t) = c_v\,.\vartheta(t).m(t) \tag{8}
$$

$$
\dot{H}_{in}(t) = c_p\,.\vartheta_{in}(t).\dot{m}_{in}(t) \tag{9}
$$

$$
\dot{H}_{out}(t) = c_p\,.\vartheta(t).\dot{m}_{out}(t) \tag{10}
$$

Substitution into equation

$$
\frac{d}{dt}p(t) = \frac{k.R}{V}[\dot{m}_{in}(t)\,.\vartheta_{in}(t) - \dot{m}_{out}(t).\vartheta(t)] \tag{11}
$$

$$
\frac{d}{dt}\vartheta = \frac{\vartheta\,.R}{p\,.V\,.c_v}\left[c_p\,.\dot{m}_{in}\,.\vartheta_{in} - c_p\,.\dot{m}_{out}\,.\vartheta - c_v(\dot{m}_{in} - \dot{m}_{out}).\vartheta\right] \tag{12}
$$

$$
\frac{d}{dt}p(t) = \frac{R\,.\vartheta(t)}{V}\,.\,[\dot{m}_{in}(t) - \dot{m}_{out}(t) \tag{13}
$$

$$
\frac{d}{dt}P_m \;= K_m(\dot{m}_{ai} - \dot{m}_{ao})(\mathrm{T}) = \frac{R.T_m\,(t)}{V_m}(\dot{m}_{ai} - \dot{m}_{ao})(t) \tag{14}
$$

After going through the intake manifold, air will enter into the combustion chamber or cylinder. Engine mass flow is the mass flow of air is injected into the combustion chamber with the air pressure of the calculation in a part of the intake manifold. Based on the air system, the engine characteristics can be assumed as a volumetric pump. And the amount of mass flow of air needed for combustion in the cylinder can be defined based on the speed-density equation.[5]

$$
q_{cyl} = \frac{\eta_v P_{out} V_d \omega}{2\pi NRT} \tag{15}
$$

Where :

qcyl = Mass Flowrate Air Entering The Combustion Chamber (kg / s)

ηv = Volumetric Efficiency (%)

Pout = Pressure Output of The Intake Manifold (Pa)

ω = Speed (Radians per minute)
R = Gas Constant (J / kg K)
N = Number Revolution Per Cycle (2 for 4 stroke)
T = Temperature (K)

A modern fuel injection system set up so that the comparison mass flowrate of air and fuel mass flowrate is kept constant to achieve volumetric efficiency without changes. Density and mass flow measurement of air is used to control the fuel is injected. So that the power generated by the engine is obtained by the following equation.[5]

$$P = \frac{\eta_c \eta_v P_{out} V_d \omega H\left(\frac{F}{A}\right)}{2\pi N R T_{in}} \quad (16)$$

Where :
ηc = The Combustion Efficiency (%)
ηe = Efficiency Electric Generator (%)
F / A = Fuel to Air Ratio
Vd = Displacement Volume (m^3)

Spark ignited engine, the combustion process occurs because there is a mixture of fuel in this case is a gas and biogas with air in the cylinder engine and ignited by a spark plug, causing sparks and combustion occurs as described in the previous chapter. The combustion process will drive the piston and crankshaft so that it will turn a generator to the torque produced by the engine speed and can produce output power. This is the equation of torque generated from the combustion of air-fuel ratio according to the stoichiometric equation.[5]

$$\dot{\omega} = \frac{1}{J}\left[\left(\frac{\eta_c \eta_v \eta_e P_{out} V_d H\left(\frac{F}{A}\right)}{2\pi N R T_{in}} - T_e\right) - bw\right] \quad (17)$$

Where
J = Inertia
tm = Torque Produced by The Engine Speed (Nm)
Te = Torque Electrically or Load (Nm)

4. Relation Between Power and Mass Flowrate

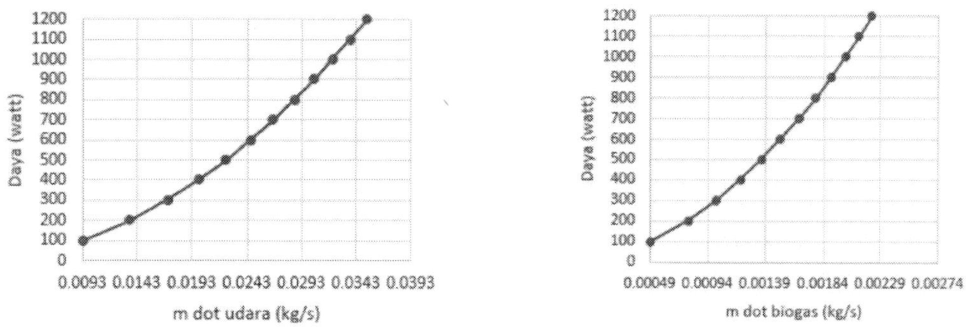

(a) (b)

Figure 2. (a) Relation Graph Between Power and Air Mass Flowrate, (b) Relation Graph Between Power and Biogas Mass Flowrate

Figure 2 is relation between power and mass flowrate. The higher the required power it needs to air and fuel will increase. Because the engine requires a greater supply of energy to produce a greater output power anyway. From the graph, it can be seen that the change in the air mass flowrate and mass flowrate is not the same biogas or not linear at each change of set point power. The most substantial change of air mass flowrate and biogas that is when the change set point 100 watts to 200 watts with a value of 0.259 kg / s and 0.018 kg / s. While for the most minor changes from the mass flowrate of air and biogas that is when the change set point power 1100 watts to 1200 watts with a value of 0.093 kg / s and 0.007 kg / s.

4.1. Tracking Set Point Up 33%

In testing of the set point tracking is raising the set point of 33% of the initial set point. On the change of set point values obtained for settling time (Ts) of 1.25 sec, rise time (Tr) of 1.25 seconds, Td 0.03 seconds, maximum overshoot (Mp) 0% and error steady state (Ess) 0% , The value of the rise time and settling time together for the first time reached the set point directly system stable. In this test indicate that a change set point capable followed by process variables. The response of tracking the set point up 33% shown in the figure 3.

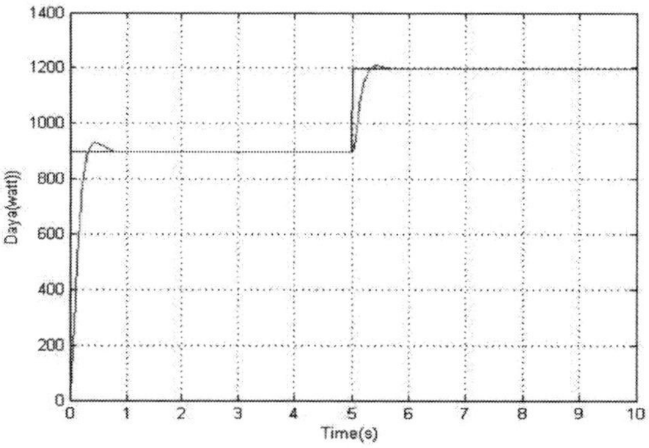

Figure 3. Graph of Output Power Response with Set Point Tracking Up 33%

4.2. Tracking Set Point Down 33%

In testing the set point tracking lowers the set point of 33% of the initial set point. On the change of set point values obtained for settling time (Ts) of 1.2 seconds, the rise time (Tr) of 1.2 seconds, Td 0.018 seconds, maximum overshoot (Mp) 0% and error steady state (Ess) 0%. The value of the rise time and settling time together for the first time reached the set point directly system stable. In this test indicate that a change set point capable followed by process variables. The response from the set point tracking down 33% found in Figure 4.

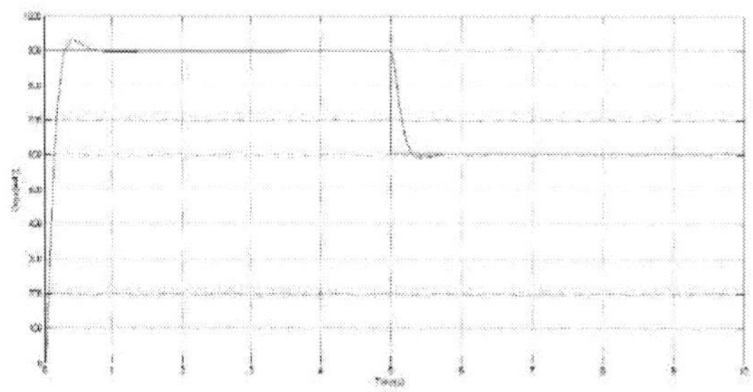

Figure 4. Graph of Output Power Response with Set Point Tracking Down 33%

4.3. Tracking Set Point Up 10%

In testing the set point tracking is raising the set point of 10% of the initial set point. On the change of set point values obtained for settling time (Ts) of 1.45 sec, rise time (Tr) of 1.45 seconds, Td 0.01 seconds, maximum overshoot (Mp) 0% and error steady state (Ess) 0% , The value of the rise time and settling time together for the first time reached the setpointdirect system stable. In this test indicate that a change set point capable followed by process variables. The response of tracking the set point up 10% found in Figure 5.

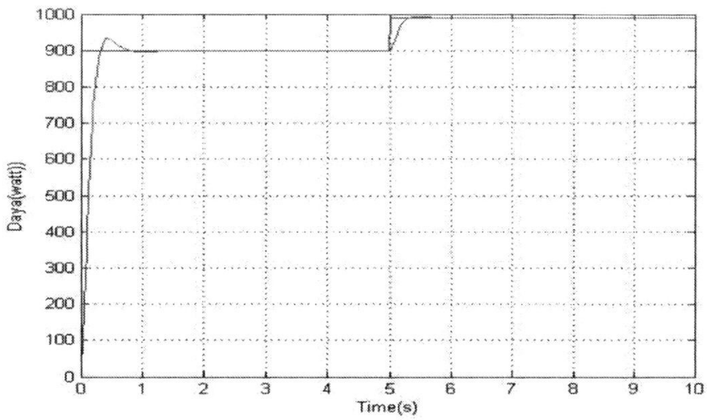

Figure 5. Graph of Output Power Response with Set Point Tracking Up 10%

4.4. Tracking Set Point Down 10%

In testing the set point tracking lowers the set point of 10% of the initial set point. On the change of set point values obtained for settling time (Ts) of 0.7 seconds, the rise time (Tr) of 0.7 seconds, Td 0.005 seconds, maximum overshoot(Mp) 0% and error steady state (Ess) 0%. The value of the rise time and settling time together for the first time reached the set point directly system stable. In this test indicate that a change set point capable followed by process variables. The response of tracking down 10% set point shown in the figure 6.

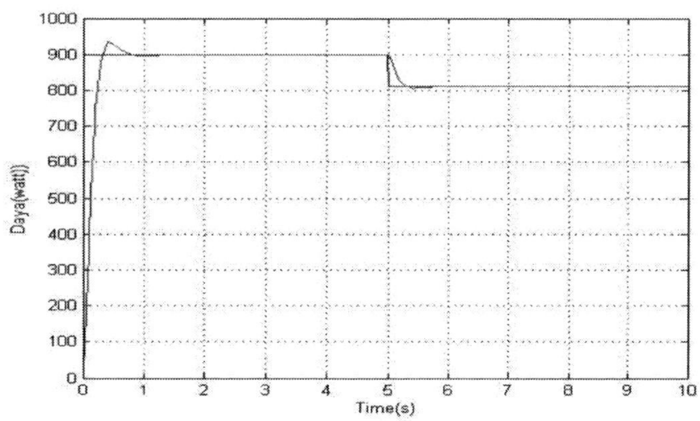

Figure 6. Graph of Output Power Response with Set Point Tracking Down 10%

5. Result of System Modelling

To test the results of modelling in simulink by entering the set point value required power ranged at 100-1200 watts. If the higher power is required, then the mass flowrate of air and biogas will be great as well, but in this study gas mass flowrate is constant.

Table 1. Relation between Mass flowrate Power of Air and Fuel

Power (Watt)	Gas Mass Flowrate (Kg/s)	Biogas Mass Flowrate (Kg/s)	Air Mass Flowrate (Kg/s)
100	0,01	0,029	0,560
200	0,01	0,047	0,819
300	0,01	0,060	1,022
400	0,01	0,071	1,192
500	0,01	0,081	1,341
600	0,01	0,090	1,476
700	0,01	0,098	1,601
800	0,01	0,106	1,716
900	0,01	0,113	1,825
1000	0,01	0,120	1,928
1100	0,01	0,126	2,025
1200	0,01	0,133	2,118

Table 1 is relation between power and mass flowrate. The higher required power it needs to air and fuel will increase. Because the engine requires a greater supply of energy to produce a greater output power anyway. The most substantial change of air mass flowrate and biogas that is when the change set point 100 watts to 200 watts with a value of 0.259 kg / s and 0,018 Kg / s. While for the most minor changes from the air mass flowrate and biogas that is when the change set point power 1100 watts to 1200 watts with a value of 0.093 kg / s and 0.007 kg / s.

Table 2. Comparison Between Fuel Mass Flowrate From Simulation Result and From Plan

Power (Watt)	Biogas Mass Flowrate (Simulation) Kg/s	Biogas Mass Flowrate (Plan) Kg/s
100	0,029	0,032
200	0,047	0,049
300	0,060	0,064
400	0,071	0,075
500	0,081	0,086
600	0,090	0,095
700	0,098	0,103
800	0,106	0,111
900	0,113	0,117
1000	0,120	0,125
1100	0,126	0,133
1200	0,133	0,142

Table 2 represents the relation of the mass flowrate of fuel and the real plant simulation results when the value of a certain power. While Table 3 is the relation of mass air flowrate simulation results and real plant during a certain power value. From these data it appears that the greater the power output is required, then the mass flowrate of fuel and air will also increase. The relation of power, fuel, and air are comparable. However, there are differences between the mass flowrate of fuel and air from the plant simulation and real data. This is due to the results of the simulation are several engine parameters taken in ideal conditions, such as pressure on the intake manifold and the others. In addition, because there are some parameters that use the approach through a number of journals and the results of previous research.

Table 3. Comparison Between Air Mass Flowrate From Simulation Result and From Plan

Power (Watt)	Air Mass Flowrate (Simulation) Kg/s	Air Mass Flowrate (Plan) Kg/s
100	0,560	0,601
200	0,819	0,871
300	1,022	1,092
400	1,192	1,255
500	1,341	1,419
600	1,476	1,571
700	1,601	1,682
800	1,716	1,789
900	1,825	1,896
1000	1,928	1,995
1100	2,025	2,099
1200	2,118	2,224

6. Conclusion

Based on the results of this research, it was found that a stable output power response with overshoot maximum value averaging below 20% and the error value below 2% in each of the set point value 100-1200 watts. In addition, the PID parameter values didiapatkan with Kp= 120, Ki = 0.1, and Kd = 15.2. While the PID parameter values for flowrate control of biogas that is Kp = 5, Ki = 3, and Kd = 1. In addition, there is a difference between the simulation and the mass flow rate of the average data plan for biogas mass flow rate at 6.94% and the mass air flow rate of 7.36%.

The higher power that generated by generator set, the mass flowrate of air and fuel will increase. For maximum power is 1200 watts, obtained air mass flowrate amounted to 2.118 Kg / s, mass flowrate of biogas amounted to 0.133 Kg / s, and gasoline mass flowrate amounted to 0.01 kg / s. As for the minimum power, obtained air mass flowrate amounted to 0.560 Kg / s, mass flowrate of biogas amounted to 0,029 Kg / s, and the gas mass flowrate amounted to 0.01 kg / s.

In this research, value of Air Fuel Ratio (AFR) for complete combustion on dual fuel system that is 15.06.

References

[1] The Agency for The Assessment and Application of Technology (BPPT) 2015 *Energy Development to Support Sustainable Development* Indonesia Energy Outlook

[2] A Abdurrakhman, T Soehartanto and B Sudarmanta 2013 *Physic Eng* **12**

[3] A I Setiawan 2002 *Harnessing The Livestock Manure Utilization Problem Solution Environment and Alternative Energy* (Jakarta : Third Printing Spreaders Swadya)

[4] J Stewart, A Clarke and R Chen 2007 *Autom. Eng. (Proc. IMEchE)* **221** 943

[5] H Bastida, E Carlos, and M Abeyesekerra 2017 *Sch. Eng.* **142** 1282-1287

IOP Conf. Series: Materials Science and Engineering **588** (2019) 012003 doi:10.1088/1757-899X/588/1/012003

Varition Analysis Of Collector Shift Time Period On Solar Water Heater Efficiency

Mustafa[1], Sudarno[1] and Wahyudi[1]

[1] Mechanical Engineering, Merdeka University of Madiun, Indonesia

Abstract. The working principle of a solar collector is to move solar radiation to the working fluid. Solar radiation that falls on the partial glass cover is partially reflected, some will be absorbed and some will be passed to the absorbent plate. The radiation that reaches the absorbent plate will be absorbed by the absorbent plate. The heat absorbed by the absorbent plate will be used to heat the fluid that works in the form of a flowing fluid. The purpose of this study was to design and make a wave plate solar collector with solar tracker, with an interval time of collector shift every 20 minutes, 40 minutes, 60 minutes. From the test results obtained the average efficiency in testing with the collector shift time interval of 20 minutes is 23.07%, with an interval of 40 minutes is 22.07%, and with an interval of 60 minutes is 21.31%.

1. Introduction

1.1. Background
Solar radiation energy was one form of alternative energy that could be used for various purposes to replace the energy produced by petroleum. For a long time, solar energy had become a source of lighting during the day, and was also widely used to dry agricultural produce, clothing, wood, heating water and others. An Indonesian gift located in the equator, where solar energy was available at all times.

As one alternative energy source, solar energy had become increasingly widespread. One of its uses was to use a device called a solar collector. At present, flat plate or wave shape solar collectors for water heaters had been developed. This collector, using wave shaped plate that serves to absorb the incoming solar energy and move the heat received to the working fluid, namely water. At the top of the collector, a transparent cover was used to cause a greenhouse effect. In order to avoid heat loss to the environment at the bottom and side of the solar collector, an insulator was provided (Made et al., 2010).

1.2. Identification of the problem
The problem of this research was to identify the influence of the variation of the time period of the collector shift at a short interval of 20 minutes, 40 minutes, 60 minutes to the efficiency of solar water heater.

1.3. The purpose of the study
The purpose of the study was knowing the effect of variations in the period of time the collector shifts to the efficiency of solar water heater.

IOP Conf. Series: Materials Science and Engineering **588** (2019) 012003 doi:10.1088/1757-899X/588/1/012003

1.4. Benefits of research
The benefits of the research was could find out the process of designing and manufacturing solar collectors with tracker. Could be used for further research with various variations so as to get better efficiency. Could be used as one of the utilization of environmentally friendly energy sources and never runs out.

1.5. Scope of problem
The scope of problem was : the testing process uses a solar water collector with a wave plate heater with a tracker. Variation in the time period of the collector shift time 20 minutes, 40 minutes, 60 minutes. Measurement results was taken for each time interval 15 minutes. Fluid capacity was set 30 liters. Flow rate was set at 0.011 liters/second. Volume of fluid inside the collector was 15 liters. Collector's tilt position was 15° facing north.

2. Literature Review
Previous research

Mustofa (2008), an experimental study of the comparison of double plate collectors and conventional collectors on the performance of solar water heater. The research resulted, the heat absorption efficiency of solar double plate heater was higher than the efficiency of conventional solar heater heat absorption.

Farid and Ismail (2010), tested the double absorber plate of wave models which produced an average heat absorption efficiency value in the double wave solar heater model with an average of 19.81%, while the value of the average heat absorption efficiency in solar absorbent plate heater flat at 12.43%. Most solar collectors were permanently installed with fixed elevating angles. This causes the solar collector to be unable to absorb solar radiation optimally because the sun was always moving, namely in the east-west direction (called the sun's pseudo-daily motion) and north-south (called the annual pseudo-motion of the sun). Absorption of solar radiation will be optimal if the direction of solar radiation was perpendicular to the surface of the field of the solar collector. Therefore, efforts were needed to direct the surface of the solar collector to be perpendicular to sunlight. The method for directing solar collectors to always follow the direction of the sun's motion was known as the method of tracking the sun (Huang et al, 2009).

Determine efficiency

The efficiency of the collector panel was the ratio between the useful heat level (QU) transferred to the liquid divided by solar radiation in the cover plate. Efficiency could be shown at equation Duffie JA and Beckman W.A. (1980), as follows:

$$\eta = \frac{q_u}{A_C G_T} \tag{1}$$

From the above equation could also use the efficiency of the collector efficiency as follows:

$$\eta = F_R(\tau\alpha) - F_R U_L \frac{(Ti-Ta)}{G_T}$$

$$\eta = \frac{\dot{m}\ C_p(T_0 - T_i)}{G_t . A_c} \tag{2}$$

where :

Qu = The energy absorbed by collector, (W/m^2)
Ac = Collector Size (m^2)
FR = Collector heat loss factor
UL = Overall heat loss (W/m^2. ° C)
Gt = Total intensity of solar radiation (W/m^2)
Ta = ambient temperature (° C)
Tin = Entering water temperature (° C)

IOP Conf. Series: Materials Science and Engineering **588** (2019) 012003 doi:10.1088/1757-899X/588/1/012003

T_{out} = water temperature out (° C)
Cp = saturated water/liquid properties (kJ/(kg.°C))
τ = Transmissivity cover glass
α = Absorbtivity of the absorbent plate
$m^{\cdot}$ = Fluid flow rate (liters/second)

Fluid mass flow rate

The fluid mass flow rate was the amount of fluid mass that flows per unit of time and could be stated as follows:

$$\dot{m} = \frac{m}{t_u} = \frac{V\rho}{t_u} \quad (3)$$

where :

m = fluid mass
tu = Time of fluid movement from one end to the other solar heat collector (liter/second)
Vρ = Volume of fluid flowing from collector exit pipe (liters).

3. ResearchMethod

The research starts from 10.00 WIB until 14.00 WIB in March 2018 until completion. Testing the variation of the time period of the collector shift was carried out in Madiun City, East Java, Indonesia A subsection

Equipment and Materials

From testing requires the following tools and materials: 3 mm thick zinc absorbent plate, 10 mm thick PVC plate, 3 mm thick glass, 25 liters water tank, framework for solar tracker collectors, circulation pipe with 25.4 mm in diameter, 3 mm thick insulator (stereofoam), aquarium pump, 1/2 " stop faucet, water hose, precision pyranometer, computer/laptop, programmable peripeal interface (PPI), analog digital converter (ADC), measuring glasses, roll cable and LM35 temperature measuring instrument.

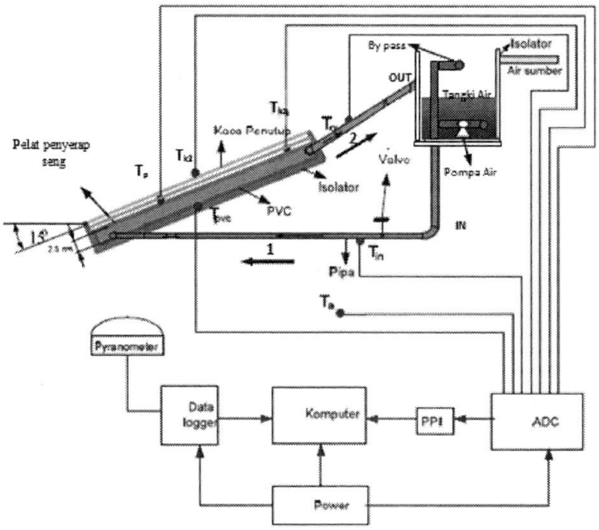

Figure 1. Scheme of solar water heating test equipment

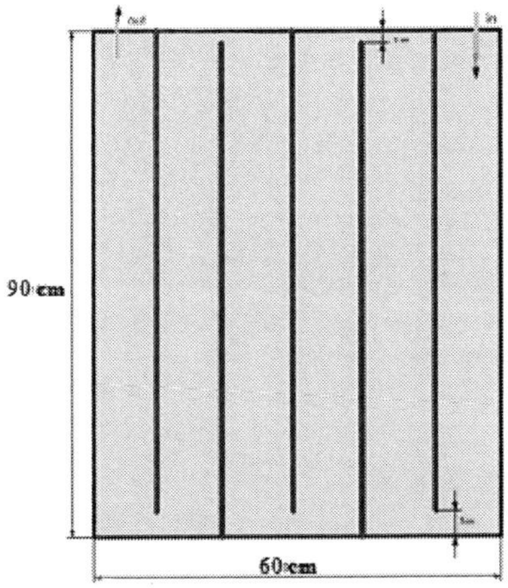

Figure 2. Scheme of collector absorber with zig zag grooves

Research Procedure

The steps taken in researching the performance of solar water heater were:

- Data collection was done by placing the collector in the sun.
- The collector was tilted at a fixed angle of 15° north.
- Install a series of measuring instruments to measure the temperature of the inlet water, absorbent plate temperature, glass temperature 1, glass temperature 2, temperature of the outgoing water, ambient temperature and solar radiation.
- Record the measurement results in accordance with the shift variation of the collector angle at 20 minutes, 40 minutes and 60 minutes.
- Direct the collector with the direction of falling sun radiation by 90 degrees by following the axis perpendicular to the shadow of the direction of the sun.
- Collect the data in table.
- Analyze the results of testing.

4. Result and Discussion

From the research results obtained data and graphs could be made as follows:

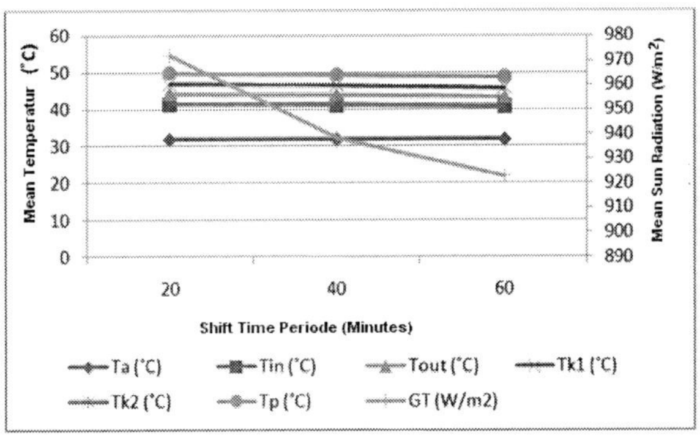

Figure 3. Relationship between the average temperature and the intensity of solar radiation

At the inlet temperature, glass temperature 1, glass temperature 2, absorbent plate temperature, and the exit temperature at each test the price was the same as the range 30°C to 50 .C. While at the ambient temperature the price was the same as the time variation of 20 minutes, 40 minutes, 60 minutes the price was lower than the inlet temperature, glass temperature 1, glass temperature 2, the temperature of the absorber plate, and the exit temperature range at 30°C. In solar radiation there was a significant decrease with a price range of 1000 W/m^2 to 800 W/m^2. This was due to the time of the same collector shift. At 20 minutes the shift time was shorter so that the displacement of the sun's upright position to the collector was faster.

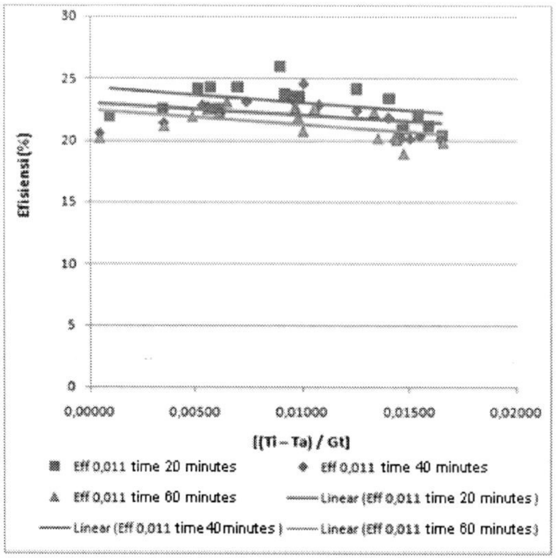

Figure 4. Graph of the relationship between the three efficiencies with (Ti-Ta / Gt)

In the test graph above it could be concluded that on the three tests the time period of the collector shift every 20 minutes, 40 minutes and 60 minutes at the same flow rate of 0.011 liters / second shows that the period of time the collector shifts greatly affects the intensity of solar radiation falling on the collector. the faster the collector's shift time toward the direction of falling solar radiation, the greater the solar radiation intensity, and the greater the temperature, the higher the efficiency obtained. From the three tests get results that show that the test using the shortest period of time the collector shift obtained the best efficiency in a period of 20 minutes while the longer the time period used, the radiation that falls to the collector and the absorption of heat gets smaller. So that the shift time affects the absorption of solar radiation in the collector which causes an increase in efficiency.

5. Conclusion

In small time interval, the collector shift was used, the greater the efficiency obtained. The best efficiency results for a time interval of 20 minutes with an average efficiency of 23.07%.

6. Ackowledgement

The research was fully funded by the Ministry of Research and Technology of Higher Education of the Republic of Indonesia on PTUPT scheme in the year 2018.

References

[1] Arismunandar W 1995 *Solar engineering technology, first edition* (Jakarta : PT. Pradnya Paramita)

[2] Burhanuddin A 2006 *Characteristics of flat plate solar collectors with variations in cover glass distance and collector tilt angle*. Essay. Faculty of Mathematics and Natural Sciences Sebelas Maret University, Surakarta.

[3] Duffie and Beckman 1991 *Solar Engineering of Thermal Processes* (Wisconsin : John Willey and Sons Inc)

IOP Conf. Series: Materials Science and Engineering **588** (2019) 012003 doi:10.1088/1757-899X/588/1/012003

[4] Farid and Ismail 2010 The effect of the double wave absorber model on the performance of a simple solar water heater *Widyateknika Journal* March 2011.

[5] Huang, YJ, Kuo, TC, Chen, CY, Chang, CH, Wu, PC, and Wu, TH, 2009 *The Design and Implementation of a Solar Tracking Generating Power System, Engineering* Letters, **17**: 4, EL _17_4_06, Advance online publication.

[6] Kristanto P and San Y K 2001 Effect of Thickness of Plates and Inter-Pipe Distance on Perfection Performance of Surya Pelat Datar Collector *Mechanical Engineering Journal*, Petra Christian University

[7] Made Sucipta, I Made Suardamana, Ketut Astawa 2010 Analysis of the Performance of Plate Solar Collectors Finned with Fin Surface Surface Variations *Scientific Journal of M Chakra Mechanical Engineering* **Vol. 4** No. 2. October (88-92)

[8] Mustafa 2008 *Experimental Study of Comparison of Double Plate Collectors and Conventional Collectors on Solar Water Heater Performance*. Thesis Mechanical Engineering Major – Universitas Brawijaya

[9] Mustofa and Ismail NR 2013 *International Journal of Research in Engineering and Technology*, **Vol. 02,** Issue.09, September 2013

[10] Rahardjo Tirtoatmodjo 2005 *Performance of Water Heaters Solar Collector Type Flat Plate With One And Two Cover Glass* Department of Mechanical Engineering - Petra Christian University

[11] Sutrisno 2002 Testing Water Heater Solar Collectors using Wave Absorber Plates with and Without Honeycomb (Surabaya : ITS)

IOP Conf. Series: Materials Science and Engineering **588** (2019) 012004 doi:10.1088/1757-899X/588/1/012004

Calorific value analysis, reduction of period weight, reaction rate, activation energy of old coconut, young coconut waste briquette burning, cocoa

W Nuriana[1], A Suryanto[2] and M Kamal [1]

[1] Faculty of Engineering, University of Merdeka Madiun
[2] Faculty of Agriculture, University of Merdeka Ponorogo

nuriana@unmer-madiun.ac.id

Abstract. Indonesia was a country with considerable biomass energy potential. Biomass energy could be explored from agricultural waste, abundant and easily obtained biomass, for example: old coconut waste, young coconut waste and cocoa waste. This study aims to obtain the heating value of charcoal briquettes from biomass of old coconut waste, young coconut waste and cocoa waste. The study was conducted on a laboratory scale, and used a complete randomized design with 3 repetitions. The procedure of this study begins with sorting raw materials, cutting and drying. Next carbonize, grind, filter, add tapioca glue, then print, and press at 115 kg/cm^2. Carbonize at 400°C, for 30, 60, 90 and 120 minutes. The briquettes produced would be analyzed for density analysis, calorific value, and highest calorific value. Then analyzed for reducing the mass and rate of the combustion reaction. The test results of the old coconut briquette density were 0.937 g/l, the highest calorific value was 6.927 kal/g with a 90 minute carbonization time. Old coconut briquettes had the lowest or lowest combustion reaction rate compared to the level of combustion reaction rate of young coconut briquettes and cocoa briquettes. The effect of a higher density value, the higher the briquette heat value and the slower or lower the combustion rate. In the Thermogravimetry test using Thermal Gravimetry Analyzer (TGA), the highest decrease in mass of old coconut waste briquettes compared to young coconut waste briquettes and cacao waste briquettes was 55.34% from 590.5°C to 599.2°C. In Thermal Gravimetric Differentials. (DTG) shows the rate of combustion reaction in old coconut waste briquettes is 0.10%/minute with an air speed of 5 ml/minute.

1. Introduction

Energy was a basic need for human life, the need for fuel was increasing. As an example of increasing energy consumption in all sectors, among others in the fields of transportation, electricity, industry and domestic. Total energy consumption in 2013 amounted to 1.1 billion BOE (Barel of Equivalent), the use of conventional fuel was quite dominating such as petroleum and coal (ESDM , 2014) The potential of biomass in Indonesia, especially agricultural waste biomass (old coconut skin waste, young coconut and cocoa) was available in large quantities and cheaply, the utilization was not maximized. Beside from being a fuel, these wastes could be used as raw materials for activated charcoal, carbon paper, battery stones and others. In general, agricultural waste was only used as fuel that was directly burned, so that it could cause environmental pollution (Jati, et al., 2005; Nuriana, 2013). Conversion of young coconut skin waste, old coconut and cocoa into briquettes will increase the density, thus increasing economic value. Fuel in the form of briquettes in addition to having high heat values could ignite for a long time (Lusia, 2008).

This research was to obtain the characteristics of the combustion reaction and its application, it was necessary to reduce the mass weight, reaction rate and its relationship with density, heating value in young coconut skin waste briquettes, old coconut and cocoa.

2. Research Method

The research was carried out and analyzed in a laboratory manner, the treatment was repeated three times. The design of the experiment using a completely randomized design with the variable effect of carbonization time and response variable was density, heating value. Weight reduction period and rate when the combustion reaction process.

2.1. Materials and tools

The materials used in this study were: old coconut waste (skin, coir, shell), young coconut (skin, coir, shell), cocoa, starch, water, aluminum foil as a wrap when oven. The equipment used in this study were ovens, knives, trays, mortars, desiccators, printing devices, briquette presses, measuring cups, picnometers, erlenmeyers, volume flasks, electric scales, 80mesh sieves, bomb calorimetry, Thermo Gravimetry Analayzer (TGA) units , Differential Themo Analayzer (DTA).

2.2. Making Briquettes

Old coconut waste, young coconut waste and cocoa waste were sorted, dried in the sun for 3 days to reduce water content. Continued carbonization at 400°C for 30, 60. 90, 120 minutes using an oven. Then pounded, sieved with 80 mesh sieve, followed by mixing with starch glue/stirred until homogeneous, printed and pressed with a pressure of 115 kg / cm^2. The printed briquettes were then dried to reduce the water content contained in the glue, dried in the sun for 2 days. Briquettes were analyzed for water content, density, heating value, combustion reaction rate, reduction in mass weight in the combustion process. From carbonization with a variation of time 30, 60, 90, 120 minutes the highest heating value of each briquette was selected and then analyzed the reaction rate and reduction in mass weight in the combustion reaction process.

3. Result and Discussion

The research was carried out and analyzed in a laboratory manner, the treatment was repeated three times. The design of the experiment using a completely randomized design with the variable effect of carbonization time and response variable was density, heating value. Weight reduction period and rate when the combustion reaction process.

3.1. Old Coconut Waste Briquette Density Test, Young Coconut waste and Cocoa waste

In figure 4.1 shows the graph of the effect of carbonization time at 400°C on the density of old coconut waste, young coconut and cocoa. The manufacture of carbonized briquettes was basically to increase the density of a raw material and obtain a greater heat value than the material before making a charcoal. Increased carbonization time increases density.

In the graph below the price of density increases significantly with the length of carbonization time (Tirono, 2011; Nuriana, 2013).

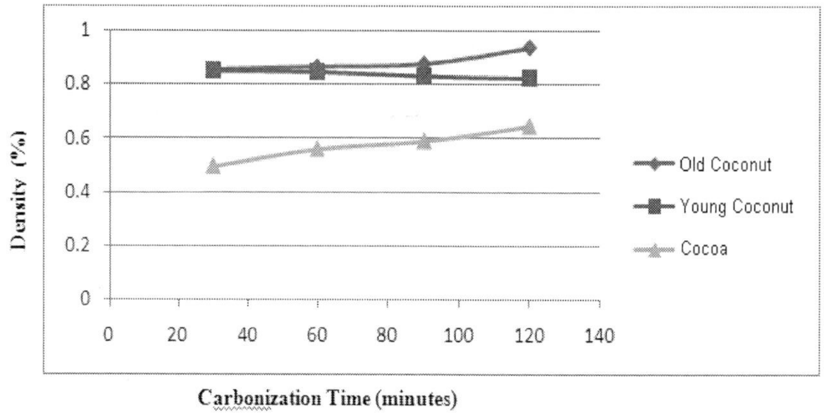

Figure 1. Effect of carbonization times 30, 60, 90, 120 minutes at 400°C on the density of old coconut waste, young coconut waste and cocoa waste.

3.2. Calorific Value Test for Old Coconut Waste Briquettes, Young Coconut Waste and Cocoa Waste

After the carbonization process was carried out at a temperature of 400°C with a time variation of 30, 60, 90, 120 minutes and briquette was made and then pounded and sifted to a size of 100 mesh, to facilitate the heat value test in combustion with oxygen in bomb calorimetry (Syamsiro, 2007). The smaller the particle size will increase the surface so that the material burns faster (Subroto, 2006; Nuriana, 2013). Calorific value was the main parameter in the assessment of a fuel or energy (Scroder E., 2006; Tirono, 2011;

Table 1. The calorific value of old coconut, young coconut and cocoa waste briquettes at a carbonization temperature of 400°C with variations of carbonization time of 30, 60, 90, 120 minutes.

No.	Waste	Time (minutes)	Heating value (cal / g)
	old coconut	30	6.635
		60	6.882
		90	6.927
		120	6.879
2	old coconut	30	5.457
		60	5.846
		90	5.328
		120	5.740
3	cocoa	30	4.925
		60	5.059
		90	4.967
		120	4.787

Source: Nuriana, 2018

The results were tested by TG, DTG in reducing the weight and rate of combustion reaction at the highest calorific value, also in testing the activation energy with Differential Scan Calorimetry (DSC). Of the three old coconut waste briquettes, young coconut, cocoa each had a value of 6,927 cal/g with

IOP Conf. Series: Materials Science and Engineering **588** (2019) 012004 doi:10.1088/1757-899X/588/1/012004

90 minutes carbonization time, 5,846 cal/g with 60 minutes carbonization time, 5,059 cal/g 60 minutes long.

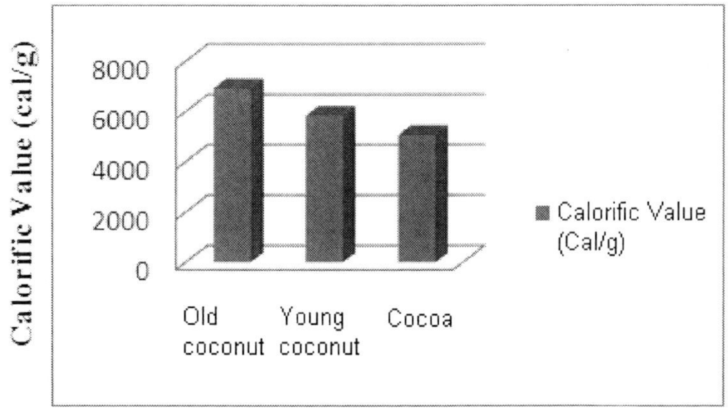

Figure 2. Calorific value of old coconut waste briquettes 6,927 cal / g with carbonization time 90 minutes, young coconut calorific value, 5,846 cal / g with 60 minutes carbonization time, cocoa 5,059 cal/g with 60 minutes long.

3.2.1. The results of the reduction in briquette weight and the reaction rate of burning old coconut waste briquettes were tested by TG and DTG. Changes in the weight of briquettes when starting burning 40°C to 100°C reduced weight was 1.1%, at a temperature of 100 °C to 262°C reduced weight 6.04 and at a temperature of 262°C to 590.5°C reduced weight 37.51% and temperature 590,5°C to 599.2 °C 55.34%. The DTG was shown as an endothermic process, and at the end of the combustion there was a faster reduction of the charcoal period. The rate of combustion reaction in Figure 4.1 was 0.10% / minute. with an air rate of 5 ml / minute.

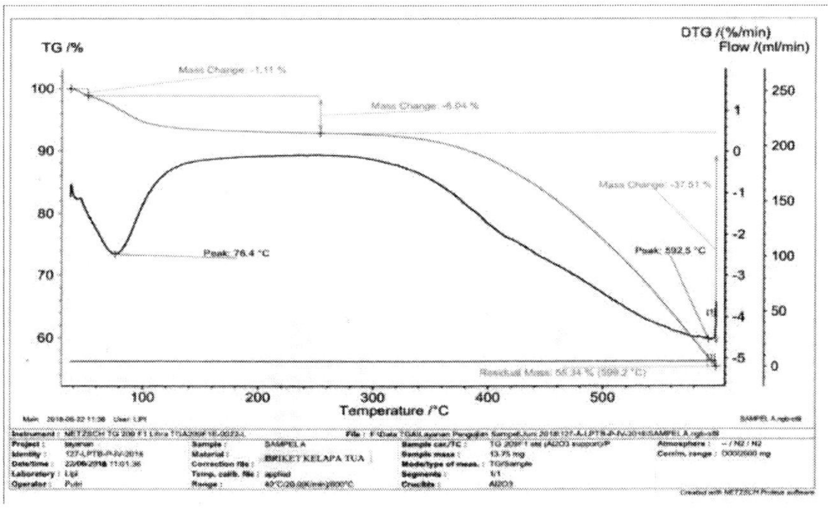

Figure 3. TG and DTG curves for old coconut waste briquettes with heating value of 6,927 cal/g, weight of 13.75 mg briquette sample, temperature of 40-600°C, air rate of 5ml/minute and heating rate of 10°C.

3.2.2. The results of reduction of briquette weight and reaction rate of burning young coconut waste briquettes with testing with TG and DTG. Changes in the weight of briquettes at the start of combustion temperature of 40°C to 79.5°C reduction in weight of 6.04%, at a temperature of 79.5°C to

188°C weight reduction of 6.50%, at a temperature of 258°C to 590°C and a reduction in weight of 42.10% . at a temperature of 590°C to 599.2°C weight reduction drastically 50.09%. The rate of combustion reaction in young coconut waste briquettes was 0.22% / minute.

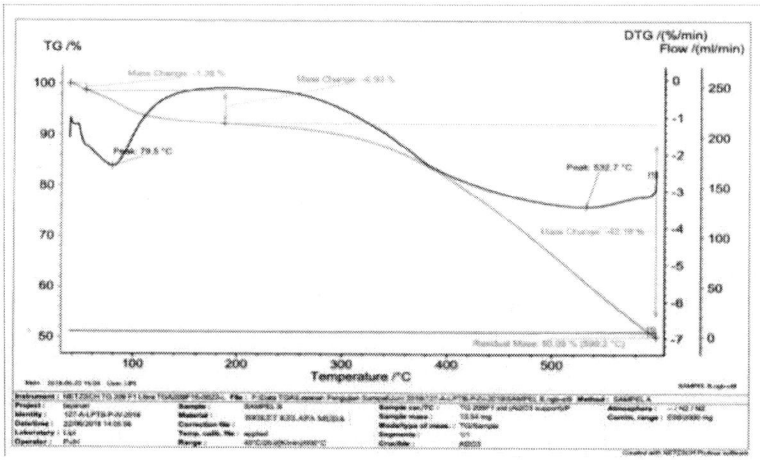

Figure 4. TGA curve and DTA for young coconut waste briquettes with heating value of 5.846 cal/g, weight of briquette samples of 13.54 mg, temperature of 40-600°C, air rate of 5ml/minute and heating rate of 10°C.

3.2.3. The results of the reduction of briquette weight and reaction rate of combustion of cocoa waste briquettes testing with TGA. Changes in the weight of briquette when starting burning temperature of 40°C to 50.5°C reduction in mass weight 0.84%, at a temperature of 50.5°C to 188°C weight reduction of 8.15%, at a temperature of 188°C to 590°C there was a reduction in weight 55.46% . At a temperature of 590°C to 599.2°C there was a reduction in weight of 35.53%. The rate of combustion reaction in young coconut waste briquettes was 0.24% / minute.

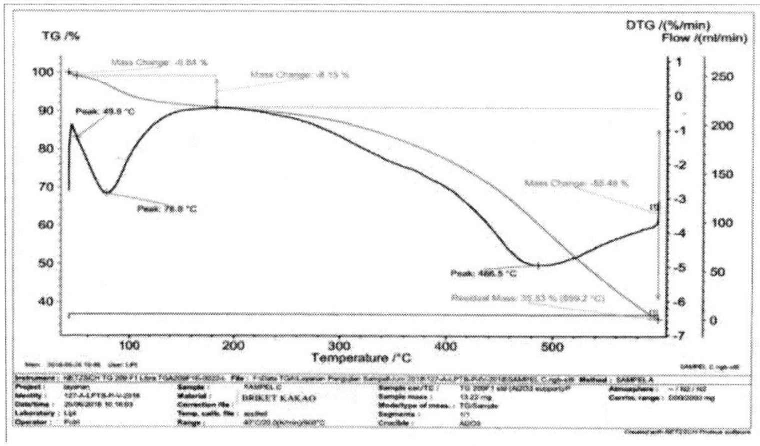

Figure 5. TGA curve and DTA for cocoa waste briquette with heating value of 5,059 cal/g, weight of briquette sample 13,22 mg, temperature 40-600°C, air rate of 5ml / minute and heating rate of 10°C.

The results of this study, the calorific value of old coconut briquettes has a slow or low combustion reaction rate compared to the rate of burning of young coconut briquettes and cocoa. The effect of the higher density value, the higher the briquette heating value and the slower or lower combustion rate according to the study of Afif, et al., 2014; Kurniawan, et al., 2012).

The highest mass weight reduction in old coconut waste briquettes compared to young coconut waste and cocoa briquettes was 55.34% from 590.5°C to 599.2°C in the analysis by Thermogravimetry (Himawanto, DA, et al., 2011; Aries D. Himawanto, 2013).

3.2.4. Activation Energy Results. Activation energy results by testing using DSC (Differential Scaning Calorimetry) tools on old coconut briquettes, young coconut and cocoa each of different weights performed with a variation of air flow rate of 20.30.40 ml / min was stated in table 4.1 below.

Table 2. Results of activation energy in old coconut briquettes, young coconut and cocoa with variations in air flow rates of 20,30,40 ml / minute.

Air flow.	Activation of briquettes (j/ g)		
(ml/minutes)	Old Coconut	Young Coconut	Cocoa
20	163,80	195,40	216,60
30	150,50	182,4	207,10
40	133,20	138,4	185,60

The results of comparison of the activation energy of old coconut briquettes, young coconut and cocoa with variations in airflow rate of 20.30.40 ml / min and the weight of each sample in the DSC test were different, could be seen in the graph below.

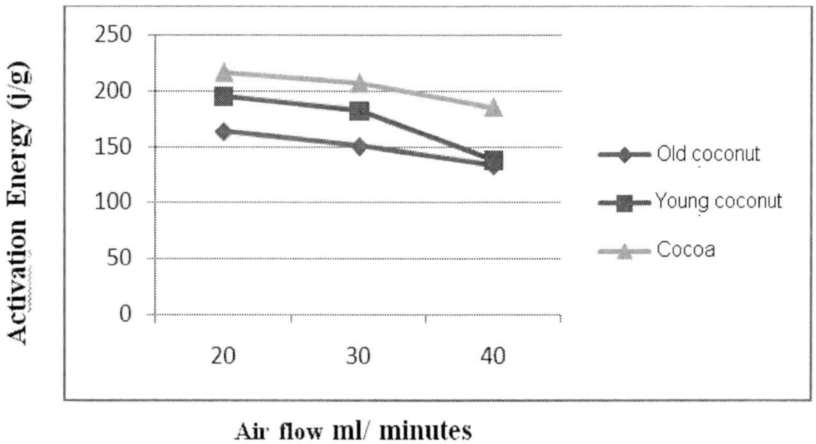

Figure 6. Energy Chart of activation of old coconut briquettes, young coconut and cocoa on variations in air flow rates of 20,30,40 ml / minute.

The higher the air flow rate was 20,30,40 ml / minute, the activation energy in old coconut briquettes, young coconut and cocoa decreases significantly. This was because the higher the air flow rate, the activation energy required for the combustion reaction was smaller. Activation energy depends on the calorific value of the briquette, the activation energy will be low if the fuel has a high fuel value, this occurs in young coconut waste briquettes. Higher air flow speeds have more oxygen content that could help make changes more perfect. Because it has burned more completely, it will require low activation energy (Aries, et al., 2013).

4. Conclusion

The use of sections to divide the text of the paper was optional and left as a decision for the author. Where the author wishes to divide the paper into sections the formatting shown in table 2 should be used.

- The highest calorific value obtained in the old coconut waste briquette was 6,927 cal/g at 400°C carbonization temperature for 90 minutes compared to young coconut waste and cocoa;
- The price of briquette density was high, the heating value has a high price;
- The reaction rate of old coconut waste briquette burning was lower (0.10% / minute) compared to young coconut waste briquettes and cocoa.

Acknowledgement

We thank the honorable ones:

- Directorate General of Strengthening Research and Development, Ministry of Research, Technology and Higher Education with letter No. SP DIPA A-04206.1.401516 / 2017 and Letter of Agreement Number: 035 / SP2H / LT / K7 / KM / 2018 Between Coordinator of Kopertis Region VII and Chancellor of Merdeka University Madiun on February 26, 2018
- Chair of the Polymer Test Laboratory, P2F-LIPI Bandung.
- Chair of the Engineering Physics Department, ITS Surabaya Physics Engineering.
- Head of School and Head of the Analytical Chemistry Laboratory at Madiun University.

References

[1] Afif M.Almu, Syahrul, Yesung AP 2014 Analisa Nilai Kalor Dan Laju Pembakaran Pada Briket Campuran Biji Nyamplung (Calophyllm Inophyllum) Dan Abu Sekam Padi *Jurnal Dinamika Teknik Mesin* **4** No 2m Juli 2014

[2] Aries D Himawanto 2013 Penentuan Energi Aktivasi Pembakaran Bruket Char Sampah Kota Dengan Menggunakan Metode Thermogravimetry Dan Isothermal Furnace. *Jurnal Rotasi* **15** No 3 Juli 2013.

[3] ESDM 2004 *Kebijakan Pengembangan Energi Terbarukan Dan Konservasi Energi. Keputusan Menteri ESDM No.0002 Tahun 2004*

[4] Himawanto, DA Indarto, Saptoadi, H Rohmat TA 2011 Karakteristik Dan Pendekatan Kinetika Global Pada Pirolisis Lambat Sampah Kota Terseleksi Reaktor *ISSN 0852 0798* **13** Juni 2011 141-146

[5] Himawanto DA, Sidhi PR, Indarto, Saptoadi H, Rohmat TA 2011 Analisa Thermogravimatry Pembakaran Briket Char Bambu *Jurnal Teknik* **11** No 1 77-84

[6] Himawanto DA, Indarto, Saptoadi H, Rohmat TA 2012 Thermogravimetric Analysis and Global Kinetics of Segregated MSW Pyrolysis *Modern Applied Science* **6** (1) 120-119

[7] Kurniawan Eddy, Budi Wahyudi S, Muslichin H 2012 Karakteristisasi Dan Laju Pembakaran Biobriket Campuran Sampah Organik Dan Bungkil Jarak (Jatropha curcas L.) *Jurnal Rekayasa Proses* **6** No. 2 2012

[8] Kurniawan Eddy 2012 *Model Matematis Laju Pembakaran Biobriket Campuran Sampah Organik Dan Bungkil Jarak (Jatropha curcas L.)*

[9] Mahajoeno Edwi 2005 *Energi Alternatif Pengganti BBM Potensi Limbah Biomassa Smart Sebagai Sumber Energi Terbarukan.*

[10] Nuriana W 2013 Synthesis Preliminary Studies Durian Peel Bio Briquettes As an Alternative Fuels *Energy Procedia Journal* **47** 295-302

[11] Nuriana W 2018 *Karakteristik Biobriket Limbah Kelapa (Tua, Muda) Dan Kakao Sebagai Energi Terbarukan Pada Variasi Waktu Karbonisasi. Kementrian Hukum Dan Hak Asasi*

Manusia Surat Pencatatan Ciptaan, Nomor EC00201800774, Tanggal Permohonan: 19 Januari 2018

[12] Scroder Eliabeth 2006 *Experiment on the Generation of Activated Carbon from Biomass* (Germany : Institute for Niclear and Energy Technologies Forschungs Karlsruhe 106-111)

[13] Tirono M, Sabit, 2011 Efek Suhu Pada Proses Pengarangan Terhadap Tempurung Kelapa *Jurnal Neutrini* **3** No. 2. Fakultas Sains dan Teknologi UIN Maulana Malik Ibrahim Malang

[14] Syamsiro M. Harwin Saptoadi 2007 *Pembakaran Briket Biomassa Cangkang Kakao: Pengaruh Temperatur Udara Preheat* Seminar Nasional Teknologi Teknik Mesin UGM Yoyakarta.

[15] Sulistyanto Amin 2005 *Pengaruh Variasi Bahan Perekat Terhadap Laju Pembakaran Biobriket Campuran Batubara Dan Sabut Kelapa*

[16] Subroto 2006 Karakteritik Pembakaran Biobriket Campuran Batubara, Ampas Tebu, Jerami. *Media Mesin* **2** Juli 2006 47-54

IOP Conf. Series: Materials Science and Engineering **588** (2019) 012005 doi:10.1088/1757-899X/588/1/012005

Characterization of Engine Performance, Combustion Process and Emission of Diesel/CNG Dual Fuel Engines with Pilot Injection Timing Variation at Low Load

Dori Yuvenda[1,2, a], Bambang Sudarmanta[1,b], Arif Wahjudi[1,c] and Jose da Silva[1]

[1] Department of Mechanical Engineering, Faculty of Industrial Technology, Institut Teknologi Sepuluh Nopember. Jalan Raya ITS, Keputih, Sukolilo, Surabaya 60111, Indonesia

[2]Department of Automotive Engineering, Faculty of engineering, Universitas Negeri Padang. Jalan Prof. Dr. Hamka Air Tawar Barat, Padang 25171, Indonesia

[a] doriyuvenda@ft.unp.ac.id, [b]bsudarmanta@gmail.com, [c]arid_w@me.its.ac.id

Abstract. The use of CNG on dual fuel diesel engines causes a decrease in engine performance, combustion and increased emissions, especially at low loads, which is affected by the addition of ignition delay period so that the maximum of peak combustion away from the top dead center (TDC). Pilot injection timing acts as an ignition delay controller. This research was carried out using an experimental method on diesel/CNG dual fuel engine a single-cylinder to determine the engine performance, combustion process and emissions characteristics with a variation of pilot injection timing from 11^0 to 19^0 BTDC with an interval of 2^0. The results showed an increase in thermal efficiency of 3.006%, maximum cylinder pressure up to 4.78 MPa, maximum heat release rate of up to 52.05 kJ /m^3/^{0}CA, and decrease hydrocarbons (HC), carbon dioxide (CO) and particulate matter (PM) emissions at pilot injection timing 17^0 BTDC under low load. Likewise, combustion performance has increased marked by a reduction in combustion duration and maximum of peak combustion approaching TDC by advancing the pilot injection timing on diesel/CNGdual fuel engine.

Keyword. Engine performance, combustion process, emissions, diesel/CNG dual fuel, pilot injection timing, ignition delay period

1. Introduction

The dual fuel engine operates using two fuels that are burned simultaneously. Alternative fuels become the main fuel while diesel fuel as a pilot diesel[1]. One of alternative fuel that is widely used in dual fuel diesel engine is compressed natural gas (CNG)[2][3]. This is because CNG has the characteristics of high auto-ignition temperature (± 650^0C)[4][5], so it can be used on engine that have a high compression ratio. The combustion process that occurs in dual fuel diesel engines is in sequence, where diesel fuel is used as a flame to burn CNG. This is because CNG cannot burn by itself. Moment a diesel fuel is injected, a mixture of diesel fuel and air-CNG mixture is called the ignition delay period. The ignition delay period occurs in two phases, physical delay and chemical delay phases. Both of these phases will affect the duration or shortness of ignition delay. The dual fuel diesel engine has a longer ignition delay period compared to single fuel diesel engines. This is due to a decrease in partial pressure from oxygen and the intake temperature of the CNG-air mixture so that it affects the pressure and temperature of the compression stroke[6][7]. The addition of ignition delay time causes a decrease in combustion performance and an increase in HC and CO emissions in dual fuel diesel engines[8]. Prolong ignition delay causes the maximum of peak pressure away from the position of the TDC so that combustion tends to occur at the step of the business which results in a decrease in engine performance, especially at low loads [4]. Several studies related to dual fuel diesel engines to overcome these problems have already been carried out. Yuvenda and Sudarmanta[9], analyze the pressure reducer by optimizing the strength of the spring as a pressure reducing gas and supplying CNG fuel to the combustion chamber and they find that by increasing the spring constant at the pressure reducer can increase the amount of CNG gas supply into the combustion chamber on the diesel dual fuel engine so that it can increase CNG gas substitution as the main fuel. Sayin, et al.[10] investigate the effect of pilot injection timing on the performance and emissions of diesel dual fuel engines under low and high load and they found that by advancing the pilot injection timing it increases thermal efficiency and decreases CO and HC emissions under low load but, moment pilot injection timing further advances the decrease in thermal efficiency. Liu, et al.[11] investigated the effect of pilot injection timing and diesel pilot quantity on dual fuel diesel exhaust emission characteristics under low, medium and high load and they found that by advancing the pilot injection timing and increasing diesel pilot quantity can reduce NO_x and PM emissions significantly but HC and CO emissions increased slightly in all loading. Yang, et al.[12][13] investigated the effect of pilot injection timing on combustion performance and reduction of emissions especially HC and CO emissions on diesel dual fuel engines at low loading and they found that by advancing the pilot injection timing it could increase cylinder pressure and heat release rate as well as significantly reduce HC and CO emissions but NO_x emissions increase. Wang, et al.[14] investigate the effect of pilot injection timing on combustion and emissions characteristics on dual fuel diesel engines under low load, and they found that by advancing the pilot injection timing can increase cylinder pressure, heat release rate and thermal efficiency and significantly reduce HC emissions but NO_x emissions increase, but if it is further advanced the injection time of the diesel pilot then it reduces the performance of the diesel dual fuel engine. Guerry, et al.[15] investigated the effect of pilot injection timing on combustion performance and emissions of diesel dual fuel engines with the method of low temperature combustion (LTC) under low and high loads, and they found that by advancing the injection time of diesel pilots can increase cylinder pressure and the rate of heat release and reduce emissions of HC and CO, but if the pilot injection time of diesel is further promoted further reduce combustion performance and increase exhaust emissions under low load. Papagiannakis, et al.[16] investigated the effect of pilot injection timing on combustion performance and emissions on dual fuel diesel engines at low loading and they found that by advancing the pilot injection timing increased cylinder pressure and heat release rate and reduced exhaust emissions on dual diesel engines fuel under low load. Shu, et al.[17] investigated the effect of pilot injection timing on thermodynamics, combustion and emissions on diesel dual fuel engines at low speed and load, and they found that by advancing the pilot injection timing can increase cylinder pressure and heat release rates and thermal efficiency while NO_x and HC emissions also increase. Based on the research that has been done above, it can be concluded that the timing of pilot injection timing can affect engine performance, combustion and emissions on desel/CNG dual fuel engines under low load. This study aims to correlate engine performance, combustion and exhaust emissions by optimizing the pilot injection timing of diesel/CNG in dual fuel engines under low load.

2. Experimental apparatus ang procedure

2.1. Test engine

The diesel engine used in this study is a Diamond DI 800 single-cylinder using a water cooling system. The engine specifications used in this study are shown in Table 1. This engine is modified into diesel/CNG dual fuel engine by adding CNG as the main fuel as shown in Figure 1. CNG is injected through the port channel that is controlled using an electrical control unit (ECU) whereas diesel fuel is injected directly into the combustion chamber using a conventional system. CNG is injected at 45^0 ATDC at the suction stroke, while the standard diesel pilot injection timing is 13^0 BTDC at the end of the compression stroke.

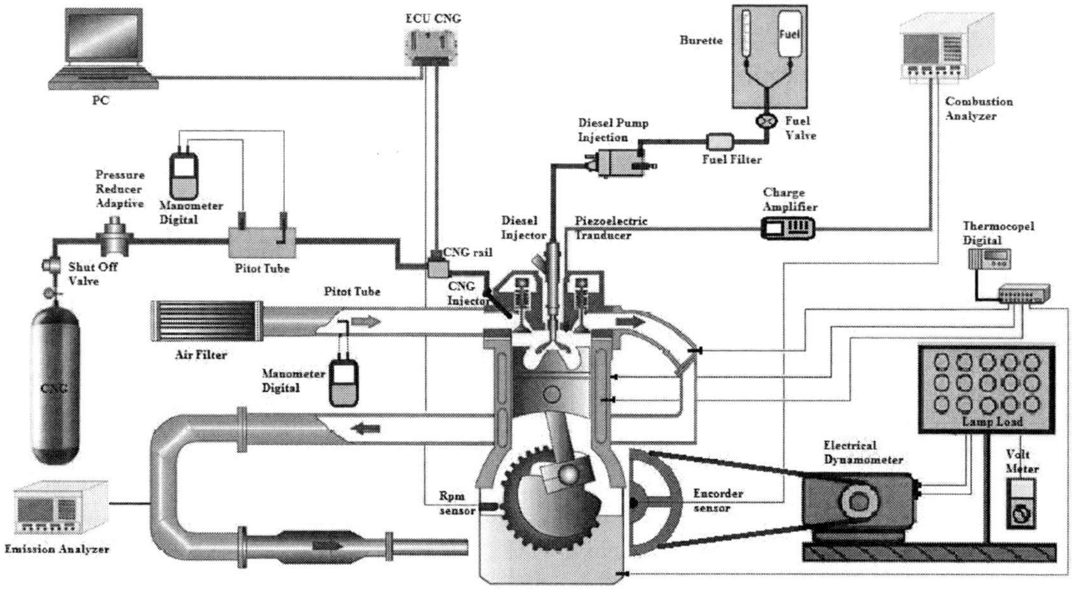

Figure 1. Experimental setup

2.2. Fuel and supply system

In this study, diesel fuel is used as a pilot for diesel while CNG is the main fuel for diesel/CNG dual fuel engine. The diesel fuel and CNG properties used in this study are shown in tables 2 and 3 below[18]. Experimental research on diesel/CNG dual fuel engines using a dual fuel system. Diesel fuel uses conventional fuel systems that already exist on the engine. CNG fuel is supplied from a gas tank that has a pressure of 250 bar which is equipped with a safety valve, then the pressure is lowered using an adaptive reducer pressure[19] to a working pressure of 2 bar. CNG is injected into the intake port using a CNG injector.

Table 1 The poperties of fuel.

Fuel Poperties	Diesel	CNG
Low heating value (MJ/kg)	42,8	48,6
Cetane number	52,5	-
Octane number	-	130
Auto-ignition temperature (^{0}C)	316	650
Stoichiometric air–fuel ratio (kg/kg)	14,69	17,2
Carbon content (%)	87	75

Table 2 Composition of natural gas.

Fuel Poperties	Volumetric concentration (%)
Methane	96.160
Ethane	1.096
Butane	0.136
Iso-Butane,n-butane	0.021
Iso-Pentane, n-pentane	0.006
N2	0.001
H2S	0.0002
H2O	0.006

2.3. Instrument arrangement and data acquisition
Diesel/CNG dual fuel engine coupled with an electrical generator used for loading the engine. Kistler piezoelectric pressure sensors are mounted on the cylinder head to measure cylinder pressure and heat release rate with a 0.20 crank angle (CA) resolution. CNG mass flow rate and air are measured using a digital manometer with an accuracy of ± 0.001. The machine is given a low load using a lamp with a capacity of 1500 watts (33.3%) connected to an electric generator. The consumption of diesel fuel is measured using a burette meter with an accuracy of 1 ml. The gas star 898 emission analyzer is used to measure HC and CO exhaust emissions while the PM is measured using a smoke opacity meter parallel to the emission analyzer.

2.4. Test procedure and conditions
The test engine is heated for 30 minutes until it reaches a temperature of ± 80^0 at ambient temperature of 25^0. Experiment testing is carried out on dual fuel diesel engines at low, medium and high loads at rotation 1500. The test is divided into two stages, first, testing the engine with standard fuel as control variable. Second, testing the engine using two fuels with a variation on the time of injection of a diesel pilot. Pilot injection timing is advanced from 110 BTDC to 190 BTDC. Advancing the pilot injection timing is done by reducing the thickness of the shim plate on the diesel injection pump holder. Reducing the thickness of the shim plate 0.1 mm can advance the pilot injection time of 1^0 CA.

3. Results and discusstions
This discussion analyzes engine performance, combustion process and emissions in diesel/CNG dual fuel engines under low loads with varying pilot injection timing. Power, specific fuel consumption (SFC) and thermal efficiency are engine performance while the combustion process includes cylinder pressure, heat release rate (HRR), ignition delay and combustion duration, then emissions include HC, CO and PM. Combustion include cylinder pressure, heat release rate, ignition delay, combustion duration, while. Ignition delay is defined as the start of the pilot injection until the start of combustion while the duration of combustion is the start of combustion (SOC) to the end of combustion (EOC). Generally, the ignition delay period on a diesel/CNG dual fuel engine is longer than a single fuel engine[6] while the duration of combustion is affected by the quality of premix combustion and diffusion of diesel/CNG fuel. The combustion process in diesel/CNG dual fuel engines has three stages. The first stage of combustion is premix from diesel fuel and a small part of the gas that starts to mix with diesel fuel, the second stage is premix combustion of CNG and the third stage is the combustion of diffusion from diesel fuel and CNG [13].

3.1. Engine performance characteristics of diesel/CNG dual fuel engine
In figure 2, shows the power, SFC and thermal efficiency under variations in pilot injection timing at low loads. Figure 2. It is seen that thermal efficiency has a significant difference with the variation of pilot injection timing compared to power and SFC. Advancing the pilot injection time from 11^0 to 19^0

BTDC can increase thermal efficiency with a maximum value of up to 14.03% at pilot injection time 17^0 BTDC under low load. This is because by advancing the pilot injection timing prolongs the mixing time of diesel fuel with CNG-air mixture so that the mixture is more homogeneous, so that it initiates a large number of hotspots around the combustion chamber which influence the premix combustion speed so as to improve diffusion combustion quality better and cause an increase in thermal efficiency. However, if the pilot injection timing is further advanced, the efficiency decreases slightly. This is because the pilot injection timing is too fast in the compression stroke with the pressure and temperature that is still low so that the combustion performance decreases which affects the decrease in thermal efficiency. This proves that the pilot injection timing affects the thermal efficiency so that optimization is needed to get the right pilot injection timing. In figure 2, also displays the power and SFC on diesel/CNG dual fuel engine under different pilot injection timing. Pilot injection timing does not have a significant effect on power and SFC, this is due to the given load and constant engine speed (1500 rpm) so that the voltage and current generated by the electricity generator are the same, as well as the engine fuel consumption.

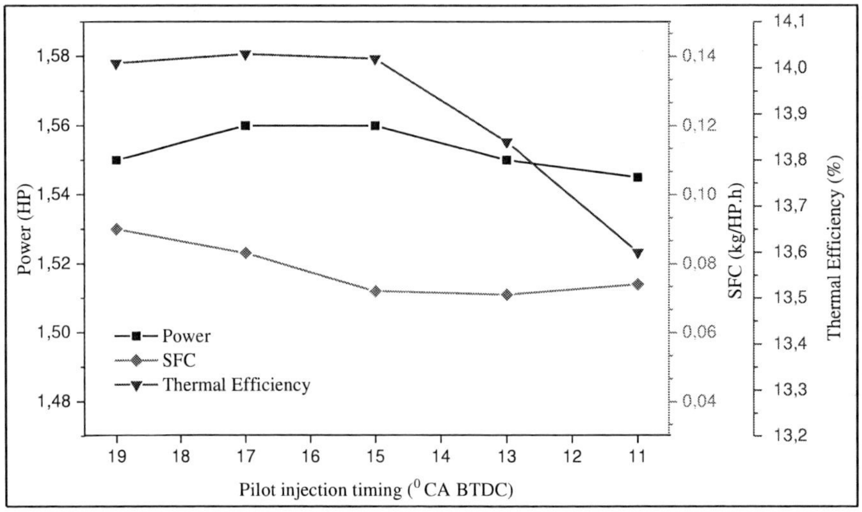

Figure 2. Engine performance versus pilot injection timing under low load

3.2. The cylinder pressure and heat release rate characteristics of diesel/CNG dual fuel engine

In figure 3, shows cylinder pressure and HRR below the variation of diesel pilot injection timing at low loads. Figure 3 explains that the cylinder pressure has two peaks, the first peak of which is the premix combustion of diesel and CNG while the second peak is the diffusion combustion of diesel fuel and CNG. Advancing the diesel pilot injection timing can increase cylinder pressure and also the cylinder pressure curve tends to lead to one peak. A significant increase occurred at the pilot injection timing of 17^0 BTDC with a value of 4.87 MPa compared to other variations. The peak of diffusion combustion occurs at an angle of 12^0 ATDC near TDC at the pilot injection timing of 17^0 BTDC. This shows that by burning the peak near TDC will produce maximum cylinder pressure. However, moment the pilot injection timing is further advanced at 19^0 BTDC there is a decrease in cylinder pressure. This is due to the injection that is too early at compression pressure and the temperature is too low so that the combustion of the fire speed that develops becomes slow. Likewise with pilot injection timing retard there is a decrease in combustion pressure, especially at the second peak in diffusion combustion. This is affected by a long ignition delay period on dual fuel engines so that maximum combustion occurs at the expansion stroke away from TDC. Figure 3. Also explains the rate of heat release (HRR) affected by pilot injection timing under low load. Advancing the pilot injection timing causes an increase in HRR, especially at pilot injection timing of 17^0 BTDC with a maximum value of 52.05 $kJ/m^3/^0CA$. Peak heat release rate occurs during diffusion combustion. This is because the cylinder pressure and high temperature due to the energy released at the compression step increases so that the peak heat release occurs at TDC. Likewise, advancing the pilot injection timing causes the ignition delay period to be longer so that mixing between the pilot fuel and the CNG-air mixture becomes more homogeneous[13]. Thus the number of ignition kernel will form more in the area of the combustion chamber so that the propagation of the fire becomes faster, resulting in increased premix and diffusion combustion. However, if the pilot injection timing is further promoted causing a decrease in HRR, due to compression pressure and the temperature is still low so that the combustion quality decreases. Besides that, by pilot injection timing retard too far there was a decline in HRR, especially during the pilot injection timing of 11^0 BTDC. This is affected by the ignition delay on the diesel/CNG dual fuel engine for longer so that the maximum of peak cylinder pressure away from the TDC so that diffusion combustion occurs more in the expansion stroke. This is indicated by the high heat release rate in the expansion stroke.

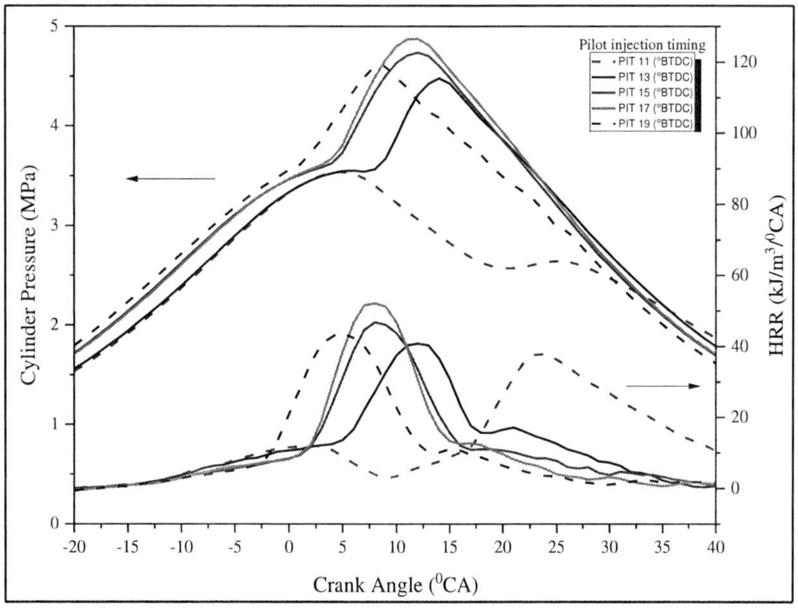

Figure 3. The cylinder pressure and HRR versus pilot injection timing under low load

3.3. Combustion performance characteristics of diesel/CNG dual fuel engine

The combustion performance described in this study is ignition delay period and combustion duration both under low load. Figure 4, shows ignition delay period and duration of combustion with different pilot injection timing under low load. Advancing the pilot injection timing increases the ignition delay period. The results show an ignition delay with a pilot injection timing of 19^0 BTDC longest compared to other variations. The addition of ignition delay period is affected by physical and chemical delay. Physical delay process is caused by the process of atomization, evaporation and mixing of fuel, while chemical delay is caused by chemical reactions that occur in the formation of ignition kernel. CNG fuel input causes a decrease in charger temperature due to the high specific heat ratio of CNG-air mixture and partial pressure reduction from oxygen, so the chemical delay process becomes long[6], and when the pilot injection timing is advanced, the delay becomes even longer. However, this addition gives a positive value for mixing the pilot fuel with the CNG-air mixture in the combustion chamber, so that the pilot fuel mixture with gas-air becomes more homogeneous and increases the formation of ignition kernel, thus increasing premix combustion from diesel fuel. On the other hand it gives a negative effect if the pilot injection timing is further advanced, where the pilot fuel is injected too early at low temperatures and pressures so that the flame propagation speed becomes slower so the diffusion combustion becomes lower. However, if the pilot injection timing is delayed the CNG-air mixture density in the combustion chamber becomes higher so that the atomization process becomes disrupted which results in fewer ignition kernel which appear. The consequence is the peak of combustion away from the TDC that occurs in the expansion stroke. In Figure 4, it also shows that the duration of combustion on diesel/CNG dual fuel engines under low load has a different phenomenon. The results show that by advancing the pilot injection timing 17^0 BTDC it can accelerate the duration of combustion so that it affects the rate of heat release. This is because the energy burned is more at the maximum cylinder pressure and also the peak position of the cylinder pressure is at TDC so that the combustion temperature is still high. Likewise with a more homogeneous mixture state produces more fire points appear in each area of the combustion chamber so that the time needed to burn the mixture of fuel and air becomes less. However, if the injection time is further advanced there is an additional duration of combustion due to the initial injection on the pressure and low temperature. Another phenomenon, moment the pilot injection timing is retard at the pilot timing injection 11^0 BTDC there is a significant increase in the duration of combustion. This is affected by the peak of the cylinder pressure that occurs at the expansion stroke so that the duration of combustion becomes longer.

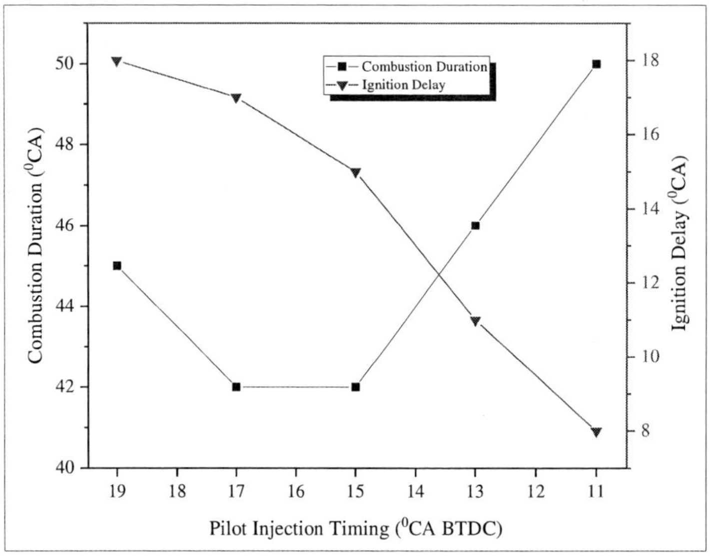

Figure 4. Combustion performance versus pilot injection timing under low load

3.4. Emissions characteristics of diesel/CNG dual fuel engine

In Figure 5, shows emission characteristics on diesel/CNG dual fuel engines at various pilot injection timing under low load. Figure 5, shows that HC, CO and PM emissions have the same chart trend under different pilot injection timing. The results show that the pilot injection timing can significantly reduce HC and CO emissions. One of the causes of CO emissions is incomplete combustion due to low cylinder pressure and temperature. Especially under low pressure and temperature loads are more likely to be lower than medium and high loads. By advancing the pilot injection timing increase the cylinder pressure and the temperature to be higher so that the combustion becomes more perfect. Likewise with HC emissions on diesel/CNG dual fuel engines more likely to increase under low load. HC emissions in 80% from diesel/CNG dual fuel are caused by unburned of CNG. The results show that by advancing the pilot injection timing can prolong the fuel and air mixture to become more homogeneous so that the unburned fuel decreases and also when the ignition starts earlier so that the fuel will burn on TDC with high temperature conditions. Moment the pilot injection timing is retard the combustion occurs more in the expansion stroke so that the pressure and temperature drop so that the fuel is more unburned and wasted through the exhaust manifold. Figure 5, also shows that PM emission has a significant decrease by advancing the pilot injection timing, because by advancing the high cylinder pressure and temperature and homogeneous mixture so that the PM emission is reduced.

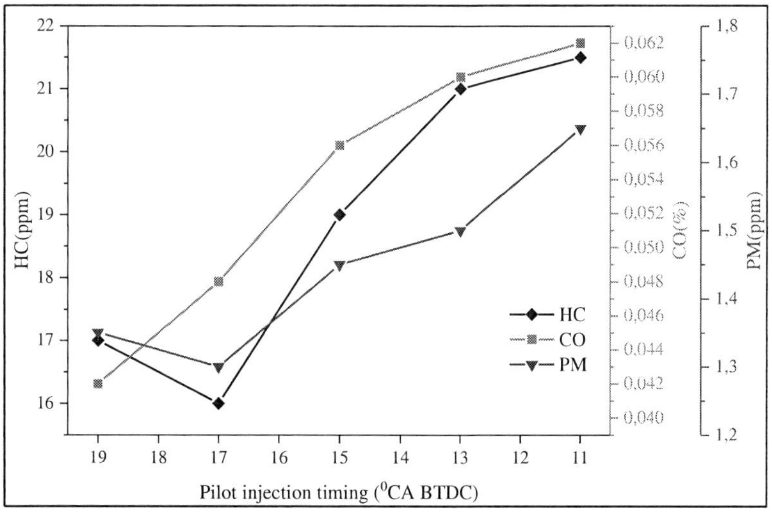

Figure 5. Emission diesel/CNG dual fuel versus pilot injection timing under high load

4. Summary

This study investigates the characteristics of engine performance, combustion and emissions of diesel/CNG dual fuel engine with varying pilot injection timeing under low load at constant speed. The Cylinder pressure, heat release rate, ignition delay and combustion duration were measured using a combustion analyzer, while emissions of HC, CO and PM were measured by using four gas analyzers. Engine performance includes power, SFC and thermal efficiency calculated by the voltage and current data obtained from the electrical generator. Based on the parameters that have been analyzed, we can conclude that the engine performance, proses pembakaran and emissions on diesel/CNG engine dual fuel is affected by the pilot injection timing under low load. Advancing the pilot injection timing of can increase cylinder pressure up to 4.87 MPa, heat release rate up to 52.05 kJ/ $m^3/^0$CA and reduce emissions of CO, HC and PM. Likewise, combustion performance has increased marked by a reduction in the duration of combustion and the maximum of peak combustion approaching TDC by advancing the pilot injection timing.

5. References

[1] M. Y. E. Selim, "Sensitivity of dual fuel engine combustion and knocking limits to gaseous fuel composition," *Energy Convers. Manag.*, vol. 45, no. 3, pp. 411–425, 2004.

[2] M. Y. E. Selim, "Pressure-time characteristics in diesel engine fueled with natural gas," *Renew. energy*, vol. 22, no. 4, pp. 473–489, 2001.

[3] G. . Abd Alla, H. . Soliman, O. . Badr, and M. . Abd Rabbo, "Effect of injection timing on the performance of a dual fuel engine," *Energy Convers. Manag.*, vol. 43, no. 2, pp. 269–277, 2002.

[4] J. B. Heywood, *Internal Combustion Engine Fundementals*, vol. 21. 1988.

[5] B. Yang, X. Wei, C. Xi, Y. Liu, K. Zeng, and M. C. Lai, "Experimental study of the effects of natural gas injection timing on the combustion performance and emissions of a turbocharged common rail dual-fuel engine," *Energy Convers. Manag.*, vol. 87, pp. 297–304, 2014.

[6] M. M. Abdelaal and A. H. Hegab, "Combustion and emission characteristics of a natural gas-fueled diesel engine with EGR," *Energy Convers. Manag.*, vol. 64, pp. 301–312, 2012.

[7] R. G. Papagiannakis and D. T. Hountalas, "Experimental investigation concerning the effect of natural gas percentage on performance and emissions of a DI dual fuel diesel engine," vol. 23, pp. 353–365, 2003.

[8] R. G. Papagiannakis and D. T. Hountalas, "Combustion and exhaust emission characteristics of a dual fuel compression ignition engine operated with pilot Diesel fuel and natural gas," vol. 45,

pp. 2971–2987, 2004.

[9] D. Yuvenda, "KARAKTERISASI PERFORMA MESIN SISTEM DUAL FUEL MENGGUNAKAN PRESSURE REDUCER ADAPTIVE DENGAN VARIASI KONSTANTA (k) PEGAS HELIX TEKAN DAN TEKANAN," 2015, no. 2014, pp. 1–8.

[10] C. Sayin and M. Canakci, "Effects of injection timing on the engine performance and exhaust emissions of a dual-fuel diesel engine," *Energy Convers. Manag.*, vol. 50, no. 1, pp. 203–213, 2009.

[11] J. Liu, F. Yang, H. Wang, M. Ouyang, and S. Hao, "Effects of pilot fuel quantity on the emissions characteristics of a CNG/diesel dual fuel engine with optimized pilot injection timing," *Appl. Energy*, vol. 110, no. x, pp. 201–206, 2013.

[12] B. Yang, C. Xi, X. Wei, K. Zeng, and M. C. Lai, "Parametric investigation of natural gas port injection and diesel pilot injection on the combustion and emissions of a turbocharged common rail dual-fuel engine at low load," *Appl. Energy*, vol. 143, pp. 130–137, 2015.

[13] B. Yang, L. Wang, L. Ning, and K. Zeng, "Effects of pilot injection timing on the combustion noise and particle emissions of a diesel/natural gas dual-fuel engine at low load," *Appl. Therm. Eng.*, vol. 102, no. x, pp. 822–828, 2016.

[14] Z. Wang *et al.*, "Impact of pilot diesel ignition mode on combustion and emissions characteristics of a diesel/natural gas dual fuel heavy-duty engine," *Fuel*, vol. 167, pp. 248–256, 2016.

[15] E. S. Guerry, M. S. Raihan, K. K. Srinivasan, S. R. Krishnan, and A. Sohail, "Injection timing effects on partially premixed diesel-methane dual fuel low temperature combustion," *Appl. Energy*, vol. 162, no. x, pp. 99–113, 2016.

[16] R. G. Papagiannakis, S. R. Krishnan, D. C. Rakopoulos, K. K. Srinivasan, and C. D. Rakopoulos, "A combined experimental and theoretical study of diesel fuel injection timing and gaseous fuel/diesel mass ratio effects on the performance and emissions of natural gas-diesel HDDI engine operating at various loads," *Fuel*, vol. 202, pp. 675–687, 2017.

[17] J. Shu, J. Fu, J. Liu, L. Zhang, and Z. Zhao, "Experimental and computational study on the effects of injection timing on thermodynamics, combustion and emission characteristics of a natural gas (NG)-diesel dual fuel engine at low speed and low load," *Energy Convers. Manag.*, vol. 160, no. January, pp. 426–438, 2018.

[18] B. Yang, L. Ning, W. H. Chen, B. Wang, K. Zeng, and W. Dong, "Parametric investigation the particle number and mass distributions characteristics in a diesel/natural gas dual-fuel engine," *Appl. Therm. Eng.*, vol. 127, pp. 402–408, 2017.

[19] D. Yuvenda, B. Sudarmanta, and E. Alwi, "Analisis Kekuatan Pegas Pressure Reducer Sebagai Penurunan Tekanan Pada Mesin Duel Fuel," vol. 17, no. 2, 2017.

IOP Conf. Series: Materials Science and Engineering **588** (2019) 012006 doi:10.1088/1757-899X/588/1/012006

Potential of Biomass Utilization in Rotary Kiln of Nickel Processing Plant

B C Pamungkas[1,a] and H Hadi[1,b]
[1] Engineering Physics Department, Faculty of Industrial Technology, Sepuluh Nopember Institute of Technology

[a] bcpamungkas@gmail.com, [b] hadiharsono00@gmail.com

Abstract. Global energy-related CO2 emissions rose by 1.4% in 2017 and reached a historic high of 32.5 Gigatonnes (Gt). One of the biggest contributors to CO2 emission is coal combustion that emits 96.1 kg CO2 for each GJ energy production. Top consumers of coal in the world are the iron-steel industry (28.5%) and the non-metallic minerals (21.4%). Rotary kiln is one of the most energy consuming equipment in nickel pyro-metallurgy processing plant. The specific energy consumption for calcine production is determined to be 2.26 GJ/ton calcine. In the rotary kiln, coal is used as fuel and a reducing agent. Charcoal from palm kernel shell has similar characteristics with coal and very suitable to substitute coal. Charcoal from PKS has many advantages compared to other biomass: high fixed carbon content, high heating value, and low ash content. Charcoal from palm kernel shell can be used as fuel and a reducing agent. Potential 9.9% reduction of CO2 emission can be achieved by using 20% PKS charcoal as fuel and 5% PKS charcoal as a reducing agent with minimum impact on the current operation.

1. Introduction

Global energy-related CO2 emissions rose by 1.4% in 2017 and reached a historic high of 32.5 Gigatonnes (Gt) (International Energy Agency, 2018). One of the biggest contributors to CO2 emission is coal combustion that emits 96.1 kg CO2 for each GJ energy production. Top consumers of coal in the world are the iron-steel industry (28.5%) and the non-metallic minerals (21.4%). Rotary kiln process is widely used in the iron-steel industry, cement industry, and non-ferrous mineral processing plant. Production of nickel by pyrometallurgical process considered energy-intensive because of high intensity of heat required for all stages: ore drying, calcination, smelting, and purifying. One of the highest energy consumption is the chemical reaction for calcination and reduction process that takes place inside the rotary kiln. At this stage, semi-dried laterite ore heated up to temperature 800-900 degree Celcius. Currently, the main source of energy in most kilns are fuel oil and coal that emit a high amount of greenhouse gas emissions. To reduce the pollution two type of improvements can be made: the first option is using more environmental-friendly fuels, such as natural gas, biofuel, biomass, solar, and the second option is by improving the energy efficiency of the kiln. From an economic point of view, renewable resources such as biomass and solar are more promising for future energy because fossil fuel price will increase when the resources exhausted in the future.

Biomass has been applied in the iron and steel metallurgical processes and cement manufacturing process [1,2]. Biomass mainly comprises of five organic elements, such as carbon (C), hydrogen (H), oxygen (O), nitrogen (N) sulfur (S); and inorganic elements such as aluminium (Al), silica (Si), potassium (K), calcium (Ca), sodium (Na), etc in its ash. The objective of this paper is to study the potential of biomass utilization in the rotary kiln of nickel processing plant.

2. Methodology and mass and energy flows in rotary kiln process

2.1. Process description

The main function of the rotary kiln is to remove all water content of semi-dried laterite ore and to perform the calcination process and reduction reaction of metallic oxide. Semi-dried ore from rotary dryer contained water in the form of residual free moisture, crystal water, and chemically bonded water. Evaporation of free moisture occurred immediately in the front section of the kiln where the temperature is lowest. Removal of crystal water and chemically bonded water require higher temperature and occurred afterward in the pre-reduction process. Reduction process of nickel oxide and other metallic oxides occurred near the outlet section of the kiln where carbon monoxide formed because of the oxygen-starved combustion environment. Oxygen required for complete combustion delivered into kiln from four air pipes installed in the middle section of the kiln to burn out remaining carbon monoxide and other combustible gases.

The rotary kiln for which this paper studied is 135 m long cylinder fabricated steel-plate with 6 m outer diameter and 0.8-0.9 rpm rotation speed. The kiln can process 241 ton per hour of semi-dried lateritic ore with an average output of 3500 ton calcine per day. The kiln has four supports with 3.1% inclination with the horizontal plane between inlet and outlet. Support number #1 is the one has the lowest elevation and located near the discharge end of the kiln.

Fossil fuel is the main source of energy input for the rotary kiln with consumption of 9140 kg per hour and 7224 kg per hour, respectively. High sulfur fuel oil used as fuel at the main burner and lance burner located on the outlet side of Kiln, meanwhile medium to high-grade coal used as reducing agent that mixed with semi-dried lateritic ore before conveyed into Kiln inlet. The heating value of high sulfur fuel oil and coal are 40,271 kJ/kg and 25,995 kJ/kg, respectively. Energy consumption of the Kiln influenced by several factors: lateritic ore chemistry and moisture, temperature distribution, rotational speed, and condition of kiln internal components.

2.2. Mass and energy balance

Mass and energy balances were carried out on the system as shown in Figure 1. Mass and energy balances implemented around the control volume of the inlet side, outlet side and kiln shell. According to the rotary kiln process flow diagram, there are six material inflow and two material outflows. Material mass balance in the rotary kiln can be expressed in material conservation equation below:

$$M_{K,in} = M_{K,fo} + M_{K,st} + M_{K,bc} + M_{K,ca} + M_{K,ap} + M_{K,rkf} \quad (1)$$

$$M_{K,out} = M_{K,c} + M_{K,ogk} \quad (2)$$

where $M_{K,in}$ and $M_{K,out}$ denote the total inflow and outflow material in rotary kiln, respectively; $M_{K,fo}$, $M_{K,st}$, $M_{K,bc}$, $M_{K,ca}$, $M_{K,ap}$, $M_{K,rkf}$, $M_{K,c}$, $M_{K,ogk}$ are the mass of high sulfur fuel oil, steam, bituminous coal, combustion air, air pipe, rotary kiln feed ore, calcine and off-gas kiln, respectively. Refer to process flow diagram for kiln feed rate of 241 ton per hour, the result of material flow analysis of rotary kiln are shown in mass balance in Table 1.

Energy balance in rotary kiln can be expressed in energy conservation equation below:

$$E_{K,in} = E_{K,fo} + E_{K,st} + E_{K,bc} + E_{K,ca} + E_{K,ap} + E_{K,rkf} \quad (3)$$

$$E_{K,out} = E_{K,c} + E_{K,ogk} + E_{K,cr} + E_{K,ev} + E_{K,hl} \quad (4)$$

where $E_{K,in}$ and $E_{K,out}$ denote the total inflow and outflow energy in rotary kiln, respectively; $E_{K,fo}$, $E_{K,st}$, $E_{K,bc}$, $E_{K,ca}$, $E_{K,ap}$, $E_{K,rkf}$, $E_{K,c}$, $E_{K,ogk}$, $E_{K,cr}$, $E_{K,ev}$, $E_{K,hl}$ are the energy of main oil, lance oil, bituminous coal, combustion air, air pipe, rotary kiln feed, calcine, off-gas kiln, chemical reaction, evaporation and heat loss from outer surface of kiln shell, respectively. Refer to process flow diagram for kiln feed rate of 241 ton per hour, the result of energy flow analysis of rotary kiln are shown in energy balance in Table 2.

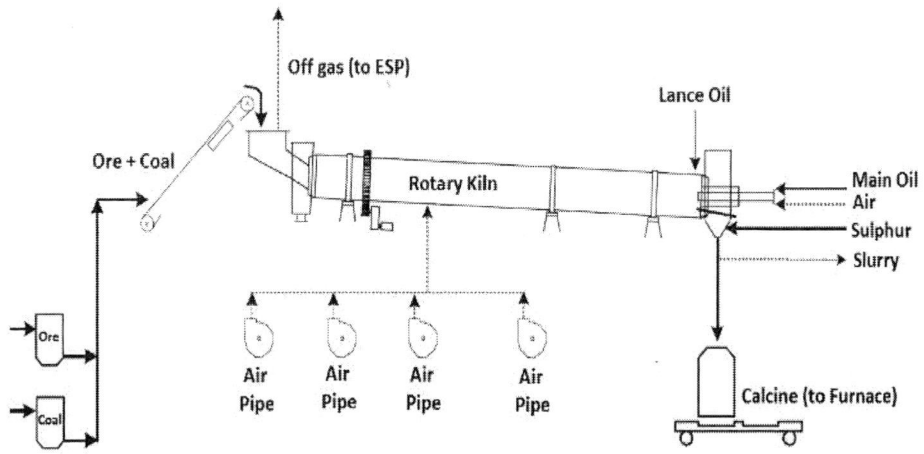

Figure 1. Flow diagram of rotary kiln

The main reference to this study is operating design data in the process flow diagram from kiln manufacturer that complemented with plant data. Plant data were obtained from the continuous measurement value recorded in the process information management system and manual measurement using appropriate tools. Analysis of mass and energy flows was conducted based on the following assumptions: rotary kiln operates at steady state conditions, the composition and the average temperature of material can be considered constant, outer shell temperature of the kiln can be divided into 44 sections and average temperature of these sections can be considered constant, ambient temperature of 27 °C can be considered constant and air supply to the rotary kiln and off-gas kiln are assumed to be ideal gases.

Table 1. Rotary Kiln Mass Balance

	Mass In [kg/h]		Mass Out [kg/h]
Ore	241,000	Calcine	148,931
HSFO	9,140	Off-gas	249,500
Coal	7,224	Dust in off-gas	27,726
Steam	980		
Combustion air	21,300		
Air pipe	146,521		
Total in	426,165	Total out	426,157

Table 2. Rotary Kiln Energy Balance

	Energy In [GJ/h]		Energy Out [GJ/h]
Ore	4.82	Calcine	83.76
HSFO	368.08	Evaporation heat	150.59
Coal	176.36	Off-gas	138.93
Steam	2.72	Heat loss	24.94
Combustion air	6.39	Chemical reaction	204.12
Air pipe	43.96		
Total in	602.33	Total out	602.33

3. Results and Discussion

3.1. Coal and biomass properties comparison

Coal specification that currently used in the rotary kiln is shown in Table 3. Coal that used as a reducing agent mixed directly with kiln feed, meanwhile coal for fuel must be ground and pulverized in the coal mill system before injected as fuel to the kiln burner system. As a reducing agent, coal addition to kiln feed influenced by nickel grade and chemical composition of laterite ore. Volatile matter portion of the coal will be burned out initially and release heat for drying process inside the rotary kiln. Subsequently, in the remaining length of the rotary kiln, fixed carbon content will form carbon monoxide and react with metallic oxide in the laterite ore. High fixed carbon content is required in the rotary kiln to ensure reduction process accomplished properly, otherwise many valuable metals will be discharged with furnace slag and wasted.

Table 3. Coal Specification For Rotary Kiln

Parameters	Unit	Test Results
Total Moisture	%wt	8.9
Total Moisture ADB	%wt	3.5
Ash Content, ADB	%wt	6.6
Volatile Matter, ADB	%wt	38.6
Fixed Carbon, ADB	%wt	51.3
Total Sulfur, ADB	%wt	0.42
Gross Caloric Value, ADB	kcal	7288
Gross Caloric Value, AR	kcal	6880
Ultimate Analysis		
Carbon ADB	%wt	76.2
Hydrogen ADB	%wt	5.6
Nitrogen ADB	%wt	1.6
Oxygen ADB	%wt	9.5

As a fuel, coal must be dried and pulverized in the coal mill system to burn completely. Primary factor for this application is heating value. Lower heating value will increase quantity of material required for same amount of energy. Secondary factor to be considered is milling requirements that indicated by Hardgrove Grindability Index (HGI) and product consistency.

Palm kernel shell charcoal selected as biomass source in this study because of leading properties compared to other biomass: high carbon content, high heating value, low ash, standard mill requirement and the finally, its abundance resources in Indonesia.

Main characteristics of biomass is high moisture and high volatile matter. High moisture will reduce heating value. Palm kernel shell as waste from palm oil mill facility contain approximately 15% of moisture and 75% volatile matter. These characteristics are not suitable for utilization in the rotary kiln as a reducing agent or fuel. Palm kernel shell need to be transformed to charcoal via carbonization or pyrolysis to increase heating value (lower moisture) and higher fixed carbon content (low volatile matter). Typical specification of palm kernel shell charcoal is shown in Table 4.

Indonesia is the largest palm oil producer in the world. Total crude palm oil (CPO) production in 2017 has reached 41.98 million ton. From one unit of fresh fruit bunch only 22.5% wt.% extracted to crude palm oil, the remaining mass will be discharge as waste unless reused as biomass energy or other utilization. Palm kernel shell portion is about 6 wt.% of fresh fruit bunch or approximately 10 million ton in 2017. Palm oil mill facility can utilize about half of the biomass as waste and remaining half will be discharged as waste.

3.2. Biomass utilization in rotary kiln

Fuel combustion characteristics, such as ignition performance, thermal reactivity and kinetic, influenced by ratio of blending between pulverized coal and biomass. Composition of 60 w.t% pulverized has the lowest activation energy and better combustion characteristics for the co-combustion of pulverized coal and palm kernel shell charcoal (Wang, et al., 2018).

Currently high sulfur fuel oil is the main source of fuel for rotary kiln and raw coal is used as the reducing agent. Pulverized coal can be used as fuel for the kiln burner up to 50% but at the moment the coal mill capacity is limited. Energy input for kiln main burner and for reducing agent is 368.08 GJ/h and 176.36 GJ/h, respectively. CO2 emission from the fossil fuel is 118.21 kg-CO2/ton-calcine and 70.32 kg-CO2/ton-calcine, respectively, which give total 188.54 kg-CO2/ton-calcine. CO2 emission with current fuel scheme is shown in Table 5.

Table 4. Typical Palm Kernel Shell Charcoal Specification

Parameters	Unit	Test Results
Total Moisture	%wt	1-5
Ash Content, ADB	%wt	0.8-3
Volatile Matter, ADB	%wt	10-12
Fixed Carbon, ADB	%wt	80-87
Gross Caloric Value, AR	kcal	6500-7000

Biomass utilization for fuel requires co-combustion with pulverized coal in ratio 60:40 of pulverized coal and pulverized PKS charcoal. Proposed fuel configuration with biomass utilization for the main burner of the rotary kiln is shown in Table 6. High sulfur fuel oil supply still maintained at the proportion of 50% input energy according to current process guideline. Pulverized coal and pulverized PKS charcoal will supply 30% and 20% of input energy, respectively. Proposed reducing agent configuration with biomass utilization is 95:5 of coal and PKS charcoal. CO2 emission from the oil, coal, and biomass is 59.11 kg-CO2/ton-calcine, 110.84 kg-CO2/ton-calcine, and zero, respectively. Total CO2 emission from the rotary kiln with biomass utilization is 169.95 kg-CO2/ton-calcine, or 9.9% lower compared to existing current fuel configuration.

Table 5. CO2 Emission With Current Fuel

Fuel type	GJ/h	kg CO2 / ton-calcine
Fuel oil	368.08	118.21
Reductant Coal	176.36	70.32
Total	544.44	188.54

Table 6. CO2 Emission With Biomass Utilization

Fuel type	GJ/h	kg CO2 / ton-calcine
Fuel oil	184.04	59.11
Pulverized coal	110.42	44.03
Pulverized charcoal	73.62	0.00
Reductant Coal	167.54	66.81
Reductant charcoal	8.82	0.00
Total	544.44	169.95

4. Conclusion

Potential of biomass utilization in the rotary kiln of nickel processing plant is promising in order to reduce CO2 emission. In the rotary kiln, coal is used as fuel and a reducing agent. Charcoal from palm kernel shell has similar characteristics with coal and very suitable to substitute coal. Charcoal from palm kernel shell can be used as fuel and a reducing agent. Potential 9.9% reduction of CO2 emission can be achieved by using 20% PKS charcoal as fuel and 5% PKS charcoal as a reducing agent with minimum impact on the current operation.

5. References

[1] International Energy Agency 2018 Global Energy & CO2 Status Report 2017

[2] Mousa E , Wang C, Riesbeck J and Larsson M 2016 Biomass application in iron and steel industry: An overview of challenges and opportunities *Renewable and sustainable energy reviews* **65** 1247-1266

[3] Rahman A, Rasul M G, Khan M, and Sharma S 2015 Recent development on the uses of alternative fuels in cement manufacturing process *Fuel* **145** 84-99

[4] Wang P, Wang G, Zhang J, Lee J Y, Li Y and Wang C 2018 Co-combustion charateristics and kinetic study of anthracite coal and palm kernel shell char *Applied Thermal Engineering* **143** 736-745

[5] Wei R, Zhang L, Cang D, Li J, Li X and Xu C C 2017 Current status and potential of biomass utilization in ferrous metallurgical industry *Renewable and Sustainable Energy Reviews* **68** 511-524

IOP Conf. Series: Materials Science and Engineering **588** (2019) 012007 doi:10.1088/1757-899X/588/1/012007

Risk Assessment on Liquefied Petroleum Gas (LPG) Handling Facility, Case Study: Terminal LPG Semarang

T A Akbar[1,2,a], A A B Dinariyana[1,2] , A Baheramsyah[1] , F I Prastyasari[1,2] and P D Setyorini[2,3]

[1] Department of Marine Engineering, Faculty of Marine Technology, Institut Teknologi Sepuluh Nopember Surabaya, Indonesia, 60111
[2] Centre of Excellences of Maritime Safety and Marine Installation (PUI-KEKAL) Institut Teknologi Sepuluh Nopember Surabaya
[3] Department of Marine Engineering, Faculty of Engineering and Marine Sciences, Hang Tuah University Surabaya, Indonesia

[a] thariqakb@gmail.com

Abstract. LPG plays important role in Indonesia as household fuel. With the rapid growth of LPG usage, its handling facility is built across the country. The handling of LPG requires various activity prior to supporting its supply chain. One of the vital facility is Terminal LPG (or on this case Terminal LPG Semarang) which serve as a hub in LPG distribution because it handles a relatively huge amount of LPG. It is exposed to operational risk especially fire risk. This paper is intended to assess the risk of LPG Terminal by using combined both qualitative and quantitative methods. The initial step is to identify the hazard by using HAZOP Study. The next step is to determine the likelihood of the risk by using FTA and ETA resulted in event frequency. The severity level is determined using consequences analysis which involved fire modeling of the intended scenarios. The result from both frequency and consequences will be presented in the risk matrix to determine the unacceptable risk scenario to develop a mitigation plan by using Fire Risk Card to assess the capability of the Terminal firefighting equipment.

1. Introduction

1.1. Background

Solar radiation energy was one form of alternative energy that could be used for various purposes to replace the energy produced by petroleum. For a long time, solar energy had become a source of lighting during the day, and was also widely used to dry agricultural produce, clothing, wood, heating water and others. An Indonesian gift located in the equator, where solar energy was available at all times.

As one alternative energy source, solar energy had become increasingly widespread. One of its uses was to use a device called a solar collector. At present, flat plate or wave shape solar collectors for water heaters had been developed. This collector, using wave shaped plate that serves to absorb the incoming solar energy and move the heat received to the working fluid, namely water. At the top of the collector, a transparent cover was used to cause a greenhouse effect. In order to avoid heat loss to the environment at the bottom and side of the solar collector, an insulator was provided (Made et al., 2010).

1.2. Identification of the problem
The problem of this research was to identify the influence of the variation of the time period of the collector shift at a short interval of 20 minutes, 40 minutes, 60 minutes to the efficiency of solar water heater.

1.3. The purpose of the study
The purpose of the study was knowing the effect of variations in the period of time the collector shifts to the efficiency of solar water heater.

1.4. Benefits of research
The benefits of the research was could find out the process of designing and manufacturing solar collectors with tracker. Could be used for further research with various variations so as to get better efficiency. Could be used as one of the utilization of environmentally friendly energy sources and never runs out.

1.5. Scope of problem
The scope of problem was : the testing process uses a solar water collector with a wave plate heater with a tracker. Variation in the time period of the collector shift time 20 minutes, 40 minutes, 60 minutes. Measurement results was taken for each time interval 15 minutes. Fluid capacity was set 30 liters. Flow rate was set at 0.011 liters/second. Volume of fluid inside the collector was 15 liters. Collector's tilt position was 15° facing north.

2. Literature Review

Previous research
Mustofa (2008), an experimental study of the comparison of double plate collectors and conventional collectors on the performance of solar water heater. The research resulted, the heat absorption efficiency of solar double plate heater was higher than the efficiency of conventional solar heater heat absorption.
Farid and Ismail (2010), tested the double absorber plate of wave models which produced an average heat absorption efficiency value in the double wave solar heater model with an average of 19.81%, while the value of the average heat absorption efficiency in solar absorbent plate heater flat at 12.43%. Most solar collectors were permanently installed with fixed elevating angles. This causes the solar collector to be unable to absorb solar radiation optimally because the sun was always moving, namely in the east-west direction (called the sun's pseudo-daily motion) and north-south (called the annual pseudo-motion of the sun). Absorption of solar radiation will be optimal if the direction of solar radiation was perpendicular to the surface of the field of the solar collector. Therefore, efforts were needed to direct the surface of the solar collector to be perpendicular to sunlight. The method for directing solar collectors to always follow the direction of the sun's motion was known as the method of tracking the sun (Huang et al, 2009).

Determine efficiency
The efficiency of the collector panel was the ratio between the useful heat level (QU) transferred to the liquid divided by solar radiation in the cover plate. Efficiency could be shown at equation Duffie JA and Beckman W.A. (1980), as follows:

$$\eta = \frac{q_u}{A_c G_T} \tag{1}$$

From the above equation could also use the efficiency of the collector efficiency as follows:

$$\eta = F_R(\tau\alpha) - F_R U_L \frac{(Ti - Ta)}{G_T}$$

$$\eta = \frac{\dot{m}\ C_p(T_0 - T_i)}{G_t.A_c} \tag{2}$$

where :

Qu = The energy absorbed by collector, (W/m^2)
Ac = Collector Size (m^2)
FR = Collector heat loss factor
UL = Overall heat loss (W/m^2. ° C)
Gt = Total intensity of solar radiation (W/m^2)
Ta = ambient temperature (° C)
Tin = Entering water temperature (° C)
T_{out} = water temperature out (° C)
Cp = saturated water/liquid properties (kJ/(kg.°C))
τ = Transmissivity cover glass
α = Absorbtivity of the absorbent plate
$\dot{m}$ = Fluid flow rate (liters/second)

Fluid mass flow rate

The fluid mass flow rate was the amount of fluid mass that flows per unit of time and could be stated as follows:

$$\dot{m} = \frac{m}{t_u} = \frac{V\rho}{t_u} \tag{3}$$

where :

m = fluid mass
tu = Time of fluid movement from one end to the other solar heat collector (liter/second)
Vρ = Volume of fluid flowing from collector exit pipe (liters).

3. ResearchMethod

The research starts from 10.00 WIB until 14.00 WIB in March 2018 until completion. Testing the variation of the time period of the collector shift was carried out in Madiun City, East Java, Indonesia A subsection

Equipment and Materials

From testing requires the following tools and materials: 3 mm thick zinc absorbent plate, 10 mm thick PVC plate, 3 mm thick glass, 25 liters water tank, framework for solar tracker collectors, circulation pipe with 25.4 mm in diameter, 3 mm thick insulator (stereofoam), aquarium pump, 1/2 " stop faucet, water hose, precision pyranometer, computer/laptop, programmable peripeal interface (PPI), analog digital converter (ADC), measuring glasses, roll cable and LM35 temperature measuring instrument.

Research Procedure

The steps taken in researching the performance of solar water heater were:

- Data collection was done by placing the collector in the sun.
- The collector was tilted at a fixed angle of 15° north.
- Install a series of measuring instruments to measure the temperature of the inlet water, absorbent plate temperature, glass temperature 1, glass temperature 2, temperature of the outgoing water, ambient temperature and solar radiation.
- Record the measurement results in accordance with the shift variation of the collector angle at 20 minutes, 40 minutes and 60 minutes.
- Direct the collector with the direction of falling sun radiation by 90 degrees by following the axis perpendicular to the shadow of the direction of the sun.
- Collect the data in table.
- Analyze the results of testing.

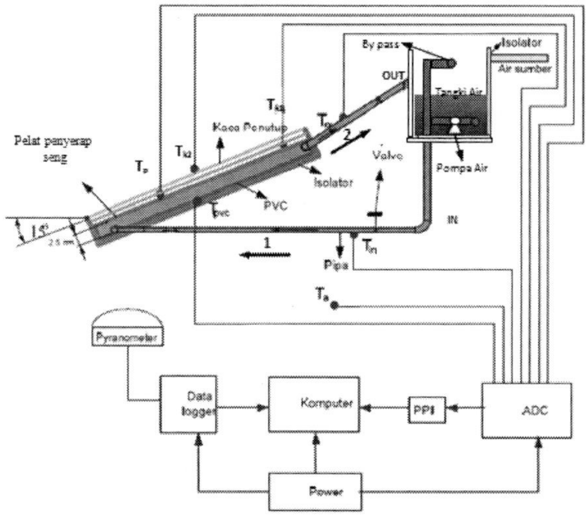

Figure 1. Scheme of solar water heating test equipment

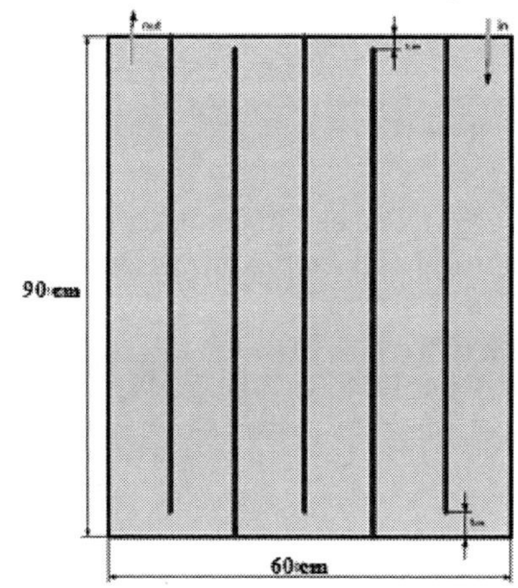

Figure 2. Scheme of collector absorber with zig zag grooves

4. Result and Discussion

From the research results obtained data and graphs could be made as follows:

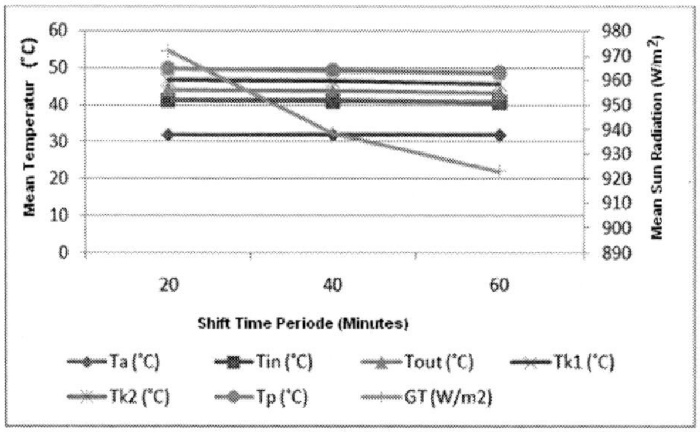

Figure 3. Relationship between the average temperature and the intensity of solar radiation

IOP Conf. Series: Materials Science and Engineering **588** (2019) 012007 doi:10.1088/1757-899X/588/1/012007

At the inlet temperature, glass temperature 1, glass temperature 2, absorbent plate temperature, and the exit temperature at each test the price was the same as the range 30°C to 50 .C. While at the ambient temperature the price was the same as the time variation of 20 minutes, 40 minutes, 60 minutes the price was lower than the inlet temperature, glass temperature 1, glass temperature 2, the temperature of the absorber plate, and the exit temperature range at 30°C. In solar radiation there was a significant decrease with a price range of 1000 W/m^2 to 800 W/m^2. This was due to the time of the same collector shift. At 20 minutes the shift time was shorter so that the displacement of the sun's upright position to the collector was faster.

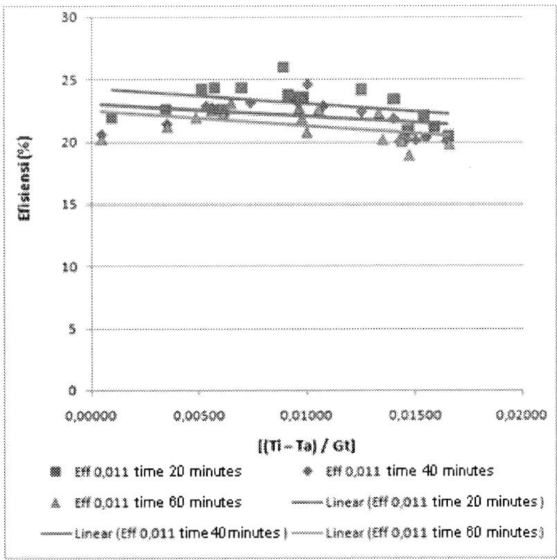

Figure 4. Graph of the relationship between the three efficiencies with (Ti-Ta / Gt)

In the test graph above it could be concluded that on the three tests the time period of the collector shift every 20 minutes, 40 minutes and 60 minutes at the same flow rate of 0.011 liters / second shows that the period of time the collector shifts greatly affects the intensity of solar radiation falling on the collector. the faster the collector's shift time toward the direction of falling solar radiation, the greater the solar radiation intensity, and the greater the temperature, the higher the efficiency obtained. From the three tests get results that show that the test using the shortest period of time the collector shift obtained the best efficiency in a period of 20 minutes while the longer the time period used, the radiation that falls to the collector and the absorption of heat gets smaller. So that the shift time affects the absorption of solar radiation in the collector which causes an increase in efficiency.

5. Conclusion

In small time interval, the collector shift was used, the greater the efficiency obtained. The best efficiency results for a time interval of 20 minutes with an average efficiency of 23.07%.

6. Ackowledgement

The research was fully funded by the Ministry of Research and Technology of Higher Education of the Republic of Indonesia on PTUPT scheme in the year 2018.

References

[1] Arismunandar W 1995 *Solar engineering technology, first edition* (Jakarta : PT. Pradnya Paramita)
[2] Burhanuddin A 2006 *Characteristics of flat plate solar collectors with variations in cover glass distance and collector tilt angle*. Essay. Faculty of Mathematics and Natural Sciences

Sebelas Maret University, Surakarta.

[3] Duffie and Beckman 1991 *Solar Engineering of Thermal Processes* (Wisconsin : John Willey and Sons Inc)

[4] Farid and Ismail 2010 The effect of the double wave absorber model on the performance of a simple solar water heater *Widyateknika Journal* March 2011.

[5] Huang, YJ, Kuo, TC, Chen, CY, Chang, CH, Wu, PC, and Wu, TH, 2009 *The Design and Implementation of a Solar Tracking Generating Power System, Engineering* Letters, **17**: 4, EL _17_4_06, Advance online publication.

[6] Kristanto P and San Y K 2001 Effect of Thickness of Plates and Inter-Pipe Distance on Perfection Performance of Surya Pelat Datar Collector *Mechanical Engineering Journal*, Petra Christian University

[7] Made Sucipta, I Made Suardamana, Ketut Astawa 2010 Analysis of the Performance of Plate Solar Collectors Finned with Fin Surface Surface Variations *Scientific Journal of M Chakra Mechanical Engineering* **Vol. 4** No. 2. October (88-92)

[8] Mustafa 2008 *Experimental Study of Comparison of Double Plate Collectors and Conventional Collectors on Solar Water Heater Performance*. Thesis Mechanical Engineering Major – Universitas Brawijaya

[9] Mustofa and Ismail NR 2013 *International Journal of Research in Engineering and Technology*, **Vol. 02,** Issue.09, September 2013

[10] Rahardjo Tirtoatmodjo 2005 *Performance of Water Heaters Solar Collector Type Flat Plate With One And Two Cover Glass* Department of Mechanical Engineering - Petra Christian University

[11] Sutrisno 2002 Testing Water Heater Solar Collectors using Wave Absorber Plates with and Without Honeycomb (Surabaya : ITS)

IOP Conf. Series: Materials Science and Engineering **588** (2019) 012008 doi:10.1088/1757-899X/588/1/012008

The development of the IHL wave-current turbine to convert the ocean energy from both wave and current

Erwandi[1], D Rahuna[2] and C S J Mintarso[2]

[1] Technology Center of Maritime Industrial Engineering BPPT, Jl. Hidrodinamika BPPT, Kompleks Kampus ITS, Sukolilo Surabaya 60112
[2] Indonesian Hydrodynamic Laboratory (IHL) BPPT, Jl. Hidrodinamika BPPT, Kompleks Kampus ITS, Sukolilo Surabaya 60112

erwandi@bppt.go.id

Abstract. Indonesia has abundant of renewable sources of energy, among others: geothermal, solar, water, wind and marine therefore the promotion of the utilization of the renewable and sustainable energy is urgent. Since 2006 Indonesian Hydrodynamic Laboratory (IHL) BPPT has conducted research and developed a technology converts the ocean energy resources into electricity, the kinetic energy of marine current and the potential energy of wave. The prototype of marine current turbine, 2 kW and 10 kW respectively, had been installed in Larantuka Strait in 2010 and 2011 (http://www.youtube.com/user/erw4ndi). Since mostly the maximum speed of the marine current of many straits in Indonesia are in order 1.0 - 2.0 m/s, we developed the turbine prototype that can convert not only the kinetic energy of marine current but also the potential energy of the wave, called as wave-current rotor turbine, in 2013. The current usually mixes with wave height with order 0.4 - 0.8 m. A 2 kW prototype was installed under Suramadu (Surabaya - Madura) Bridge close to pier no. 56. The prototype showed a good performance especially in converting the kinetic energy of marine current. The research was continued in 2014-2016 under Ministry of Research and Technology (INSINAS RISTEK) budget scheme. Some innovations have been done to improve the performance of turbines. .

1. Introduction

Indonesia has plentiful of renewable and sustainable energy resources. Those are: geothermal, solar, water, wind and marine. Some feasibility studies of geothermal resources has been finished on 2017. On July 2018, the Indonesian Government has launched the 75 MW peak wind turbine farm at Sidrap, Sulawesi Island. Many solar power plants have being built in the islands of East Nusa Tenggara Province. However, the marine is the least exploited. As archipelagic country the sources of ocean renewable energy, theoretically, are beyond the other ones [1].

The power of hydrokinetic energy of the speed of the marine current that pass through many straits in Indonesia has attracted many investor companies to exploit it [13]. The straits in East Nusa Tenggara Province are the examples where the flow speed of the marine currents between the islands typically have strong velocities in the range of 2-4 [m/s] or more. It mixes with wind waves and swells come from the open sea surrounding the straits.

Since 2006 fiscal year, the IHL of BPPT has been establishing research on development of the turbine of the marine current. It is initiated with the predicting of the strength of the marine current in the several straits at Nusa Tenggara Province. Then a 3 blades Darrieus turbine was designed and tested in tank facilities of IHL. Manufacturing of 2 kW prototype was started in 2009 and installed in Larantuka Strait in 2010. In 2011 the capacity of turbine was increased to 10 kW [2] [3].

In 2013 the turbine was moved to Madura Strait installed under Suramadu Bridge close to pillars no. 56. Suramadu Bridge is the bridge that connects Surabaya City and Madura Island. Since the maximum speed of tidal stream in Madura Strait is only 1.28 m/s and there is a tendency that the water under Suramadu Bridge sometimes is wavy water, we changed the blade position of Darrieus Turbine not perpendicular to the direction of the flow of tidal stream but has an angle 60° to the direction of the flow. The turbine gave a good performance especially in converting tidal stream into electricity. It is appreciated by the visitation of Minister of Research and Technology to the site to see the prototype.

The design process was continued in 2014 - 2016 under INSINAS research budget scheme managed by Ministry of Research and Technology. Some improvement designs are carried out to enhance the performance of the turbine. This paper has purpose to show the development process of the design wave-current rotor. Firstly the ocean modeling by numerical technique is utilized to predict the speed of current flow under Suramadu Bridge. Secondly the foil is designed to catch the kinetic energy of current speed. Next the physical model is tested in Maneuvering Ocean Engineering Basin (MOB) of IHL. The prototype is built and installed in Suramadu Bridge. The turbine can work well with cut-in speed 0.6 m/s.

2. Materials and Methods

This research follows the following method: firstly mesoscale numerical simulation is conducted to predict the velocity of tidal stream at Madura Strait, secondly based on the mesoscale numerical results, the foil of the wave-current rotor is designed with assumption that the category of current speed is slow. Next the prototype of wave-current rotor is tested in MOB basin of IHL, furthermore the prototype of wave-current rotor is manufactured and finally it is installed under Suramadu Bridge close to pillar no. 56.

The Geophysical Fluid Dynamic Navier Stokes equations and continuity equation are discretized with staggered grid system. First order upstream differencing scheme is applied to the advection term, second order centered differencing scheme is applied to diffusion term. Equations are integrated by forward-backward scheme [7].

At the open boundaries, the tidal constituent data at Karang Jamuang (North Madura Strait) and Karang Kleta (East Madura Strait) produced by DISHIDROS Indonesian Navy are fed to computational domain [8]. The sea elevations calculated by the tidal components are given at every time step. Velocity gradient normal to the open boundary is assumed to be 0.

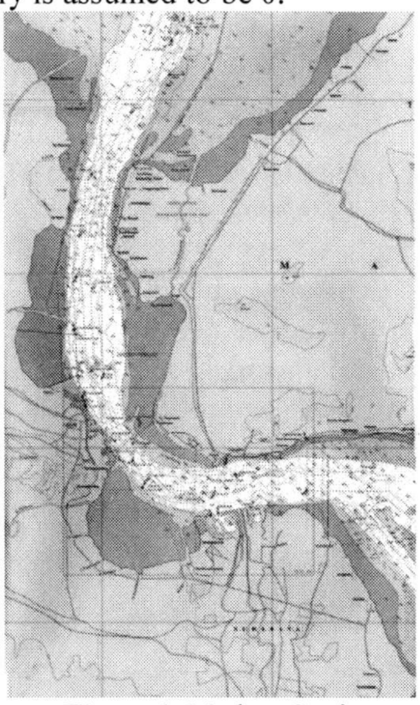

Figure 1. Madura Strait

2.1. Numerical Prediction of the Current Speed at Madura Strait

The MEC Ocean Model (Ocean modeling program of Marine Environment Committee of the Society of Naval Architects of Japan) [4] has been utilized as the initial approach for the prediction of the strength of the marine current at Madura Strait. Madura Strait (Selat Madura) is situated between East Java and Madura Island, as shown in Figure 1. In the east of Madura Strait there is Suramadu Bridge, the longest bridge in Indonesia. Numerical prediction was carried out intensively using variable mesh. Erwandi [5][6] showed the variable mesh technique to predict the speed of current distribution in Alas Strait.

2.2. Grid Generation of Madura Strait

The MEC-Model uses the orthogonal constant grid system and variable grid system. Since the bathymetry data are very coarse, , for the first approximation, the 50 meter accuracy in horizontal direction and 0.5 meter accuracy in vertical direction of the orthogonal constant grid system is generated to discretize the bathymetry contour of the strait. The variable mesh is concentrated in the site where the Suramadu Bridge exists. The variable mesh in bird view is shown in Figure. 2.

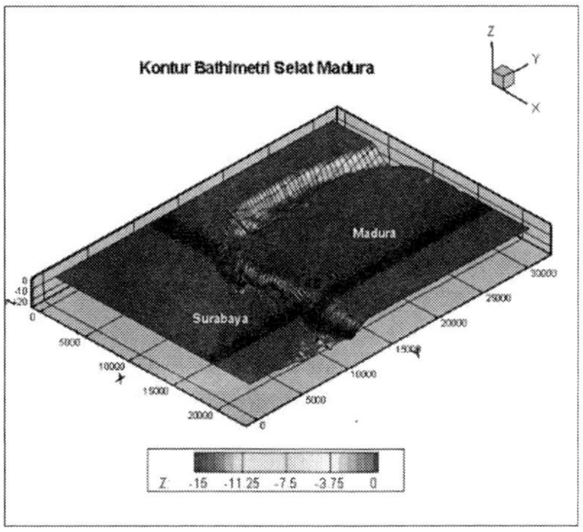

Figure 2. Variable mesh at Madura Strait

2.3. Designing the Foil of Wave-Current Rotor Turbine

Since the rate of tidal stream characteristic under Suramadu Bridge is slow, then special design of foil should be considered. We had been designed special foil named as LHI 166 [9]. LHI is abbreviation of Laboratorium Hidrodinamika Indonesia (Indonesian Hydrodynamic Laboratory) and 166 is the thickness code, means 16.6% of chord length. LHI 166 foil is designed for special purpose. It gives high lift in low Reynolds Number. The full inverse and mixed inverse formulation panel method was employed to design the shape of the foil. The lift and drag force of LHI 166 was calculated numerically using commercial CFD program with angle of attack 0° - 358°. Fig. 4 shows the mesh when CFD is utilized to find the lift and drag force coefficient.

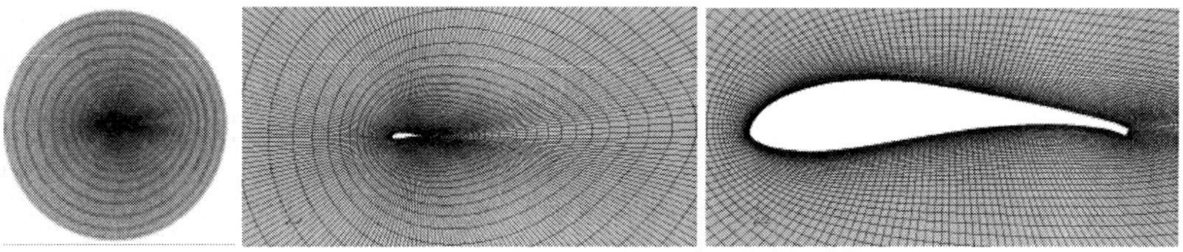

Figure 3. Mesh generation of LHI 166 foil

2.4. Design and Hydrodynamic Test of Wave-Current Rotor Turbine

Three blades wave-current rotor turbine ver. 1.0 was designed originally from the research of performance Darrieus Turbine designed in IHL. The blade position was changed. In Darrieus turbine the blade position is perpendicular to the direction of the stream flow of marine current whilst the wave-current rotor turbine has angle 45° and 60° to the direction of the flow. Figure 4 shows the shape of wave current rotor turbine.

Three aluminum model of wave-current rotor turbine blade with sectional area foil LHI-166 shape was manufactured using CNC machine. The length of the blade and the length of the chord were 1 meter and 10 cm respectively. The blade can move freely up to -20° at a hinge close to the leading edge of the foil. The passive variable pitch is applied to control the movement of the blade. The hydrodynamic test showed that this application can increase the efficiency of turbine up to 38%. However there are disadvantages of it, the torsional vibration increases when the velocity of marine current increases [11]. The wave-current rotor turbine was tested both in towing tank and wave tank (Maneuvering and Offshore engineering Basin / MOB). Figure 5 showed the testing process in MOB.

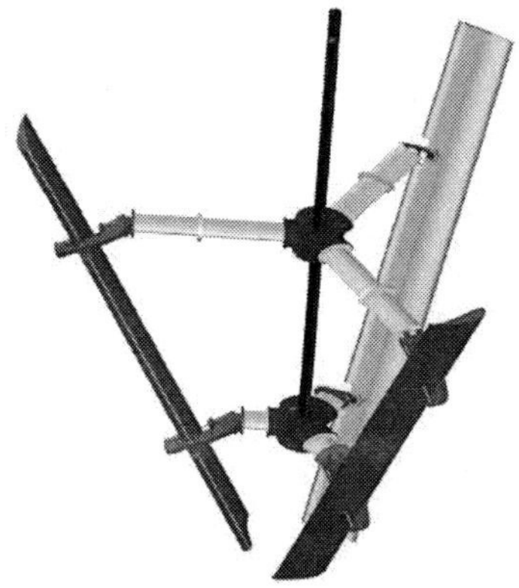

Figure 4. Wave-current rotor turbine

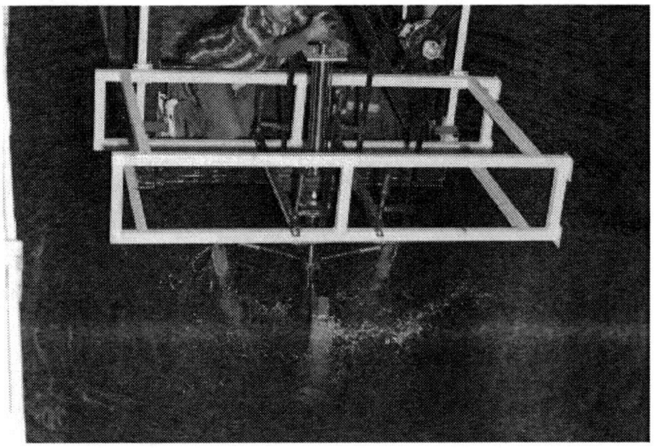

Figure 5. Testing of wave-current rotor turbine at MOB

2.5. Wave-Current Rotor Installation at Suramadu Bridge

A 3.5 kW wave-current turbine prototype was manufactured between two floating pontoon of fiber reinforced plastic. The length of the pontoon is 8 meter, the width is 1 meter, and draft is 0.5 meter. Total width of the prototype is 6 meter. The blade of the turbine has length 2.6 meter and diameter 3.6 meter. Four nylon mooring ropes diameter 7 cm and length 80 meter were connected to each bow and aft of the pontoons respectively. The two pontoons are connected with a cremona bridge where one of the edge of cremona can be raised and lowered easily using hoist and leverage at the other edge as a hinge. Figure 6 shows the schematic view the system to raise and to lower the wave-current rotor turbine. Figure 7 is the drawing of the prototype of wave current-rotor.

In the deck, we placed bevel gear from vertical axis to horizontal axis, the hydraulic brake, a 1 : 20 horizontal gearbox, and permanent magnetic generator (PMG). The maximum rotation rate of PMG is 250 rpm. The fluctuated AC electric current as output of PMG is converted to DC electric current through rectifier and then goes to inverter where the standard voltage and frequency are produced. A handmade system of dump load is utilized to transfer the electricity when the electricity produced by PMG is bigger than the load. Figure 8 is the prototype of wave-current rotor turbine when it was assembled at Pantai Ria Kenjeran Surabaya. The total displacement of the prototype is 8 ton.

The prototype was installed close to the bridge pillar no. 56, the anchor and concrete block for mooring system was not needed. The rope was directly tied to the pillar. Figure 9 showed the prototype installed in bridge pillar no. 56. The installation system of the prototype is well known as floating type installation system. Since it is floated then the performance of the turbine in converting the potential energy of the wave cannot be measured and observed. The prototype only works to convert the kinetic energy of tidal stream.

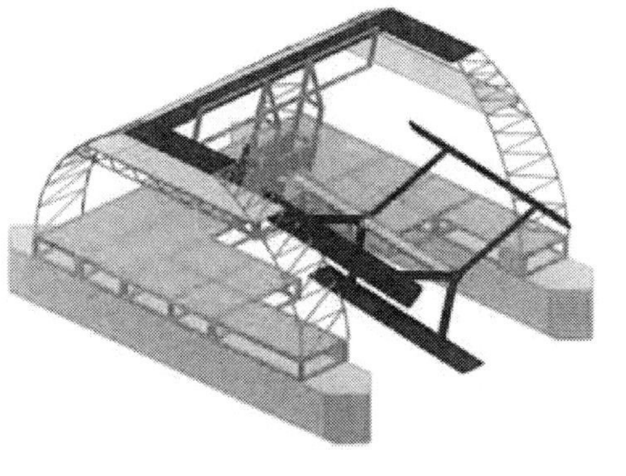

Figure 6. System to raise & lower the turbine

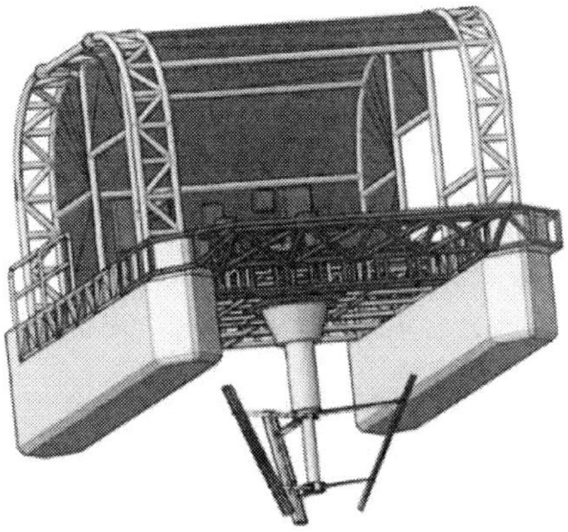

Figure 7. The drawing of wave-current rotor

Figure 8. Prototype of wave current rotor

Figure 9. Prototype installed in bridge pillar no. 56

3. Results and Discussions

3.1. Results of Numerical Simulations

The tide type of the Madura Strait is mixed predominant semi-diurnal, therefore we assumed only semi-diurnal tide at the open boundary and we considered the amplitude of sea elevation is the summation of the amplitude of the tidal constituents M2, S2, K1, and O1. Phase lag is set by M2 tide. The variability of spring-neap cycle is reproduced by using amplitude M2+S2 for mean spring tide, and M2-S2 for mean neap tide. Both cases use phase lag M2. Fig. 11 (a) and (b) show the vector and power distribution of current velocity at high tide and low tide respectively. The position of Suramadu Bridge is indicated in rectangle box. The maximum velocity of current at high tide at point where the Suramadu Bridge pillar no. 56 is 1.38 m/s whilst the maximum velocity of low tide current at same position is 1.1 m/s. The prediction of those maximum values using numerical simulation was confirmed to the direct measurement reported by Haryo Dwi Atmono et. al. 2005 [10]. They have a good agreement between numerical results and direct measurement.

As shown in Fig. 11, the power produced from the kinetic energy of the marine current speed can be determined by following equation:

$$P = 0.5\rho A V^3 \eta \qquad [\mathrm{W}] \tag{1}$$

Where:
P: power [watt]
ρ : density of water [1024 kg/m3]
A : turbine cross sectional area [m2]
V : current speed [m/s]
η : the efficiency of the turbine (0.593)

The efficiency value that is set to 0.593 is theoretical efficiency. The real efficiency of the turbine, actually, is lower than 0.593, however in this study it is assumed that the mechanical turbine efficiency is 1. Since the fastest current speed in vertical direction is near the surface thus the turbine should be installed close to the surface. The cross sectional area A is set to 1 m2.

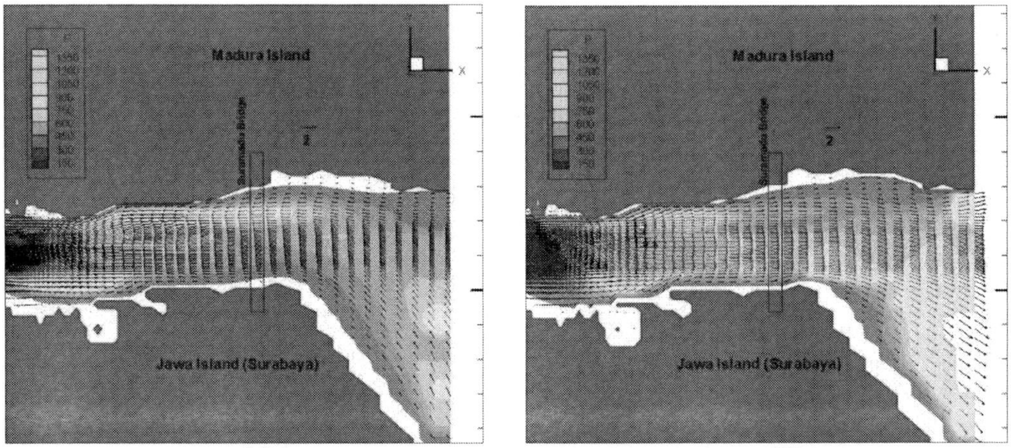

(a) (b)

Figure 10. Vector distribution of Marine Current Speed at Suramadu Bridge Madura Strait (a) High tide (b) low tide

3.2. Results of Wave-Current Blade Foil Numerical Simulations
Based on the mesoscale numerical simulation, two dimensional detail numerical simulation was carried out. Figure 11(a) shows the sample result numerical calculation of LHI 166 at angle of attack 14° whilst Figure 11(b) is the pressure coefficient distribution of Figure 11(a). The knowledge of flow characteristic at every position in the certain angle of attack will give a fruitful information to design the performance of turbine when the foil passes upstream and downstream area. Since the tangential force produced by foil is very essential to rotate the turbine system then a careful design of passive variable pitch will increase the efficiency of the turbine. The results of numerical analysis is then verified by testing the physical model not only in towing tank but also in MOB. The foil LHI 166 gave a good performance since the average efficiency is 38.8%. It worked well at tip speed ratio (TSR) 2.25. The passive variable pitch can remove the phenomena of negative torque and give a high efficiency at slow velocity of the stream. It provides the very low cut-in speed of the turbine. The turbine started rotating at 0.25 m/s.

However the application of passive variable pitch gave disadvantage when the towing speed was increased. The turbine system started shaking at towing speed 1.3 m/s. The passive variable pitch produced very high lift force in higher speed. High lift force means high tangential force. Since the resultant of tangential force is bigger than normal force perpendicular to radial arm, then the radial arm will experience fluctuated bending moment. The fluctuated bending moment is conveyed to the shaft and the whole turbine system [11][12].

The performance of turbine can be represented in Figure 12. The maximum rotation rate of the turbine transfer to PMG was 167 rpm. The maximum electric power produced by PMG was 2003 Watts at 1.21 m/s. The electricity generated by PMG was fluctuated depend on the turbulent flow fluctuation of the tidal stream and the shaking phenomena of wave-current rotor turbine system. In Figure 12 the electricity output measured by power meter is in blue symbol, whilst the red dot represents the rpm power diagram taken from the PMG technical specification sheet.

From Figure 12, we also conclude that the selection of the gearbox rate seems too low. The gearbox installed in wave-current rotor turbine has rate 1:20. The analysis of the power generated by PMG, the torque in the shaft is still bigger than the torque of PMG. The change of gear box rate will increase the power produced by PMG since the maximum power of PMG is 3500 Watts at 250 rpm.

The prototype was installed for 6 months. As continuation of the research, we are designing new turbine called wave-current rotor turbine ver. 2.0 to enhance the performance of the turbine in converting the potential energy of the wave height.

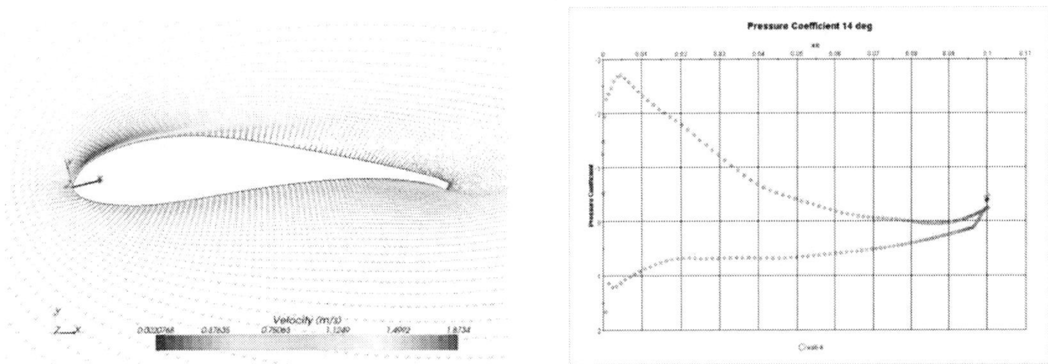

(a) (b)

Figure 11. Sample result of LHI 166 foil at angle of attack 14°

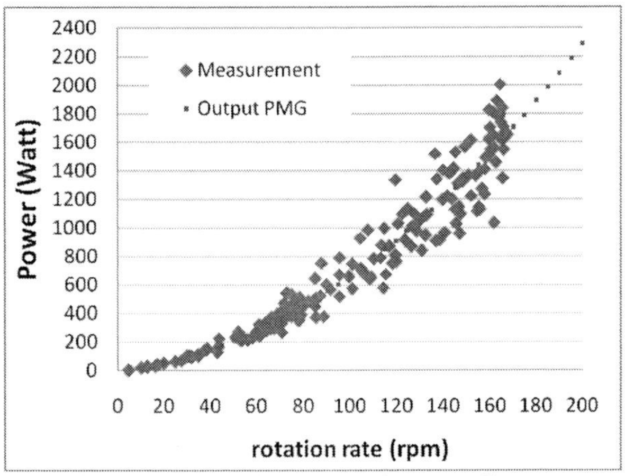

Figure 12. Rotation rate and power diagram

4. Conclusion

The numerical ocean modeling (MEC model) is very useful to predict the power of the hydrokinetic energy from marine current speed in Madura Strait and the tidal current under Suramadu Bridge. Moreover the detail numerical simulation can also predict the performance of the foil LHI 166. Finally the prototype of wave-current rotor can work well in converting kinetic energy of tidal current under Suramadu Bridge.

Acknowledgements
This research is supported by Ministry of Research and Technology, under INSINAS research budget scheme 2014-2016 entitles **The Development of the Wave-Current Rotor Converter Prototype to Convert Both Marine Current Hydrokinetic Energy and Wave Potential Energy to Electricity**. The authors would like to expressed their gratitude to Dr. Agoes Rustiana Hoetman for the encouragement and support for the research on the Development of Ocean Renewable Energy. This research is dedicated to the late Prof. Said D. Jenie, that we direct the most sincere thanks for his idea, support, encouragement, patience, and enthusiasm.

References
[1] Erwandi August 29 2005 Marine Current Energy: The Alternative Energy that Ecological Friendly and Sustainable (in Bahasa Indonesia), *Article on Kompas Newspaper*
[2] Erwandi et al 2011 *Proc. Int. Conf. on Sustainable Energy and Advanced Material* (Solo Indonesia) p. 15
[3] Erwandi 2011 *Proc. Int. Conf. on World Renewable Energy* (Bali Indonesia)
[4] Erwandi, Toda Y, Asakura H, Iwata Y, Wayama T 2004 J. *of Kansai Society of Naval Architects of Japan* **241**
[5] Erwandi et al 2007 *Proc. Nat. Sem. on Theory and Application on Marine Technology* p. A137 - A148
[6] Erwandi 2008 Numerical Simulation of Ocean Current for Tidal Current Power Plant in Indonesia *Symp. on Marine Environment and Modeling for East Asian Seas* University of Tokyo, Japan
[7] Toda Y, Erwandi, Nishioka K 2004 Preliminary Computation Using Body Force Distribution in Full-3D Region for Accounting for the effect of the structures, *Proc. the 5th MEC Model Workshop The Society of Naval Architects of Japan*, Osaka
[8] DISHIDROS 2005 *Table of Indonesian Tidal 2005* (DISHIDROS Indonesia Navy)
[9] Erwandi 2007 *Turbine Rotor Model: Engineering and Manufacturing*, BPPH BPPT Internal Report
[10] Atmono H. et al 2005 *Sedimentation Study on Approach Bridge of Suramadu Bridge, Project Contract no : 04/Spk-Jasa/Cic-Srmd/2005* (Suramadu Bridge Project Consortium Indonesian Contractor Surabaya Report)
[11] Hantoro R, Utama I, Sulistyono A, Erwandi 2011 *Australian J. of Basic and Applied Sciences* **5**(7) p. 121 - 128
[12] Hantoro R, Utama I, Erwandi 2009 *Proc. Int. Symp. On Ocean Science, Technology, And Policy, World Ocean Conference* (Manado Indonesia)
[13] Singgih V April 3 2018 Indonesia to Build World's Biggest Tidal Power Plant, *Article in The Jakarta Post Newspaper*

IOP Conf. Series: Materials Science and Engineering **588** (2019) 012009 doi:10.1088/1757-899X/588/1/012009

Subsurface Analysis Using Gravity Data at Lili Sepporaki Geothermal Area

Jubilate Sihombing[1, a] **Wien Lestari** [1,b] **Mariyanto** [1,c] **and Wiwid Joni** [2,a]

[1] Geophysical Engineering Department, Faculty of Civil Enviromental and Geo Engineering, Institut Teknologi Sepuluh Nopember, Indonesia
[2]Center of Mineral Resources, Coal and Geothermal, Ministry of Energy and Mineral Resources, Indonesia

[a]jubelnababan29@gmail.com , [b] wien.geofisika@gmail.com, [c]mariyanto@geofisika.its.ac.id

Abstract. Gravity survey plays an important role in early stage of subsurface imaging in geothermal system. The aim of this study is to understand the subsurface structure and possibly the heat source based on rock density distribution with three dimensional inversion modeling (3D). In 2010, Center of Mineral Resources, Coal and Geothermal of Indonesia was performed gravity survey at Lili Sepporaki geothermal area with 205 stations within a 10 x 10 km coverage area. The complete Bouguer anomaly was obtained with the density of 2.8 g/cm3 and this density was used to do 3D modelling. Bouguer anomalies ranging from 40 mgal to 62 mgal were observed in the study area with low anomalies values located in the northern part and high anomalies located in the southern part of the study area. Based on the 3D inversion modeling results, there are low density rocks with values between 2.0 - 2.3 g/cm3 around the hot springs of the research area with a depth of <1 km. These rocks are interpreted as rocks that associated with fractures forming a fault zone that controls the surface manifestation of the study area. High density rocks with values between 3.0 - 3.3 gr/cm3 are located south of research area with depth >2 km. This rocks are interpreted as intrusive igneous rocks and are thought as the heat source of geothermal systems in the study area.

1. Introduction

Geothermal potential in Sulawesi is the third largest after Java and Sumatra, which is around 3229 MWe spread in 76 locations. Most of the geothermal potential in Sulawesi Island is associated with non-volcanic environments. Non-volcanic geothermal is a geothermal system that is not related to Quaternary volcanic activity, found in sedimentary, plutonic, and metamorphic environments associated with tectonic processes with manifestations characterized by the presence of hot springs (Gupta and Roy, 2006). This research area is one of the geothermal potential areas, but until now it has not been used indirectly for electricity generation. The existence of geothermal systems in this area is characterized by geothermal manifestations in the form of hot springs with temperatures of 84.20C to 970C. Geothermal Lilli Sepporaki is located in a non-volcanic environment which is dominated by tertiary volcanic brick bounded by the active Palu Koro fault.

One of the geophysical methods used in geothermal exploration is the gravity method. Gravity method is a geophysical method based on differences in gravitational fields due to differences in the density of constituent rock layers beneath the earth's surface (Kearey et al., 2002; Pradana et al., 2017). The

IOP Conf. Series: Materials Science and Engineering **588** (2019) 012009 doi:10.1088/1757-899X/588/1/012009

physical magnitude measured in the gravity method is the acceleration of the earth's gravity to get the physical quantity of density. The acceleration data of gravity obtained during measurement is processed into an anomaly of gravity acceleration of the earth which is called an anomaly bouguer. From the results of data processing, modeling both forward modeling and inversion modeling is carried out to determine differences in rock density, so that the data can be used to determine subsurface geological structures in the study area (Telford et al., 1990).

In geothermal exploration, gravity methods are generally used to describe subsurface structures that control geothermal systems (Porkhial et al., 2015), and also to find out which magma or intrusive rock bodies are related to heat sources from geothermal systems (Represas et al., 2013). Separation of regional and residual anomalies using filtering butterworth. The butterworth filter provides a maximum flat response price on the passband frequency and a zero price at the stopband (Shaker and Oweis, 2016). To delineate rock density contrasts associated with structural zones and heat sources, 3D inversion modeling is done using UBC-GIF Grav3D software using the algorithm from Li and Oldenburg (1998b) (Martakusumah et al., 2015). This inversion modeling is not unique or not single meaning that an anomaly response can be produced by several subsurface models (Grandis, 2009).

2. Geology of Lili Sepporaki Geothermal Area

Lili-Sepporaki geothermal area is located in the region of West Sulawesi, which is dominated by volcanic rocks. Based on stratigraphy (Figure 1), the formation that existence in the Lilli geothermal area consist of Walimbong Volcanic (Tvw), Andesite Feldspatoid (Tf), Andesite Porphy (Tp), Indivisible Volcanic (Tvt), Basaltic Basaltic Lava Bobongbatu (Tlbb), Lava Andesite Buttu Sawergading (Tls), Lava Andesite Buttu Butu (Tlb), Lava Andesite Buttu Talaya (Tlt), Lava Andesite Buttu Dambu (Tld), Lava Andesite Buttu Kamande (Tlk), Lava Andesite Buttu Sawergading (Tls) Alluvium (Qal) (Setiawan et al. 2010).

Lilli's geological structure pattern is clarified into seven structure patterns. Based on the seven structure patterns, the most important that controlled the existence of geothermal surface manifestation is the structure pattern N10-200^{0}E / N190-200^{0}E and N110-120^{0}E / N290-300^{0}E. The intersection of both of this faults form the media that flowing hydrothermal fluid to the surface. There is a depression structure that located in the Lilli geothermal area. It is characterized by the rest of the escarp depression that formed a curve up to half radial. Based on morphology and regional structure patterns, the forming depression is thought to be the result of collapse of the previously formed axes. Geothermal systems in the research area is thought to be limited by this depression and characterized by the surface manifestations that formed hot springs (Setiawan et al. 2010).

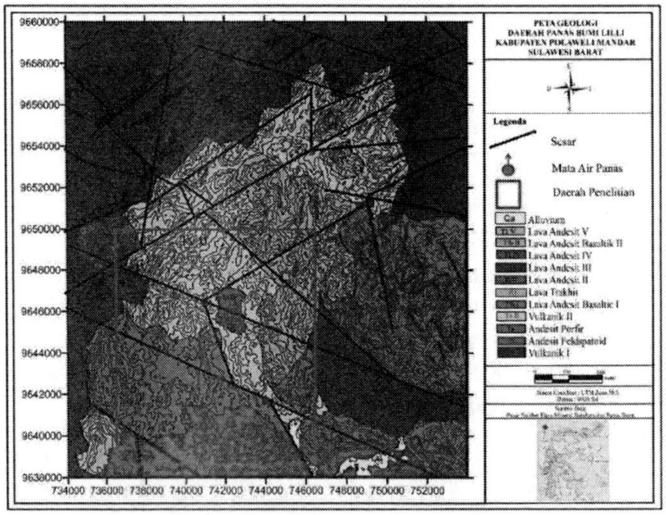

Figure 1. Geology of Lili geothermal area (Center of Geological Resources, 2011)

3. Gravity

3.1. Complete Bouguer Anomaly

In 2010, Center of Mineral Resources, Coal and Geothermal of Indonesia (PSDMBP) was performed gravity survey at Lili Sepporaki geothermal area with 205 stations within a 10 x 10 km coverage area. The measurements were made with the LaCoste & Romberg type G-802 gravitimeter at 205 stations on an area of approximately 100 km2. Figure 2 shows distribution of the gravity stations in the region. The spacing between the stations varies from 250 m to 450 m. In this study, estimation of rock density was carried out using the parasnis method. The parasnis method is based on the Bouguer anomaly equation assuming the Bouguer anomaly is zero. In this method, the variable x (tide correction to latitude correction) and y (bouguer and terrain correction) are plotted on the cartesian coordinate, then a linear line equation can be searched by the least square method. Figure 2 shows the distribution of data from curve (0.04185h-terrain) vs (g_Obs-g_theoritis + 0.3085h) with the density used to make terrain correction is 1 g / cm3. Linear regression (least square) on the distribution of data shows rock density of 2.80 g / cm3. In complete bouger anomaly calculation, for bouguer correction and terrain correction using a density of 2.80 g / cm3.

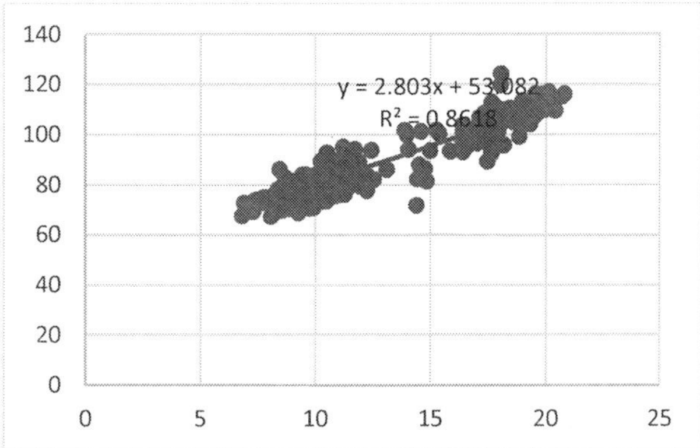

Figure 2. Parasnis method to estimate the average density at Lili Sepporaki geothermal area.

Figure 3 shows the complete Bouguer anomaly map of the study area. Bouguer anomalies ranging from 40 mgal to 62 mgal were observed in the study area. There is low value (40 mgal to 54 mgal) anomaly spreads to the north dominantly, northwest and northeast from the study area. These negative anomaly areas are considered related with the fault that controlled the geothermal system in the region. There is a high value anomaly (55mgal t0 62 mgal) that spread in the southern part of the dominant, southwest and southeast of the study area. coinciding with the high-density igneous rocks.

3.2. Residual Anomaly

The complete Bouguer anomaly that has been obtained contains all possible anomalous sources that are below the surface, namely regional anomalies and residual anomalies (Hinze et al., 2013). The regional Bouguer gravity anomaly is resulting from the presence of deeper and broader structures. The residual Bouguer gravity anomaly is resulting from the presence of smaller and more superficial ones. To get the target that has been determined, the anomalies need to be separated (Setianingsih et al., 2013). In this study, the separation of the complete Bouguer anomaly was carried out using the Butterworth filter. The butterworth filter provides a maximum flat response price on the passband frequency and a zero price at the stopband. This filter can be carried out a high pass filter and filter passes through low (low pass filter) (Shaker and Oweis, 2016).

IOP Conf. Series: Materials Science and Engineering **588** (2019) 012009 doi:10.1088/1757-899X/588/1/012009

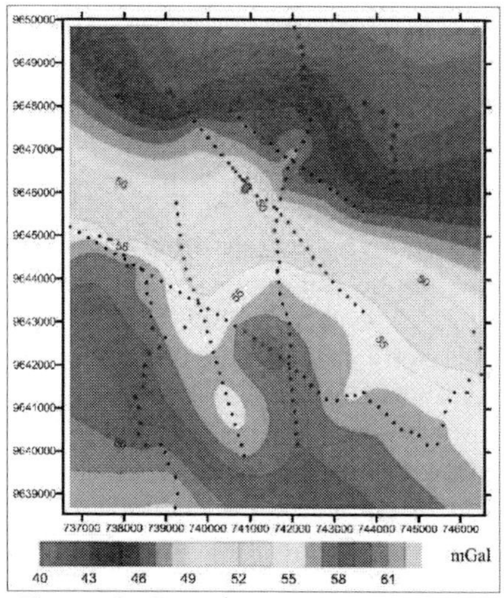

Figure 3. Complete Bouguer anomay of Lili-Sepporaki geothermal area

Figure 4 shows the residual Bouguer anomaly map of the study area. Residual Bouguer anomalies ranging from -2.4 mgal to 2.4 mgal were observed in the study area. There is low value (-2.4 mgal to 0 mgal) anomaly spreads to the north dominantly and in middle of southern from the study area with the direction northwest to southwest. The low value anomaly probably related to the fault zone in the subsurface and and the fault that controlled and formed surface manifestations. There is a high value anomaly (0 mgal to 2.4 mgal) that spread in the southern part of the dominant, southwest and southeast of the study area. coinciding with the high-density igneous rocks.

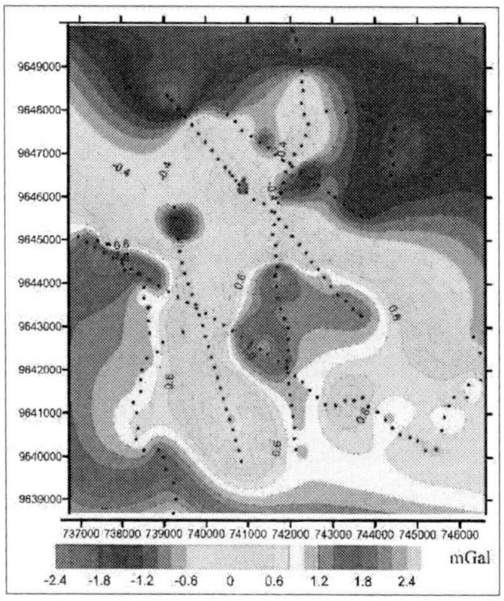

Figure 4. Residual Bouguer anomay of Lili-Sepporaki geothermal area

3.3. *3D Inversion Modelling*

The 3-D inversion model is processed use Li and Odenburg algorithm. The result of 3d inversion modelling has varies values from 2.0 gr/cm3 to 3.2 gr/cm3. 3-D inversion model shows that the low density value 2,0 - 2,3 gr/cm3 is located around the hot springs of the research area. The high density value 2,8 – 3,0 gr/cm3 is located the southern part of area which has more than 2000 m depth. To analyze and interpretation the three dimensional inversion model, the cross section to show the two dimensional that imaging the subsurface condition to interpretated is created from 3-D model (Parapat et al. 2017). The cross section with the direction from southwest to northeast (Figure 4).

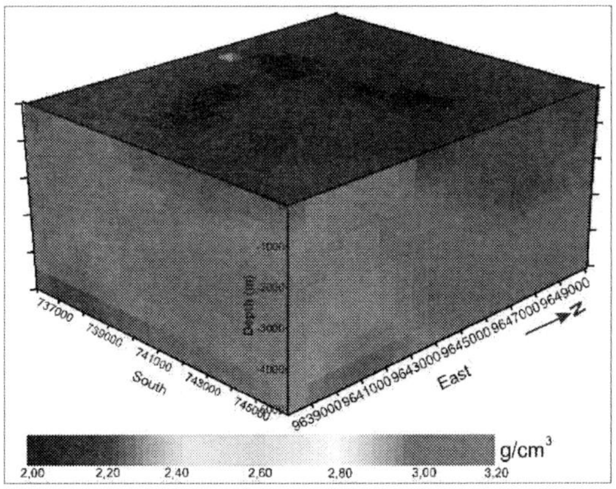

Figure 5. 3-D inversion model of Lili-Sepporaki geothermal area.

This 2-D cross section is located in the dominant the high Bouguer anomaly that area leads to the southeast, where as the medium Bouguer anomaly is in the middle to the northwest and the low Bouguer anomaly is to the northwest. This 2-D cross section model shows that there are the low density rocks with a range of values 2.0-2.3 gr/cm3 at 5000 to 6000 and 9000 to 12000 toward the northeast at depths of <2000 m. This rocks are thought to be associated with a weak that zone associated with fault. There is surface manifestion in this rocks, that matter support the indication that this rocks is associated with fault that controlled the manifestation. While the high density rocks with a range of values from 2.8 to 3.0 gr/cm3 at 0 to 5000 toward the northeast with a depth of >2000 m. This rocks are thought to be associated to intrusive igneous rocks that associated with the heat source of Lili Sepporaki geothermal system. That matter is supported by the location of high Bouguer anomaly values that located in southern part of the research area (Martakusumah et al. 2015; Parapat et al. 2017).

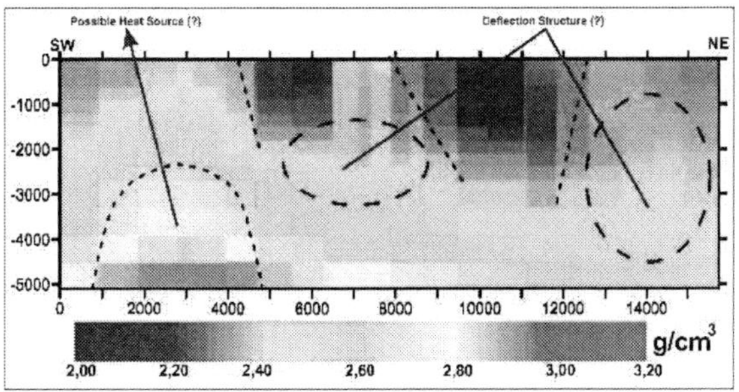

Figure 6. 2-D model cross section of 3-D inversion model of Lili-Sepporaki geothermal area

4. Conclusions

Based on 3-D distribution model of Lili-Sepporaki geothermal area, we can conclude.:

There is a low density value with value 2.0-2.3 g/cm3 which has a thickness of 500-1000 meters, related with the fault zone and high density value with value 2.8-3.0 g/cm3 which has a thickness more than 2000 meters, asummed as igneous rock intrusion.

IOP Conf. Series: Materials Science and Engineering **588** (2019) 012009 doi:10.1088/1757-899X/588/1/012009

5. Acknowledgement
We would like acknowledge to Institut Teknologi Surabaya which also gave us to participate .

References

[1] Pradana F, Lestari W and Warnana D D Subduction Density Data Value Modelling of Java Island *Geosaintek* **Volume 2** Number 6
[2] Grandis H 2009 *Pengantar Pemodelan Inversi Geofisika* (Bandung : Himpunan Ahli Geofisika Indonesia)
[3] Gupta H K and Roy S 2006 Geothermal Energy: An alternative resource for the 21st Century (Amsterdam : Elsevier) pp. 292
[4] H Zahra S and Oweis H T 2016 Application of high-pass filtering techniques on gravity and magnetic data of the eastern Qattara Depression area, Western Desert, Egypt *NRIAG Journal of Astronomy and Geophysics* **Volume 5** Number 1 pp 106-123(18).
[5] P Kearey, M Brooks and I Hill 2006 *An Introduction to Geophysical Exploration* (UK : Blackwell Publishing)
[6] P Represas , F A M Santos , and J Ribeiro 2013 Interpretation of gravity data to delineate structural features connected to low-temperature geothermal resources at Northeastern Portugal *Journal of Applied Geophysics* **Volume 92** pp. 30-38.
[7] R Martakusumah , W Srigutomo , Suryantini, A B Pratama, Trimadona, and A Haans 2015 *Gravity analysis for hidden geothermal system in Cipanas, Tasikmalaya Regency, West Java,* (Melbourne : Proceedings of the World Geothermal Congress) pp.9
[8] S Porkhial , J Nouraliee , M R Rahmāni and D Ebrāhimi 2015 *Study on Conceptual Model of Arabdizaj Prospect in West Azabaijan Province, NW. Iran* (Melbourne : World Geothermal Congress Proceeding)
[9] W M Telford, L P Geldar and R E Sheriff 1990 *Gravity Method in Applied Geophysics : 2nd Edition* (UK : Cambridge University Press) pp. 6-61.
[10] Y Li and D W Oldenburg 1998 3-D inversion of gravity data *Geophysics* **Volume 63**, Number 1, pp. 109–119

Effect of jet velocity on LPG to flame stability and flame temperature distribution on Inverse Difusion Flame (IDF)

Pramoda Agung Sumadhijono[1,2,a], **Bambang Sudarmanta**[1,b], **and Arif Wahjudi**[1,c]
[1]Departement of Mechanical Engineering, Faculty of Industrial Technology, Sepuluh Nopember Institute of Technology. Jalan Raya ITS, Keputih, Sukolilo, Surabaya 60111, Indonesia
[2]Departement Of Mechanical Engineering, Faculty of Engineering, University of 17 Agustus 1945 Surabaya, Jalan Semolowaru No. 45, Surabaya 60118, Indonesia

[a]pramoda_as@untag-sby.ac.id, [b]sudarmanta@me.its.ac.id, [c]arif_w@me.its.ac.id

Abstract. Liquid Pretolium Gas (LPG) experimental investigation used to Inverse Diffusion Flame (IDF) burners is presented. The effect of fuel jet velocity on the flame stability and flame temperature distribution of LPG Inverse Diffusion Flame (IDF) was investigated. Fuel jet speed (Vf) = 0.154 m/s and 0.330 m/s was chosen for analysis. Air jet velocity varies from Va = 7.431 m/s to Va = 15.924 m/s for Vf = 0.154 m/s and Va = 8.493 m/s up to Va = 21.231 m/s for Vf = 0.330 m/s . The results of the observation show that the velocity of the fuel jet affects the ratio of the velocity of the air-fuel jet which will ultimately have a significant effect on the flame. It was found that the stability of LPG IDF flame at Vf = 0.154 m/s occurred at Va = 11.677 m/s, the equivalence ratio (ϕ) = 1.99. The stability of LPG IDF flame at Vf = 0.330 m/s occurs at Va = 14.862 m/s with an equivalence ratio (ϕ) = 1.56. It was observed that the temperature along the IDF centerline increased when the distance from the burner (Z) exit increased. The nature of this temperature distribution shows that central cold air in the air jet is gradually heated towards the IDF downstream. The highest flame temperature is seen in Z between 90-110 mm. It was found that the length of the flame increased with increasing flow of fuel jets. At a constant fuel jet speed the flame length increases but will return again to a certain fuel-air ratio. At Vf = 0.154 m/s the highest flame height is 111 mm and at Vf = 0.330 m/s the highest flame height is 143 mm.

1. Introduction

Energy is the driving force for various human activities. Energy production in the industrial world is largely sourced from combustion, where combustion is defined as the process of chemical reaction between fuel and oxidizer accompanied by the production of heat energy in the form of light or flame. Fuel is a substance capable of releasing heat when an oxidation reaction occurs, while an oxidizing agent is an element that contains oxygen (eg pure oxygen, air). Processes that may occur during the combustion process include: chemical and physical interaction processes accompanied by the release of heat energy derived from the energies of chemical (exothermic) bonds, heat transfer processes, mass transfer processes, and fluid particle movements. The biggest type of fuel used today comes from the remains of natural fossils with the basic elements of hydrocarbons in the form of: solid as coal, liquid such as gasoline and diesel, and in the form of gases such as methane, propane, butane and others.
Various steps to meet energy needs are important topics along with the decreasing number of fossil energy sources. The current energy system is still largely produced from the fuel combustion process in

steam, gas and steam and gas power plants. Likewise with the transportation system, as well as various processes in the industry that require heat sources from the results of burning fuel in the combustion chamber. The study of combustion greatly affects the increasingly scarce use of fuel. Experimental studies and theoretical studies of combustion have been carried out to better understand the phenomenon of combustion. Simulations with various numerical methods are increasingly being carried out supported by advances in information technology. These steps are basically aimed at obtaining a new method of burning fuel with more savings, cleaner and more stable, so that in the end the fulfillment of the energy needs obtained from the combustion process can be more guaranteed.

Various methods have been examined both in terms of the quantity and quality of the flow of air and fuel mixtures as well as burner equipment engineering. From the aspect of the flow of mixture of air and fuel, one method for obtaining clean combustion is combustion in conditions of mixture with more air value (excess water) that is high or rich in oxygen or poor in fuel. And from the burner equipment engineering aspect, the use of swirler is still an alternative that is often found to improve combustion.

Inverse Diffusion Flame (IDF) is a special type of diffusion flame. This differs from the normal diffusion flame-Normal Diffusion Flame (NDF) is related to the relative position of the fuel outlet and the air outlet / oxidant. At the IDF, a fuel outlet (external) surrounds the air outlet (internal). The low-speed fuel jet drags in because of the high speed of the air and mixes with air. NDF has a wide flammability range. Premixed flame has the characteristics of low soot formation. IDF is a feature of NDF and premixed fire, hence the advantages of both fires in terms of various flame stability, less pollutant emissions and operational safety.

Wu and Essenhigh [1] presented a fire map for IDF using methane. Six different types of fire are proposed for IDF. Mikofski et al. [2] examined the effects of various air flow rates on ethylene and methane IDFs using laser-induced hydroxyl fluorescence and induced laser soot. They observed that fire ethylene has a greater tendency to form soot. Sobiesiak and Wenzell [3] examined the effects of dynamic mixing between natural gas and air in a reverse flame structure. Mikofski et al. [4] examined the relationship between high ethylene flame and methane using luminosity images. They found that the glowing flame height was greater than the height of the reaction zone. Mahesh and Mishra [5] conducted an experimental study on IDF fire structures using LPG behind the burner. They reported variations in fire appearance and centerline temperature distribution at different air-fuel ratios.

Zhen et al. [6] conducted an experiment to study the effect of nozzle length on IDF LPG. The structure of fire and CO concentration are compared in multi jet fuel burners with long and short nozzles. They observed that the potential core and IDF fire base were shorter in the short nozzle compared to the long nozzle case. Mahesh and Mishra [7] conducted a comparative study of IDF LPG behind and coaxial burners. The results show that a compact flame with little luminosity is observed in the backstep burner. Stelzner et al. [8] presented experimental and numerical investigations on the IDF structure of rich methane.

Elbaz and Roberts [9] try to understand the role of entrainment and mixing in IDF methane. Elbaz and Roberts [10] attempted to characterize methane IDF with air and annular fuel jets. They observed that an increase in the air-fuel speed ratio increases the entrainment of the fuel to the air flow; therefore, the air-fuel mixture is increased. Bhatia et al. [11] numerically investigates IDF and NDF. They reported that oxygen enrichment and gravity variation had a significant effect on the NDF flame structure compared to IDF. Barakat et al. [12] presented experimental and theoretical studies of the effects of jet dynamics and geometry parameters of burners on the behavior of IDF fuel LPG. The energy associated with IDF is strongly related to the distribution of temperature and temperature levels. Takagi et al. [13] the temperature characteristics that are numerically examined are derived from preferential diffusion effects on NDFs and IDFs.

Sze et al. [14] presented a detailed study of IDFs in two different burners, one with coaxial settings and the other with circumferentially regulated ports. They observed a higher flame temperature in the harbor arranged in a circular manner because of a more intense air fuel mixture. Dong et al. [15] examined the effect of air port diameters on the characteristics of heat transfer from butane IDF. Zhen et al. [16] reported the effect of equivalence ratios and Reynolds numbers on circulating temperature, fire

IOP Conf. Series: Materials Science and Engineering **588** (2019) 012010 doi:10.1088/1757-899X/588/1/012010

appearance and CO 2 emissions. Zhen et al. [17] compared the IDF thermal field with flame and the flame that was fitted with fire using LPG as fuel.

Kamal [18] compared the characteristics of IDF and NDF soot. Patel et al. [19]; Examining experimentally the appearance of fire and emission characteristics of LPG-fueled IDF with coaxial burners. Long flame, CO and NOx emissions are checked at the equivalent ratio and the angle of the different swirler propellers. IDF with swirl, shows wider, shorter and more stable compared to IDF without swirler. Adequate investment and swirl effects for 30 ° propeller angles produce an environment that matches low CO and NOx emissions.

Through a literature review, it is known that although a large number of papers are available on IDF, the effect of fuel jet velocity on the distribution of fire, fire stability, equivalence and altitude ratios has not been found. Therefore, our knowledge of the effect of fuel jet velocity on fire characteristics is relatively rare and incomplete. The main purpose of this study was to determine that the effect of LPG fuel jet speed on the characteristics of the flame produced by the IDF burner. Then an experimental test was performed to find out the performance of the IDF burner at the chosen fuel jet speed.

2. Experimental apparatus and procedure

2.1. Experiment Settings

Figure 1 shows the geometry details of the coaxial burner. The burner consists of two coaxial tubes with a inner tube 10 mm in diameter and an outer tube with a diameter of 30 mm. The burner is operated in such a way that the central air jet is surrounded by an annular fuel jet. To investigate the effect of fuel jet velocity on the characteristics of the variable setting flame as shown in table 1.

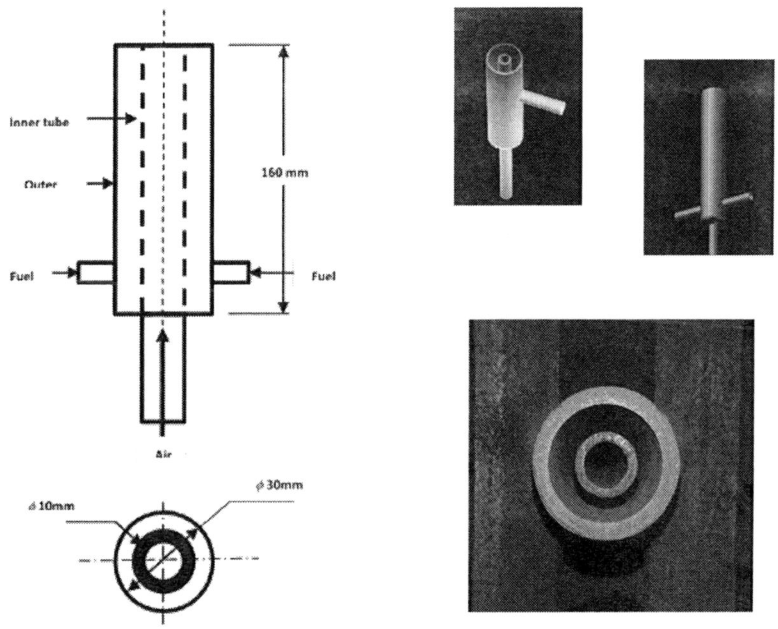

Figure 1. Geometrical details of the coaxial burner

Figure 2 shows the experimental test setting scheme used in this work. The air is supplied in a coaxial burner from the blower. Valves are used to control air flow. Air flow is measured with a calibrated rotameter (± 2.0% accuracy of full scale) installed in the air duct. Experiments carried out using LPG as fuel. LPG contains 70% C_4H_{10} (butane), 30% C_3H_8 (propane). The needle valve is used to control the

IOP Conf. Series: Materials Science and Engineering **588** (2019) 012010 doi:10.1088/1757-899X/588/1/012010

flow of fuel. The fuel flow is measured with a calibrated rotameter (± 2.0% accuracy of full scale) mounted on the fuel line. The shape and size of the fire are captured with the help of a high resolution digital camera (24.2 megapixels, ISO 12800 sensitivity) which has video recording capability at 30 fps.

Tabel 1. Research variable

Speed Jet Fuel (V_f) (m / s)	Air Speed (V_a) (m / s)
0.154	7.431
	9.554
	11.677
	13.800
	15.924
0.330	8.493
	11.677
	14.862
	18.047
	21.231

The length of flame (Ht) is defined as the distance between the burner and the tip of the flame along the vertical center line of the IDF. The tip of the fire means the point where the fire is visible to the naked eye. Ten images were recorded for each IDF configuration. The length of the last flame is considered as the average flame length of ten images under the same conditions. Equivalent ratio, Φ is defined as:

$$\Phi = \frac{AFR_{stoic}}{AFR_{act}} = \frac{\left(Ma/_{Mf}\right)_{stoic}}{\left(Ma/_{Mf}\right)_{act}} \tag{1}$$

where (Ma / Mf) stoic is the stoichiometric air-fuel ratio and (Ma / Mf) is the actual air-fuel ratio. Variations in flame appearance, flame length, flame distribution and centerline temperature distribution as a function of fuel jet speed, air jet speed, equivalence ratio, are analyzed in the following sections.

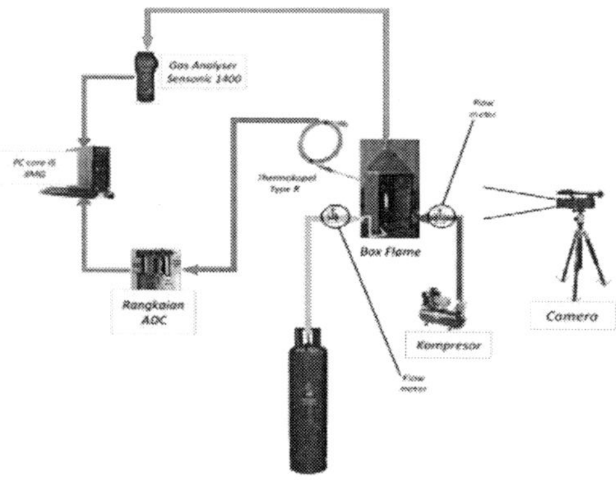

Figure 2. Scheme in research

2.2. *Instrument arrangement and data acquisition*

Retrieval of flame temperature data was carried out with 8 R type thermocouples. The diameter of the thermocouple wire was 0.3 mm and the diameter of the ceramic was 4 mm. Thermocouples are mounted vertically on a fire. The temperature is taken simultaneously (8 thermocouples) at each altitude. Data from thermocouple is used as digital data using Analog Digital Converter, using PLC software. Digital

IOP Conf. Series: Materials Science and Engineering **588** (2019) 012010 doi:10.1088/1757-899X/588/1/012010

data is transferred to a computer and forms a temperature data table. This data is then processed in MATLAB software for temperature distribution. The measurement mechanism is shown in Figure 3:

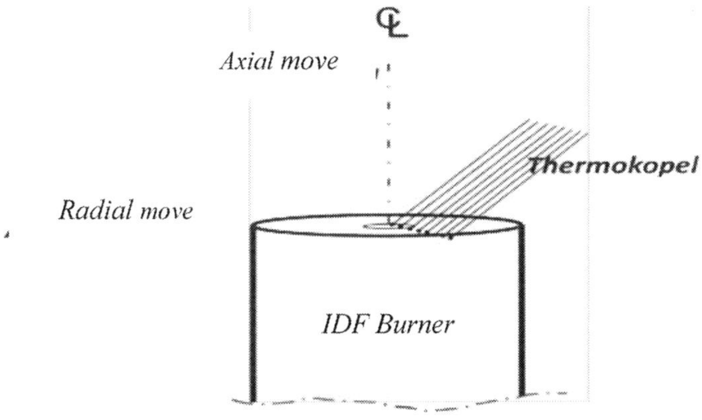

Figure 3. Acquisition Data

3. Results and discusstions

3.1. Flame appearance and distribution Flame temperature

Wu and Essenhigh [1] identified six types of IDF regimes based on the form of flames. The speed and stoichiometry of air-fuel flow affects the shape and structure of the IDF. The emergence of the IDF reflects the quality of the physical process that is completed during burning. In this study, the flame display of IDF on coaxial burners was investigated for different air flow and fuel conditions. Figure 4 shows the IDF structure. IDF structure has been characterized as a double structure (Zhen et al. [6]). The IDF dual structure consists of a fire base (entrainment zone) at the root of the flame and the next flame torch. The fire neck connects the base of the fire and the flame torch. Fire neck, which is yellow, indicates the presence of soot (soot). The flame torch consists of an inner blue zone and an external luminous zone as shown in Fig. 4

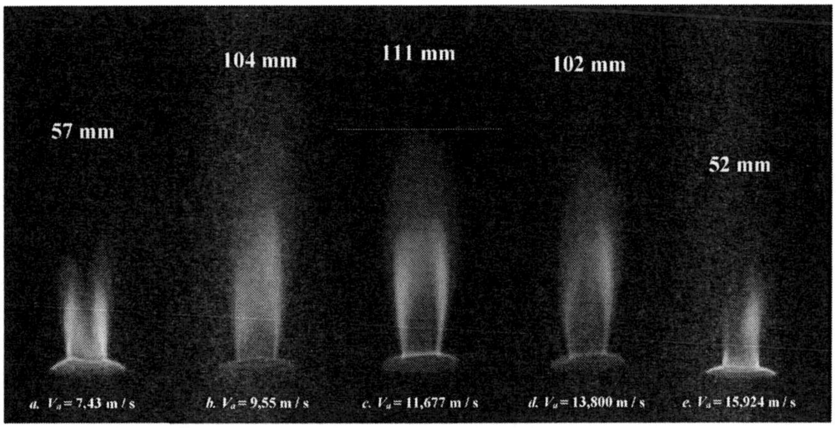

Figure 4. IDF structure and flame length for V_f = 0,154 m / s

IOP Conf. Series: Materials Science and Engineering **588** (2019) 012010 doi:10.1088/1757-899X/588/1/012010

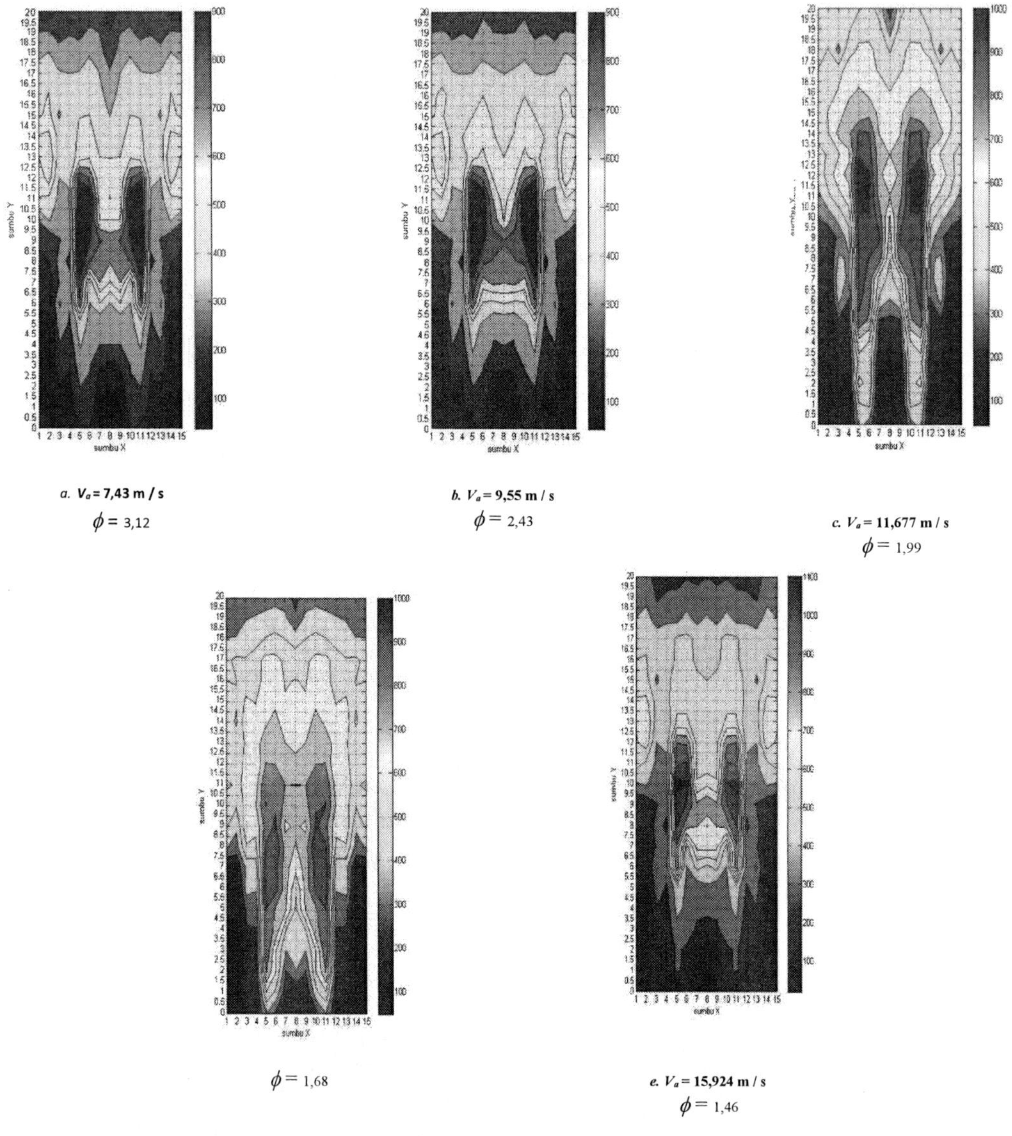

Figure 5. IDF structure and flame length for (ϕ) ; $V_f = 0{,}154$ m / s

Figure 4. shows variations in fire appearance for increased air jet speeds, Va = 7.43 m / s to 15.92 m / s, with fixed fuel jet speeds (Vf = 0.154 m / s). It was seen that the length of the flame continued to increase with the increasing speed of the air jet. An increase in IDF length can be associated with sufficient entrainment of fuel in a central air jet. High-speed air jets are accelerated to react with fuel jets and fuel-rich mixtures form longer IDFs. However, at IDF Va = 7.43 m / s, so weak entrainment results in so very rich and results in soot formation, fires are shorter than higher air Va.

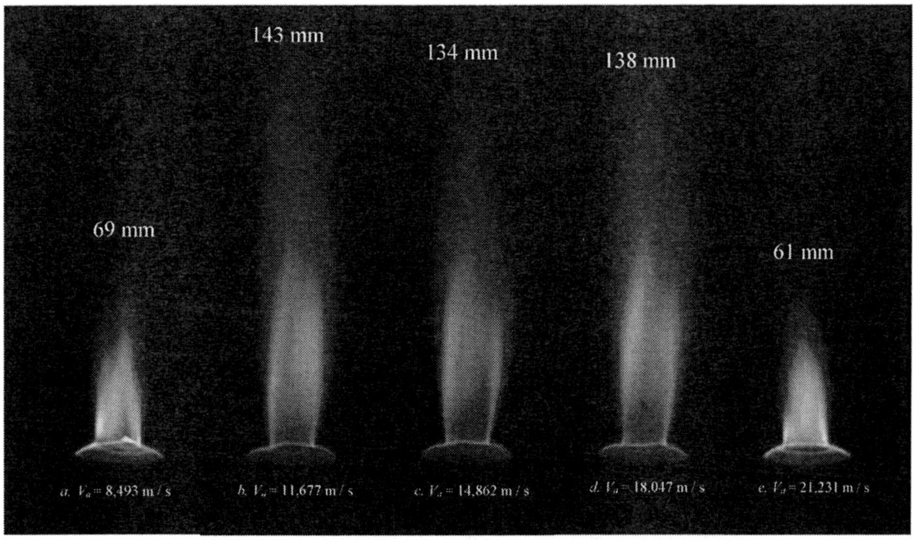

Figure 6. IDF structure and flame length for V_f = 0,330 m / s

Figure 7a. Temperature distribution and equivalence ratio (ϕ) ; V_f = 0,330 m / s

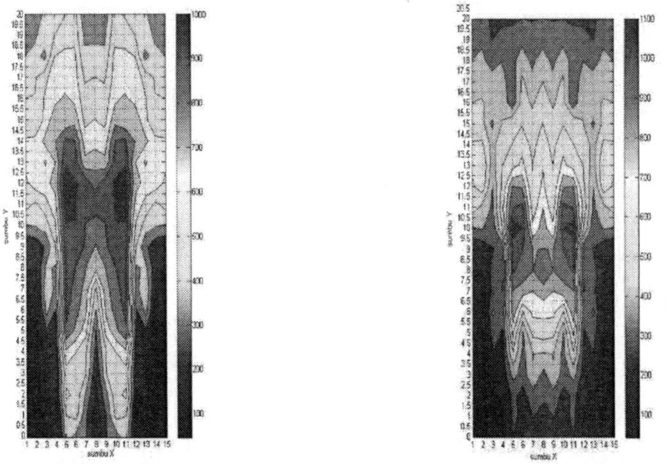

d. V_a = 18,047 m / s
ϕ = 1,38

e. V_a = 21,231 m / s
ϕ = 1,09

Figure 7b. Temperature distribution and equivalence ratio (ϕ) ; V_f = 0,330 m / s

Likewise, Va = 15,924 m / s, so the high difference between the speed of the air jet and the speed of the fuel jet and the increase in air mass results in the high flame return decreasing. But there is no visible formation of excess soot. As shown in Figure. 4, the length of the blue zone (Hb) has a quadratic trend with Vf. Figure 5. Shows the distribution of flame from the speed of an increasing air jet. The area of fire with high temperature will increase with increasing air jet speed, but will also experience a decrease if in certain compositions jet speed continues to increase.

As shown in Figure. 4, when Va is small, the length of the blue zone (Hb) increases linearly with an increase in Va up to 11,677 m / s. When Va increases, the blue zone decreases slightly as shown in Figure 4 (d) - (e). The velocity of the air jet increases from 7.431 m / s to 15.924 m / s, resulting in a shorter IDF with a thinner flame torch. In addition, the reduction in IDF length was observed with increasing air jet speed. In normal diffusion flames, the fuel is burned with entrained air from the surroundings; therefore, the color is yellow.

At the IDF, the central air jet withstands the fuel jet. The amount of fuel trapped in an air jet controls the extent to which the fuel is burned in pre-mode or diffusion mode. IDF color changes indicate the entrainment effect of the fuel jet. It can be seen that the increase in the speed of the air jet increases the entrainment of the fuel jet, which leads to more fuel being burned in pre-mixed mode. The blue zone represents the initial combustion. At high enough air jet speeds, all fuel is burned in pre-temp mode. Sze et al. [14] have reported the same effect of air jet speed on the length of flame and color of IDFs. This study clearly shows that the appearance of the IDF may depend on the speed of the air jet and the speed of the fuel jet.

The effect of Φ on the appearance of the flame, the length of the flame and the non-rotating IDF luminosity can be observed qualitatively through direct visualization of the flame (Fig. 4 and 6). The increase in length and luminosity of non-rotating IDF can be associated with inadequate entrainment of fuel due to increased fuel rate. Because the equality ratio with the air jet speed continues to increase, the fuel momentum also increases. Therefore, a number of fuels do not fall under the entrainment effect of a central air jet. This unburned fuel, burning in diffusion mode on the outside of the flame causes the long flame to elongate with soot formation. The length of the flame is an important characteristic, which helps to qualitatively evaluate the effect of air-fuel mixing. The variation of flame length as a function of Φ and velocity is presented in Figure 4,5,6,7 along with the temperature distribution.

IOP Conf. Series: Materials Science and Engineering **588** (2019) 012010 doi:10.1088/1757-899X/588/1/012010

3.2. Center line temperature

Figure 8 shows the temperature at the Center line, for each flame observed. It appears that with the increasing speed of the fuel jet the higher the highest average temperature can be achieved at the center line. The highest temperature at the center line is reached at altitudes between 90 mm to 110 mm. For Vf = 0.154 m / s the highest temperature at Va = 15.924 m / s this is because the equvalence ratio has a value of ϕ = 1.46 so that the fuel mixture - air between stoichiometry and actual is not much different. Likewise with Vf = 0.330 m / s the highest temperature at the center line is at Va = 21.231 m / s with the equvalence ratio has a value of ϕ = 1.09.

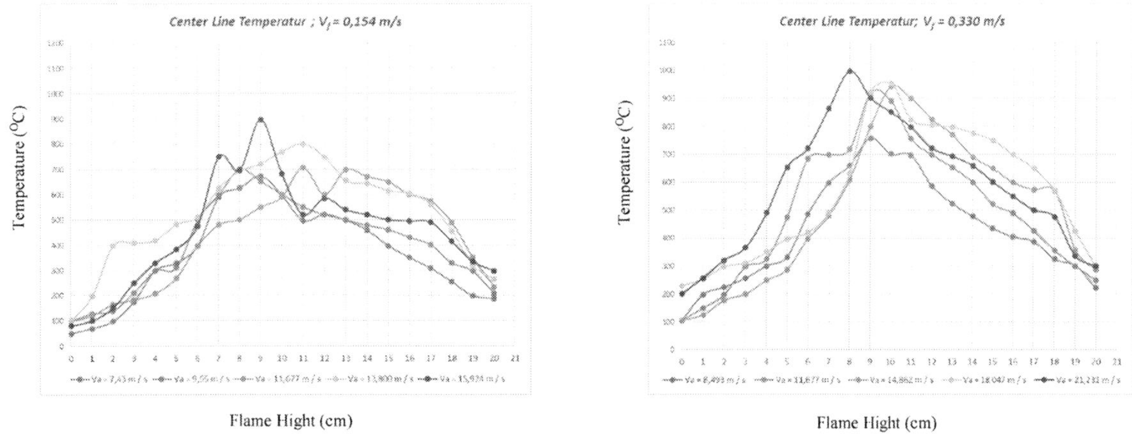

Figure 8. Center line temperature

4. Summary

An experimental study was conducted to investigate the effects of different fuel jet speeds and air on the appearance of IDFs. The effect of the fuel jet speed on the appearance of the fire was investigated. The effect of the equivalence ratio on the flame length is checked. The main conclusions from this study are summarized below.

1. The length of the IDF flame below the speed of a constant air jet continues to increase with the speed of the air jet. The length of the flame that decreases with the speed of an increased air jet is observed because of better fuel entrainment which increases the mixture of air and fuel. For V_f = 0,154 m / s the highest flame is 111 mm at V_a = 11,677 m / s and V_f = 0,330 m / s at V_a = 11,677 m / s the highest fire height reaches 143 mm.
2. Temperature distribution is seen in more and more high temperature areas as air velocity increases and jet fuel speeds are constant. High temperature areas are in flames with Va = 9.55 m / s; ϕ = 2.43; V_f = 0.154 m / s and V_a = 14.862 m / s; ϕ = 1.56; V_f = 0.330 m / s. But the increasing speed of air jets in hot areas will also decrease.
3. The highest temperature at the centerline between 90 mm to 110 mm in temperature between 800 ^{O}C and 1000 ^{O}C. It can be concluded that the higher the speed of the highest temperature air jet in the line will be higher. This is due to the effective entrainment of the fuel, which leads to a better increase in fuel and air to produce an elongated IDF blue zone.

5. Acknowledgments

This work has been supported by the Combustion and Fuel Laboratory managed by the Department of Mechanical Engineering, Faculty of Industrial Technology, Sepuluh November Institute of Technology, Surabaya, Indonesia and the IDF Burner Development Project managed by the Departement Of Mechanical Engineering, Faculty of Engineering, University of 17 Agustus 1945 Surabaya, Indonesia.

IOP Conf. Series: Materials Science and Engineering **588** (2019) 012010 doi:10.1088/1757-899X/588/1/012010

References

[1] K T Wu and R H Essenhigh 1984 *Mapping and structure of inverse diffusion flame of methane*, twentieth symposium (International) on combustion, Combustion Institute 1925–1932.

[2] M A Mikofski, T C Williams, C R Shaddix and L G Blevins 2004 *Effect of varied air flow on flame structure of laminar inverse diffusion flame,* Combustion Institute

[3] A Sobiesiak, J C Wenzell 2005 Characteristics and structure of inverse diffusion flames on natural gas *Proceed. Combustion Institute* 30 743–749.

[4] M A Mikofski, T C Williams, C R Shaddix and L G Blevins 2006 Flame height measurement of laminar inverse diffusion flames *Combust Flame* 146 63–72.

[5] S Mahesh and D P Mishra 2010 Flame structure of LPG-air inverse diffusion flame in a backstep burner, *Fuel* 89 2145–2148.

[6] H S Zhen, YS Choy, CW Leung and CS Cheung 2011 Effects of nozzle length on flame and emission behaviours of multi-fuel-jet inverse diffusion flame burner, *Appl. Energy* 88 2917–2924.

[7] S Mahesh and D P Mishra 2011 Study of the turbulent inverse diffusion flame in recessed backstep and coaxial burners *Combust. Explos. Shock Waves* 47 (3) 274–279.

[8] B Stelzner, F Hunger, S Voss, J Keller, C Hasse and D Trimis 2013 Experimental and numerical study of rich inverse diffusion flame structure *Proceed. Combust. Institute* 34 1045–1055.

[9] A M Elbaz and W L Roberts, 2014 Flame structure of methane inverse diffusion flame *Exp. Therm. Fluid Sci.* 56 23–32.

[10] A M Elbaz and W L Roberts 2014 Experimental characterization of methane inverse diffusion flame, *Combust. Sci. Technol.* 186 1249–1272.

[11] P Bhatia, V R Katta, S S Krishnan, Y Zheng, P B Sunderland and J P Gore 2012 Simulations of normal and inverse laminar diffusion flames under oxygen enhancement and gravity variation, *Combust. Theory Modell* 16 774–798.

[12] H Z Barakat, M R Salem, A Morgan and H E Hany 2013 Study of effects of burner configuration and jet dynamics on characteristics of inversed diffusion flames, *J. Mech. Eng. Res.* 5 (7) 128–144.

[13] T Takagi, Z Xu and M Komiyama 1996 Preferential diffusion effects on the temperature in usual and inverse diffusion flames, *Combust. Flame* 106 252–260.

[14] L K Sze, C S Cheung and C W Leung, 2006 Appearance, temperature, and NOx emission of two inverse diffusion flames with different port design, *Combust. Flame* 144 237–248.

[15] L L Dong, C S Cheung and C W Leung 2013 Heat transfer optimization of an impinging port-array inverse diffusion flame jet *Energy* 49 182–192.

[16] H S Zhen, C W Leung and C S Cheung 2010 Thermal and emission characteristics of a turbulent swirling inverse diffusion flame, *Internat. J. Heat Mass Transf.* 53 902–909.

[17] H S Zhen, C W Leung and C S Cheung 2011 A comparison of the thermal, emission and heat transfer characteristics of swirl-stabilized premixed and inverse diffusion flames, *Energy Convers. Manage* 52 1263–1271.

[18] M M Kamal 2007 Soot formation and oxidation in normal and inverse diffusion flames, *J. Power Energy* 221 (4) 481–495.

[19] Patel V and Rupesh S 2017 Experimental and Numerical Investigation of LPG Fuelled Inverse Diffusion Flame in a Coaxial Burner, *International Journal of Advanced Thermofluid Research (ijatr)* **Vol. 3** No. 1.

IOP Conf. Series: Materials Science and Engineering **588** (2019) 012011 doi:10.1088/1757-899X/588/1/012011

Co – pyrolysis of Empty Palm Fruit Bunch (EPFB) and Low Quality Coal Under CO_2 Atmosphere

S Zullaikah[1,a], R Y Naulina[1], L D Puspitasari[1], U Nadliroh[1] and H Niawanti[2]

[1] Department Of Chemical Engineering, Institut Teknologi Sepuluh Nopember (ITS) Jl. Arief Rahman Hakim, Surabaya 60111 Indonesia
[2] Deparment of Chemical Engineering, Mulawarman University, Samarinda 75119 Indonesia

[a] szulle@chem-eng.its.ac.id

Abstract. The objective of this research was to study blending ratio effect on co-pyrolysis of Empty Palm Fruit Bunch (EPFB) and low quality coal using CO_2 gas purge on the characteristics of gas, solid and liquid products. Characteristics of gas products consisting of H_2 and CO_2 gas concentration. Solid product characteristics consist of pore diameter, surface area, total pore volume, solid structure. Meanwhile, the characteristics of liquid products consist of liquid composition. EPFB and low quality coal were reduced to a maximum particle size of about 16 meshes at pre treatment process. The blending ratio of low quality coal and EPFB were 100/0, 80/20, 60/40, 40/60, 20/80 and 0/100, temperature at 500 °C. The result obtained that H_2 gas increased between 0 % - 0.423 % due to the higher blending ratio. The concentration of CO_2 gas was fluctuated. Meanwhile, pore diameter and total pore volume decreased due to the higher blending ratio. The pores diameter were 61.7316 nm and 18.8361 nm obtained at 0 % and 80 % EPFB, respectively. Total pore volumes were 0.0111 cm^3/ and 0.0063 cm^3/g obtained at 0 % and 80 % EPFB, respectively. Surfaces area were increased, they were 0.72 m^2/g and 1.3480 m^2/g at 0 % and 80 % EPFB, respectively. Physical analysis was obtained that the higher of blending ratio (EPFB 100%), the more concentrated of reddish brown color. Liquid products identified contain 41 compounds with the dominant compound of the alkene group. The largest compound were hexadecane (23.69%) and heptacosane (22.61%).

1. Introduction

Energy demands is increasing along with growth in the industrial and transportation sectors. During this time, most of the energy are derived from fossil fuels and minerals that are non-renewable resources. In fact, the amount of fossil fuels and minerals is limited. So it takes effort to reduce dependence on the use of fossil fuels and minerals as well as efforts to optimize the use of new alternative energy sources. Biomass is a source of renewable energy that is environmentally friendly when compared to coal. EPFB is one of the most common biomass in Indonesia. Indonesia is the largest producer of oil palm in the world with an area of 6.17 million Ha in 2013 (BPS). EPFB is a part of palm oil that is classified as untargeted lignocellulosic waste. EPFB is the largest solid waste generated by oil palm plantations. The availability of EPFB is quite abundant reaching 26,697,464 tons per year[1]. During this time, the utilization of EPFB only as fuel boiler, compost and also as hardener in oil palm plantation.

Pyrolysis is a technology to produce alternative renewable energy using biomass materials. It is a high temperature process without present of O2. In general, the material for the pyrolysis process is only biomass. It will cause some disadvantages, such as biomass having high content moisture content, low calorific value, low density and so on[5]. To overcome this problem, biomass and coal co-pyrolysis methods were used. The combination of biomass and coal in the co-pyrolysis process can complement each other's deficiencies. The addition of coal to biomass pyrolysis may affect product composition and distribution. While the high content of H/C in biomass can be used as a H2 donor in the pyrolysis process. The hydrogen can stabilize the large radical structures produced by coal pyrolysis. Co-pyrolysis process produces three kinds of products: solid, liquid and gas. These three components can be further utilized.

In this study used CO2 as purge gas due to CO2 gas contributed to global warming. The use of CO2 gas as a purge gas in co-pyrolysis will help reduce the emissions of CO2 gas in the atmosphere. In co-pyrolysis using CO2, tar is likely to crack and become gas, otherwise tar will also be polymerized and secondary char formation occurs[2]. The use of CO2 gas as a purge gas is also capable of enlarging the specific surface area of the char and increasing the yield of the gas produced[3]. The solid product formed can be used for fuel or used as activated carbon. Liquid product as bio-oil can be used as an additive or mixture in fuel. Meanwhile, the gas formed can be burned directly or as a source of syngas[7].

This study was expected to know the effect of blending ratio on co-pyrolysis between EPFB and low quality coal using CO2 purge gas in terms of gas, solid and liquid product characteristics. It was analyzed in terms of quantitative (product yield) and qualitative (product composition analysis). In addition, this study was also expected to optimize the use of resources in the form of low quality coal and EPFB in order to produce alternative energy products.

2. Experimental

2.1 Material

EPFB were obtained from Malang, East Java, coal was obtained from Semen Gresik Tuban Factory and CO_2 purge gas was obtained from Genta Prima Gas Surabaya.

2.2 Co-Pyrolysis of EPFB and Low Quality Coal Using CO_2 Purge Gas

Co-pyrolysis was used free fall reactor that operated at atmospheric pressure. The equipment scheme can be seen in Figure 1. In the preparation stage, there were smoothing and sieving process and also drying process. Coal and EPFB were smoothed and sieved with a maximum size of 16 mesh or 1.19 mm. For the drying process, the finely fed feeds were fed into the oven at 110 °C for 3 hours or until constant weight was achieved. The drying process was needed due to the pyrolysis process was based on the standard drybasis material. Then, the low quality coal and EPFB were mixed according to the variable blending ratio determined by 100 gram. Before the sample was put into the reactor, the reactor was heated with a constant temperature of 500 °C while the CO_2 gas was flowed. After the temperature was reached and constant, sample was fed into the reactor through the hopper using gas flow conditions discharged so that no sample is pushed out due to pressure from the CO_2 gas. Feed for pyrolysis was dropped slowly into the reactor by adjusting the hopper valve. During the pyrolysis process, the temperature was also kept constant and the CO_2 gas carrier fed into the reactor at a constant rate of 1 L/min. This co-pyrolysis process lasts for 1 hour. From the results of co-pyrolsis, produced three types of products were gas, liquid and solid. After the pyrolysis was complete, the reactor was cooled till reached the room temperature. Char was a solid product that collected using a beaker glass and placed at the bottom of the reactor. Liquid or primary gas product was a condensed gas product that turns phase into liquid and collected into erlenmeyer. While, the gas product or secondary gas was an uncondensable gas that collected in a gas holder.The hot solid products, cooled at room temperature and then weighed to obtaine a solid mass. Liquid product that was collected in erlenmeyer, then filtered to got solid liquid solid-free product, then afterwards weighed to obtained the mass of liquid. The mass of gas obtained by reducing the mass of initial feed by the total solid and liquid mass (assuming the input material was equal to the output).

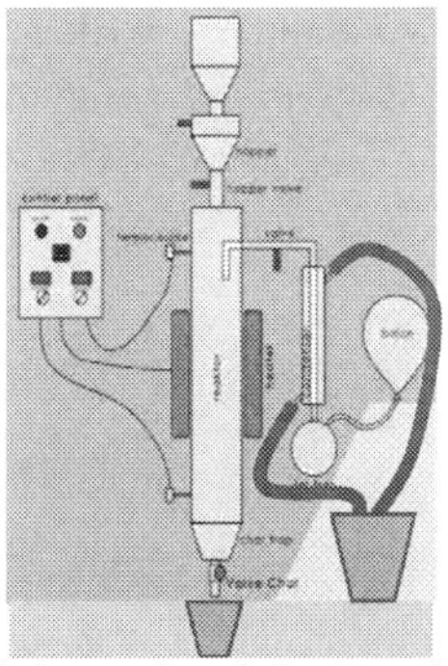

Figure 1. Schematic of Co-Pyrolysis Equipment

The equation below was used to calculated the yield of solid, liquid and gas in product :

$$\text{Yield of solid} = \frac{Mass\ of\ solid}{Mass\ of\ raw\ material} \text{ x } 100\ \% \tag{1}$$

$$\text{Yield of liquid} = \frac{mass\ of\ liquid}{Mass\ of\ raw\ material} \text{ x } 100\ \% \tag{2}$$

$$\text{Yield of Gas} = \frac{Mass\ of\ gas}{Mass\ of\ raw\ material} \text{ x } 100\ \% \tag{3}$$

2.3 Analysis Methods

In this study was conducted several analysis methods for each product. GC (Gas Chromatography) was used to analyze the composition of CO_2 and H_2 gases in gas product, the equipment was available in Biochemistry Laboratory, Chemical Engineering Department, ITS. The solid product was analyzed using BET (Brunauer-Emmett-Teller) method to determine the total surface area, pore diameter and pore volume performed in the Electrochemistry laboratory, Chemical Engineering Department, ITS. In addition to using BET, solid product was also analyzed using SEM (Scanning Electron Microscopy) method to find out the solid morphology structure of the product conducted in Integrated Physics Laboratory, State University of Malang. While liquid product was analyzed using GC-MS (Gas Chromatography-Mass Spectrometry) to found out the liquid product composition conducted in Gelora Djaya Laboratory, Surabaya.

3. Results And Discussion

3.1 Characteristics of Raw Material

EPFB were obtained from Malang-East Java, while coal obtained from Semen Gresik Tuban Factory. Before use, the feedstock was dried and chopped to a maximum particle size of 16 mesh or 1.19 mm. The characteristics of each raw materials were analyzed by proximate analysis including total moisture, volatile matter, ash content and fixed carbon. ASTM method was used for proximate analysis. The results of the proximate analysis of coal and EPFB were presented below.

Table 1. Result of coal and EPFB proxymate analysis

Parameter	Analysis Result (%)		Method
	Coal	EPFB	
Total Moisture	27.01	13.02	ASTM D3173
Volatile Matter	17.70	58.17	ASTM D3175
Ash content	11.73	10.43	ASTM D3174
Fixed Carbon	43.56	18.38	-

Based on the results of the proximate analysis in Table 1, the coal used in this research was categorized as low rank coal (subbituminuous). The fixed carbon range for subbituminuous coal is about 43.6 %. EPFB that used in this study had a lower total moisture and higher volatile matter than coal, obtained about 58.17% and 13.02 %, respectively.

3.2 Pyrolysis Result

3.2.1 Blending ratio on EPFB and Low Quality Coal Co-pyrolysis at 500°C

This research was focused on the effect of blending ratio of coal and EPFB on co-pyrolysis at 500 °C. Time of pyrolysis was 1 hour, the pyrolysis process was purified using 1L/min of CO_2 gas. The products from pyrolysis process were liquid, solid and gas products. The yield of data obtained from the acquisition of solid, liquid, and gas products is presented in Table 2.

Table 2. Result of pyrolysis product of EPFB and low quality coal at Low Temperature 500 °C

Coal	Liquid (%)	Solid (%)	Gas (%)
100 %	2.7053±2.5	70.9358±0.5	26.3589±1.5
80 %	2.7200±1.03	69.5830±1.04	27.6970±2.4
60 %	3.2113±1.08	67.7850±0.1	29.0037±0.13
40 %	3.7186±1.7	65.1030±0.2	31.1784±0.61
20 %	4.4427±1.03	48.1288±9.2	47.4285±10.74
0 % (EPFB 100 %)	5.1744±2.13	30.7161±3.15	64.1095±1.74

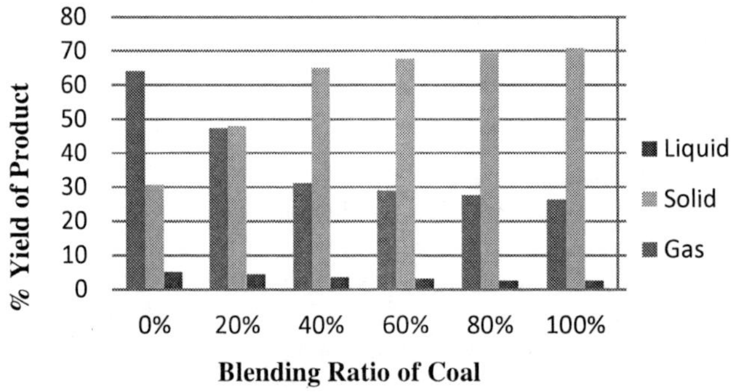

Figure 2. Effect of Blending Ratio on Yield of Products

It can be concluded from Figure 2 that the greater blending ratio of the coal, the more yield of liquid and gas obtained, meanwhile the less yield of solid product obtained. It was happened due to the highest blending ratio of coal (100 %), the components in coal were more difficult to evaporate and decompose than the components in EPFB. It can be proved by looking at the volume of volatile matter of coal (17.70 %) less than the volatile matter of EPFB (58.17 %). The yield of solid product was larger compared to liquid and gas due to during the pyrolysis process it was possible to formed a second char. In addition, CO_2 gas reacted with liquid products cause the decreasing of liquid products while gas products increased. The increasing of micropore caused the rich of solid product in carbon elements. Other factors that influence product yield were particle size, temperature, heating rate, residence time, steam/biomass ratio, catalyst and others[2]. However, this study more focused on the influence of blending ratio.

3.3 Analysis Result

3.3.1 Effect of Blending Ration on BET Analysis Result

Based on Table 3, it was found that blending ratio had an effect on char pore diameter, char surface area and char total pore volume. With high blending ratio condition, pore diameter and total pore volume were decreased while surface area was increased. Based on the theory, the effect of CO_2 gas in pyrolysis was to increase the surface area of char[2]. The most important factor affecting particle size is temperature, the higher of temperature, the smaller of particle size obtained[8]. It was also obtained the decreasing result of pore diameter and total pore volume. Based on the theory, the decreasing of particle char size had an effect on the increasing of gas yield and decreasing of liquid and solid yield. Particle size not only affecting heat and mass transfer but also the possibility of a second reaction within a char particle. The volatile component in char can trigger a second reaction (eg cracking, condensation and polymerization) inside the char particles. The polymerization of some volatile components due to the accumulation of large molecules on the pore wall, resulting in increasing of yield char and decreasing of volatile components, is occurring more frequently if the particle size of char is larger[4].

Table 3. Result of BET Analysis

Parameter	Coal : EPFB			
	100:0	**80:20**	**40:60**	**20:80**
Pore diameter (nm)	61.7316	30.3056	46.5352	18.8361
Surface area (m^2/g)	0.7230	2.7190	0.7090	1.3480
Total pore volume (cm^3/g)	0.0111	0.0206	0.0082	0.0063

Table 4. BET analysis result of feedstock

Parameter	Feedstock	
	Coal	**EPFB**
Pore diameter (nm)	74.4808	51.9846
Surface area (m^2/g)	0.2150	0.1480
Total pore volume (cc/g)	0.0040	0.0019

Table 4 is the result of BET analysis on feedstock. It can be seen that the difference of pore diameter on coal material before and after pyrolysis process, which was approximately 12.7492 nm. The pore diameter of the coal feedstock before pyrolysis was greater than after pyrolysis. Meanwhile, the surface area of coal feedstock was larger after pyrolysis. It indicates that pyrolysis was capable of enlarging the material surface area. Similarly, the total pore volume becomes larger after pyrolysis.

IOP Conf. Series: Materials Science and Engineering **588** (2019) 012011 doi:10.1088/1757-899X/588/1/012011

3.3.2 Effect of Blending Ratio on SEM Analysis Result

Figure 3 shows the SEM analysis of co-pyrolysis solid product at 500 °C. Images were taken with 1500 times of magnification. The three images were almost identical, having irregular, heterogeneous and irregular structures. However, if observed by naked eye, EPFB had more pores (more porous) than coal.

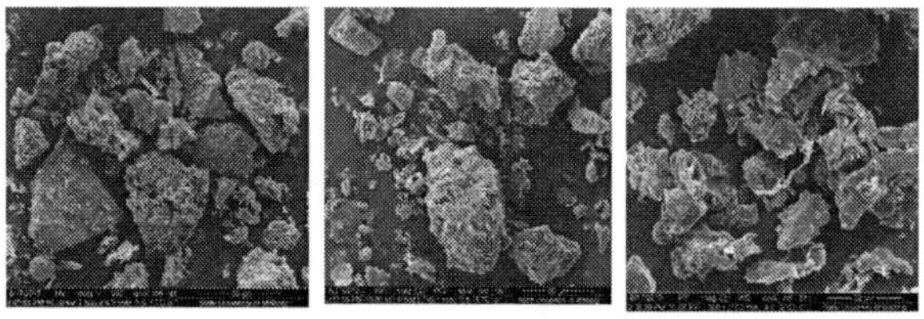

a b c

Figure 3. SEM Analysis (a) Coal: EPFB = 40:60
(b) Coal : EPFB = 100:0 (c) Coal: EFFB = 0: 100

3.3.3 Effect of Blending Ratio on GC Analysis of Gas Product

Pyrolysis gas process was a product that can not be condensed into liquid products. During the pyrolysis process, the resulting gas is accommodated in a gas trap (balloon). GC analysis was used to know the level of H_2 and CO_2 on gas. Char was converted into gas by thermal cracking process in pyrolisis process. Char was very reactive with CO_2 gas.

3.3.4 Effect of Blending Ratio on Physical and Chemical Condition of Gas Product

Liquid product that obtained from this research was called as bio-oil. Bio oil is the main product desired from any pyrolysis process because it has many advantages. One of them is as a fuel mixture. Bio oil in the liquid phase was obtained by condensing the output gas from the reactor. Not all amounts of gas phase changed into a liquid. The condensed gas was called the primary gas, while the non-condensed gas was called secondary gas. The physical differences of liquid products in each blending ratio are shown in Table 6.

Shortly after the liquid product was obtained, the comparison of color was done by naked eye. In general, bio oil was reddish brown color. Based on Table 6, it is represent the difference between each blending ratio. In the variable coal/ EPFB = 100/0 obtained a bright reddish brown liquid product. As the blending ratio increased, the density of color and texture of liquid products was sharper till the dark brown color obtained from 100 % EPFB. EPFB gave a sharper and thicker brown color due to cellulose, hemicellulose and lignin content contained in EPFB.

Based on the GC results in Table 5 can be concluded that the higher of blending ratio (EPFB 100%), the higher of H_2 gas concentration obtained. It was happenned due to the fact that biomass is rich in H_2. So as to be possible in this pyrolysis process, EPFB functions as a H_2 donor and allows for synergistic effects[8]. However, the highest concentration of H_2 gas was obtained in blending ratio of coal: EPFB = 20:80, it was equal to 0.8258 %. This condition were optimal conditions to produce H_2 gas during pyrolysis process.

The concentration of CO_2 gas was fluctuated due to the pretreatment of different raw materials. So, H_2O content in the raw materials also varies as well. The amount of H_2O caused more CO_2 formation, because H_2O was also an oxygen supply (O_2) which reacted with carbon (C) and produced CO_2. In addition, CO_2 used as a purge gas was also possible to react with hydrogen supplied by biomass and form H_2O and CO[2].

Table 5. GC Result of Gas Product

Coal/	Surface area	Gas volume in syringe (mL)	Concentration (v/v)

IOP Conf. Series: Materials Science and Engineering **588** (2019) 012011 doi:10.1088/1757-899X/588/1/012011

EPFB	H_2	CO_2	H_2	CO_2	Total (H_2+CO_2)	H_2	CO_2
100/0	-	66.320	-	0.237783	0.237783	-	118.891399%
20/80	4955	58378	0.001652	0.209308	0.210959	0.825833%	104.653831%
40/60	892	45.588	0.000297	0.163451	0.163748	0.148667%	81.725288%
60/40	1.134	61.609	0.000378	0.220892	0.221270	0.189000%	110.446022%
80/20	-	52.645	-	0.188753	0.188753	-	94.376322%
0/ 100	2.538	86.304	0.000846	0.309433	0.310279	0.423000%	154.716575%

The volume of liquid products obtained was varying due to less optimal condensation process during pyrolysis. The efficiency of the condenser in the course of the pyrolysis process was less than the maximum due to several things, the gas contact with the cooling medium was minimum and the blockage of the condenser by feedstock.

In addition to the effect of equipment unefficiency, the volume of liquid products was also affected by the type of reactor used. Since the reactor type was a free fall reactor, with a height reactor design of about 90 cm enabling when the pyrolysis process occurs, the reaction between feedstock and CO_2 purge gas had some problems. High reactors and a small volume of feedstock caused the volume of gas spaces was high. It cause temperature instability in the reactor, the temperature panels often showed a temperature drop about 20 °C. In addition, the distance between the reactor heights with a condenser length prevented the condensing of gas; it was affected to the formed liquid product. In addition to known physical conditions of liquid products were formed, chemical conditions also needed to analyze using GC-MS. The obtained liquid was an unhomogeneous oil phase, tar and water was not homogen. The content of water in tar was not separated due to many compounds contained in tar dissolved in water. The following Figure 4 is a liquid chromatogram of liquid product and Table 7 is a component contained in liquid products.

Table 6. Physical differences of liquid product on each blending ratio

Coal/ EPFB	Physical Differences			
	Color	Odor	Volume (mL)	Density (g/mL)
100/0	Bright Reddish brown	Oil scent	3	0.9
80/20	Reddish brown slightly dark	Oil scent	2.9	0.94
60/40	Darker reddish browns	Oil scent	4.2	0.76
40/60	Darker reddish browns	Oil scent	3.9	0.95
20/80	Almost solid reddish brown	Oil scent	5.8	0.77
0/100	Solid brown	Oil scent	4.7	1.1

IOP Conf. Series: Materials Science and Engineering **588** (2019) 012011 doi:10.1088/1757-899X/588/1/012011

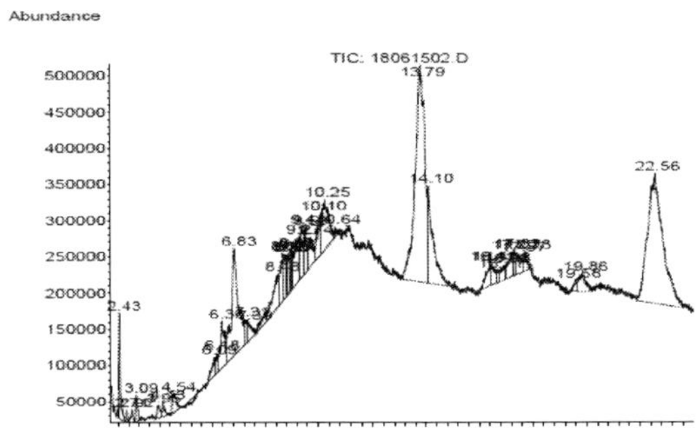

Figure 4. Chromatogram result of GC-MS analysis on blending ratio Coal/EPFB = 80/20

Based on the results of analysis using GC-MS obtained 41 compounds that were identified. With the largest area were hexadecane (23.69 %) and heptacosane (22.61 %).

3.3.5 Synergis Effect

Coal and EPFB have complex structures, which may indicate differences in thermal decomposition. It was enable the synergistic effect between the two raw materials both in quantity and quality. Synergistic effects were defined as an increase or decrease in the results of co-processing experiments with the results of calculation[6]. Based on the approach taken by [6],[8]:

$$Ycal = Yc \times BRc + Yb \times (1 - BRc) \quad (4)$$

Where: Y_{cal} was the yield of the calculated result, Yc and Yb were yields of individual experimental measurements of coal and EPFB, and BRc was the blending ratio of coal. If $Y_{exp} > Y_{cal}$ there was a synergistic effect with the addition of EPFB. The following were the Y_{exp} and Y_{cal} calculations of each blending ratio at 500 °C.

Table 7. Y_{exp} and Y_{cal} calculation of co-pyrolysis products at 500 °C

Coal :EPFB	Solid		Liquid		Gas	
	Exp	Cal	Exp	Cal	Exp	Cal
100/0	70.935	70.9358	2.705	2.7053	26.358	26.358
80/20	69.583	62.8918	2.72	3.1991	27.697	33.909
60/40	67.785	54.8479	3.211	3.6929	29.003	41.459
40/60	65.103	46.8039	3.718	4.1867	31.178	49.009
20/80	48.128	38.7600	4.442	4.6805	47.428	56.559
0/100	30.716	30.7161	5.174	5.1744	64.109	64.109

Based on Table 7 it can be concluded that there was a synergistic effect on solid product where $Y_{exp} > Y_{cal}$. Synergistic effects occured when there was a chemical interaction between two materials. Synergistic effect occured when the process was in a condition above the heating temperature, ie 500 °C. According to Pan YG, under low heating conditions, two processes will be separated and no synergetic effects are found. Factors that can increase the synergistic effect not only temperature and blending ratio but also rate of heating and reaction time[8]. High heating rates and short reaction residence times prevented the formation of secondary reactions. So, the more primary gas was produced, which was condensable gas becomes tar. However, short residence time made reaction contact between two raw materials was not guaranteed.

4. Conclusion

IOP Conf. Series: Materials Science and Engineering **588** (2019) 012011 doi:10.1088/1757-899X/588/1/012011

It can be concluded that the higher blending ratio (EPFB 100 %) caused the concentration of H_2 gas increased, the CO_2 gas concentration fluctuated and gas yields increased. The influence of blending ratio on solid characteristic was higher blending ratio (EPFB 100 %), char pore diameter and total pore volume tend to decreased, surface area tend to increased and yield of solid decreased. While, the effect of blending ratio on the liquid characteristic was the higher blending ratio (EPFB 100 %), the darker of reddish-brown color, the density increased and the liquid yield increased.

References

[1] Erlan R 2012 *Konversi Minyak Nabati Menjadi Green Diesel dan Green Gasoline Dengan Proses Hydrocracking dan Hydrotreating Pada Katalis Nimo/Al_2O_3, Nimo/ Al_2O_3- SiO_3, Nimo/ SiO_2 dan Nimo/ Zeoli* Disertasi ITS Surabaya

[2] C Guizani, FJ Escudero Sanz and S Salvador 2013 *Effect of CO_2 on Biomass Fast Pyrolysis : Reaction Rate Gas Yield and Char Reactivity Properties* **116** 310 – 320

[3] Li Jing, Dai Jianjun, Liu Guangqing, Zhang Hedong, Gao Zuopeng, Fu Jie, He Yanfeng and Huang Yan 2016 Biochar from Microwave Pyrolysis of Biomass: *A Review Biomass and Bioenergy* **94** 228 – 244

[4] Semino D and Tognotti L 1998 Modelling and Sensitivity Analysis of Pyrolysis of Biomass Partcles in A Fluidised Bed *Computers Chem.Engng* **22** 699 – 702

[5] Shuaidan Li, Xueli Chen, Aibin Liu, Li Wang and Guangsuo Yu 2015 Co-Pyrolysis Characteristic of Biomass and Bituminous Coal Bioresource *Technology* **179** 414 – 420

[6] Suelves I, Moliner R and Lazaro M J 2000 Synergetic Effects in The Co-Pyrolysis of coal and Petroleum Residues : Influences of Coal Mineral Matter and Petroleum Residues Mass Ratio *Journal of Analytical And Applied Pyrolysis* **55** 29 -41

[7] Sukseswati D D 2010 *Karakteristik Sifat Fisik Dan Kimia Minyak Hasil Pirolisis Lambat Campuran Sampah Kertas dan Daun* Skripsi Universitas Sebelas Maret Surakarta

[8] Z Li, X Shaoping, Z Wei and L Shuqin 2007 Co-pyrolysis of Biomass and Coal in a Free Fall Reactor Fuel **86** 353 – 359

IOP Conf. Series: Materials Science and Engineering **588** (2019) 012012 doi:10.1088/1757-899X/588/1/012012

Passive design (building envelope) impact to cooling load of researh centre building ITS

T. Soehartanto[1] , Matradji[2] and Roekmono[2]
[1] Department of Instrument Engineering, Faculty of Vocation, Sepuluh Nopember Institute of Technology, Surabaya, Indonesia
[2] Department of Physical Engineering, Faculty of Industrial Engineering, Sepuluh Nopember Institute of Technology, Surabaya, Indonesia

totokstf@yahoo.com

Abstract. The ITS Research Center (RC) building as one of the new buildings in ITS has a unique architecture compared to ITS building architecture in general. The shape of this architecture is interesting to study because the side of the building facing north and south is the majority of tinted glass, which is given a frame on the left and right sides with a thickness of 4 meters each. The wall protrudes inside with a certain slope angle, so that it protects the glass window from solar radiation (from the east and from the west) directly. The 11th floor sheath is tinted glass that curves downward with a certain tilt angle, so it can provide a shade effect on the glass floor below. This type of building design is a passive design (one of the green building methods) to reduce thermal transfer through the building envelope. Thermal transfer through the envelope of the Research Center building will be presented in this paper, through the calculation of overall thermal transfer value (OTTV) covering the north and south side of the building from the second floor to the 11th floor. The covering from the 3rd to the 10th floor consists of spandrel, tinted glass and parapet, while the sheath on the 2nd and 11th floors is tinted glass. The calculation results show that the total OTTV from the north and south side of the building envelope is 6.18 Watt / m2, this shows that the design of ITS research center building architecture is effective in reducing the thermal transver through the sheath.

1. Introduction

40% of the cooling load in buildings comes from OTTV (Overall Thermal Transfer) of the building envelope, so the design of the building envelope and the direction of the building envelope play an important role in reducing cooling load from OTTV (passive design method). The ITS Research Center (RC) building as one of the new buildings in ITS has a form of modern architecture (applying the Green Building concept). The envelope of the ITS Research Center building facing North and South is a glass window with a frame on the left side and on the right side with a width of 4 meters each. The shape of the frame protrudes into a certain slope angle so that the glass window does not get direct sunlight radiation from the east or from the west. For this reason, the impact of the RC ITS building's passive design on the cooling load will be explained through the calculation of thermal transmittance per floor (from the 2nd floor to the 11th floor). Sheath design per floor (from the 2nd floor to the 10th floor) is in the form of a spandrel, tinted glass and parapet, while the 11th floor sheath is a tinted glass that protrudes out with a certain angle of brightness. This design is intended to provide a shade effect on the glass window below

2. Methodology

OTTV calculations according to SNI-03-6389-2000, are shown in the following flow chart:

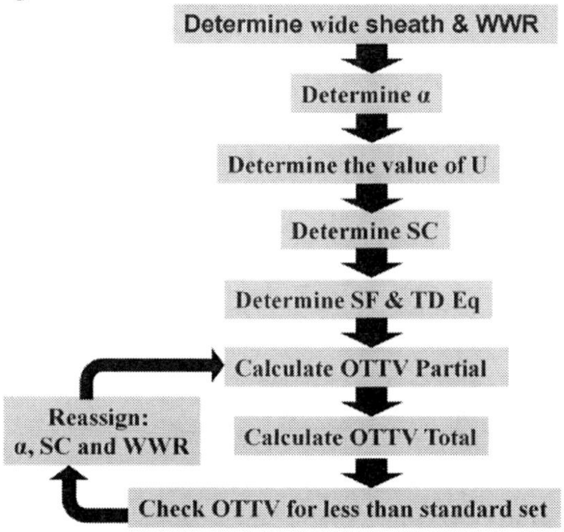

Figure 1. OTTV Flow Chart

OTTV for each area of the outer wall of a building with a certain orientation :

$$OTTV = \alpha[(UW \times (1 - WWR)] \times TDEk + (SC \times WWR \times SF) + (Uf \times WWR \times \Delta T) \quad (1)$$

Where :

OTTV = Overall thermal transfer value (W/m^2)
α = absorbance of solar radiation.
UW = Thermal transmittance of opaque walls ($W/m^2.K$)
WWR = Comparison of window area with the width of the entire outer wall at the specified orientation
TDEk = Equivalent temperature difference (K).
SF = Solar radiation factor (W/m^2)
SC = The shade coefficient of the phenetration system.
Uf = Thermal transmittance phenetration ($W/m^2.K$)
ΔT = Temperature difference between the outside and the inside

OTTV throughout the outer wall, the following equation is used :

$$OTTV = \frac{(Ao1 \times OTTV1) + (Ao2 \times OTTV2) + \dots\dots\dots + (Aoi \times OTTVi)}{Ao1 + Ao2 + \dots\dots\dots\dots Aoi} \quad (2)$$

Where :

Aoi = Wall area on the outer wall i (m^2).
This total area includes all opaque wall surfaces and the surface area of the windows found on these walls
OTTVi = The total thermal displacement value for the i wall as a result of the calculation using the equation 2

IOP Conf. Series: Materials Science and Engineering **588** (2019) 012012 doi:10.1088/1757-899X/588/1/012012

2.1. Passive Design of RC ITS Builiding

Figure 2 shows a photo of the ITS RC building, this picture shows a glass window facing the north side with a wall frame on the east side and west side, so that sunlight from the east and from the west cannot illuminate directly into the window.

Figure 2. Passive Design of RC ITS Building

RC ITS building consists of 11 floors, where the second floor of the sheath is the same colored glass as the 11th floor, while the 3rd floor to the 10th floor consists of spandrel, tinted glass, and parapet with width and height as shown in figure 3 (scale 2.2 cm = 3.6 m).

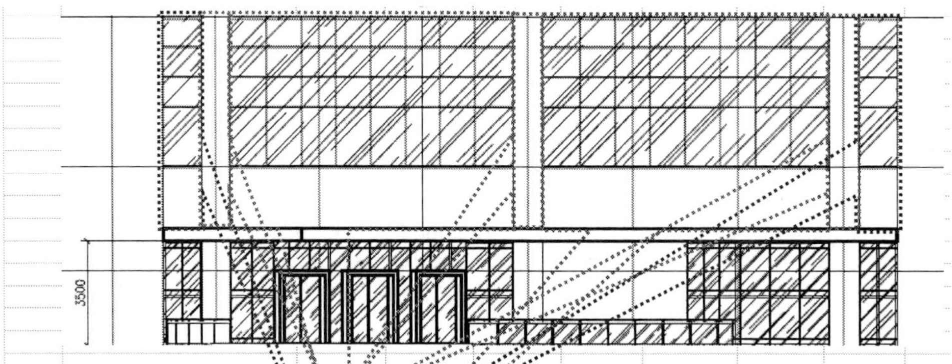

				Wide (m)
1	building bone concrete	Wide @ 0.6 cm	skala 2.2 cm = 3.6 m	2.95
2	glass on the left and right side	Wide 1.6 cm		2.62
3	left-right center side glass	Wide 21.36 cm		34.95
	WIDE BUILDING OF RC ITS			40.52

Figure 3. Wide sheathing of the north side of the RC ITS building

The size of the sheath (width and height) per floor of the RC ITS building on the north and south side is shown in table 1, this size will be used to calculate the wall area per floor.

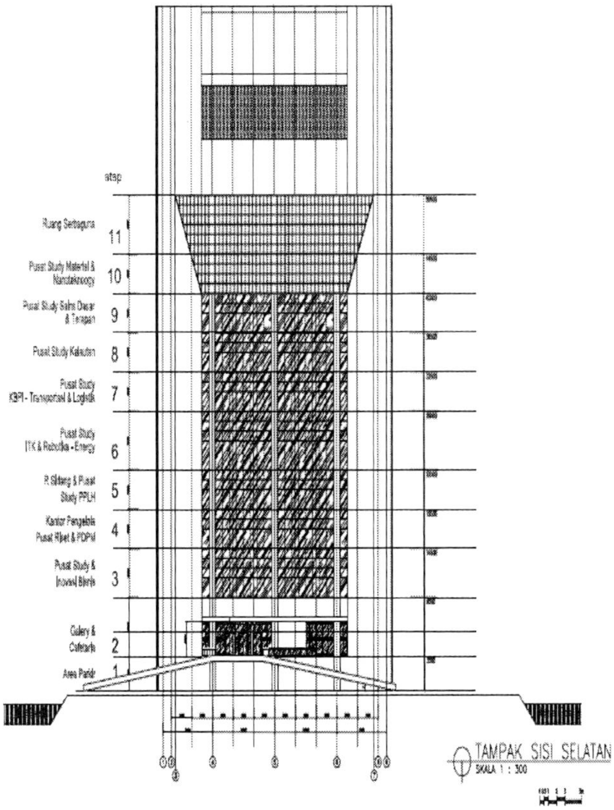

Figure 4 . Looks north side of RC ITS building

Table 1 . Size (width and height) of the sheath per floor of the RC ITS building north and south

Floor	Function	Width (m)	Height (m)
1	Parking	40,52	3,50
2	Gallery & Cafetaria	38,72	6,00
3	Business Study & Innovation Centre	38,72	5,00
4	Office of Research & PDPM Centre Management	38,72	4,00
5	PPLH Study Room & Study Centre	38,72	4,00
6	Centre for ITK & Robotics - Energy	38,72	6,00
7	Centre of Study-Transportation & Logistics KBPI	38,72	4,00
8	Centre of Marine Study	38,72	4,00
9	Centre & Applicable Science Study Centre	38,72	4,00
10	Centre of Study Material & Nano Technology	38,72	4,00
11	Multipurpose Room	38,72	6,00

Floors 3 to 10 on the north and south sides consist of spandels, tinted glass and parapets with high sizes as shown in table 2. The sheath of floors 3 to 10 has the same width.

Table 2. Spandrel Height, Glass Height and Parapet each level of RC ITS North and South Side

North Side			South Side		
Upper Spandrel Height (m)	**Glass Height (m)**	**Lower Parapet Height (m)**	**Upper Spandrel Height (m)**	**Glass Height (m)**	**Lower Parapet Height (m)**
	6,00			6,00	
0,50	3,00	1,50	0,50	3,00	1,50
0,50	2,00	1,50	0,50	2,00	1,50
0,50	2,00	1,50	0,50	2,00	1,50
0,50	4,00	1,50	0,50	4,00	1,50
0,50	2,00	1,50	0,50	2,00	1,50
0,50	2,00	1,50	0,50	2,00	1,50
0,50	2,00	1,50	0,50	2,00	1,50
0,50	2,00	1,50	0,50	2,00	1,50
	6,00			6,00	

The results of the calculation of the sheath area per floor facing the north and south sides and the calculation of the area of the colored glass windows north and south side, are used to calculate the WWR (Wall Window Ratio).

Table 3. WWR per floor of RC ITS building on the north and south side

Floor	**North Side**			**South Side**		
	WWR Spandrel (%)	**WWR Glass** (%)	**WWR Parapet** (%)	**WWR Spandrel** (%)	**WWR Glass** (%)	**WWR Parapet** (%)
2		0,13			0,13	
3	0,0106383	0,06	0,03191	0,0106383	0,06	0,03191
4	0,0106383	0,04	0,03191	0,0106383	0,04	0,03191
5	0,0106383	0,04	0,03191	0,0106383	0,04	0,03191
6	0,0106383	0,09	0,03191	0,0106383	0,09	0,03191
7	0,0106383	0,04	0,03191	0,0106383	0,04	0,03191
8	0,0106383	0,04	0,03191	0,0106383	0,04	0,03191
9	0,0106383	0,04	0,03191	0,0106383	0,04	0,03191
10	0,0106383	0,04	0,03191	0,0106383	0,04	0,03191
11	0	0,13	0	0	0,13	0

3. Results and Discussion

The calculation of OTTV RC building ITS is carried out in accordance with the flow chart (Figure 1), the calculation results are as follows :

The value of α (solar radiation absortance) according to SNI-03-6389-2000 (α value for outer walls and opaque roofs) is shown in table 4.

Table 4. A α value for outer walls and opaque roofs

Bahan Dinding Luar	**α**
Beton Berat	0,91
Bata Merah	0,89
Bitumunous Felt	0,88
Batu Sabak	0,87
Beton Ringan	0,86
Aspal Jalan Setapak	0,82
Kayu Permukaan Halus	0,78
Beton Ekspos	0,61
Ubin Putih	0,58
Bata Kuning Tua	0,56
Atap Putih	0,5
Cat Alumunium	0,4
Kerikil	0,29
Seng Putih	0,26
Bata Gelazur Putih	0,25
Lembaran Alumunium yang Dikilapkan	0,12

The north side and the south side of the RC ITS building are in the form of GLASS and a small portion in the form of Spandrel and parapet, so that the absorption coefficient of solar radiation for the outer wall (SPANDREL) can be assumed to be α = 0.12.

U, SC and SF values can be obtained from the glass manufacturer's data as shown in table 5, in this study only U and SC values were used

Table 5. A α value for U , SC, and SF

Type Of Glass	Standard Thickness (mm)	Energy Characteristic				Light Characteristic		Solar Factor (%)	Shading Coeficient (3"4")	U Value W/m²k
		Transmittance (%)	Reflectance (%)	Absoption (%)	Ultra Violet Transmission (%)	Transmittance (%)	Reflectance (%)			
Panasap	5	56	5	39	25	55	5	67	0.77	5.8
Bronze	6	50	5	44	20	50	5	64	0.73	5.8
(BRFL)	8	41	5	54	14	41	5	58	0.66	5.7

SF Value use SNI-03-6389-2000 for many orientation.

Tavle 6. SF Value for many orientation

Orientasi	**U**	**TL**	**T**	**TGR**	**S**	**BD**	**B**	**BL**
	130	113	112	97	97	176	243	211

Different Equivalent Temperature (TD ek) on the wall is calculated using data from SNI-03-6389-2000.

IOP Conf. Series: Materials Science and Engineering **588** (2019) 012012 doi:10.1088/1757-899X/588/1/012012

Table 7. $T_{D\ Equivalent}$ on the Wall Value

Weight per unit area (kg/m²)	$T_{D\ eqv}$
< 125	15
126 – 195	12
> 195	10

Based on Table 7, to know the equivalent TD of a wall, we must know the weight per unit area, done by multiplying the wall density against the wall thickness.

Table 8. The sheath in the form of SPANDREL

	Thick (m)	Density (kg/m³)	Weight/Area (kg/m²)
Tinted Glass	0.008		20.000
Air Space (with high emissivity)	0.010		
Gypsum Board	0.006	880.000	5.280
Glasswool	0.050	32.000	1.600
			26.880

SPRANDEL weight / area is less than 125 kg/m^2, then $T_{D\ ekivalen}$ is 15 Kelvin.

Table 9. The sheath in the form of PARAPET

	Thick (m)	Density (kg/m³)	Weight/Area (kg/m²)
ACP	0.03	2,672.00	80.16
External Wall Plester	0.01	1,568.00	20.38
Brick Wall	0.10	1,760.00	176.00
Internal Wall Plester	0.01	1,568.00	20.38
			296.93

PARAPET weight / area is more than 195 kg/m^2, then $T_{D\ ekivalen}$ is 10 Kelvin. The equivalent TD value for GLASS (according to SNI-03-639-2000) is 5 Kelvin.

3.1. Calculation U Value

The sheath of the 3rd to 10th floor consists of spandrel, tinted glass and parapet, so that the translucent thermal transmittance value (UW) and translucent thermal transmittance value (UF) are calculated as Table 10.

Table 10. The translucent thermal transmittance value (UW) and translucent thermal transmittance value (UF)

No	Type of Façade	Components of The Façade	Thickness (m)	K (W.m.K)	R (m^2.K.W)	U Value (W/M^2.K)
2	**TINTED GLASS**	External Surface			0,04	
	Penasap (Blue Green)	Tinted Glass	0,01			5,8
		Internal Surface			0,13	
					0,53	1,89
						7,69
3	**SPRANDEL**	External Surface			0,4	
	With tinted glass blue green	Glass	0,01			5,8
		Air Space (with high emissivity)	0,01		0,12	
		Gypsum Board	0,01	0,17	0,04	
		Glasswool	0,05			0,04
		Internal Surface			0,13	
					0,69	1,46
						7,29
4	**PARAPET**	External Surface			0,4	
	ACP and Brick	ACP	0,03	0,4	0,08	
		External Wall Plester	0,01	0,57	0,02	
		Brick Wall	0,1	0,77	0,13	
		Internal Wall Plester	0,01	0,57	0,02	
		Internal Surface			0,13	
					0,78	
						1,28

3.2. Calculation OTTV every floor in south and north side

The results of OTTV calculations on the north or south side on the 2nd to 11th floors are Q1, Q2 and Q3 in table 11.

Table 11. Result of OTTV

Floor	Façade Type	α	Uw	WWR	1-WWR	TD Ek		Q1
2nd Floor	Tinted Glass	0,12	0	0,13	0,87	0	0	0
3rd Floor	Spandrel	0,12	7,29	0,01	0,99	5	4,33	
	Tinted Glass	0,12	0	0,06	0,94	0	0	20,87
	Parapet	0,89	1,28	0,03	0,97	15	16,54	
4th Floor	Spandrel	0,12	7,29	0,01	0,99	5	4,33	
	Tinted Glass	0,12	0	0,04	0,96	0	0	20,87
	Parapet	0,89	1,28	0,03	0,97	15	16,54	
5th Floor	Spandrel	0,12	7,29	0,01	0,99	5	4,33	
	Tinted Glass	0,12	0	0,04	0,96	0	0	20,87
	Parapet	0,89	1,28	0,03	0,97	15	16,54	

Floor	Façade Type	α	Uw	WWR	1-WWR	TD Ek		Q1
6th Floor	Spandrel	0,12	7,29	0,01	0,99	5	4,33	
	Tinted Glass	0,12	0	0,09	0,91	0	0	20,87
	Parapet	0,89	1,28	0,03	0,97	15	16,54	
7th Floor	Spandrel	0,12	7,29	0,01	0,99	5	4,33	
	Tinted Glass	0,12	0	0,04	0,96	0	0	20,87
	Parapet	0,89	1,28	0,03	0,97	15	16,54	
8th Floor	Spandrel	0,12	7,29	0,01	0,99	5	4,33	
	Tinted Glass	0,12	0	0,04	0,96	0	0	20,87
	Parapet	0,89	1,28	0,03	0,97	15	16,54	
9th Floor	Spandrel	0,12	7,29	0,01	0,99	5	4,33	
	Tinted Glass	0,12	0	0,04	0,96	0	0	20,87
	Parapet	0,89	1,28	0,03	0,97	15	16,54	
10th Floor	Tinted Glass	0,12	0	0,04	0,96	0	0	0
11th Floor	Tinted Glass	0,12	0	0,13	0,87	0	0	0

Table 12. The sum of Q1, Q2 and Q3 is the OTTV value on a floor with a certain wind direction

SC	SF		Q2	Uf	ΔT		Q3	OTTV i	Façade Area
0,66	58,00	4,89	4,89	7,69	5,00	4,91	4,91	10,46	232,32
0,00	130,00	0,00		0,00	15,00	0,00			
0,66	58,00	2,44	2,44	7,69	5,00	2,45	2,45	5,56	193,60
0,00	130,00	0,00		0,00	10,00	0,00			
0,00	130,00	0,00		0,00	15,00	0,00			
0,66	58,00	1,63	1,63	7,69	5,00	1,64	1,64	3,93	154,88
0,00	130,00	0,00		0,00	10,00	0,00			
0,00	130,00	0,00		0,00	15,00	0,00			
0,66	58,00	1,63	1,63	7,69	5,00	1,64	1,64	3,93	154,88
0,00	130,00	0,00		0,00	10,00	0,00			
0,00	130,00	0,00		0,00	15,00	0,00			
0,66	58,00	3,26	3,26	7,69	5,00	3,27	3,27	7,19	232,32
0,00	130,00	0,00		0,00	10,00	0,00			
0,00	130,00	0,00		0,00	15,00	0,00			
0,66	58,00	1,63	1,63	7,69	5,00	1,64	1,64	3,93	154,88
0,00	130,00	0,00		0,00	10,00	0,00			
0,00	130,00	0,00		0,00	15,00	0,00			
0,66	58,00	1,63	1,63	7,69	5,00	1,64	1,64	3,93	154,88
0,00	130,00	0,00		0,00	10,00	0,00			
0,00	130,00	0,00		0,00	15,00	0,00			
0,66	58,00	1,63	1,63	7,69	5,00	1,64	1,64	3,93	154,88
0,00	130,00	0,00		0,00	10,00	0,00			

SC	SF		Q2	Uf	ΔT		Q3	OTTV i	Façade Area
0,66	58,00	1,63	1,63	7,69	5,00	1,64	1,64	3,93	154,88
0,66	58,00	4,89	4,89	7,69	5,00	4,91	4,91	10,46	232,32

According to SNI 03-6389 concerning energy conservation of building envelopes in buildings, the OTTV value for buildings must not exceed 45 Watts / m^2.

The results of OTTV calculations per floor on the north side or south side of the RC ITS envelope are 6.18 Watts / m^2, so the OTTV value of the cover of the north and south sides is 12.36 Watts / m^2. The OTTV value of the RC ITS building is still below the standard, this indicates that the RC ITS building passive design is quite effective in reducing the cooling load of buildings originating from OTTV.

4. Conclusion

The colored glass windows of the RC ITS building are placed on the north and south sides with a protruding position and protected by a frame on the east side and on the west side, so that sunlight from the east and from the west cannot illuminate directly the glass window surface colored. The envelope of the east side and west side of the RC ITS building is a brick wall that is given a small window hole on each floor. The design of this building turned out to be quite effective in reducing the burden of cooling the building due to thermal transfer from the building envelope or commonly called the overall thermal transfer value (OTTV). The results of OTTV calculations on the north and south sides obtained a value of 12.36 Watts / m2, this value is still below the SNI 03-6389 standard regarding energy conservation of building envelopes in buildings that OTTV values for buildings should not exceed 45 Watts / m^2

Acknowledgement

This work was supported by a grant (D42018 - LOKAL ITS 2018 (ABDIMAS) provided by the Institute of Research and Public Services (LPPM), Institut Teknologi Sepuluh Nopember (ITS), Indonesia. The authors thank Mr. ali Musyafa' and Mr. Tri Joko for valuable technical support.

References

[1] S Loekito *Analisis konservasi energi melalui selubung bangunan* Available ; *ced.petra.ac.id/index.php/civ/article/download/16465/16457.*

[2] Badan Standardisasi Nasional 200 *Konservasi energi selubung bangunan pada bangunan gedung* SNI 03-6389-2000 Jakarta Indonesia

[3] Panduan Pengguna Bangunan Gedung Hijau Jakarta *Selubung Bangunan* **vol. 1**. Available *https://id.scribd.com/document/337989857/Vol-1-Selubung-Bangunan*

[4] Feri Harianto *Konservasi Energi Selubung Bangunan Pada Gedung Graha Galaxy Surabaya*

[5] Jurnal Karakteristik Fasade Bangunan Factory Outlet Di Jalan Ir. H. Djianda Bandung. Hal 4-7 Available:*https://www.academia.edu/7985812/KARAKTERISTIK_FASADE_BANGUNAN_FACTORY_OUTLET_DI_JALAN*

[6] International Energy Agency 2018 *Global Energy & CO2 Status Report 2017*

IOP Conf. Series: Materials Science and Engineering **588** (2019) 012013 doi:10.1088/1757-899X/588/1/012013

Effect of acid-catalysed in one step production of biodiesel on total sugars of defatted rice bran

M Yasmin[1], S Zullaikah[1,2], A Permatasari[1], I Marita[1] and M Rachimoellah[1]

[1]Department of Chemical Engineering, Institut Teknologi Sepuluh Nopember (ITS), Kampus ITS Keputih Sukolilo, Surabaya 60111 Indonesia

szulle@chem-eng.its.ac.id, szullehakim@gmail.com

Abstract. Defatted rice bran is a major byproduct of rice bran-based biodiesel and rich in carbohydrate. The high amount of carbohydrate (starch) from its process can be utilized as energy fuel feedstock. In this work, the effect of acid catalyst amount and reaction time on the yield of oil and carbohydrate products in in-situ transesterification from rice bran was investigated. The longer reaction time was used; oil products (FAMEs yield and its content) and carbohydrate product (total sugars content and recovery) showed the opposite trendline. The longer reaction times was achieved, the more yield of FAMEs was obtained but the lower total sugars content on defatted rice bran. While, the less amount catalyst was enough to faster reaction and produce high total sugars content.

1. Introduction

An increase of emissions as a consequence of excessive consumption of fossil fuels leads to some negative effects to environment security. Therefore, a demand of renewable fuels to substitute the recent fuels increases year by year. The potential renewable and alternative diesel fuel is biodiesel. Biodiesel has attracted a wide attention as clean, biodegradable, and environmentally benign fuel. Biodiesel can be mixed with petroleum diesel in any proportion or used directly in diesel engines without modification [1].

Biodiesel can be directly converted by transesterification of oil and alcohol with the presence of catalyst. Commercial biodiesel production still relies on refined vegetable oils such as soybean oil in US, rapeseed and sunflower seed oils in Europe, palm oil in Southeast Asia, and coconut oil in the Philippines [2,3]. However, those renewable sources arise 70-80% cost in biodiesel production [3]. The economic competitiveness in biodiesel production can utilize low-cost raw materials such as rice bran, a byproduct of rice milling. It can be a potential feedstock due to high lipids content (10-25%) and available in large quantity [4].

In recent years, in-situ process has drawn considerable attention as a way to reduce the production cost further [5]. Rice bran oil (RBO) based biodiesel under in situ acid esterification [6], rice bran-based biodiesel with ultrasound assisted in-situ esterification [2], in-situ transesterification under supercritical condition [7], two steps in-situ process of rice bran-based biodiesel [5], and in situ subcritical transesterification [4] had been conducted as the economic competitiveness to produce biodiesel.

Most of those researches still focus on yield of biodiesel can be achieved and its optimization. While, rice bran-based biodiesel production yielded a high amount of defatted rice bran as a byproduct. Defatted rice bran is rich in carbohydrate and can be used as bioethanol feedstock. However, information of total sugars on defatted rice bran as a byproduct of rice bran-based biodiesel is rarely found. Therefore, the objective of this work is to investigate the effect of acid-catalyst amount and reaction time on FAME and total sugars from in situ transesterification process.

2. Materials and methods

2.1. Materials

Rice bran was obtained from local rice mill in Sidoarjo, Indonesia. Fresh rice bran was immediately separated from its impurities by sieving method (<0.6 mm mesh) and stored in a sealed container at 4°C to prevent the formation of free fatty acids (FFAs) by lipase enzyme. Rice bran was put at 80°C in an oven for 2 h to reduce its water content before it is used as feedstock in the process. Standard fatty acid methyl esters (FAMEs) were obtained from Sigma Chemicals Company (St. Louis, MO, USA). All solvents and reagents were either of high-performance liquid chromatography (HPLC) grade or of analytical reagent grade were obtained from commercial sources.

2.2. Extraction of rice bran

Rice bran (10 g) was extracted by Soxhlet extraction method using extraction thimble filter (35 x 120 mm) and methanol (150 ml) as a solvent. The extraction was conducted at 65°C in predetermined time (0.5, 1, 3, and 5 h). The liquid product was distilled to separate methanol from crude rice bran oil (CRBO). Then, the CRBO was analyzed and weighed to know the initial FFAs content and yield of the extraction. The yield of CRBO was determined as the weight ratio of CRBO to the rice bran and the FFAs analysis was determined according to Rukunuddin et al [8]. The solid product (defatted rice bran) was dried at 65°C for an hour and analyzed to know the total sugars content. Total sugars content was determined by Luff-Schoorl method according to Egan et al [9]. The initial total sugar in rice bran was 41.82%.

2.3. In-situ production of biodiesel

Soxhlet extraction apparatus equipped with magnetic stirrer was employed in this work [3]. In general, rice bran (10 g) was put in the extraction thimble filter (35 x 120 mm) while methanol (150 ml) as solvent and reactant with sulfuric acid as catalyst was put in round-bottom flask. The in-situ process was conducted at 65°C in predetermined time and amount of catalyst. The liquid product was distilled to recover methanol from the oil product. Then, the oil product was extracted by n-hexane for three times (50 ml each) and washed by distilled water until neutral pH. The mixture formed two layers, upper layer (organic phase) and bottom layer (aqueous layer). The aqueous layer was removed and discarded while the organic layer was distilled to recover hexane from its FAMEs product. The FAMEs product was dried at 80°C for an hour to remove the remaining water. The FAMEs were weighed and analyzed to know its yield and purity. The yield of FAMEs was determined as the weight ratio of FAME to its crude oil while the FAMEs content was analyzed by gas chromatography (GC). The solid product (defatted rice bran) was dried at 65°C for an hour and analysed to know total sugars content. Total sugars content was determined by Luff-Schoorl method according to Egan et al [9].

2.4. Product analysis

Total sugars were determined by Luff-Schoorl method, which is based on iodometric determination of the unreduced Cu (II) ions remaining after the reaction with reducing sugars [9]. Titratable acidity expressed by anhydrous citric acid was determined by titrating sample with 0.1 M NaOH. The recovery of total sugars was determined as the weight of total sugars in defatted rice bran to the initial weight of the total sugars in rice bran.

FAMEs content of the sample was determined by gas chromatography. The sample with n-hexane dilution (0.5 μL) was injected into the GC. External standard calibration curves were obtained by using 0.2-20 mg pure standard. Oleic acid methyl ester was selected for the determination of FAMEs calibration factor and was used for all FAMEs. Chromatographic analysis was performed in a HP 6890 (Hewlett-Packard Inc., Avondale, Pennsylvania, USA) gas chromatograph equipped with a flame ionization detector. The column used was HP-1 crosslinked methyl siloxane column (60m x 0.25mm i.d. x 1μm film thickness, Hewlett-Packard Inc., Avondale, Pennsylvania, USA). The operating conditions were: the injector and detector temperature were set at 250°C, the column temperature was held at 200°C for 2 min, and then was raised to 300°C at 15°C/min and was held for 10 min. Helium was

IOP Conf. Series: Materials Science and Engineering **588** (2019) 012013 doi:10.1088/1757-899X/588/1/012013

used as the carrier gas with a linear velocity of 40 cm/s at 200°C. The yield of FAMEs was obtained by dividing of the weight of the FAMEs in the oil product by that of the rice bran.

3. Results and discussion

3.1. Effect of extraction time on yield of CRBO and total sugars

In this work, ratio of rice bran to methanol used was 1:15 (g/ml). Table 1 shows the effect of extraction time on yield of CRBO and FFAs content. The yield of CRBO increased along with increasing of extraction time. At 0.5 h, yield of CRBO was 6.10% and it increased to 19.46% for extended time of extraction (5 h). Shiu et al [5] stated that extraction lipid from rice bran was a slow process so that it needs longer time to extract the lipids. However, the increment of extraction time affected the extraction rate. The longer extraction time used, the rate will be slower. Lei et al [10] reported that time to complete extraction of rice bran with methanol was 5 h.

The FFAs content in CRBO depended on extraction time significantly. When extraction was prolonged to 5 h, the FFAs content increased rapidly. The increasing of FFAs content can be caused by hydrolysis of the glycerides such as triglycerides (TG), diglycerides (DG), or monoglycerides (MG) into FFAs [3].

Table 1. Effect of extraction time on yield and FFAs content of CRBO.

Time (h)	Yield of CRBO (%)	FFAs content (%)
0.5	6.10±0.28	32.27±1.99
1	14.97±0.47	35.07±1.98
3	17.90±0.85	36.47±1.98
5	19.46±0.36	53.53±3.63

Extraction conditions: mass of rice bran = 10 g, T=65°C, and rice bran to methanol = 1:15 (g/ml).

Table 2. Effect of extraction time on total sugars in defatted rice bran[a].

Time (h)	**Total sugars**
0.5	37.97±0.18 (90.79±0.44)
1	32.25±0.83 (77.12±2.00)
3	28.02±0.56 (66.99±1.34)
5	27.45±0.56 (65.63±1.34)

Extraction conditions: mass of rice bran = 10 g, T=65°C, and rice bran to methanol = 1:15 (g/ml).
[a] Values reported as content (wt. %) with recoveries (%) in parenthesis

Total sugars on defatted rice bran was also affected by extraction time. The longer of extraction time, both of content and recovery of total sugars decreased. The initial total sugars content in rice bran was 41.82% and it decreased until 27.45% for 5 h of extraction. The highest loss of sugars in rice bran was 14.37% after 5 h of extraction. The decline of total sugars can be caused by utilization of methanol as polar extraction solvent. The more polar the solvent is, the more carbohydrate can be preferentially extracted [1,5].

3.2. Effect of reaction time

In this study, biodiesel was obtained through in-situ transesterification process, in which the oil extraction process and the reaction was carried out simultaneously with the addition of acid catalyst [6,10]. The selection of sulfuric acid as catalyst was due to high level of FFAs (>30%) in rice bran and to prevent saponification reaction between base catalyst and FFAs [3,11].

Effect of reaction time on FAMEs yield and content is shown in Fig 1. Both of FAMEs yield and content increased along with increasing of time. The FAMEs yield was 8.71% (at 0.5 h) and it increased to 17.47% as the reaction time increased to 5 h. The highest increasing of FAMEs yield (5.6%) was obtained when reaction time was increased from 0.5 h to 1 h. Gunawan et al [1] stated that the conversion FAMEs from rice bran oil was rapid at the first hour and slower for further hours. Therefore after 1 h of reaction, an increment of FAMEs yield was not too significant. FAMEs yield also depended on the oil extraction rate rather than reaction rate of transesterification [4].

Total sugars content and recovery were also affected by reaction time. Figure 2 shows that total sugars content in defatted rice bran slightly decreased from 5 min to 30 min of reaction. The trend of total sugars content was opposite with FAMEs yield. The decline of sugars content in defatted rice bran can be caused by acid condition. Acidic condition as a result of acid catalyst presence in the system might hydrolyze starch (as the major carbohydrate in rice bran) into monomer sugars [4]. Therefore, the content of starch decreased along with increasing of reaction time. Recovery of total sugars in defatted rice bran has the same trendline with its content. The longer reaction time used, the recovery of total sugars decreased gradually.

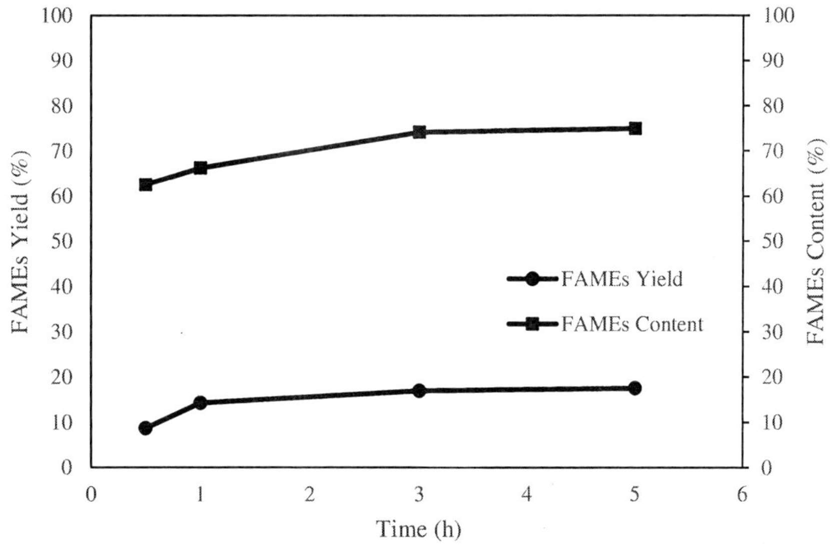

Figure 1. Effect of reaction time on FAMEs yield and content (T=65oC, ratio rice bran to methanol = 1:15 (g/ml), volume of catalyst = 0.5 ml).

IOP Conf. Series: Materials Science and Engineering **588** (2019) 012013 doi:10.1088/1757-899X/588/1/012013

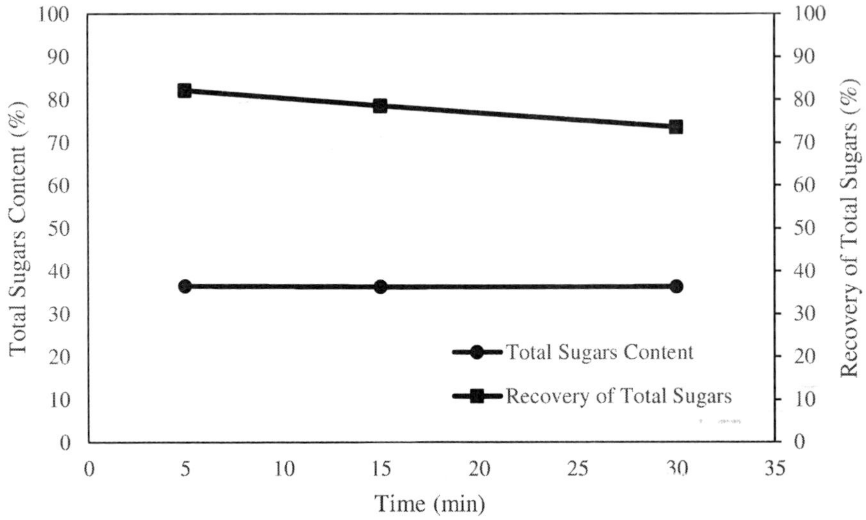

Figure 2. Effect of Reaction Time on Total Sugars Content and Recovery (T=65oC, Ratio rice bran to methanol =1:15 (g/ml), Vol. sulfuric acid = 0.5 ml)

3.3. Effect of catalyst amount

The reaction between methanol and oils catalyzed by acid is very fast [4]. Figure 3 shows that adding more catalyst did not increase FAMEs yield significantly and the highest FAMEs yield (17.99%) was obtained by using 0.1 ml of catalyst. Zullaikah et al [3] stated that adding less catalyst amount was enough to accelerate the reaction. Therefore, adding less amount of catalyst was appropriate to faster the reaction and prevent darker color of FAMEs [3,12].

Total sugars content did not affect significantly when amount of catalyst was added. Figure 4 shows that the trendline is similar with FAMEs yield. The more catalyst was added, both of FAMEs yield and total sugars content was not affected significantly. Less amount of catalyst used in the system will decrease total sugars content so that utilization of less amount of sulfuric acid was enough to produce high total sugars content in defatted rice bran. Recovery of total sugars decreased along with more sulfuric acid added into the system. Therefore, less amount of sulfuric acid was enough to prevent loss of total sugars in defatted rice bran.

IOP Conf. Series: Materials Science and Engineering **588** (2019) 012013 doi:10.1088/1757-899X/588/1/012013

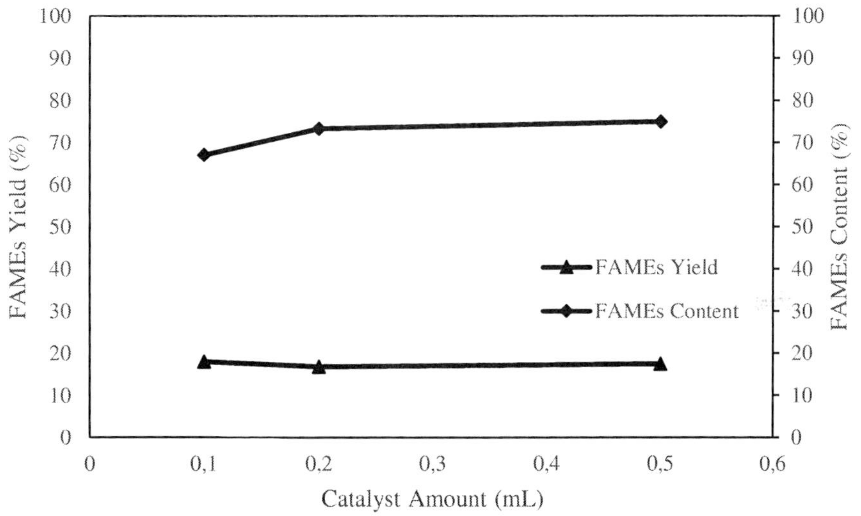

Figure 3. Effect of Catalyst amount on FAMEs Yield and Content (T=65oC, Ratio rice bran to methanol =1:15 (g/ml), Reaction time = 5 h)

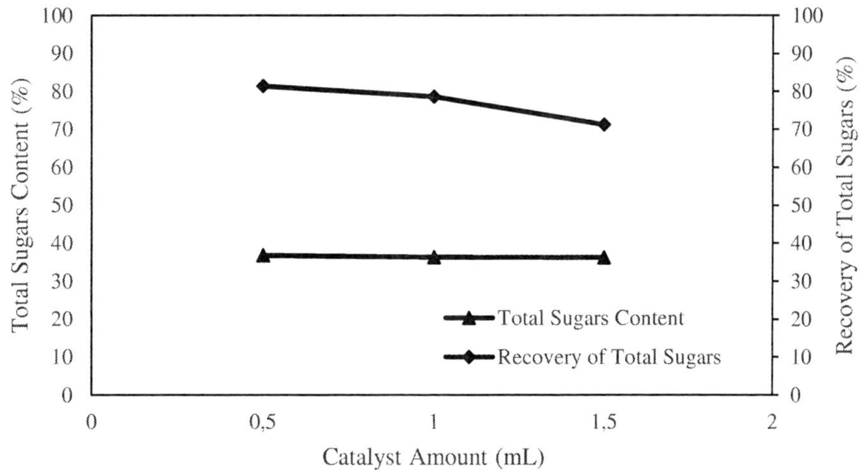

Figure 4. Effect of Catalyst amount on Total Sugars Content and Recovery (T=65oC, Ratio rice bran to methanol =1:15 (g/ml), Reaction time = 5 h)

4. Conclusion

Reaction time and acid catalyst amount affected total sugars on defatted rice bran as major byproduct from catalytic in situ biodiesel production. The longer reaction time can lower total sugars on defatted rice bran even though increasing biodiesel yield. Therefore, less amount of catalyst was enough to increase FAMEs yield and did not lower total sugars in defatted rice bran significantly.

Acknowledgments

This research was financially supported by Directorate General of Higher Education, Ministry of National Education, Indonesia and Institut Teknologi Sepuluh Nopember through project No. 0392/IT2.7/PM/2012.

References

[1] Gunawan S, Maulana S, Anwar S and Widjaja T 2011 *Ind Crops Prod* **33** 624-28
[2] Yustianingsih L, Zullaikah S and Ju Y H 2009 *J. Energ. Inst* **82** 133-37
[3] Zullaikah S, Rakhadima Y T, Rachimoellah M, Widjaja T and Sumarno 2014 *IPTEK. Journal of Proceeding Series* **1** 351-54
[4] Zullaikah S, Rahkadima Y T, and Ju Y H 2017 *Renew. Energy* **111** 764-70
[5] Shiu P J, Gunawan S, Hsieh W H, Kasim N S and Ju Y H 2010 *Bioresour. Technol.* **101** 984-89
[6] Ozgul-Yucel S and Turkay S 2003 *J. Am. Oil Chem. Soc.* **80** 81-4
[7] Kasim N S, Tsai T H, Gunawan S and Ju Y H 2010 *Bioresour. Technol.* **100** 2007-11
[8] Rukunudin I, Whitea P, Bernb C, and Baileyc T 1998 *J. Am. Oil Chem. Soc.* **75** 563-68
[9] Egan H, Kirk R, and Sawyer R 1981 *Sugars and preserves* 152-53
[10] Lei H, Ding X, Zhang H, Chen X, Li Y, Zhang H, and Wang Z 2010 *Fuel* **89** 1475-79
[11] Ju Y H and Vali S R 2005 *J. Sci. Ind. Res.* **64** 866-82
[12] Ju Y H and Zullaikah S 2013 *J. Taiwan Inst. Chem. Eng.* **44** 924-8

IOP Conf. Series: Materials Science and Engineering **588** (2019) 012014 doi:10.1088/1757-899X/588/1/012014

Experimental Validation of the Piezoelectric Energy Harvester for Wearable Devices

Akhmad Afrizal Rizqi[1], Ariprihartai[1],i[3], *, I Wayan Dasnai[2], Purnomoi[1], Arif Nur Afandyi[1]

[1]Departement of Electrical Engineering, Faculty of Engineering, Universitas Negeri Malang, Indonesia

[2]Departement of Chemistry, Faculty of Mathematics and Natural Sciences, Universitas Negeri Malang, Indonesia

[3]Centre of Advanced Materials for Renewable Energy, Indonesia

E-mail: `ariprihartai.ft@um.ac.id`

Abstract. This paper presents the experimental validation of the piezoelectric array in harvesting the vibration energy from human activity. This experiment is aimed to choose the best array topology of piezoelectric energy harvesting (PEH) for replacing the battery on our wearable devices. Based on our experiments, the no-load characteristic of PZT array shows the big hope for future implementation on wearable devices.

1. Introduction

Energy harvesting device used to collect ambient energy resources and convert those into electrical energy. The most commonly used ambient energy is light, wind, thermal gradients and vibrations [1, 2]. For example [3], Author generates electrical energy from combining piezoelectric and electromagnetic in one unit. They used some topologies to produce electrical energy efficiently from two sources. Author [4] examined nanoscale energy harvesting which resulted in the higher energy by adjusting the surface material

Energy harvesters using piezoelectric are one of the most popular ones. Piezoelectricity can convert vibration energy into electrical energy. Vibration energy provides in surroundings, for example, in mechanical equipment, buildings, roads, railways, and human body [5]. Research in [6] had investigated and updated the Spring-comprised Piezoelectric Energy Generator (SPEG) by varying spring strength and support excitation. In [7], Authored examined the implementation of piezoelectric energy harvesters in the galloping phenomenon. Researchers utilize in aerodynamic instability for energy harvesters. The results of the proposed system show a significant increase compared to the energy harvesting system proposed in the literature.

The existing research on piezoelectric energy harvesting was focused on the piezo material. In this paper, we proposed an experimental validation of piezoelectric energy harvesters for wearable devices. The purpose of this study is to examine the feasibility of PEH for powering the wearable sensors. We conducted several scenarios to choose the best array topologies for our application.

The rest of this paper organized as follow. Section 2 is related works and Section 3 is the Experimental Method. Section 4 and Section 5 describe the Result and Analysis, and Conclusion, respectively.

2. Related Works

In[8], piezoelectric power generation technology for pavements have two methods: piezoelectric material composite and transducer embedded pavement piezoelectric power generation. Transducer embedded pavement technology for piezoelectric power plants can obtain higher and more controlled power output.

There are three types of mechanical-electric conversion modes in the piezoelectric transducer: d31, d33, and d15. D33 mode comes from vertical compressive pressure, whereas d31 mode electricity is generated from horizontal compressive stress. In the d15 mode, it is rarely used on the pavement because it requires the effect of shear stress to produce electrical energy. The d33 mode has a higher piezoelectric coefficient. Therefore, this study uses the d33 mode of stacked piezoelectric transducers. The wireless sensor is installed on the bridge to ensure the vehicle runs safely and monitors the structural state of the bridge [9-12]. The battery is needed to supply the system, but using a battery has the disadvantage of repeated replacements, very expensive and not feasible in some special situations. This potential use of piezoelectric energy harvesting systems for civil infrastructure is gaining attention [13,14].

One way to harvest vibration energy on bridges is to use piezoelectric. In the study [15], tested the piezoelectric energy harvester from the vehicle bridge vibration. Based on the vibration characteristics of the clutch platform, two piezoelectric energy harvesters (PEHs) with different basic frequencies. PEH-1 is designed with the natural frequency of the bridge and PEH-2 is designed with the vibration frequency of the vehicle-bridge coupling. PEH voltage output was measured in nine different cases and the energy harvesting performance of PEHs was discussed. The results show that the PEH-1 voltage has a peak when the vehicle enters or leaves the bridge and is almost zero when the vehicle moves on the bridge. While the PEH-2 voltage when the vehicle moves on the bridge. Experimental results also show that more energy can be harvested when PEH is installed in the middle of the bridge.

To utilize renewable energy in the transportation sector, research on the application of thermoelectric and piezoelectric effects on the pavement of energy harvest has grown significantly. The study [16] provides a literature review of the application of thermoelectric and piezoelectric to produce electricity on the pavement. In the previous literature, piezoelectric with piezoelectric transducers (PZTs) showed limited amounts of electrical output, while pipe systems that worked with thermoelectric (TEG) generators could produce more electricity. So that on the sidewalk has the potential to harvest energy. The results of a study on the Florida highway, if the entire road network is covered by a proposed system (PP-TEG system) gets 55 GWh of electricity per day, while one covered by PZTs only produces 4.04 MWh of electricity per day. Based on the cost-effectiveness analysis of the two systems, unless the PZT system is only paved in parts of the road with very high traffic volumes, the PP-TEG system is more cost-effective than the PZT system. Compared with the above existing works, our work has two distinction:

- We test the feasibility of the piezoelectric arrays in generating low power electrical energy.
- We measured the vibration rather than force.
- We used the round type of Piezoelectric with diameter 20mm.

3. Experimental method

We conducted experiments for analyzing the feasibility of the piezoelectric (PZT) for harvesting pressure energy from human feet, see in figure 1. This experiment was divided into 2 scenarios: (1) no load test; (2) loading test. In both scenarios, we measured the current and voltage on the output terminal of the bridge rectifier. Each test was applied to 5 PZT arrays as shown in Figure 2. The data collected from 5 samples. That sample has a different gender, weight, and height as shown in Table 1.

IOP Conf. Series: Materials Science and Engineering **588** (2019) 012014 doi:10.1088/1757-899X/588/1/012014

Table 1. sample information.

Gender	Age	Height(cm)	Weight(kg)
male	21	166	45
male	21	169	68
maler	20	170	55
female	20	155	45
male	20	175	75

Then, the output current and voltage for each experiment is averaged. Besides, we also record the data using an oscilloscope to capture the working frequency of piezoelectric. To meet the requirement for powering the wearable wireless sensors, we used resistors with resistance 100 Ω, and 1 kΩ. Moreover, we used linear regression for data analysis.

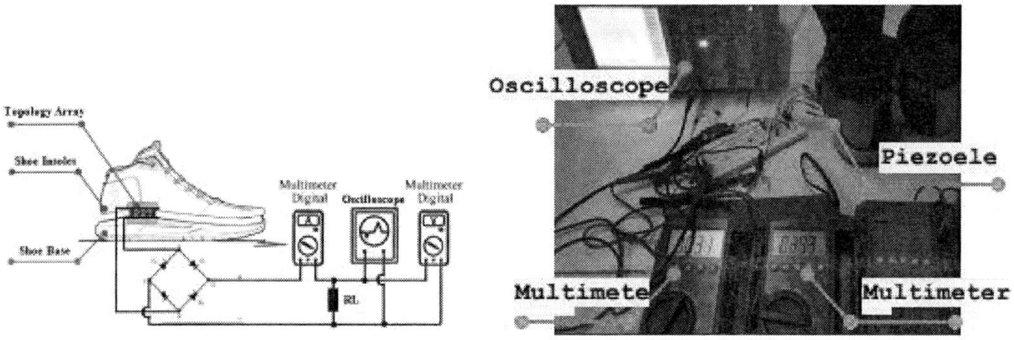

Figure 1. Experiments for analyzing

4. Result and Analysis

Experiments had done and results were collected and plot into graphs. The measurement results for no-load test for all PZT array topologies are shown in Figure 3 and Figure 4. As the vibration increased, the open circuit voltages and short-circuit current are increased, except the PZT array #5 or topology 5.

Figure 5 and Figure 6 depicts the effect of vibration on voltage and current generated by the PZT arrays which loaded by 100ohm and 1Kohm. As the vibration is increased, then the resulting voltage and current on PZT terminals are increased. The voltage dropped to 0.004925V and the current reached 2.9785A when the vibration is 1m/s2 at a load of 1Kohm.

The experimental results regarding the most efficient topology are Topology 5 (PZT array in series-parallel connection). The topology 1 (single PZT) draw a current value of 3.033mA, a voltage value of 0.00184V and a power value of 0.005581mW as depicted in Figure 7.

Figure 8 depicts the result of the topology 2 that is PZT array in series connection. The current value generated by this array is 1.4166mA, the voltage is 0.00843V and the power is reached to 0.011942mW. Figure 9 shows the current value of 3.075mA, the voltage value of 0.002045V and the power value of 0.006288mW.

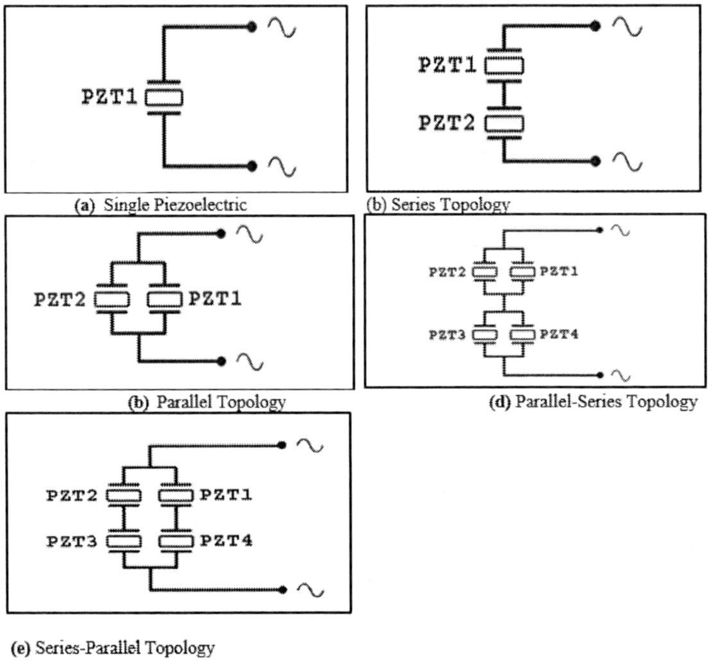

Figure 2. Pizoelectric Array.

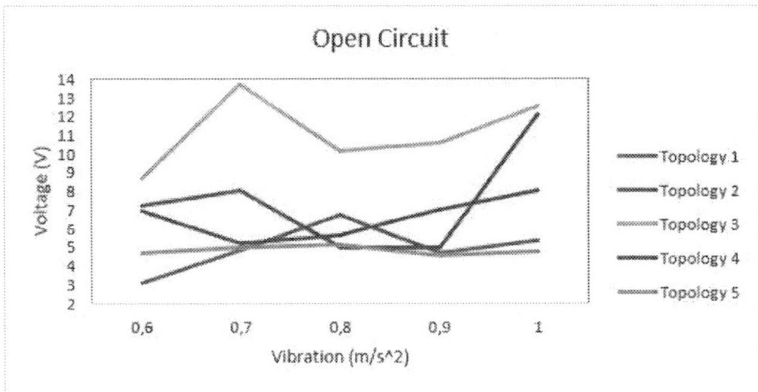

Figure 3. Open circuit Voltage.

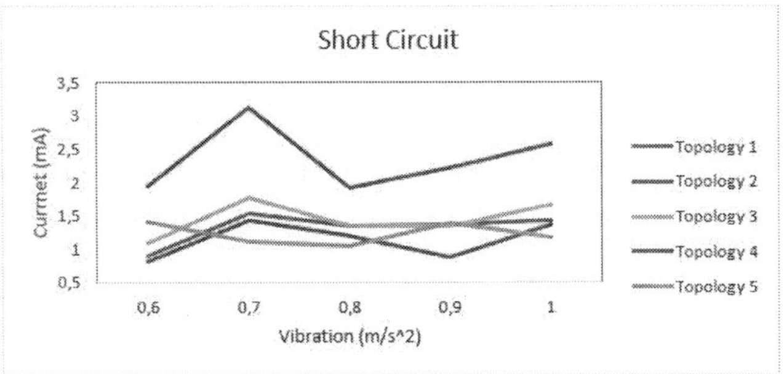

Figure 4. Short Circuit Current.

Figure 10 shows measurement results for topology 3 where 2 PZTs connected in parallel. The 1 kΩload was drawn current for about 2.9785mA when the output voltage is 0.004925V. Thus, the generated power by this PZT array is 0.014669mW. Figure 11 shows the output current, the output voltage, and the power value of the PZT array 4 are 2.166mA, 0.00257V, and 0.005567mW, respectively.

Based on the obtained results, the best array PZT is 4 PZTs connected in series and parallel.

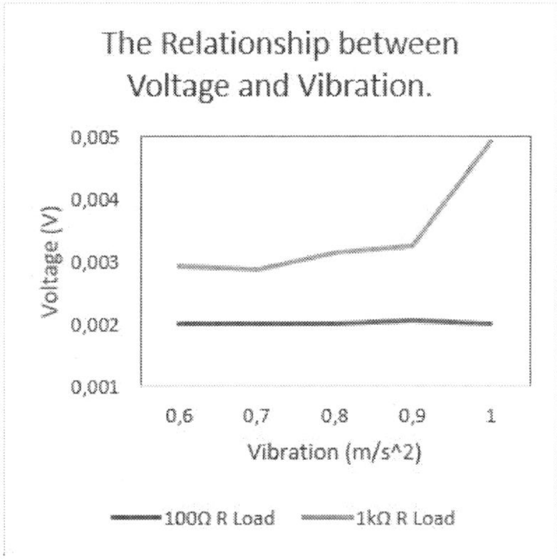

Figure 5. Graph of Voltage with Vibration.

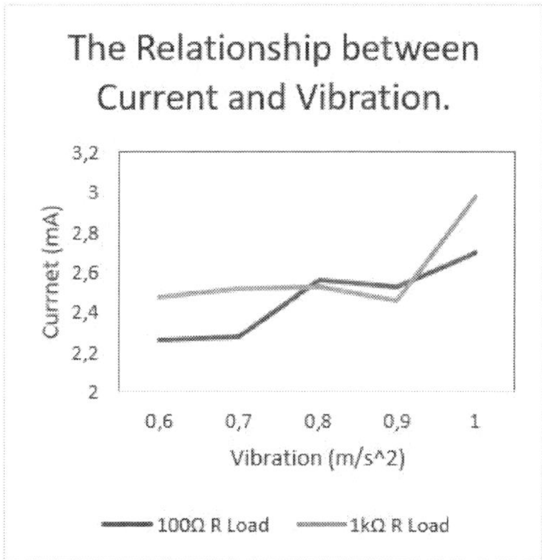

Figure 6. Graph of Current with Vibration.

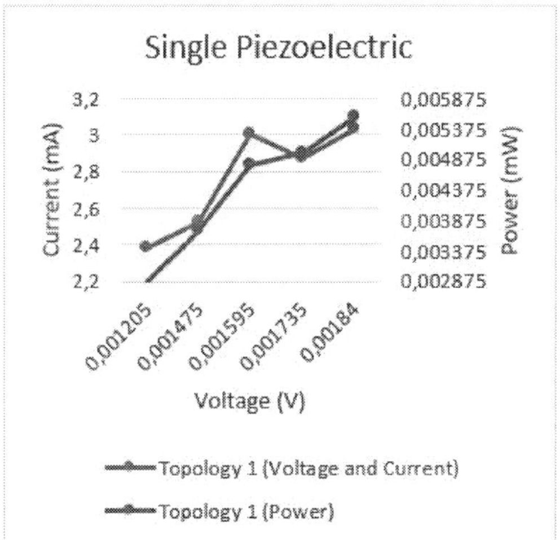

Figure 7. Output characteristic of PZT array #1.

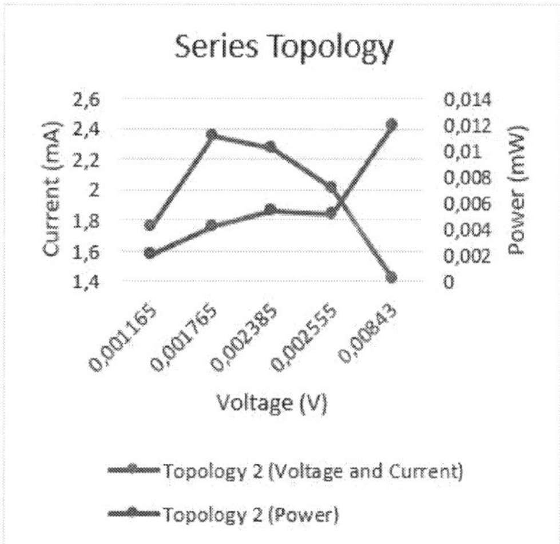

Figure 8. Output characteristic of PZT array #2.

However, direct implementation of those arrays to powering the microcontroller, sensors, and transmitter is impossible. It because the voltage dropped too much from several volts to mV. By the recent technology of dc to dc converter or PMIC, the generated voltage by PZT array could be increased and provide the necessary amount of voltage and current to the output. In addition, the supper capacitor may provide energy while there is no vibration or during harvesting.

5. Conclusion

Experimental validation of the piezoelectric energy harvester for wearable devices had tested. By using several topologies, the highest efficiency produces by topology 5. Based on microcontroller specifications, the power, which produces by topology 5 is not sufficient for giving supply to the wearable devices directly. According to the no-load characteristic of PZT array, the PEH has

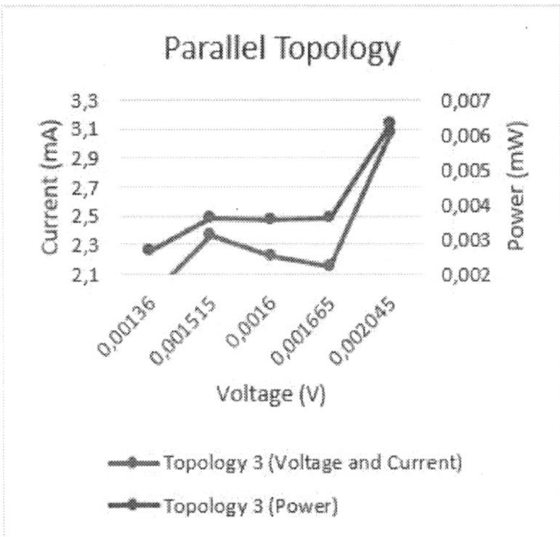

Figure 9. Output characteristic of PZT array #3.

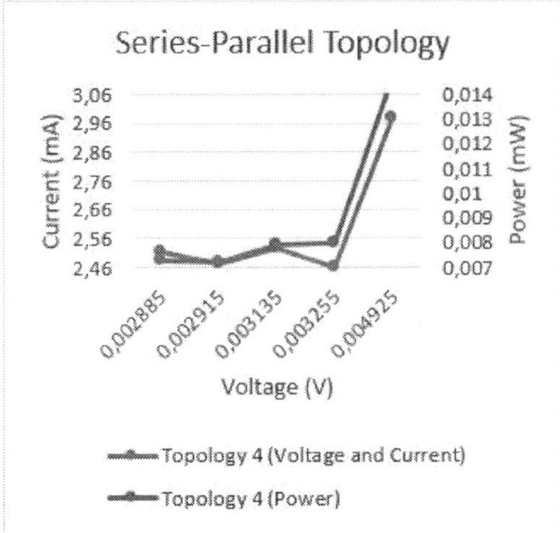

Figure 10. Output characteristic of PZT array #4.

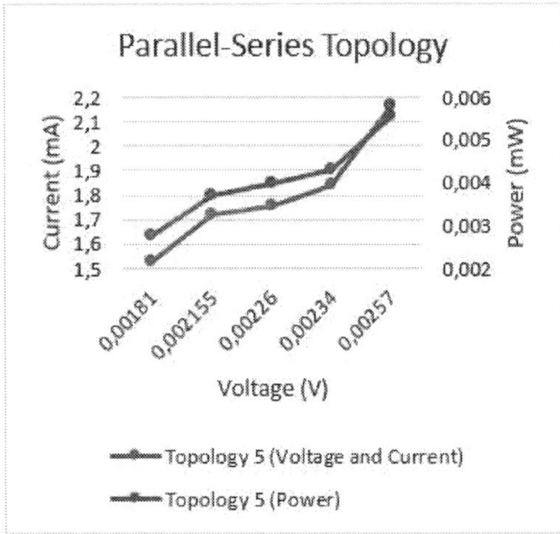

Figure 11. Output characteristic of PZT array #5.

great opportunities for replacing batteries on wearable devices. In the future, we developed the power management IC to maintaining the no-load characteristic of the PZT array.

6. Acknowledgment

These work are a part of Camry Programme 2018, Universitas Negeri Malang, Indonesia.

References

[1] Roundy, S., Wright, P. K., & Rabaey, J. (2003). A study of low level vibrations as a power source for wireless sensor nodes. Computer Communications, 26(11), 1131–1144.

[2] Beeby, S. P., Tudor, M. J., & White, N. M. (2006). Energy harvesting vibration sources for microsystems applications. Measurement Science and Technology, 17(12), R175–R195.

[3] Hamid, R., & Yuce, M. R. (2017). A wearable energy harvester unit using piezoelectric–electromagnetic hybrid technique. Sensors and Actuators A: Physical, 257, 198–207.

[4] Elvin N, Erturk A. Advances in energy harvesting methods. Springer Sci & Bus Media; 2013.

[5] Fan, T. (2018). Nano-scale energy harvester of piezoelectric/piezomagnetic structures with torsional mode. Mechanical Systems and Signal Processing, 112, 147–153.

[6] Allamraju, K. V., & Korla, S. (2015). Novel Design, Analytical and Experimental Studies of Spring Comprised Piezo Electric Energy Harvester. Materials Today: Proceedings, 2(4-5),
[7] Usman, M., Hanif, A., Kim, I.-H., & Jung, H.-J. (2018). Experimental validation of a novel piezoelectric energy harvesting system employing wake galloping phenomenon for a broad wind spectrum. Energy, 153, 882–889.
[8] Wang, C., Zhao, J., Li, Q., & Li, Y. (2018). Optimization design and experimental investigation of piezoelectric energy harvesting devices for pavement. Applied Energy, 229, 18–30.
[9] Pakzad, S. N., & Fenves, G. L. (2009). Statistical Analysis of Vibration Modes of a Suspension Bridge Using Spatially Dense Wireless Sensor Network. Journal of Structural Engineering, 135(7), 863–872.
[10] Asadollahi, P., & Li, J. (2017). Statistical Analysis of Modal Properties of a Cable-Stayed Bridge through Long-Term Wireless Structural Health Monitoring. Journal of Bridge Engineering, 22(9), 04017051.
[11] Zhang, Y., Kurata, M., & Lynch, J. P. (2017). Long-Term Modal Analysis of Wireless Structural Monitoring Data from a Suspension Bridge under Varying Environmental and Operational Conditions: System Design and Automated Modal Analysis. Journal of Engineering Mechanics, 143(4), 04016124.
[12] Spencer, B. F., Jo, H., Mechitov, K. A., Li, J., Sim, S.-H., Kim, R. E., … Agha, G. (2015). Recent advances in wireless smart sensors for multi-scale monitoring and control of civil infrastructure. Journal of Civil Structural Health Monitoring, 6(1), 17–41.
[13] Maruccio, C., Quaranta, G., Lorenzis, L. D., & Monti, G. (2016). Energy harvesting from electrospun piezoelectric nanofibers for structural health monitoring of a cable-stayed bridge. Smart Materials and Structures, 25(8), 085040
[14] Takeya, K., Sasaki, E., & Kobayashi, Y. (2016). Design and parametric study on energy harvesting from bridge vibration using tuned dual-mass damper systems. Journal of Sound and Vibration, 361, 50–65.
[15] Zhang, Z., Xiang, H., Shi, Z., & Zhan, J. (2018). Experimental investigation on piezoelectric energy harvesting from vehicle-bridge coupling vibration. Energy Conversion and Management, 163, 169–179.
[16] Guo, L., & Lu, Q. (2017). Potentials of piezoelectric and thermoelectric technologies for harvesting energy from pavements. Renewable and Sustainable Energy Reviews, 72, 761–773.

IOP Conf. Series: Materials Science and Engineering **588** (2019) 012015 doi:10.1088/1757-899X/588/1/012015

Fatigue Calculation of Offshore Jacket Structural For Monitoring

Y W Wu[1], H T Wu[1], C H Chung[1], J R Lyu[1], S H Yang[1], S Y Chen[1] and Y Y Chang[1]

[1] Ship and Ocean Industries R&D Center (SOIC). 14F., No.27, Sec. 2, Zhongzheng E. Rd., Tamsui Dist., New Taipei City 251, Taiwan (R.O.C.)

Abstract. Based on the IEC 61400-3 offshore wind turbine(OWT) design guidelines. The NREL 5MW wind turbine with Taiwan domestic jacket support structure in 18 meters water depth was selected to perform the fatigue strength calculation for the support structure. Commercial software Bladed and SACS are used to perform the required structural responses and fatigue strength calculations. The stress concentration factors(SCF) and S-N curves for the stress calculations of tubular joints are adopted based on the recommendation of DNV GL guidelines. The magnitude of the stress variation range and the corresponding number of counts are obtained by the rainflow counting algorithm. Finally under the assumption of Palmgren-Miner linear cumulative damage rule, the cumulative damage of the structure component can be calculated and then the fatigue life can be estimated. The numerical evaluation results can also be available for the further reference on the determination of sensor locations for the structural monitor system.

1. Introduction

Due to the focus on the issues of environmental protection and warming, the exploitation of renewable energy become more and more important for all the nations in the world. The application of wind energy is one of the favourable way for renewable energy utilization. In general, the designed service life of offshore wind farm is about 20 to 25 years, where both the maintenance and performance of the wind farm should be taken into account. Thus, it is necessary to assess structural fatigue life during the design stage. In this paper, the NREL 5MW wind turbine with Taiwan domestic jacket support structure in 18 meters water depth was selected to perform the required analysis. Where the software Bladed is used to calculate the aerodynamics load of wind turbine and then the marine structure software SASC is adopted to calculate the responses of support structure.

According to the IEC61400-3 design standard specification, there are several design load conditions should be taken into account during the design stage. Such as DLC-1(Power Production), DLC-2(Power Production plus occurrence of fault), DLC-3(Start up), DLC-4(Normal shut down), DLC-5(Emergency shut down), DLC-6(Parked), DLC-7(Parked and fault conditions) and DLC-8(Transport assembly maintenance and repair), etc. The purpose of this paper is to evaluate the fatigue life of the support structure. Therefore, only the design load condition of the DLC1.2 combined probability distribution is chosen for the analysis.

Sea bed condition and environmental data of an offshore site near Taiwan Strait are selected as boundary and load conditions for the analysis. In this paper, the time domain loads are calculated under the specified weather conditions by using the software "Bladed", and then the local stresses are obtained by using the marine structural analysis software "SACS". The magnitude of the stress variation range and the corresponding number of counts are obtained by the Rainflow counting algorithm. Finally, the cumulative fatigue damage local joints are evaluated by fitting the appropriate SN curve and the Palmgren-Miner's rule. And then the fatigue life of the support structure can be evaluated.

2. Wind turbine system

2.1. Wind turbine

The NREL 5MW turbine design[1] is chosen for this paper, include the material, tower and blade structural design together with generator itself show in the **Table 1**. The bottom diameter of the tower is 5 m with thickness of 0.038 m, while the top diameter connecting to nacelle is 3.12 m with thickness of 0.03 m. The basic dimension, component mass and rotor speed are shown in the **Table 2.**

2.2. Support structure

In general, all dead loads must be included in the design with nominal densities of the materials chosen. For loads/structures not covered by the load cases stated in this design basis, or DNV-OS-J101 [2], the environmental loads must be determined based on a return period of 50 years. Fatigue design of the jacket foundation is made as per [2] and DNVGL-RP-0005[3]. A typical jacket foundation arrangement and its main dimension are illustrated in **Figure 1.**

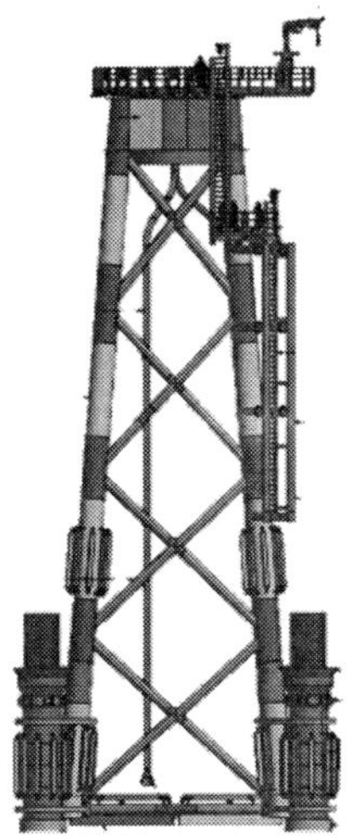

Figure 1. Jacket foundation arrangement.

Table 1. Main dimension of the wind turbine

Rated power(MW)	5
Rotor, Hub diameter(m)	126 ,3
Hub height(m)	90
Cut-in, Rated, Cut-Out wind speed(m/s)	3, 11.4, 25
Cut-In, Rated Rotor speed(rpm)	6.9, 12.1
RNA mass(t)	350
Tower mass(t)	347
Transition piece(t)	666
Jacket support structure(t)	462

Table 2. Material properties

Young's modulus (GPa)	210
Shear modulus (GPa)	81
Density (kg/m^3)	7850

2.3. Natural frequency

The offshore wind turbine encounters cyclical loads of wind, wave and rotor for full operation life. In the design stage, the natural frequency of the overall structure including wind turbine and jacket foundation should be checked to avoid the resonance with the wind and wave as well as the vibration caused by the rotation of the rotor.

The rotor speed ranges between 6.9 and 12.1 rpm. The speed can be converted to times per second, from 0.115 Hz to 0.202 Hz. For a three-blades wind turbine, the corresponding exciting frequency range is fall between 0.345 and 0.606 Hz. which can be converted to times per second, from 0.115 to 0.202 Hz. With three blade mounted on the wind turbine, the affecting frequency range is also triple, from 0.345 to 0.606 Hz. The wave and wind frequency distribution are respectively analysis by JONSWAP and Kaimal spectrum. All possible affecting frequencies are sorted in **Figure 2**. The only remained safe range, in which the natural frequency shall be designed, is between 0.233 and 0.25 Hz. **Figure 3**. Wind turbine mode shapes shows the calculated first and second side-to-side and fore-to-aft modes respectively.

IOP Conf. Series: Materials Science and Engineering **588** (2019) 012015 doi:10.1088/1757-899X/588/1/012015

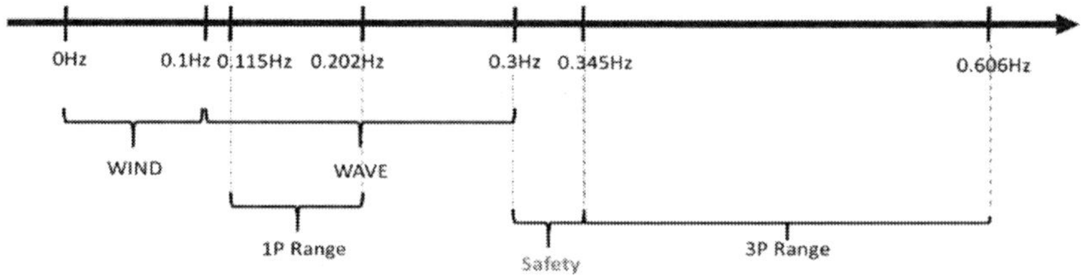

Figure 2. Safe frequency range of wind turbine structure

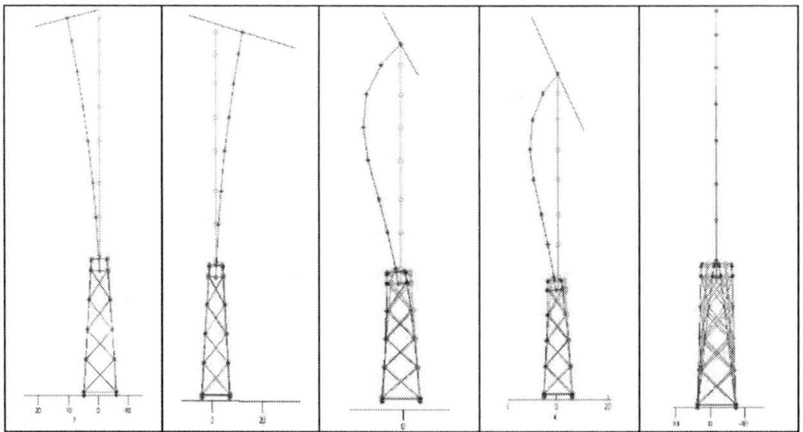

0.312Hz	0.314Hz	1.771Hz	2.180Hz	2.356Hz
1st side-side	1st fore-aft	2nd side-side	2nd fore-aft	1st torsion

Figure 3. Wind turbine mode shapes

3. Metocean

3.1. Foundation position

The environmental data comes from National Oceanic and Atmospheric Administration (NOAA). Dominant wind direction is north-northeast, 22.5 degrees clockwise rotation, for around six years at Taiwan Strait in **Figure 4.** Jacket foundation installation position In summer, the wind direction is not fixed and the average speed is relatively low except for typhoon. Prevailing wave direction is north-northwest.

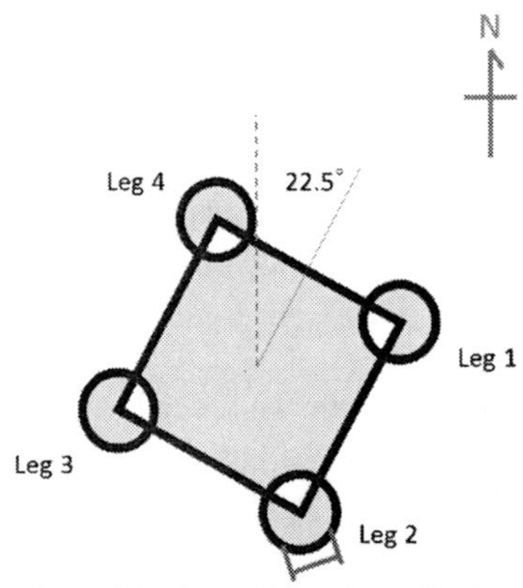

Figure 4. Jacket foundation installation position

3.2. Wind data

The wind data is given in **Table 3 Wind speed versus wind direction**, As for Wind plot show in Figure 5. (a) wind speed distribution, (b) wind rose map

Table 3 Wind speed versus wind direction

		Wind Direction (deg.)												
	From	-15	15	45	75	105	135	165	195	225	255	285	315	Probability
From	To	15	45	75	105	135	165	195	225	255	285	315	345	
Wind Speed (m/s) 1	3	1.0%	0.7%	0.5%	0.3%	0.2%	0.3%	0.3%	0.5%	0.7%	0.7%	0.7%	0.8%	6.8%
3	5	1.8%	1.5%	0.5%	0.1%	0.1%	0.1%	0.3%	0.9%	1.4%	1.1%	0.8%	1.1%	9.8%
5	7	2.3%	2.9%	0.5%	0.1%	0.1%	0.1%	0.4%	1.4%	2.1%	0.9%	0.4%	0.5%	11.6%
7	9	1.5%	4.4%	0.4%	0.0%	0.0%	0.1%	0.5%	2.3%	2.2%	0.2%	0.1%	0.2%	11.9%
9	11	0.9%	5.0%	0.3%	0.0%	0.0%	0.0%	0.5%	3.3%	1.1%	0.1%	0.0%	0.0%	11.2%
11	13	0.4%	6.2%	0.3%	0.0%	0.0%	0.0%	0.4%	2.3%	0.4%	0.0%	0.0%	0.0%	10.0%
13	15	0.2%	6.2%	0.3%	0.0%	0.0%	0.0%	0.2%	0.9%	0.1%	0.0%	0.0%	0.0%	7.8%
15	17	0.1%	6.1%	0.1%	0.0%	0.0%	0.0%	0.2%	0.4%	0.0%	0.0%	0.0%	0.0%	6.9%
17	19	0.0%	6.6%	0.2%	0.0%	0.0%	0.0%	0.1%	0.2%	0.0%	0.0%	0.0%	0.0%	7.1%
19	21	0.0%	5.3%	0.1%	0.0%	0.0%	0.0%	0.0%	0.1%	0.0%	0.0%	0.0%	0.0%	5.5%
21	23	0.0%	4.9%	0.1%	0.0%	0.0%	0.0%	0.0%	0.1%	0.0%	0.0%	0.0%	0.0%	5.1%
23	25	0.0%	3.3%	0.1%	0.0%	0.0%	0.0%	0.0%	0.0%	0.0%	0.0%	0.0%	0.0%	3.4%
25	27	0.0%	2.0%	0.0%	0.0%	0.0%	0.0%	0.0%	0.0%	0.0%	0.0%	0.0%	0.0%	2.1%
27	29	0.0%	0.7%	0.0%	0.0%	0.0%	0.0%	0.0%	0.0%	0.0%	0.0%	0.0%	0.0%	0.7%
29	31	0.0%	0.2%	0.0%	0.0%	0.0%	0.0%	0.0%	0.0%	0.0%	0.0%	0.0%	0.0%	0.3%
31	33	0.0%	0.0%	0.0%	0.0%	0.0%	0.0%	0.0%	0.0%	0.0%	0.0%	0.0%	0.0%	0.1%
33	35	0.0%	0.0%	0.0%	0.0%	0.0%	0.0%	0.0%	0.0%	0.0%	0.0%	0.0%	0.0%	0.0%
	總	8.2%	[illegible]	3.3%	0.6%	0.5%	0.5%	2.9%	12.4%	8.1%	2.9%	2.1%	2.6%	100%

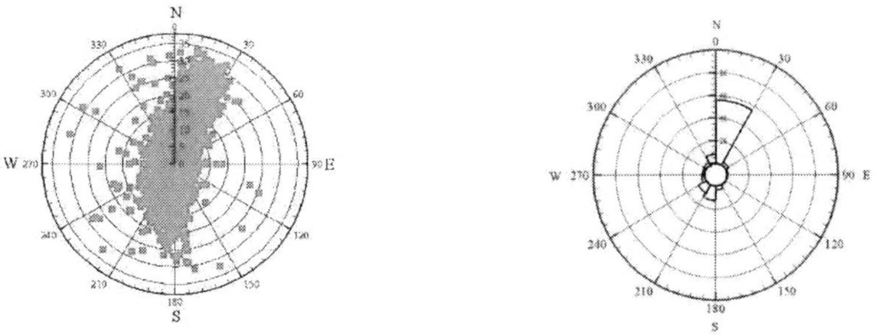

(a) (b)

Figure 5. (a) wind speed distribution, (b) wind rose map

3.2.1. Wave Condition

Wind versus wave is presented in the following as scatter diagrams for selected wind speed intervals with the distribution of significant wave height and wave peak period given as hours per year in average for the period 2010-01-01 to 2016-12-31 in the **Table 4.** Wave scatter diagramThe wave directions and misalignment are given in **Figure 6.** Wave Direction joint Probability Table

Table 4. Wave scatter diagram

			Hs(m)						
		From	0	1	2	3	4	5	Probability
	From	To	1	2	3	4	5	6	
Tp(s)	0	1	0.0%	0.0%	0.0%	0.0%	0.0%	0.0%	0.0%
	1	2	0.1%	0.0%	0.0%	0.0%	0.0%	0.0%	0.1%
	2	3	3.0%	0.1%	0.0%	0.0%	0.0%	0.0%	3.2%
	3	4	13.6%	3.9%	0.0%	0.0%	0.0%	0.0%	17.4%
	4	5	11.0%	10.4%	0.2%	0.0%	0.0%	0.0%	21.7%
	5	6	6.6%	9.2%	7.0%	0.1%	0.0%	0.0%	22.8%
	6	7	4.2%	1.4%	10.3%	3.9%	0.1%	0.0%	19.8%
	7	8	3.6%	0.1%	0.9%	[illegible]	1.0%	0.0%	11.2%
	8	9	1.8%	0.1%	0.0%	0.2%	0.4%	0.0%	2.4%
	9	10	0.6%	0.1%	0.0%	0.0%	0.0%	0.0%	0.7%
	10	11	0.5%	0.1%	0.0%	0.0%	0.0%	0.0%	0.6%
	11	12	0.1%	0.1%	0.0%	0.0%	0.0%	0.0%	0.2%
	12	13	0.0%	0.0%	0.0%	0.0%	0.0%	0.0%	0.0%
	Probability		45.0%	25.4%	18.4%	9.8%	1.4%	0.0%	100.0%

Wave Rose	1	2	3	4	5	6	7	8	9	10	11	12	All Dir
Sector	0	30	60	90	120	150	180	210	240	270	300	330	
Frequency	19.3%	50.9%	0.2%	0.1%	0.1%	0.1%	5.9%	15.2%	6.0%	1.5%	0.3%	0.5%	100%

Figure 6. Wave Direction joint Probability Table

4. Fatigue assessment

Offshore wind turbines are exposed to turbine excitation, wind and wave loads varying in time throughout their lifetime [4]. It leads to constantly varying in time stresses and making the support structure prone to fatigue. In general, fatigue damage accumulation at welded tubular joints can be estimated using the calculated local hot spot stresses. First, stress concentration factors(SCF) and S-N

IOP Conf. Series: Materials Science and Engineering **588** (2019) 012015 doi:10.1088/1757-899X/588/1/012015

curve for the tubular joint can be obtained according to the recommendation of the guidelines of DNV GL RP C203 [5]. and the hot spot stresses can then be calculated in time-domain. Then the magnitude of the stress variation range and the corresponding number of counts are obtained by using the rainflow counting algorithm as shown in **Figure 7.** Rainflow Counting Finally, the cumulative fatigue damage ratio can be calculated by using the Palmgren-Miner rule as shown in **Figure 8.** Fatigue assessment according to Palmgren-Miner rule Fatigue assessment according to Palmgren-Miner rule, and the fatigue life can then be estimated based on the calculated cumulative damage ratio.

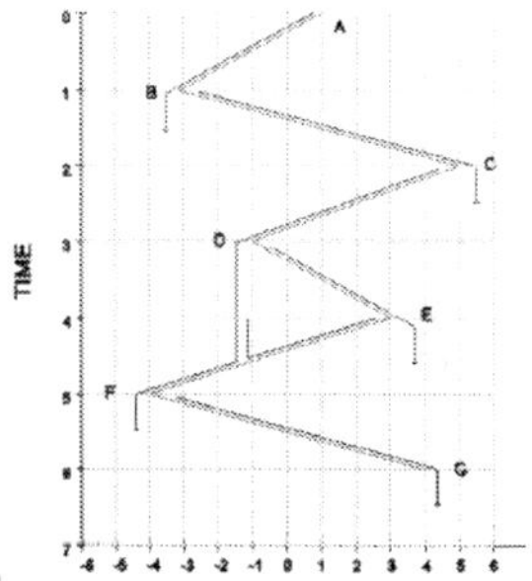

Figure 7. Rainflow Counting

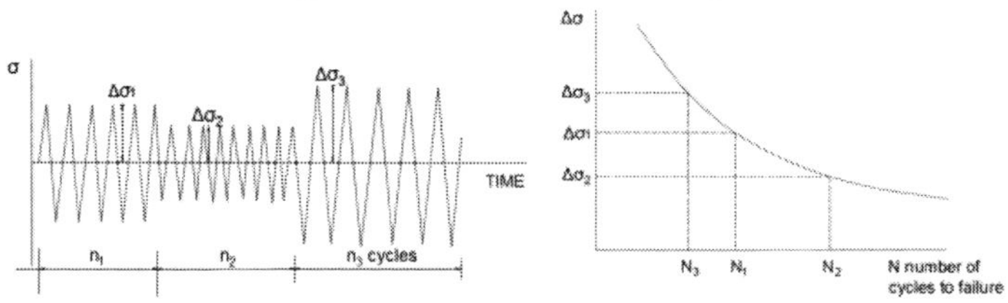

Figure 8. Fatigue assessment according to Palmgren-Miner rule

4.1. Fatigue result

Table 5. Fatigue damage life shows part of the calculated results. The analysis results show that the calculated fatigue life is more than 20 years generally for most of the joints. and the calculated fatigue life is relatively shorter while structure members are located near the pile position, it is suspected that it is caused by the large force acting on those structure members near the soil.

Table 5. Fatigue damage life

	Joint	Member	Damage ratio	Damage life
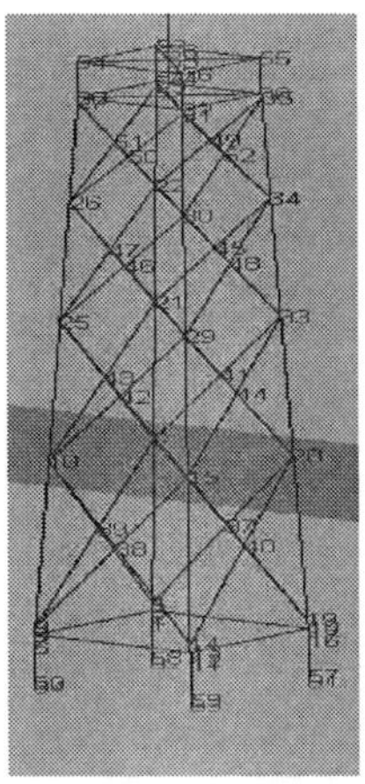	57	57 – 61	0.29	79
		57 – 16	0.91	22
	58	58 – 1	0.48	40
		58 – 62	0.16	121
	59	59 – 63	0.47	41
		59 – 11	0.15	125
	60	60 – 6	0.3	58
		60 - 64	0.95	21

IOP Conf. Series: Materials Science and Engineering **588** (2019) 012015 doi:10.1088/1757-899X/588/1/012015

5. Conclusion

In this paper, a NREL 5MW wind turbine with Taiwan domestic jacket support structure is selected as an example to perform the fatigue strength calculation for the support structure. Commercial software Bladed and SACS are used to perform the required structural responses and fatigue strength calculations. The calculated results show that the estimated fatigue life for most of the joints of supporting structure members are expected to be more than the required designed service of 20 years. And it is expected that the numerical evaluation results can also be available for the further reference on the determination of sensor locations for the structural monitor system.

6. References

[1] Jonkman J, Butterfield S, Musial W and Scott G 2009 Definition of a 5-MW reference wind turbine for offshore system development. NREL/TP-500-38060 (National Renewable Energy Laboratory, USA)
[2] Det Norske Veritas 2014 DNV-OS-J-101 Design of offshore wind turbine structures.
[3] Det Norske Veritas-Germanischer Lloyd 2016 DNVGL-RP-0005 Fatigue Design of Offshore Steel Structure.
[4] IEC,'"Wind turbine Part1:Design requirement for offshore wind turbine"' ,IEC-61400-3 ,February.
[5] DNVGL"Fatigue design of offshore steel structure, "DnvGL-RP-C203,2016

Acknowledgments

The authors would like to thank the Bureau of Energy, Ministry of Economic Affairs, Taiwan for their support under Grant Number 106-D0106.

IOP Conf. Series: Materials Science and Engineering **588** (2019) 012016 doi:10.1088/1757-899X/588/1/012016

Self-powered IoT Based Vibration Monitoring of Induction Motor for Diagnostic and Prediction Failure

Adim Firmansah[1], Aripriharta[1,3,*], Nandang Mufti [2,3], Arif Nur Affandi[1], Ilham Ari Elbaith Zaeni[1]

[1]Department of Electrical Engineering, Faculty of Engineering, Universitas Negeri Malang, Indonesia
[2]Department of Physic, Faculty of Mathematics and Natural Science, Universitas Negeri Malang, Indonesia
[3]Centre of Advanced Materials for Renewable Energy, Indonesia

E-mail: `aripriharta@gmail.com`

Abstract. The vibration of the motor provides information for failure diagnostic and prediction through signal processing. In this paper, we present an IoT-based monitoring system for recording the vibration of the induction motor. A 27mm piezoelectric is used to collect the vibration data from the motor. Each self-powered IoT node is supplied by the motor vibration its self by utilizing two 27mm piezoelectrics in series. We conducted an experiment with 3 scenarios, which aims to collect the vibration data from different positions, they are bottom, top and at the coupling. For each scenario, the speed of the induction motor is changed from low, middle to high speed. Based on our results, the proposed system is a success to monitor the motor vibration clearly. The results can display on the smartphone. Each node can send the recorded vibration to the cloud with an average delay of about 1 s.

1. Introduction

The most widely used actuator in an industry is the induction motor which used on the liquid mixing machines, conveyor drives, cranes for lifting goods. Since the operation cycle of those machines is continuous, thus the mechanical part of the motor could be failed. The monitoring system for diagnosing and predicting the condition of induction motors become a high priority task [1].

There have been many studies in diagnosis the mechanical damage of the induction motor in literature. Motor vibration can be generated from motor bearings, broken rotor bars, overloaded motors, imbalanced motors [2]. The existing monitoring systems [3] [4] was demonstrated that the specific frequency of vibration is related to kind of damage. However, those previous works were haven't used a remote monitoring system. Other former works implemented an induction motor vibration monitoring system by using IoT with the web-based interface. The web-based system [5] [6] still lacks high flexibility in its use.

The mechanical damage of the induction motor should be handled immediately to prevent the system halt. In this paper, we propose the monitoring system by using self-power IoT, where a 27mm piezoelectric is used to collect the vibration data of the induction motor. In addition, our

IOP Conf. Series: Materials Science and Engineering **588** (2019) 012016 doi:10.1088/1757-899X/588/1/012016

IoT-node is self-powered devices which powered through harvesting the vibration energy from an induction motor. The proposed system is also equipped with apps to provide a graphical interface for the user through a smartphone.

Finally, the study of this paper is arranged as follow. Section 2 describes the method that consists of hardware and software design. The experiment results and analysis are in Section 3. The conclusion of this works is described in Section 4.

2. Method

The proposed IoT-based monitoring system depicted in Figure 1. This system composes by n self-powered IoT nodes. Each node collects data from 27mm-thin piezoelectric and forward data to the cloud through the router. The vibration data of the motor collected using the nodeMCU with piezoelectric as a sensor. The microcontroller connects to wifi without requiring additional modules. The user can monitor the motor vibration from a smartphone.

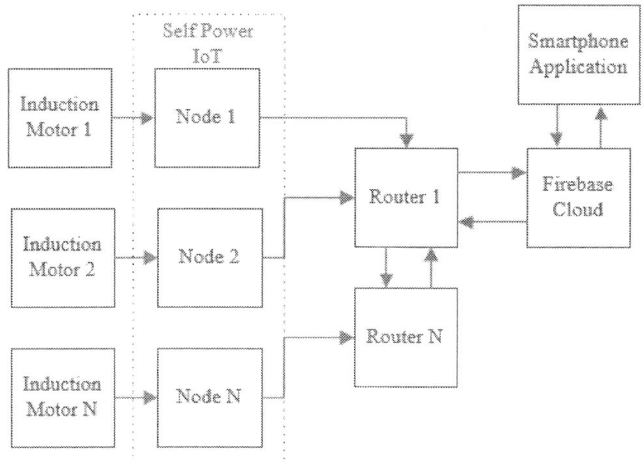

Figure 1. System block diagram.

2.1. Hardware Design

The hardware for the proposed system consists of a piezoelectric energy harvester, NodeMCU, and router. The piezoelectric energy harvester is shown in Figure 2. The vibration energy from induction motor converted by 27mm-thin piezoelectric become ac voltage. LTC3588 manages the electric power and stabilize it to the 3.3V dc voltage. A supercapacitor placed on the output terminal of LTC3588 to store the electrical energy.

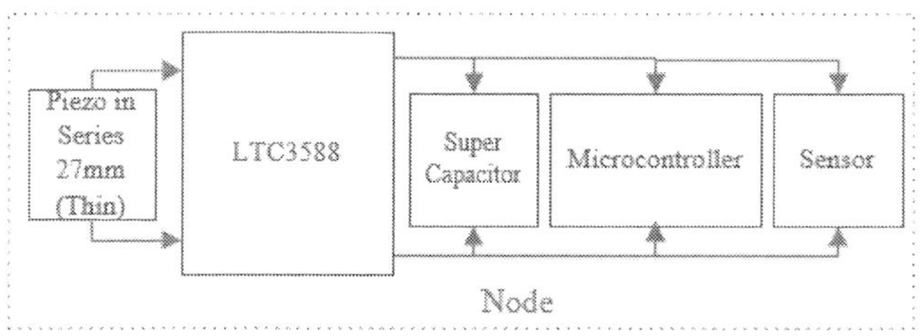

Figure 2. Piezoelectric energy harvester.

Motor vibration data sent to database periodically by the microcontroller via wifi network. The data transmission cycle from the microcontroller to the smartphone application shown in Figure 3.

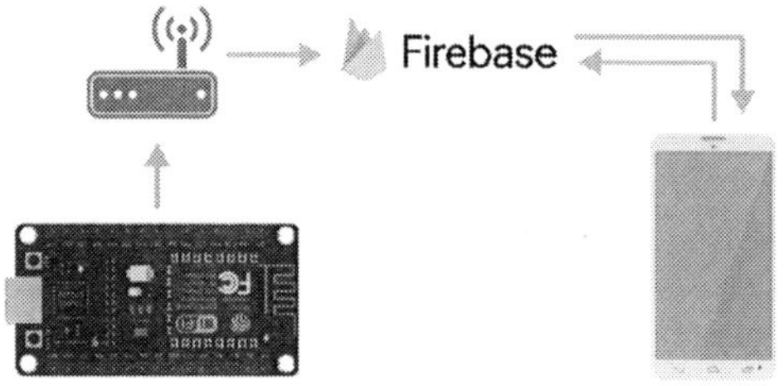

Figure 3. The cycle of sending data.

While sending data to the database, the microcontroller stops the data retrieval process. To minimize data loss during the sending process, the system created by maximizing microcontroller memory for data storage. After the data meets the microcontroller memory, all data stored uploaded to the database in once upload process.

NodeMCU is a microcontroller which designed for low power application. Its equipped with 17 GPIO pins and a WIFI transceiver [10]. This MCU operates at 3.3V and consumes low power. We used this device for collecting vibration from a sensor which connected to pin A0, data acquisition and controlling the WIFI module as depicted in Figure 4.

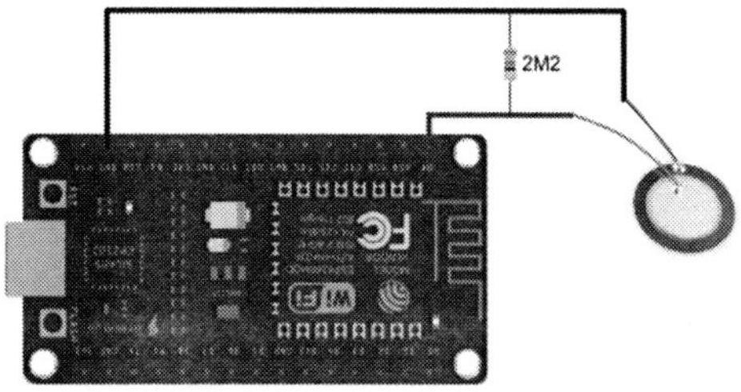

Figure 4. Connection for piezo-electric.

2.2. *Software Design*

The software design is divisible into three parts. The first part of software design is data acquisition. The measurable vibration is the analog values which range of 0-1023 (0 - 3.3V). We normalized vibration values by using the following equation:

$$V = \frac{AnalogValue}{Vmax} \tag{1}$$

The second software design is a cloud database. We used Firebase because of its free and has several Google services in the cloud, including instant messaging, user authentication, real-time databases, storage, hosting, etc [7]. Firebase supports various platforms such as Android, iOS, Linux, Arduino [8]. The compatibility is one of the considerations in database selection. Figure 5 shows the Firebase cloud system.

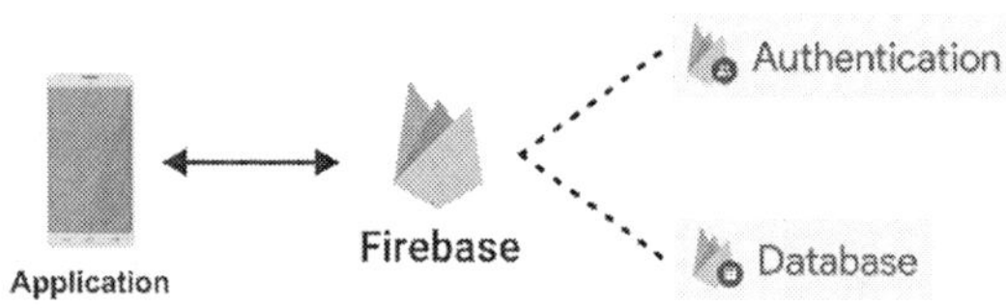

Figure 5. Firebase cloud system.

The third part in software design is Apps. Figure 6 depicts our apps for a user interface on a smartphone. Figure 6 (a) shows the design of the user authentication for entering the username and password. The authentication process aims to limit the access to our system. After logging in, the user goes directly to the main page. Figure 6 (b) shows the main page that displays data in the graph, as well as some other functions. First data and last data are variable data ranges, that searched for minimal value, maximal and average for analysis. By providing data observation time for users, data updates can be set automatic or manual. To manually update the data, the user uses the update button provided. To manage data updates automatically by marking the automatic update checkbox under the update button. The application can be used for several users at the same time on a different device.

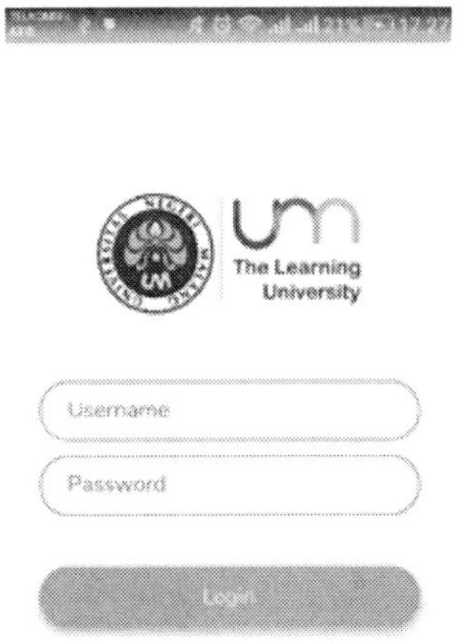

Figure 6. User authentication.

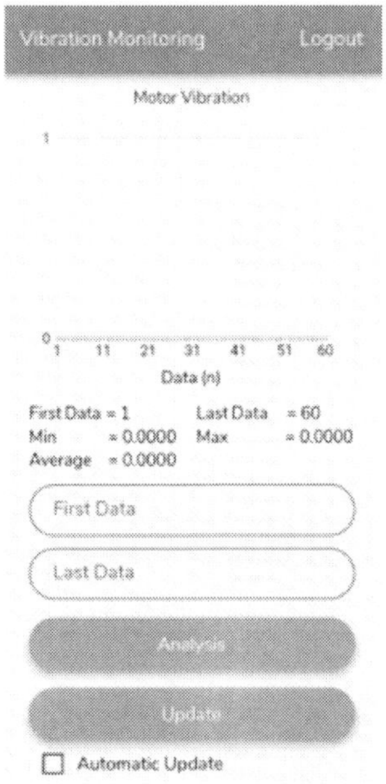

Figure 7. Main page.

2.3. Experiment Scenarios

We demonstrated three scenarios for validating the performances of our proposed monitoring system through an experiment. The sensor is placed in several positions (top, front, and bottom) to find the best point in data retrieval. Laying the sensor is shown in Figure 7-8. The magnitude of the motor vibration is detected using piezoelectric sensors. The sensor produces an AC voltage whose magnitude depends on the value of the pressure/vibration detected.

The 3-phase motors tested in this paper have the following specifications. This motor is controlled by VSD Altivar-31 to control the motor speed from low to high speed and vice versa. Each experiment was conducted for 15 minutes and the results displayed on a smartphone screen.

3. Result and Analysis

Tests carried out as shown in Figure 9 demonstrate our self-IoT can operated by harvesting the motor vibration energy. LTC-3588 gets an input voltage from two piezoelectrics with a series configuration that reaches 2.56 V (maks). The output of the LTC-3588 stored on the supercapacitor (1F, 3.3V) to powering the nodeMCU's, the measured power from the single self-IoT node is 183 mW per round.

IOP Conf. Series: Materials Science and Engineering **588** (2019) 012016 doi:10.1088/1757-899X/588/1/012016

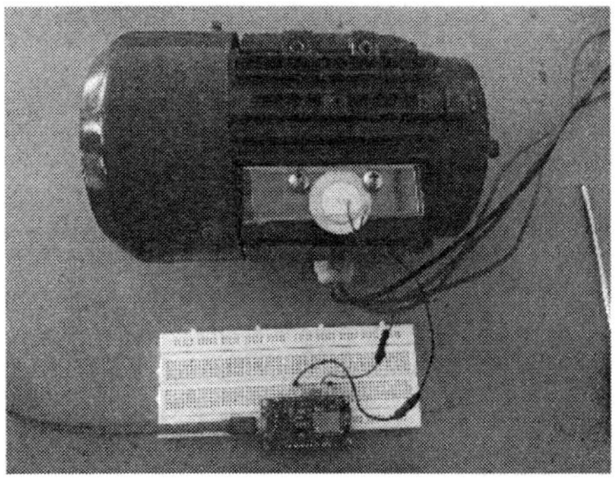

Figure 8. The sensor placed at the top.

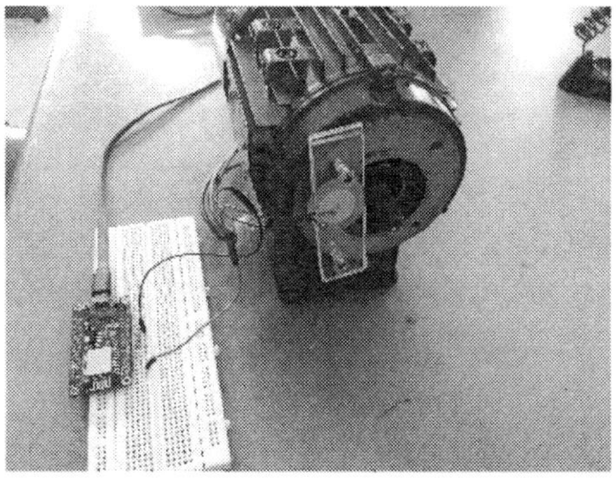

Figure 9. The sensor placed at the front.

Table 1. Motor spesification.

Spesification	Value
Power	0,09kW
Voltage	230/380V
Speed	1350 rpm
Protective grade IP	55
Frequency	50Hz
Weight	36kg

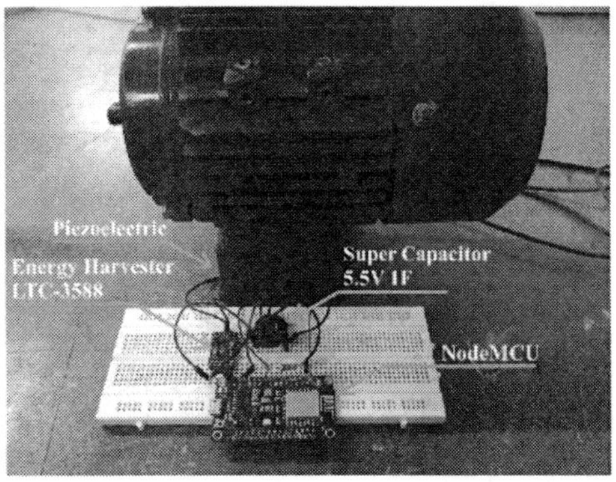

Figure 10. Experiment.

The motor vibration has taken with a piezoelectric sensor on multiple placements at the motor in the laboratory. Experiments are carried out with several variations of motor speed to observe the effect of motor speed on the vibration. Based on [9], the magnitude of the motor speed affects the vibration value. Figure 10 represents the motor vibration when the piezoelectric placed at the top of the motor. At 0Hz frequency, the sensor detects a little vibration by the sensor reading result in the range of 0.09V. At maximum speed (f = 50Hz) the highest point is at 0.46V. With the frequency of 10Hz and 25Hz, the generated vibration has a difference value that is more stable than the 50Hz frequency.

Figure 11 shows the result of vibration reading when the sensor placed in front of the motor. In the results, the vibration generated by the sensor is higher with a peak voltage of 0.53V at a frequency of 50Hz.

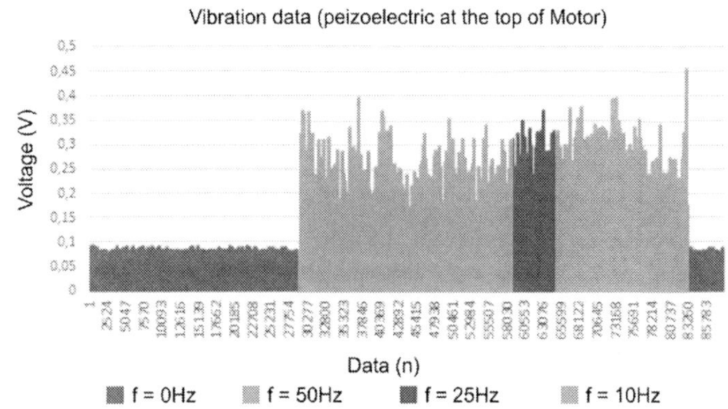

Figure 11. Vibration data (piezo-electric at top of the motor).

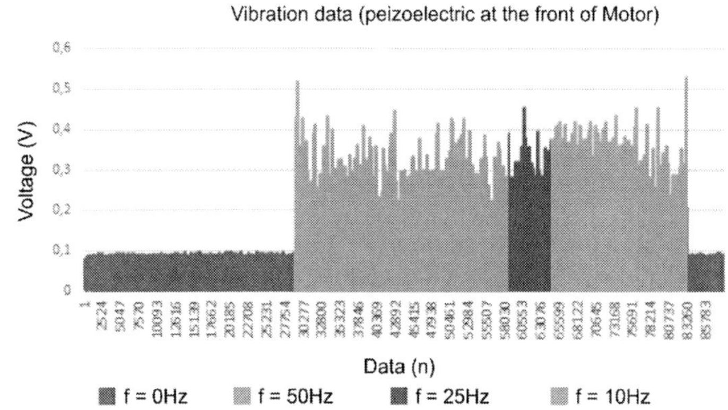

Figure 12. Vibration data (piezo-electric at front of the motor).

In the last scenario, a piezoelectric sensor placed under the motor. Figure 12 represents the results of sensor readings with a value of 0.12V while resting. At maximum speed (f = 50Hz) the peak voltage is 0.23V. From the three places of piezoelectric sensors in the retrieval of induction motor vibration data, the bottom position of the motor provides better data because it has a relatively stable value on the applied operating frequency. Vibration data used as a reference is data taken after maintenance on an induction motor. In a motor that has a problem with its condition, there will be a vibration surge or a decrease in vibration value (if the speed decreases) in the monitoring process. The data obtained can be used as a reference in planning a periodic maintenance schedule through further diagnostic and prediction algorithm. From 10 times of trials, the delay for data transmission through our system is about 1s.

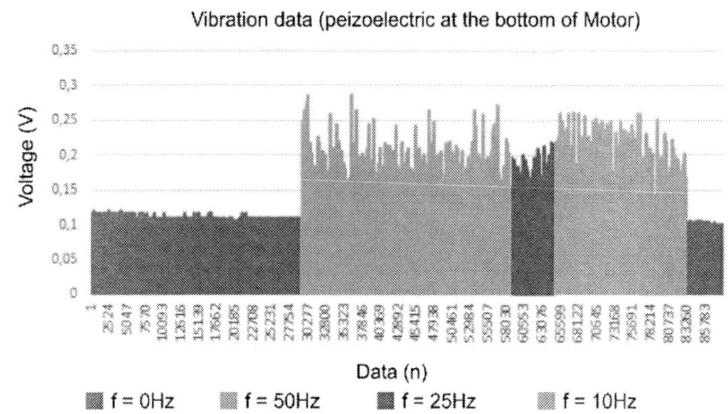

Figure 13. Vibration data (piezo-electric at bottom of the motor).

4. Conclusion

The Self powered IoT for vibration monitoring of electric motor has been developed and tested successfully. The vibration of motor uses for powering the entire monitoring system including a sensor, microcontroller, and WIFI module. By using a piezoelectric sensor, the motor vibration can read by the IoT node. The experiment results show the vibration data taken from different places. The user can access data from smartphone remotely and switch from one to another's motor.

The future project will implement the system for n number of motors, where n is larger than 5. Each node sends data to the same database. The signal processing, e.g. STFT, DWT and prediction algorithm will develop in the cloud system such as Matlab cloud for providing accuracy on diagnostic and prediction.

References

[1] Michael T 2011 Introduction motor condition monitoring: Vibration analysis technique – a practical implementation *Proc. Int. Electric Machines & Drive Conf.* pp 406-411
[2] K Agoston 2015 Fault Detection of the Electrical Motors Based on Vibration Analysis *Proc. Int. Conf. Interdisciplinarity in Engineering* pp 547-533
[3] Mariana I, Robert B 2012 *Study DC Motor Diagnosis Based on the Vibration Spectrum and Current Analysis* pp 1-4
[4] Wu-Jeng L, Chiaming Y, You-Sheng L, Shu-Chu T and ShiMiao H 2018 *JustIoT Internet of Things base in Firebase Real-time Database* (Taiwan: Kun Shang University) p 44
[5] D Gangga, V Ramachandran 2018 IoT Based Vibration Analitycs of Electrical Machines *J. Internet of Things* pp 1-12
[6] A Ajitha et al 2017 IoT Platform for Condition Monitoring of Industrial Motors *Proc. Int. Conf. on Communication and Electronics System* pp 260-265
[7] Dezheng W, Steven D P 2007 *Using Mechanical Vibration to Estimate Rotor Speed in Induction Motor Drives* (USA: Purdue University) pp 2412-2417
[8] Abdullah A, Yahya A, Mohammed A D, Ibrahim A, Faycal B, Abbes A and Guillaume A 2017 Real-Time Communication Network using Firebase Cloud IoT Platform for ECMO Simulation *Proc. Int. Conf. on the Internet of Things* pp 178-182
[9] Alfredo M R, Gonzalo N C 1994 *Mechanical Vibration of Three-phase Induction Motors Fed by Nonsinusoidal Currents* (Casilla: Universidad de Chile) pp 166-172
[10] Espressif Systems 2015 *ESP8266EX Datahseet* p 17

IOP Conf. Series: Materials Science and Engineering **588** (2019) 012017 doi:10.1088/1757-899X/588/1/012017

IoT-based Control and Monitoring for DC Motor Fed by Photovoltaic System

Kusmayanto Hadi Wibowo[1], **Aripriharta**[1,3*], **Markus Diantoro** [2,3], **Slamet Wibawanto**[1], **Kartika Candra Kirana**[1], **Fitrah Wicaksono Yunianto Saputra**[1], **M Nuzulul Chanif**[1]

[1] *Departement of Electrical Engineering, Faculty of Engineering, Universitas Negeri Malang, Indonesia*

[2] *Departement of Physics, Faculty of Mathematics and Natural Sciences, Universitas Negeri Malang, Indonesia*

[3] *Centre of Advanced Materials for Renewable Energy, Indonesia*

E-mail: aripriharta.ft@um.ac.id

Abstract. Internet of Thing (IoT) makes the monitoring and controlling process become accessible everywhere and every time. In this paper, the IoT is used to control the speed of a DC Motor remotely through smartphone apps. The proposed system uses the 50Wp solar panel and DC-DC Converter to provide 24V dc to the motor. In specific, a PID algorithm is applied to control motor speed from the measured back EMF. There are 3 experiments that have been done. The first is controlling motor speed through a smart phone, the second is the relationship between EMF and DC Motor speed, the third experiment was monitoring motor speed on Thingspeak and Blynk apps.

1. Introduction

Monitoring the real-time situation using IoT has begun to develop. In [1], they develop monitoring of electrical energy in real-time. They use Raspberry Pi to access data via a smartphone. The data monitored is the measurement of voltage, current, and neutral current. The data is used to monitor the electricity used. While the authors [2], have developed IoT-based monitoring systems for the efficiency of solar power plants. They use Raspberry pi and Message Queuing Telemetry Transfer Protocol (MQTT) to be accessed via smartphones.

In addition to the development of IoT-based monitoring, improved DC Motor control performance is also growing. The author [3], has designed a control device with a PID algorithm. The project is applied to DC Motor speed. While in [4] designing motor speed control with fast response and high precision. The two papers already have good results, but motor control has not been specifically applied.

Development of the Solar Panel as a source of energy for DC Motors has been widely developed. As in reference [5], the Solar Panel is an energy source for DC Motor water pumps. Although the results of previous studies have been significant, for applications of solar-powered pumps, PID algorithm is needed to get good results. In this paper describes the development of IoT-based dc-pump motor control, where DC Motors are supplied from the Solar panel. This project contributes to the implementation of PID algorithms and IoT-based controls for DC

IOP Conf. Series: Materials Science and Engineering **588** (2019) 012017 doi:10.1088/1757-899X/588/1/012017

Motor speed. This system will be implemented for dc pump regulation systems in remote areas of Indonesia.

The rest of this paper is organized as follow. Material and Method is described in Section 2. Section 3 describes Result and Analysis. Section 4 is conclusion.

2. Material and Method

Figure 1 presents the design of DC Motor Control based on IoT. The system consists of 4 parts. The first part is the DC Converter, this section is to provide 25V to the motor. The next section is about DC Motor control. This project uses the PID algorithm to control DC Motor speed. This DC Motor control is designed based on IoT, so the smartphone becomes the control device. IoT-Nodes need routers to connect to the internet so that nodes can connect with users. The last part is data communication per node. But this paper only discusses one node, which tests the DC Motor control based on IoT. This project uses Thingspeak to monitor the results of DC motor control and Blynk applications to control DC motors.

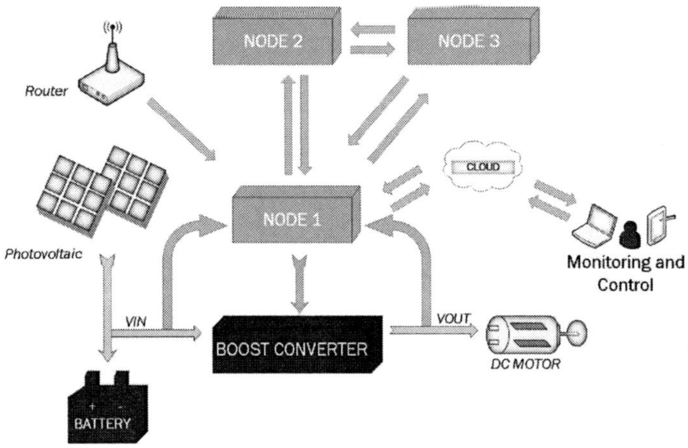

Figure 1. System block diagram

2.1. Solar Panel

This project uses solar modules to supply DC Motor loads. The solar module used for the development of this project is the Solartech SRM-50D Sunrise Solar Module. Here are the parameters of the solar module:

Table 1. Parameter of Sunrise Solartech SRM-50D

Parameter	Value
Standart Power	50 W
Short Circuit Current (Isc)	3.14 A
Open Circuit Voltage (Voc)	21.4 V
Max. Power Current (Imp)	2.88 A
Max. Power Voltage (Vmp)	17.4 V
Max. System Voltage	1000 DC
Number of cell	36

From these parameters, it can be graphed Power to Voltage on Solar Modules. The following is a graph of the PV characteristic:

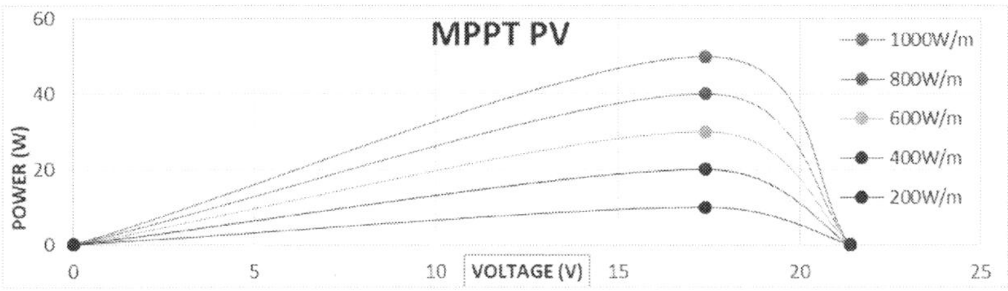

Figure 2. Graph of PV characteristic

From the above graph in Figure 2, it is explained that the lower the Radiation obtained by the Solar Module affects the decrease in power released by the Solar Module. At the same time, the voltage remains at maximum Voltage, so it can be concluded that Radiation will affect the current coming out of a Solar.

In addition to knowing about the parameters of the solar panel, the potential of the location regarding solar conditions is also very influential for the success of this project. This paper refers to the Suncalc.org site to find out the condition of the sun in Hamlet III, Poncokusumo, Malang, East Java, Indonesia. The following are the conditions of the sun at that location:

Table 2. The condition of Solar at the location.

Situation	Hour
`Dawn`	4:59
`Sunrise`	5:20
`Culmination`	11:22
`Sunset`	17:25
`Dusk`	17:46

2.2. IoT Based Control System

There are several hardware used in our IoT based control system, namely IoT-Nodes, routers, and DC Motor. Figure 3 shows the proposed system.

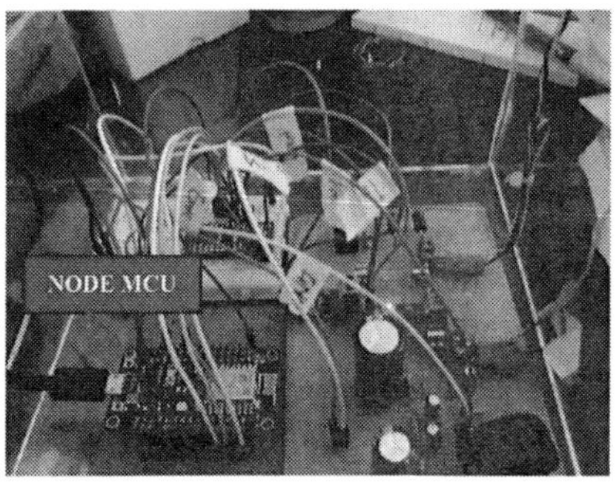

Figure 3. The proposed system

2.2.1. IoT-Nodes The IoT-Node used is the Node-MCU. The Node-MCU is supplied by the solar panel whose output voltage has been set to 5 volt DC. The Node-MCU functions to control and transmit data to the internet. Therefore, this project uses Node-MCU because it already has a wifi module in it.

2.2.2. Router The router functions to connect IoT-Node with the internet. IoT-Node can send and receive data via the internet. Users can control from the phone because the data entered by the user is connected to the IoT-Node via the internet, while the monitoring results are obtained from sensor data.

2.2.3. DC Motor DC Motors in this project can be used for water pumps. The water pump used is a 24 volt DC pump, 10 m^3/h with a speed of 4300 rpm [13]. The motor speed at the pump must continue to a stable condition even though the voltage on the solar cell decreases. Therefore, the Node-Mcu will control the speed entering the motor.

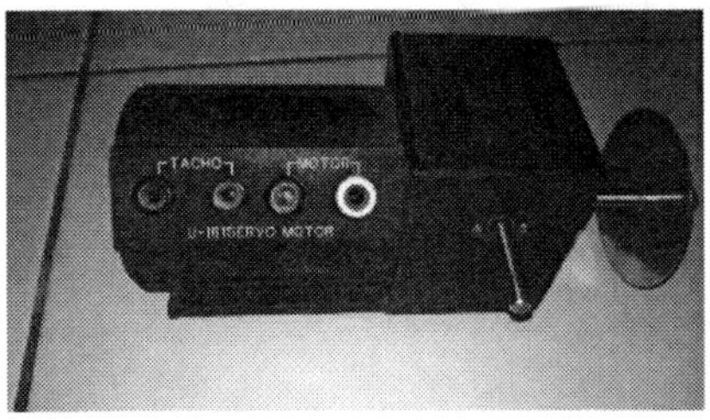

Figure 4. DC Motor

2.3. DC-DC Converter
The DC-DC converter used is Boost Converter with the circuit as shown in Figure 5. This converter is used to adjust the voltage to match the value desired by the User. In addition to increasing the voltage, the converter also has a function to stabilize the voltage generated by solar cells. The component we use is an inductor that has a size of 1mh; Capacitors with a capacity of 2200uf / 50V; Diodes 1n5402 with a maximum current of 3A and a maximum voltage of 200volt, and Mosfet with type 1RFZ44n for switching in the circuit. This component we can through the equations that we take from references [6].

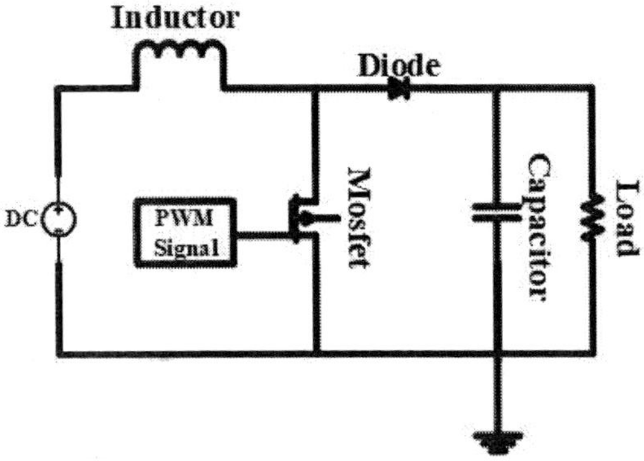

Figure 5. Boost converter circuit

To get the duty cycle value is :

$$D = 1 - \sqrt{\frac{rmp}{Ro}} \tag{1}$$

$$R_{mp} = \frac{V_{mp}}{I_{mp}} \tag{2}$$

$$R_o = \frac{V_o}{I_o} \tag{3}$$

To get the inductor value is :

$$L_{min} = \frac{R_o}{dilfs}\frac{4}{27} \tag{4}$$

$$dil = \frac{V_{mp}D}{Lf} \tag{5}$$

To get the capacitor value is :

$$C_{min} = \frac{D_{min}(1 - D_{(min)})^2}{R_{mp(min)}dV_o fs} \tag{6}$$

$$C_{min} = \frac{D_{max}(1 - D_{(max)})^2}{R_{mp(min)}dV_o fs} \tag{7}$$

$$dV_o = \frac{dQ}{C} \tag{8}$$

2.4. PID

Proportional Integral and Derivative widely used for the control system in the industry according to section 2.3 of the DC-DC Converter, the converter can increase the voltage according to the reference voltage if the PWM signal issued by the Controller matches the desired response. The PID will adjust the PWM signal to get an optimal response. Figure 7 is the code in Arduino about the PID algorithm (use the library on Arudino). Here is a block diagram of a PID [7-8]:

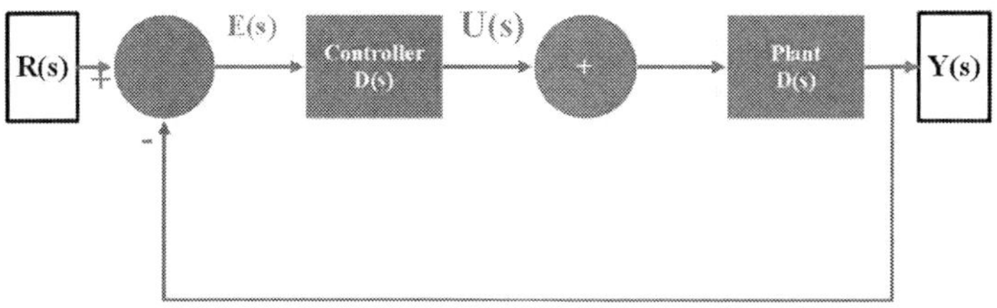

Figure 6. Block diagram of PID

A controller is set using the following formula:

$$D_{(s)} = K_p + \frac{K_i}{s} + K_D. \tag{9}$$

2.5. Data Acquisition

In this project, we do not use motor speed sensors (sensorless) through the Back-EMF method. The Back EMF captured by the voltage sensor and convert to a digital value by ADC which is onboard our IoT-Nodes. If the DC Motor is supplied by a voltage source then the DC Motor will move. When the motor moves, the voltage appears that is opposite to the flow of the permanent magnet coil. This is called the EMF method. EMF is directly related to motor speed, so knowing the emf value can calculate the speed of the motor speed. Figure 8 is the equivalent of DC Motor.

IOP Conf. Series: Materials Science and Engineering **588** (2019) 012017 doi:10.1088/1757-899X/588/1/012017

```
#include <PID_v1.h>
#define Sensor A0
#define PWM 3
Double Setpoint, Input, Output; //Define Variables we'll be connecting to
Double Kp=0.07425, Ki=0.012, Kd=0.12; //Initial tuning parameters
PID mypid(&Input, &Output, &Setpoint, Kp, Ki, Kd, DIRECT);
Int Vreff = 25; //define Reff
Void setup()
{
Serial.begin(9600);
Pinmode(Sensor, INPUT);
Mypid.setmode(AUTOMATIC);
}
Void loop()
{
Setpoint = Vreff;
Input = Sensor;
Mypid.Compute();
Analogwrite(Input, Output);
Serial.println(String(Output));
Delay(1000);
}
```

Figure 7. PID algorithm code

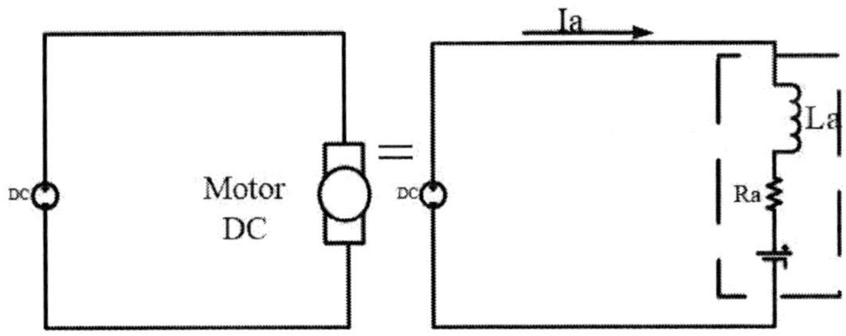

Figure 8. The equivalent circuit of DC Motor

We can represent the motor as a series connection of resistance, inductance, and the back EMF voltage source. It is this voltage source that we want to measure. Here's the equation from EMF [10]:

$$EMF = \frac{n}{k} \tag{10}$$

Speed on the motor can be represented by:

$$n = \frac{V_s - i_a R_a}{k} \tag{11}$$

Which, n is the motor speed. Vs is the input voltage of the motor, Ia is the current from the motor. While Ra is the motor resistance and k is the coefficient.

2.6. Blynk Apps

This project can control DC Motors using smartphones. Blynk is an application that we use to control DC Motors, Blynk can be downloaded for free by users. Its use is very easy, the user only needs to download the blynk application and scan the barcode from the project developer. As an important note, users need to have a connection to connect to a DC Motor. Figure 9 is the Blynk app. In the figure, there is a motor speed controller. Users only need to move the motor speed cursor (RPM Control) from left to right. In addition to control, the developer also provides facilities to monitor the value of Voltage on the motor and motor speed.

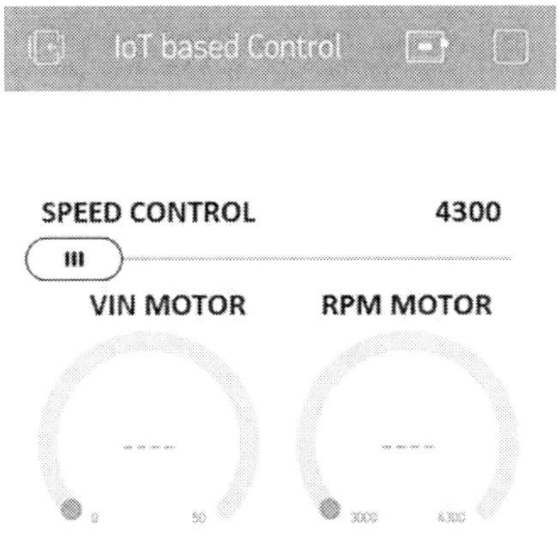

Figure 9. Blynk Apps

2.7. Cloud Systems

The cloud becomes an intermediary for communicating between users and nodes. With cloud systems, users can monitor and control DC Motors remotely. Control is done by inputting the reference speed on the smartphone. The reference speed will be sent to the node through the cloud so that the node will control the motor speed to match the reference speed.

3. Result and discussion

The proposed system experiment was successfully carried out. Table 3 shows the test results from the Boost Converter, where the output of the boost converter is a DC motor. The first experiment (time = 13.42), it shows testing without using a load at all. The next experiment DC motor brakes level 1 and the user enters the motor speed with a value of 4300 rpm (maximum speed). The speed is the same as the reference voltage (24 Volts). During the next test (time = 13.52), using DC Motor braking with level 2, the result of the output voltage and motor speed is still stable. When the output voltage is more than 24 volts, the converter cannot reduce the reference voltage because the Boost Converter only has the function of increasing the voltage value. The results from output voltage have an effect on Motor Speed. This data is a parameter of the PID algorithm. Besides data from the output voltage sensor, another important parameter is the input data from the user. Input data from the user is the value of the desired motor speed. From the motor speed data is converted into voltage data using the EMF method, so that the data becomes a set point to produce PWM signals.

Table 3. Test Results Boost Converter

Time	Vin(Volt)	Vout(Volt)	RPM
13.42	21.09	33.24	0
13.50	21.48	24.39	4300
13.52	21.38	24.49	4300
13.54	21.4	24.33	4300
13.57	21.25	25	4300

The next experiment is to test the relationship between EMF with DC Motor Speed. Figure 10 is the motor speed chart of 10 experiments. From 10 experiments, the average motor speed was 4262 RPM, with a maximum speed of 4297 rpm and a minimum speed of 4207rpm. Figure 11 is a graph of the relationship of EMF with DC Motor speed. The relationship between EMF and DC Motor speed is directly proportional.

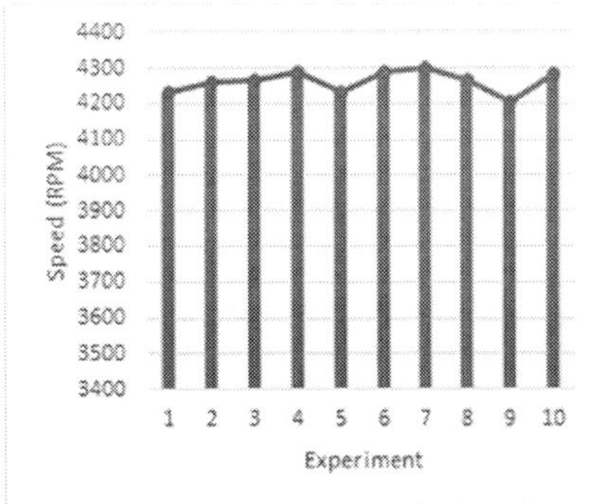

Figure 10. Graph of 10 Experiment

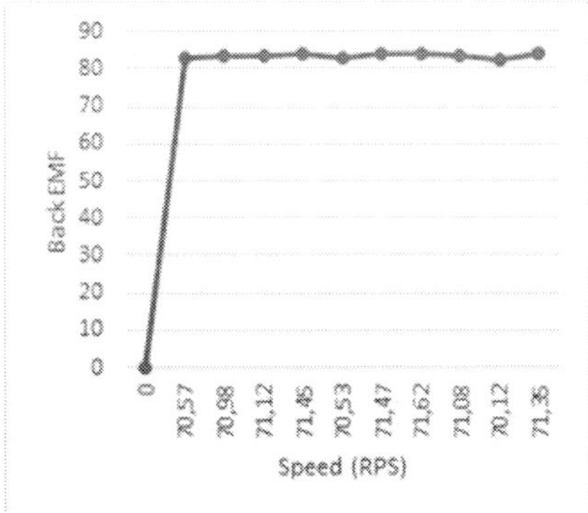

Figure 11. Graph of the relationship between EMF and motor speed

The last experiment is monitoring DC Motor speed on Thingspeak and Blynk app. Figure 12 shows the results of monitoring output voltage in the boost converter. The monitoring results are sent to the Thingspeak web every 15 seconds, while the data to be processed into PWM conducted continuously. The monitoring results from the Blynk application can be seen in Figure 13.

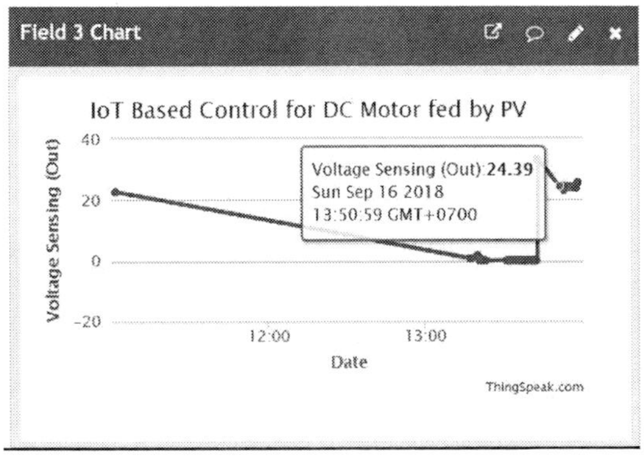

Figure 12. Monitoring using Thingspeak

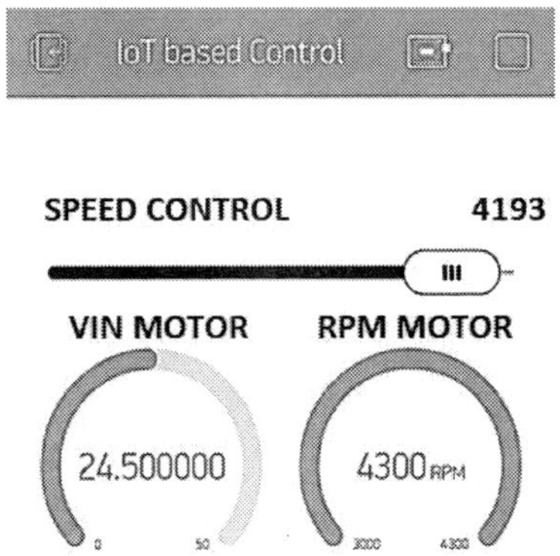

Figure 13. Monitoring using Blynk App

4. Conclusion

The IoT based DC Motor control system which is funded by PV 50Wp was built and tested. This paper describes the development of DC Motor control based on the IoT and the PID uses a Solar Module. The results of the testing tool have reached the motor speed according to the user's wishes. Motor speed is affected by the motor input voltage. When the motor input voltage is lower than the reference voltage, then the voltage automatically increase until matches the reference voltage. In additional result, the relationship between EMF with Speed dc motor is directly proportional. The results of monitoring also work well.

In the future, this project will be evaluated, and continue to be developed so that the project reaches its maximum. In addition, we will develop the certain protocol for the flat topology of n number nodes which places on the remote area. This system will be used for controlling and monitoring a large number of dc pump fed by Photovoltaic system.

5. References

[1] Phung M D, De La Villefromoy M and Ha Q 2017 *N Management of solar energy in microgrids using IoT-based dependable control* (International Conference on Electrical Machines and Systems) pp 1-6
[2] Choi C.-S, Jeong J.-D, Han J, Park W.-K and Lee I-W 2017 *Implementation of IoT based PV monitoring system with message queuing telemetry transfer protocol and smart utility network* (International Conference on Information and Communication Technology Convergence) pp 107-179
[3] Flores-Moran E, Yanez-Pazmino W and Barzola-Monteses J 2018 *Genetic algorithm and fuzzy self-tuning PID for DC motor position controllers* (19th International Carpathian Control Conference) pp 162-168
[4] Jing J, Wang Y and Huang Y 2016 *The fuzzy-PID control of brushless DC motor* (International Conference on Mechatronics and Automation) pp 1440-1444
[5] Mohamed A A S, Berzoy A and Mohammed O 2015 *Control parameters optimization for PM DC motor in photovoltaic applications* (International Electric Machines and Drives Conference) pp 1742-1745
[6] Ayop R. and Tan C W 2018 *Design of boost converter based on maximum power point resistance for photovoltaic applications* (Solar Energy vol 160) pp 322–335
[7] Elsrogy W M, Fkirin M A and Hassan M A M 2013 *Speed control of DC motor using PID controller based on artificial intelligence techniques* (International Conference on Control, Decision and Information Technologies)
[8] Viola J, Angel L and Sebastian J M 2017 *Design and robust performance evaluation of a fractional order PID controller applied to a DC motor* (Journal of Automatica Sinica vol 4(2)) pp 304–314
[9] Liu X and Sanchez-Sinencio E 2015 *A Highly Efficient Ultralow Photovoltaic Power Harvesting System With MPPT for Internet of Things Smart Nodes* (IEEE Transactions on Very Large Scale Integration (VLSI) Systems vol 23(12)) pp 3065–3075
[10] https://www.precisionmicrodrives.com/content/ab-021-measuring-rpm-from-back-emf/
[11] https://www.designnews.com/electronics-test/pic-microcontrollers-provide-2-cost-effective-methods-brushed-dc-motor-control/23988528347199
[12] Handson Technology.ESP8266 NodeMCU WiFi Module User Manual.
[13] http://huayangtech.manufacturer.globalsources.com/si/6008840396785/pdtl/Submersible-pump/1160284703/Submersible-Pump-24V-DC.htm

IOP Conf. Series: Materials Science and Engineering **588** (2019) 012018 doi:10.1088/1757-899X/588/1/012018

Effect of Vacuum Level on Heating Speed on Ethanol Destillation Process Using Coaxial Vacuum Tube Solar Collector

Sutomo[1], W Nuriana[1], Sutrisno[1] and J Triono[2]

[1] Mechanical Engineering, Merdeka University of Madiun, Indonesia
[2] Informatic Management, Merdeka University of Madiun, Indonesia

Abstract. Vacuum tube collectors that could produce heat up to a temperature of 80°C could be used to process ethanol distillation, because the evaporation process of ethanol only requires a temperature of 79°C at 1 atm. Vacuum tube collectors in terms of many types of heat absorbers, each type had a different heat absorption ability. This study will review the performance of the coaxial vacuum tube collector used for the process of ethanol distillation. The study was carried out in space using halogen lamps as solar radiation simulators. In the study conducted with three radiation variations, namely 998 W/m^2, 878 W/m^2, and 782 W/m^2. The pressure / vacuum level of the collector tube cavity varies from 0.043 atm, 0.179 atm, 0.315 atm, 0.5atm, and 1 atm. Ethanol research material with 30% content using water as a distillation cooling medium. The results of this research could be known the effect of the vacuum level on the speed of achieving temperature in the distillation process using a solar collector coaxial vacuum tube.

1. Introduction

The most abundant renewable energy source in the area around the equator was solar thermal energy. The technology applied to harvest eye heat energy today had so far two types of technology, namely photovoltaic solar energy technology and thermal solar energy technology (Agus, 2012). The vacuum tube collector, which was one of the thermal solar energy technologies, was very possible to be developed in the region around the equator because the intensity of the radiation was abundant and the technology was simpler.

Shah et al. (2004), stated in theory that in vacuum tube collectors the effect of the shadow from the adjacent tube and the angle of arrival of the beam radiation could affect the collector's performance. Vacuum tube collector (vacuum tube collector) was one of the various types of heat collector collectors, which had an area of absorption (absorber) which was effective, because in the position of the sun's orbit 2 where the sunlight was perpendicular to the absorber the percentage was the same, to prevent heat loss which had been absorbed using vacuum air insulators which had low conductivity.

According to Kim et al. (2007), vacuum tube solar collectors had high heat extraction efficiency, compared to flat plate collectors in a temperature range above 80 ° C, because the collector had a combined effect of selective surface coating and vacuum insulation that protects the absorbent element. In addition to solar thermal energy sources which were renewable energy, ethanol was also a potential source of energy to be developed. Ethanol could be produced from carbohydrates contained in plants, (cassava, corn, sorghum, etc.) or molasses (waste of sugar production). Agricultural crops were considered renewable because they get energy from the sun through photosynthesis. In the

IOP Conf. Series: Materials Science and Engineering **588** (2019) 012018 doi:10.1088/1757-899X/588/1/012018

combustion process, ethanol could increase the efficiency of combustion, especially in Otto motors because it contains approximately 35% oxygen and had a higher octane number (112), compared to gasoline whose octane value was less than 90 (Rahayu et al., 2010).
Ethanol could be produced in several ways, one of which was through biological processes by fermentation, which could produce ethanol with low levels (10-20%). To increase ethanol levels could be done by separating the water contained in it, which could be done by distillation. Distillation was the separation of components based on their boiling point. The boiling point of pure ethanol was 79 ° C, boiling point of water was 100 ° C under standard conditions (Riyan et al., 2012).
Looking at both energy potential and utilization, it was possible to combine the two renewable energy sources. This study aims to determine the effect of the vacuum on the collector tube on the speed of achieving ethanol evaporation temperature in the ethanol distillation process using a coaxial vacuum tube collector.

2. Research Method

Data was taken from the beginning of the test, due to the nature of ethanol which could evaporate at room temperature. Data obtained from the test include radiation intensity, collector tube pressure, temperature in the glass tube, absorber temperature, ethanol temperature in and out of the collector and ambient temperature. Data retrieval was done every minute for approximately 90 minutes in each test.
Coaxial vacuum tube collector for ethanol distillation in principle utilizes heat trapped in the collector tube for evaporation. To prevent heat loss trapped in the collector, use a vacuum air insulator. The ethanol that passes through the collector was heated and evaporation occurs, then the ethanol vapor was condensed for condensation. Collector heating was carried out with variations in radiation intensity of 998 W/m^2, 878 W/m^2 and 782W/m^2. Collector pressure level when heating varies 1 atmosphere, 0.5 atmosphere, 0.315 atmosphere, 0.179 atmosphere and 0.043 atmosphere.

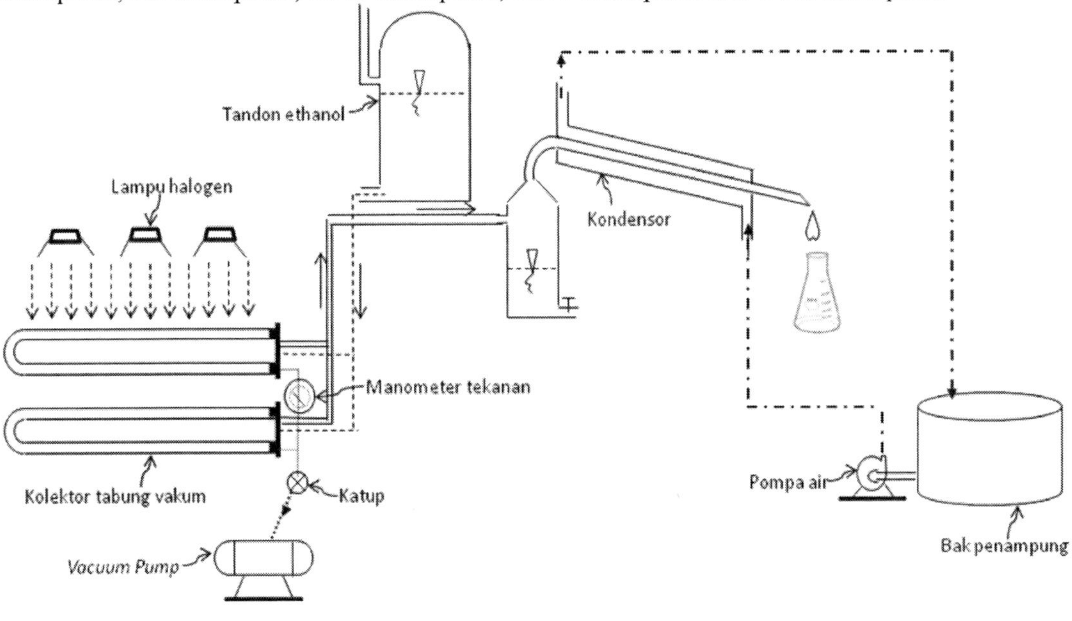

Figure 1. Schematic of a solar collector coaxial vacuum tube

3. Result and Discussion

The temperature of ethanol evaporation at each change in collector pressure was shown in Figure 2 for radiation intensity 782 W / m2, figure 3 for radiation intensity 878 W / m2 and figure 4 for radiation intensity of 998 W / m2. The picture shows, the lower the pressure on the collector the faster the evaporating temperature of ethanol was reached.

Figure 2 through 4 show a graph of the temperature achievement and ethanol heating time in the distillation process using a coaxial vacuum tube collector. The results show in the same heating period in the collector tube that the pressure was lower the temperature was higher. In this study ethanol heating was carried out at ethanol evaporation temperature which ranged from 79°C to 83°C and maintained during testing. The study carried out three variations of radiation intensity, namely 782 W/m^2, 878 W/m^2 and 998 W/m^2 with 1 atmospheric collector pressure, 0.5 atmosphere, 0.315 atmosphere, 0.179 atmosphere and 0.043 atmosphere. Figure 5 shows the time to achieve the ethanol evaporation temperature heated in the collector, at each pressure level of each radiation intensity.

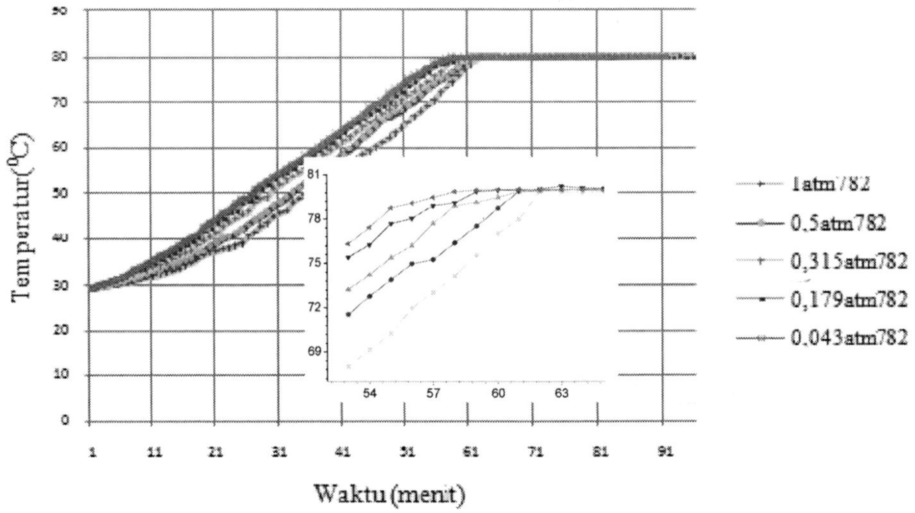

Figure 2. Evaporation temperature at a radiation intensity of 782 W/m^2.

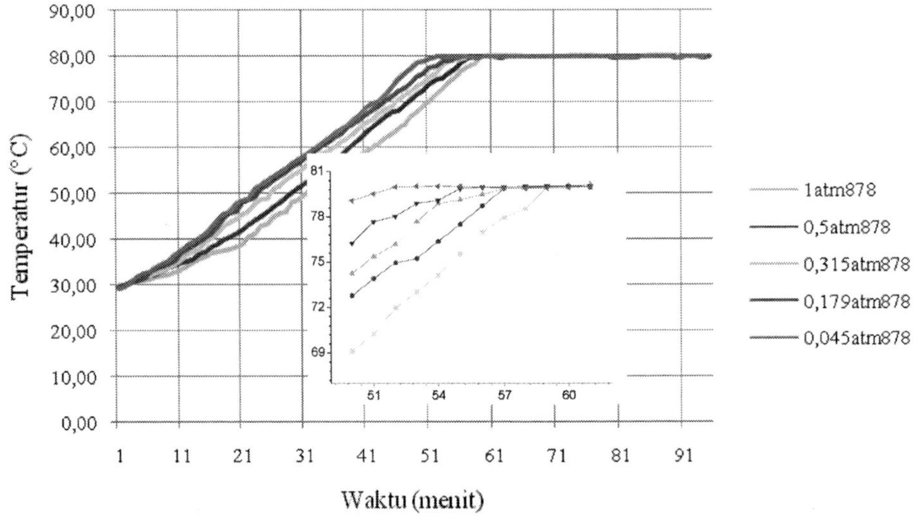

Figure 3. Evaporation temperature at radiation intensity of 878 W/m^2.

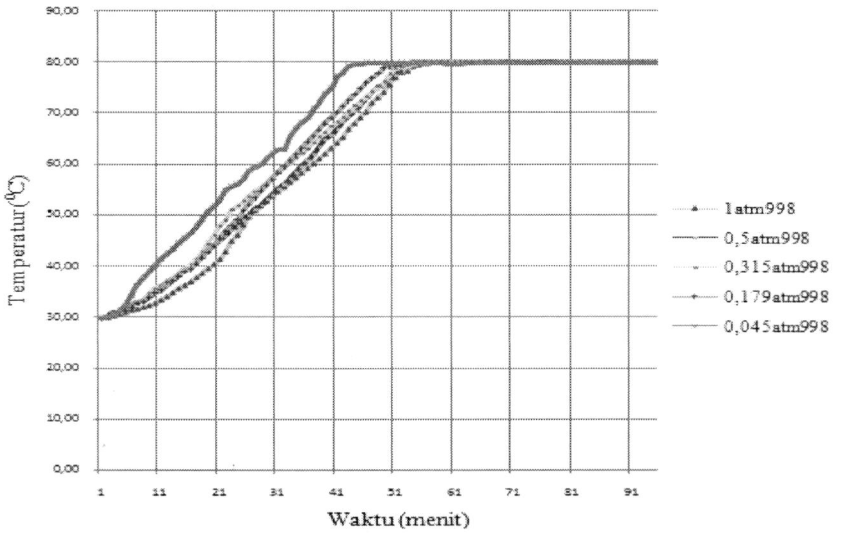

Figure 4. Evaporation temperature at a radiation intensity of 998 W/m^2.

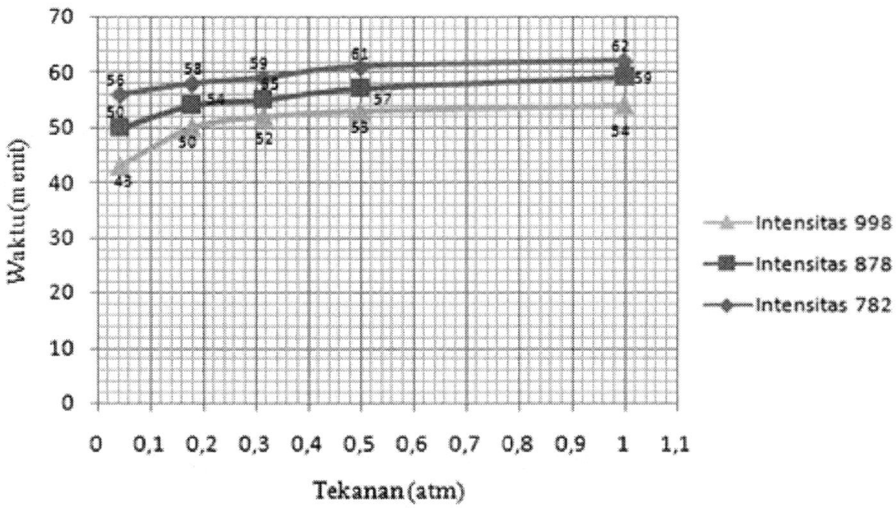

Figure 5. Time to achieve evaporation temperature.

Figure 5 shows that of the three radiation intensities, when the ethanol evaporation temperature was reached when the heating occurred in the collector showed that, the lower the collector pressure the faster the ethanol evaporation temperature was reached.

4. Conclusion

Based on the results of testing carried out on the ethanol distillation process using a solar collector coaxial vacuum tube showed that of the three radiation intensities, when the ethanol evaporation temperature was reached when the heating in the collector showed that, the lower the collector pressure the faster the ethanol evaporation temperature was achieved.

5. Acknowledgement

The research was fully funded by the Ministry of Research and Technology of Higher Education of the Republic of Indonesia on PTUPT scheme in the year 2018.

IOP Conf. Series: Materials Science and Engineering **588** (2019) 012018 doi:10.1088/1757-899X/588/1/012018

References

[1] Agus Tri Putra ID G 2012 Analysis of the use of solar energy into thermal and electrical energy in a steam engine *Jurnal Logic* **Volume 12** nomor 1

[2] Budihardjo I and G L Morrison 2009 Performance of water-in-glass evacuated tube solar water heaters *Science Direct, Solar Energi* **83** 49-56

[3] Duffie J A and Beckman W A *Solar Engineering of Thermal Processes* (New York :A Willey-Intrscience Publication)

[4] Gao Y , R Fan, X Y Zhang, Y J AN, M X Wang, Y K Gao, and Y Yu 2014 Thermal performance and parameter analysis of a U-pipe evacuated solar tube collector *Science Direct, Solar Energi* **107** 714-727.

[5] Glembin J, G Rockendorf and J Scheuren 2010 Internal thermal coupling in direct-flow coaxial vacuum tube collectors *Science Direct, Solar Energi* **84** 2010 1137-1146.

[6] Khopkar SM 2003 *Basic Concepts of Analytical Chemistry (*Jakarta : Universitas Indonesia)

[7] Marco Ricci, Enrico Bocci, Emanuele Michelangeli, Andrea Micouldgeli, Mauro Villarini, and Vincenzo Naso 2015 Experimental Tests of Solar Collectors Prototypes Systems", *Science Direct, Energi Procedia* **82** 744 – 751

[8] Michel Hayek, Johny Assaf, and William Latief 2011 Thermal analysis of evacuated solar tube collectors *Science Direct, Energi Procedia Energi* **6** 618-626.

[9] Morrison G L, I Budihardjo and M Behnia 2004 Water-in-glass evacuated tube solar water heaters *Solar Energi* **Vol.76** Issues 1–3 pp.135-140

[10] Mulyono 2010 Cavity Optimization for variations in the degree of vacuum as an insulator *Jurnal Teknik Industri* volume 11 no.1

[11] Kretzschmar, Sebastian Herrmann and Matthias Kunick, "*Zittau's Fluid Property Calculator*", Department of Technical Thermodynamics Faculty of Mechanical Engineering, Zittau/Goerlitz University of Applied Sciences

[12] Riyan Nazarudin Primy, Agus Pambudi and Pradita Anggit 2012 Synthesis and Characterization of Fe2O3 / CaF2 solids *Jurnal Sain dan Seni ITS* **Vol. 1** No.1

[13] Robert H. Perry and Don W Green 2007 *Perry's Chemical Engineers' Handbook Eighth Edition* (New York : McGraw-Hill)

[14] Ruobing Liang, Liangdong Ma, Jili Zhang and Dan Zhao 2011 Theoretical and experimental investigation of the filled-typeevacuated tube solar collector with U tube *Science Direct Solar Energi* **85** 1735–1744

[15] L J Shah and S Farbo 2004 Vertical evacuated tubular collector utilizing solar radiation from all direction *Science Direct Applied Energi* **78** 371-395.

[16] Winter Ed 1990 *Solar collectors, energi and materials* (Cambridge MA : MIT Press)

[17] Yong Kim and Teobeom Seo 2007 Thermal performances comparisons of the glass evacuated tube solar collectors with shapes of absorber tube *Science Direct Renewable Energi* **32** 772-795.

[18] Xinyu Zhang , Shijun You , Hongchuan Ge , Yan Gao , Wei Xu , Min Wang, Tao He and Xuejing Zheng 2014 Thermal performance of direct-flow coaxial evacuated-tube solarcollectors with and without a heat shield *ScienceDirect Energi Conversion and Management* **84** 80–87.

IOP Conf. Series: Materials Science and Engineering **588** (2019) 012019 doi:10.1088/1757-899X/588/1/012019

Numerical study of the effect of oxidation zone inlet air temperature variation on municipal solid waste pellet gasification process on downdraft type reactor characteristics

B Sudarmanta[1], D Hidayatulloh[1], D R Ependi[1], A R Saleh[1,2], A Setiyawan[1] and A Bachtiar[1]

[1]Department of Mechanical Engineering, Institut Teknologi Sepuluh Nopember, Indonesia
[2]Department of Mechanical Engineering, Pasir Pangaraian University, Indonesia

bsudarmanta@gmail.com

Abstract. Gasification is an efficient technology to extract energy from MSW biomass into energy in the form of syngas. One of the most influential paramerers to improve syngas quality is temperature, so the development of the design and operation of the downdraft type reactor is done by heating the inlet air of the oxidation zone through an external heater. Research with numerical studies was carried out on the effect of the inlet air in the oxidation zone to determine the temperature distribution and syngas composition along the reactor. Based on experimental research, heating the inlet air of the oxidation zone with external heaters in this study was carried out with 5 temperature variations, namely 80 °C, 110 °C, 150 °C, 180 °C, and 200 °C. The modeling used is the standard model k-epsilon, Radiation P1, the transport species model with turbulence used is finite-rate / eddy-dissipation, and Discrete Phase Model (DPM). The results obtained from this study were that the highest air temperature was obtained at 200°C at 1004 °C in the oxidation zone. The composition of syngas CO, H_2, and CH_4 at temperature increases of 80-200 °C increased from 21.64%, 9.24%, and 2.86% to 22.65%, 10.49%, and 3.10%. Then increasing the LHV syngas from 4,757.20 kJ/m3 to 5,106.02 kJ/m3. Cold gas efficiency increased from 72.17% to 79.31%.

1. Introduction

Increased energy consumption occurs every year in the 2000-2014 period. The average annual growth for the period 2000-2014 was 3.99% in 2014. Besides that, oil production in Indonesia since 2010-2014 decreased by around 4.41% [1]. If this is not considered, it will lead to a serious energy crisis, so that efforts are needed to cover the energy crisis by developing new renewable energy as an alternative energy.

Waste management with technology has great potential to become renewable energy. This is because the amount of landfill increase in Indonesia itself is known to have reached 175,000 tons/day or equivalent to 64 million tons/year [2]. The technology that makes it possible to convert waste into renewable energy is by using a thermochemical process, [3]. Municipal Solid Waste (MSW) is a solid waste produced by urban communities, obtained from landfills (TPA). The latest method of processing MSW is by turning it into a pellet shape. Changing the shape into pellets has the advantage of reducing moisture content where the moisture content of MSW pellets is more homogeneous than not being

IOP Conf. Series: Materials Science and Engineering **588** (2019) 012019 doi:10.1088/1757-899X/588/1/012019

converted into pellets [4].

Experimental research on the gasification process using downdraft type reactors has been carried out by Depi Rustam [5], Aji Triyanto [6], Rizki Wase [7] and Irfan Maulana [8] in 2017 with MSW biomass raw material. The study was carried out by [5] with gasification characterization which showed the effect of inlet air heating on the oxidation zone with external heating on the temperature distribution of the reactor, especially the oxidation zone. The study showed that the presence of air heating from 80 °C to 200 °C was able to increase the temperature of the oxidation zone from 818 °C to 942 °C and improve the quality of syngas, which increased LHV syngas from 4227.99 kJ/m^3 to 4609.99 kJ/m^3.

Gasification research has been carried out numerically before by several researchers, namely [9], who have also examined the simulation of rice husk gasification on downdraft type glue. The study only observed the 2-dimensional plane of the reactor and varied the AFR which could maximize the efficiency of the gasification process. Next is the design of a rice husk gasification reactor in a downdraft type reactor, developed by [10] with a simulation of downdraft type Computational Fluid Dynamics (CFD) gas with rice husk biomass. The study observed in 3 dimensions with variations in throat angle and inlet air. Then based on the results of the research [5] it is known that the addition of an external air heater is needed to be able to achieve the desired optimum temperature.

Therefore, in this study modeling and simulation will be carried out with variations in the addition of an external air heater before the air enters the internal heater which will be supplied to the downdraft reactor type oxidation zone with MSW pellet raw material. In this study, the gasification process that occurs in the reactor will be numerically modeled using the standard k-epsilon, Radiation P1, the transport species model with turbulence used is finite-rate/eddy-dissipation, and Discrete Phase Model (DPM). The data from the experimental research will be used as a validation of the results of gasification research using numerical models.

2. Formatting the title, authors and affiliations

Modeling and simulation in this study using CFD software. The steps taken are:

2.1. *Pre-processing*

This stage includes several sub-stages, namely making geometry, making meshing, and determining the modeling domain.

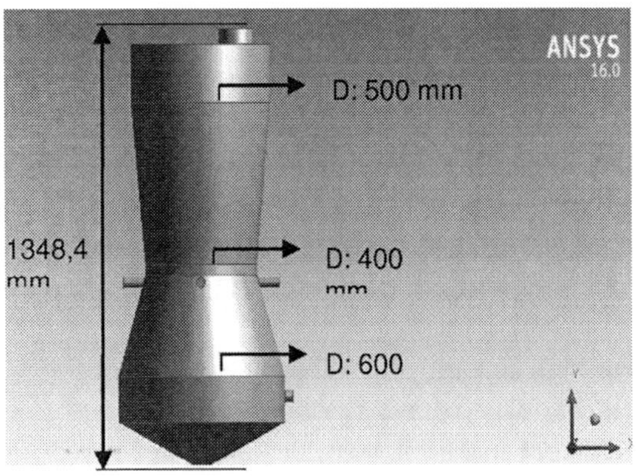

Figure 1. Reactor design

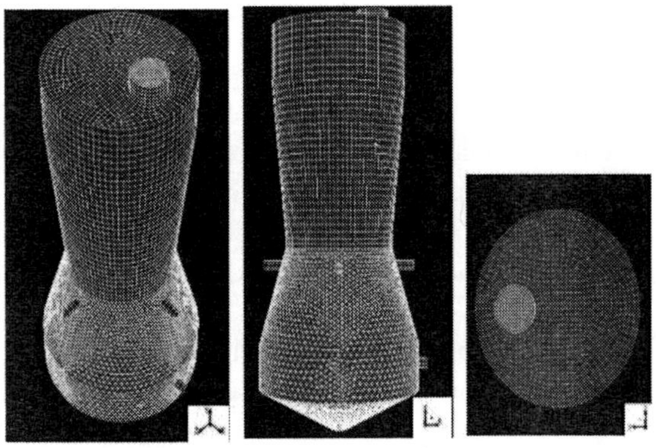

Figure 2. Meshing on the reactor

IOP Conf. Series: Materials Science and Engineering **588** (2019) 012019 doi:10.1088/1757-899X/588/1/012019

Table 1. Basic characteristics of the pellet MSW

Proximate analysis	(% wt)
Moisture content	9,82
Ash content	14,71
Volatile matter	65,78
Fixed carbon	9,69
Ultimate analysis	(% wt)
C	39,83
H	6,7
O	38,11
N	0,35
S	0,14
HHV	13.84 MJ/kg

Table 2. Boundary condition settings used in modeling

Boundary Condition	Type	Value
Air inlet	Mass flow inlet	$\dot{m}$: 0,003032 kg/s T : 80,110,150,180,200)°C
Biomass Inlet	Mass flow inlet	0,00309 kg/s
Syngas Outlet	Pressure-outlet	
Reactor	Wall	Material : Steel

2.2. Processing

Processing is setting up the Ansys Fluent 16 software. Some of the settings that will be carried out include models, materials, boundary conditions.

a) Models
The P1 radiation model was used in this study [11]. The finite rate/Eddy dissipation reaction model is used [12]. The viscous turbulence model used is the k-epsilon model, [12]. The reaction model in this study was taken from several sources summarized in table 1.

b) Materials
The input value for ultimate and proximate analysis is shown in Table 1.

c) Boundary Condition
In this study, boundary conditions are attached to Table 2

2.3. Post-processing

Observations will be made on each variation. Qualitative data in the form of contours taken in vertical sections (z-center reactor and x-center reactor) and horizontal cross section on each reactor section y (drying zone, pyrolysis zone, partial combustion zone, reduction zone)

3. Results and Discussion

Characteristics of the gasification process in a study, one of which was obtained through variations in air temperature in the oxidation zone inlet. Five variations of inlet air temperature were carried out with numerical methods in this study, namely: 80 °C, 110 °C, 150 °C, 180 °C, and 200 °C. Through proximate and ultimate test data, the composition of each element (C, H, O, N, S) and the content of moisture, ash, fix carbon, volatile, so that coal particles as input for experimental testing can be assumed as MSW particles.

3.1. Analysis of Temperature Distribution Throughout the Reactor

Temperature distribution along the reactor is shown to determine the effect of air heating temperature. The temperature value of each zone is taken using the Report-Surface integral-Area weighted average menu in the thermocouple position that corresponds to the experimental method. The Area weighted average menu is used to take the average temperature value in the Y coordinate section. To determine the boundaries between zones in the gasification process, the graph in Figure 3 is obtained with 12 starting points taken at reactor height: 0.2 m, 0.3 m, 0.4 m, 0.5 m, 0, 6 m, 0.7 m, 0.8 m, 0.9 m, 1 m, 1.1 m, 1.2 m, and 1 , 3 m.

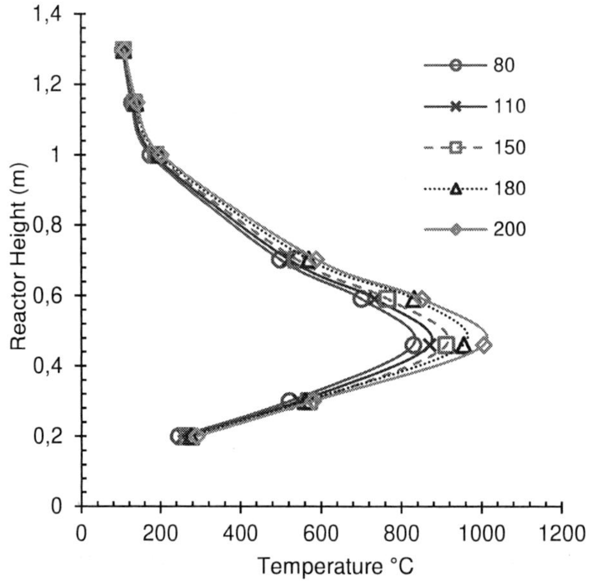

Figure 3. Temperature distribution along gasifier of each variations

Table 3. The reactions involved in the gasification

Reaction		ΔH (kJ/kmol)	A	E (J/kmol)	Source
$C(s) + 0.5O_2(g) \rightarrow CO(g)$	R1	-110,5	0,052	$6,1\times10^7$	[16]
$C(s) + CO_2(g) \rightarrow 2CO(g)$	R2	+172	0,00732	$1,125\times10^8$	[16]
$C(s) + H_2O(g) \rightarrow CO(g) + H2(g)$	R3	+131,4	0,00782	$1,15\times10^8$	[16]
C (s)+ 2H2 (g)→ CH4(g)	R4	-75	0.12	1.7×10^7	[17]
CO(g)+ 0.5 O2(g) → CO2(g)	R5	-283,1	$2,2\times10^{12}$	$1,67\times10^8$	[18]
H2(g)+ 0.5 O2(g) → H2O (g)	R6	-242	$6,8\times10^{12}$	$1,68\times10^8$	[19]
CH4(g) + 0.5 O2 (g)→CO(g) + 2H2 (g)	R7	-36	1×10^{15}	1×10^8	[14]
CO(g) + H2O (g)→ CO2(g) + H2 (g)	R8	-41	$2,75\times10^{10}$	$6,1\times10^7$	[17]
2CO(g)+ 2H2 (g) → CH4(g) + CO2 (g)	R9	-247	1×10^{10}	1×10^8	[14]

Based on Figure 3, it can be seen that the temperature in the reactor for each scenario tends to have a uniform pattern. At an altitude of 104 cm from the bottom of the reactor, the measured temperature is 150 °C, indicating that the height is at the boundary between the drying zone and the pyrolysis zone for the addition of an air temperature of 80 °C [13]. The addition of air temperature at 110 °C, 150 °C, 180 °C, and 200 °C, resulted in a shift in the drying zone at an altitude of ± 110 cm at a temperature of 150

°C. So that the boundary between the drying zone and the pyrolysis zone with the addition of an inlet air temperature of 80 - 200 °C is at an altitude of 104 cm to 110 cm. In the drying zone the water content in the fuel is evaporated into steam. The closer the temperature is 150 °C, the drying process will get better.Menurut [14] batas zona pirolisis akan terjadi pada kisaran suhu 150-700 °C. It can be seen that figure 3, the addition of air temperature variation 80 - 200 °C shows that in the height range 60 - 65 cm is the limit of the pyrolysis zone, which produces products in the form of hydrogen, carbon monoxide, carbon dioxide, hydrocarbons, and char.

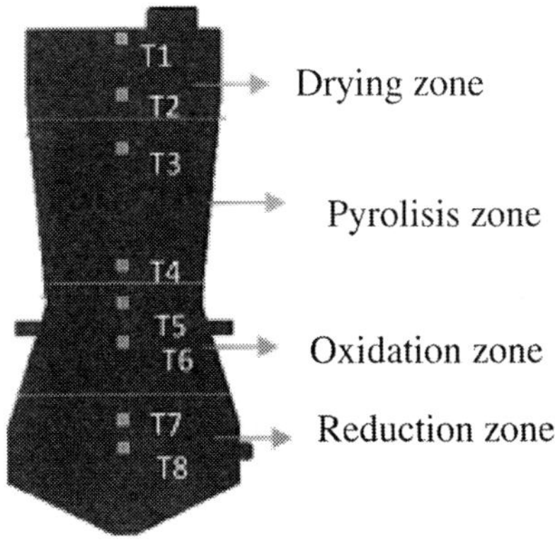

Figure 4. Zone boundaries in a downdraft reactor with a variation of 200 °C air temperature and T1-T8 thermocouple position

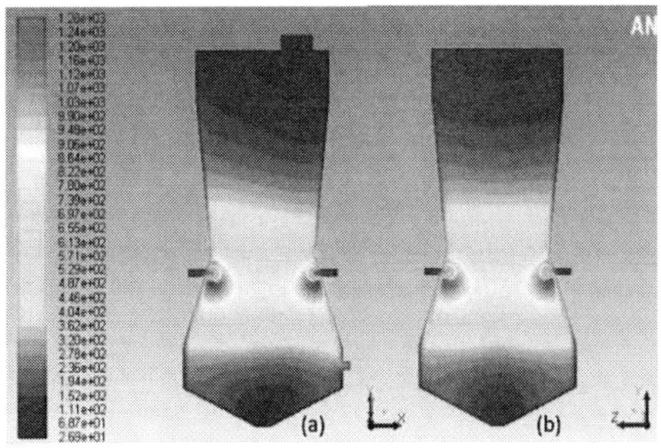

Figure 5. Temperature distribution contour at inlet air temperature variations 200 °C

At an altitude of 50 cm with air temperature variations; 80 - 200 °C, producing temperatures at the bottom of the reactor in a row of value; 830 - 1004 °C. These values show the maximum temperature in the partial oxidation zone. According to [14] the oxidation zone will occur in the temperature range of 700-1500 °C, so that the partial oxidation temperature in this study is appropriate because it is still in that range. In the partial oxidation zone exothermic reactions occur (char combustion, partial oxidation, hydrogen combustion) which produce carbon dioxide (CO2), carbon monoxide (CO), and water vapor (H2O) products. According to [15], the reduction zone operates at a temperature range of 400-700 °C. That way you can know the boundary between the partial oxidation zone and the reduction zone for air temperature variations; 80 - 200 °C, respectively in the height range; 41 - 37 cm. The reduction zone is shown at a height of 30 cm with air temperature variations; 80-200 °C, producing temperatures in a row valued at 520 - 580 °C. Thus the large maximum zone of the gasification process in the reactor is known, namely at a variation of 200 °C air temperature, among others; drying zone at an altitude between 110-130 cm, pyrolysis zone at an altitude between 65-110 cm, oxidation zone at an altitude between 37-65 cm, and a reduction zone at altitudes between 20-37 cm.

The boundaries between zones in the gasification process based on this numerical method can evaluate and provide recommendations from the experimental method to determine the position of the thermocouple with the same reactor. Thermocouple placement in the experimental method include: T1 = 1.30 m; T2 = 1.14 m; T3 = 1 m; T4 = 0.7 m; T5 = 0.6 m; T6 = 0.5 m; T7 = 0.3 m; T8 = 0.2 m. In Figure 4 shows that the placement of thermocouples in the drying zone for T1 is appropriate because to measure the temperature at the end of the reactor. The recommendations for the T2 position for experimental research are then transferred to a height of 1.2 m, where the position can be representative of the temperature in the drying zone. In the pyrolysis zone of T3 and T4 it is recommended to change the position of the thermocouple at an altitude of 0.875 m, where it can represent the pyrolysis zone temperature. T5 and T6 can be recommended for placing a thermocouple at a height of 0.51 m as a temperature alternative to the oxidation zone. T7 at a height of 0.3 m can already be used as a representative temperature in the reduction zone. While T8 is known as a temperature representative in the syngas outlet zone.

The results in the drying and pyrolysis zones in Figure 5 show that the temperature distribution is less symmetrical in the Z coordinate section as shown in Figure 5 (a). The indications that cause this asymmetry are the reactor design factors. In the drying zone, the biomass input geometry is in an asymmetrical position in the middle plane z and has a diameter that is different from the reactor wall diameter, so that the temperature distribution is not evenly distributed throughout the reactor wall. If the reactor design is replaced by making the position of the biomass input in the middle of the reactor indicated to produce a temperature difference will be very small.

Tabel 4. Syngas composition from the results of numerical methods

T_{air} (°C)	Syngas composition					
	CO (% vol)	H_2 (% vol)	CH_4 (% vol)	CO_2 (% vol)	N_2 (% vol)	O_2 (% vol)
80	21,64	9,24	2,86	2,62	61,6	2,04
110	21,8	9,45	2,9	2,3	61,6	1,94
150	22,12	9,87	2,98	1,87	61,5	1,67
180	22,43	10,28	3,05	1,52	61,4	1,32
200	22,65	10,5	3,1	1,34	61,4	1,01

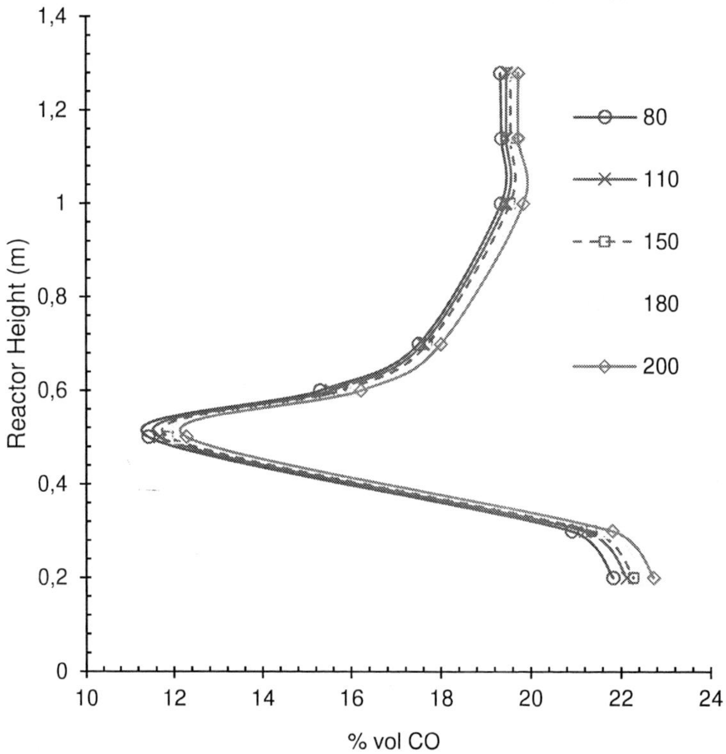

Figure 6. Average distribution of volumetric percentages CO along the reactor for each variation

IOP Conf. Series: Materials Science and Engineering **588** (2019) 012019 doi:10.1088/1757-899X/588/1/012019

3.2 Analysis of flammable syngas composition

Based on the results of Table 5 the composition of flammable syngas generated by the average numerical method has increased. In CO compounds, an increase of 1.01% was obtained from the inlet air temperature of 80 °C to 200 °C. Similarly, H_2 and CH_4 compounds obtained increased by 1.26% and 0.24%. This proves that if the inlet air temperature is increased, it will improve the quality of syngas by increasing the composition of flammable syngas, namely CO, H_2, and CH_4. The largest non-flammable syngas composition is N_2, because the gasifying agent used is free air containing O_2 and N_2 compounds. Values in CO_2 and O_2 compounds tend to be small and decrease from 80 °C to 200 °C inlet air temperature. It can be indicated that the small value of the CO_2 compound is reduced by Boudouard reaction and the reversible reaction of R8 in the reduction zone and the faster the reaction as the inlet air temperature is added. O_2 compounds look small, indicating that the gasification process works optimally. Decreasing O_2 composition in syngas as an effect of increasing the inlet air temperature of the oxidation zone shows that the increase in air temperature makes O_2 more reactive

3.2.1 *Volumetric Analysis of CO Distribution.* Temperature comparison chart along the reactor with an average percentage of volumetric CO for 5 variations in the addition of inlet air temperature, each section of Y coordinate with the thermocouple position according to the experimental method (T1-T8).

When viewed from a graph of the average distribution of CO volumetric percentages shown in Figure 6, CO compounds in the drying zone are not significant. This is because the modeling used by the reactions in table 1 in this study did not occur in the drying zone. According to [15] to find out the drying process must be modeled with the multiphase model and use UDF (user define function) on the fluent. However, this study did not use the multiphase and UDF models, so that the drying process in the reactor could not be seen.

The pyrolysis zone is quite suitable because of the devolatilization process which breaks down the dried MSW pellets into several gas products, one of which is CO gas. In Figure 6 the percentage value of volumetric CO in the pyrolysis zone (T3 and T4) is less than 20% where in the range of zones the CO produced product is 10-20 mole%, (Basu, 2012). Then in the partial oxidation zone (T5 and T6) by increasing the air temperature from 80-200 °C, the percentage value of CO volumetric decreases. This is because CO reacts with oxygen (R5) to produce CO_2 ($CO (g) + 0.5O_2 (g) \rightarrow CO_2 (g)$). Besides that in this zone CO can also be produced through partial combustion (R1), so that the product of this partial combustion will be continued with a reaction in the reduction zone. The partial combustion reaction ($C(s) + 0.5O_2 (g) \rightarrow CO (g)$) which produces this CO product is a slow reaction [2]. If we review in more detail the air temperature variation of 200 °C, the difference in the percentage value of volumetric CO between T5 and T6 is 3.92%, which is greater than the air temperature variation of 80 °C which is 3.85%. This shows that the partial oxidation process is more reactive with higher air temperature, so the reaction in the zone will be faster.

The significant increase in CO volumetric percentage value from T6 to T7 shows that in the reduction zone many produce CO. This is because there are 3 reactions that produce CO, which are (1) boudouard reaction (R1) which is to change the char that reacts with CO_2 to CO, (2) the water-gas reaction (R3) which is to change the char that reacts with H_2O to CO and H_2 , (3) steam reforming reaction (R7) which converts CH_4 to CO and H_2 because it reacts with O_2. While at T8, CO volumetric percentage value also increased but not too significant

3.2.2 *Volumetric Analysis of H_2 Distribution.* If viewed from the graph of the average distribution of volumetric H_2 percentage, the drying zone of H_2 gas is not significant. This is because the modeling reactions used in table 1 do not occur in the drying zone. In the pyrolysis zone there is a devolatilization process which breaks the dried MSW pellets into several gas products, one of which is H_2. This is appropriate, because up to a temperature of 600 °C H_2 gas will experience an increase due to the shift reaction, and will decrease when above 600 °C [14].

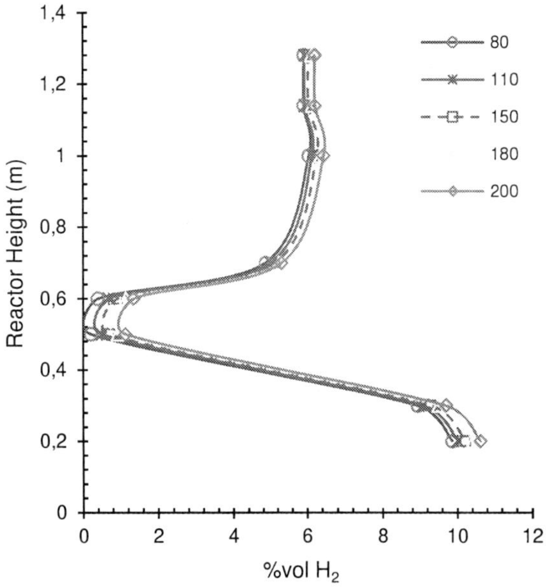

Figure 7. Average distribution of volumetric percentages H_2 along the reactor for each variation

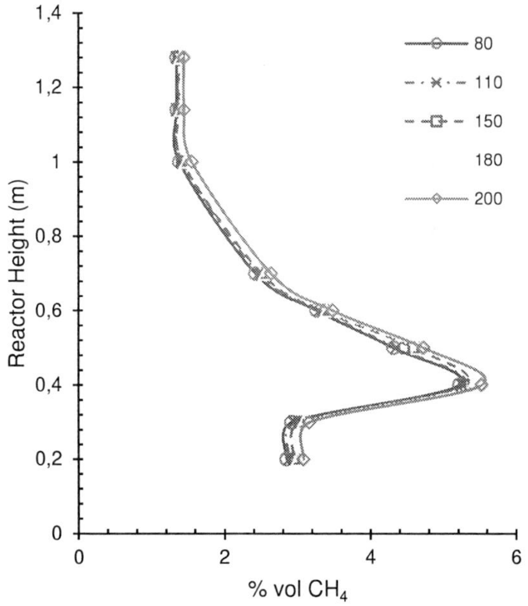

Figure 8. Average distribution of volumetric percentages CH_4 along the reactor for each variation

Then in the partial oxidation zone (T5 and T6) by increasing the air temperature from 80 to 200 °C, the percentage value of volumetric H_2 decreases. This happens because H2 reacts with oxygen (R6) to produce H_2O. H_2 also decreases to the reduction zone limit, because it reacts with CO (R9) so that it forms CH_4 and CO_2. If viewed in more detail from Figure 7 variations in air temperature 200 °C the difference in volumetric H_2 percentage value between T5 and T6 is 0.21% greater than the air temperature variation of 80 °C which is 0.18%. This shows that the H_2 combustion process is more reactive with the addition of a higher air temperature, so the reaction in the zone will be faster. A significant increase in the volumetric percentage value of H_2 from T6 to T7 indicates that the reduction zone produces a lot of H_2. There are two reactions that produce H_2 in the reduction zone. The first is the shift reaction (R8) which converts CO that reacts with H_2O to H_2 and CO_2. The second reaction is water-gas (R3) which converts the char that reacts with H_2O to CO and H_2. The third reaction is metane gasification (R7) which converts CH_4 to CO and H_2 because it reacts with O_2.

3.2.3 *Volumetric Analysis of CH_4 Distribution* . Adding T9 point at 0.4 m height to find out more details of CH_4 percentage volumetric average distribution. When viewed from the average distribution of volumetric CH_4 percentage shown in Figure 4.12, the CH_4 gas drying zone is not significant. This is because in this study with the modeling used the reactions in table 1 did not occur in the drying zone. In the pyrolysis zone, CH_4 increases. This is because in the pyrolysis zone there is a devolatilization process which breaks the dried MSW pellets into several gas products, one of which is CH_4. CH_4 compounds also increase to the limit between the partial oxidation zone and the reduction (T9). This is because at the boundary between these zones CH_4 is produced from methanation reaction (R9). That is the reaction of CO and H_2 which produces CH_4 and CO2.

3.3 *Analysis of Cold Gas Efficiency Gasification Process*

Figure 9 shows that heating the air temperature from 80-200 °C can increase cold gas efficiency by 7.34% with a relatively linear increase. This increase was not only influenced by the increase in LHV in syngas, but also influenced by the mass flow rate of biomass (MSW pellet) and the syngas mass flow rate.

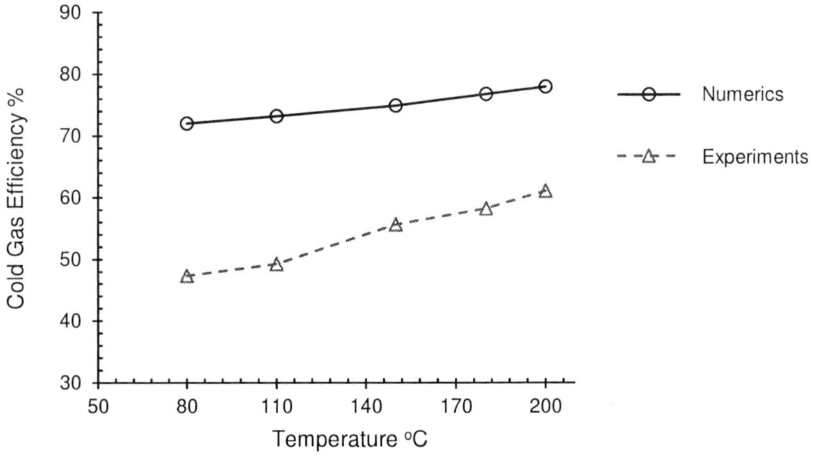

Figure 9. Effect of air temperature on cold gas efficiency

Research conducted by experimental methods shows that with an increase in air temperature from 80-200 °C able to increase cold gas efficiency can increase by 14%, from 47% to 61%. This difference is quite large because the increase is not only affected by an increase in LHV in syngas, but also influenced by the mass flow rate of biomass (MSW pellet) and syngas mass flow rate. The influencing factor is in addition to measuring the syngas mass flow rate which is carried out using the experimental method, measurements are taken from the outputs in the form of char and ash. While the limitation of the problem in the study with the numerical method of output measured only syngas.

4. Conclusion

Characterization of the downdraft type gasification reactor with the addition of inlet air temperature variations in the partial oxidation zone from 80 °C to 200 °C are as follows:

1. Temperature distribution of each gasification zone increases;
   - zone drying increases (T2) from 130 °C to 140 °C
   - the pyrolysis zone increases (T4) from 498 °C to 578 °C
   - the partial oxidation zone increases (T6) from 830 °C to 1004 °C zona reduksi meningkat (T7) dari 520 °C sampai 580 °C
2. Volumetric percentages on flammable syngas (CO, H_2, and CH_4) increased from 21.64%, 9.24%, and 2.86% to 22.65%, 10.49%, and 3.10%.
3. Cold gas efficiency increased from 72.17% to 79.51%.

Acknowledgments

We express our gratitude to LPPM ITS for providing us with financial assistance so that this research can be carried out.

References

[1] Hutagalung and Aldi Martino 2015 *Rencana Strategis 2015-2019* Kementerian Energi dan Sumber Daya Mineral Direktorat Jenderal Minyak dan Gas Bumi.

IOP Conf. Series: Materials Science and Engineering **588** (2019) 012019 doi:10.1088/1757-899X/588/1/012019

[2] Mintarsih and Tuti Hendrawati 2015 *Penanganan Sampah Plastik* Kementerian Lingkungan Hidup dan Kehutanan.
[3] Indarto 2016 *Studi Eksperimental Gasifikasi Briket Municipal Solid Waste Dengan Reaktor Gasifikasi Tipe Downdraft Berpengendali Suhu Otomatis Pada Zona Partial Combustion (*Surabaya : Institut Teknologi Sepuluh Nopember)
[4] B Nyakuma, ArshadAhmad, Anwa Johari, Tuan Abdullah, Olagoke Oladokun, Alkali, and Habib 2016 *Gasification of Oil Palm Empty Fruit Bunches (OPEFB) Briquettes for Bio-Syngas Production*" (Universiti Teknologi Malaysia : Departement of Chemical Engineering)
[5] Ependi, Depi Rustam and B Sudarmanta 2017 *Studi eksperimental pengaruh suhu udara inlet zona oksidasi pada proses gasifikasi pelet municipal solid waste terhadap unjuk kerja gasifier tipe downdraft* (Surabaya : Institut Teknologi Sepuluh Nopember)
[6] Trianto, Muhammad Aji and B Sudarmanta 2017 *Studi eksperimental pengaruh suhu proses gasifikasi pellet municipal solid waste (msw) terhadap unjuk kerja reaktor gasifikasi tipe downdraft berpengendali suhu pada zona partial combustion sistem kontinyu (*Surabaya : Institut Teknologi Sepuluh Nopember)
[7] Primawan, Rizki Wase and B Sudarmanta 2017 *Studi eksperimental pengaruh laju alir massa udara pada proses gasifikasi pelet municipal solid waste (msw) terhadap kandungan tar dan carbon conversion rate gasifier tipe downdraft (*Surabaya : Institut Teknologi Sepuluh Nopember)
[8] Ardiansyah, irfan maulana and B Sudarmanta 2017 S*tudi eksperimental pengaruh air fuel ratio proses gasifikasi pellet municipal solid waste (msw) terhadap unjuk kerja gasifier tipe downdraft sistem kontinyu (*Surabaya : Institut Teknologi Sepuluh Nopember)
[9] M Burhan Cahyo and B Sudarmanta 2012 *Simulasi Gasifikasi Sekam Padi pada Gasifier Downdraft Menggunakan Software Fluent (*Surabaya : Institut Teknologi Sepuluh Nopember)
[10] Dziyad Dzulfansyah 2014 *Perancangan reaktor gasifikasi sekam padi tipe downdraft menggunakan analisis computational fluid dynamics* (Bogor : IPB)
[11] Lu X and Wang T 2012 *Effect of radiation models on coal gasification simulation.* (Houston Texas : Proceedings of the ASME International Mechanical Engineering Congress & Exposition)
[12] Silaen A and Wang T 2010 Effects of turbulence and devolatilization models on gasification simulation *International Journal of Heat and Mass Transfer*
[13] Timo Hyppanen 2011 *A Study On Rate Correlations Of Gasification Reactions* Lappeenranta University Of Technology
[14] Prabir Basu 2010 *Combustion and Gasification in Fluidized Bed* Taylor and Francis Group.
[15] Satake 2006 *Biomass Gasification and solution for Agro Waste* Jepang.
[16] Chen C, Horio M and Kojima T 2000 *Numerical simulation of entrained flow coal gasifiers* Chemical Engineering Science.
[17] Lu X and Wang T 2014 Investigation of low rank coal gasification in a two-stage downdraft entrained-flow gasifier *International Journal of Clean Coal and Energy*
[18] Westbrook CK and Dryer FL 1981 *Simplified reaction mechanism for the oxidation of hydrocarbon fuels in flames* Combustion Science and Technology
[19] Jones WP and Lindstedt RP 1988 *Global reaction schemes for hydrocarbon combustion* Combustion and Flame

IOP Conf. Series: Materials Science and Engineering **588** (2019) 012020 doi:10.1088/1757-899X/588/1/012020

Effects of Advancing the Pilot Injection Timing on The Engine Performance, Combustion and Emissions on Diesel dual fuel Engine at High Load

Bambang Sudarmanta[1, a], Atok Setiyawan[1], Ary Bachtiar K P[1], Dori Yuvenda[1,2, b], and Jose da Silva[1]

[1] Department of Mechanical Engineering, Faculty of Industrial Technology, Institut Teknologi Sepuluh Nopember, Kampus ITS, Sukolilo, Surabaya 60111, Indonesia

[2] Department of Automotive Engineering, Faculty of engineering, Universitas Negeri Padang. Kampus UNP Air Tawar Barat, Padang 25171, Indonesia

[a] bsudarmanta@gmail.com, [b]doriyuvenda@ft.unp.ac.id

Abstract. Pilot injection timing is a very important parameter at the start of ignition which will affect engine performance and emissions on diesel dual-fuel engines, especially at high loads. This research was carried out on the Diamond DI 800 diesel engine, a single-cylinder that was modified into a dual fuel engine with the aim to determine the effect of advancing the pilot injection timing (PIT) on engine performance, combustion and emissions under high loads with variation pilot injection timing (11^0, 13^0, 15^0, 17^0 and 19^0 BTDC). The cylinder pressure, heat release rate (HRR), combustion duration, ignition delay, power, SFC and brake thermal efficiency (BTE), as well as carbon monoxide (CO), hydrocarbon (HC) and particulate matter emissions, were analyzed. The results showed that advancing the pilot injection timing increased the maximum cylinder pressure by 69.52 bar; the heat release rate was 117.79 $kJ/m^3/^0CA$, brake thermal efficiency was 25.17% and significantly reduced HC, CO and PM emissions at the pilot injection timing 17° BTDC. Improved combustion performance occurs when the combustion duration is reduced even though the ignition delay period occurs by adding pilot injection timing under high load.

1. Introduction

Currently, alternative fuels are the solution to the scarcity of fuels such as ethanol fuel on gasoline engine [1]–[4], while diesel engine use CNG [5] and biodiesel fuels [6]. The use of CNG fuel in a diesel engine is better known as diesel dual fuel (DDF). Diesel dual fuel engine still causes problems related to engine performances and emissions [7]–[9]. Where the use of pure diesel fuel (100%) is replaced in part by CNG fuel, causing a decrease in combustion performance which affects the decrease in engine performance [10][11]. Besides that, CNG utilization causes an increase in emissions of carbon monoxide (CO) and hydrocarbons (HC)[12][13], although the addition of CNG can reduce nitrogen oxide (NO_x) and particulate matter (PM) emissions[14].

The problem with diesel dual fuel engine is that there is a decrease in engine performance and increased emissions compared to a standard diesel engine, especially high load[15][16]. The reason is that at high load the engine consumption large amounts of fuel to produce maximum power in offsetting the load, so that the mixture of fuel and air is richer, allowing combustion to become imperfect. Besides that, it

IOP Conf. Series: Materials Science and Engineering **588** (2019) 012020 doi:10.1088/1757-899X/588/1/012020

is also affected by the addition of the ignition delay, resulting from a decrease in the partial pressure of the oxygen and the intake temperature of the air-gas mixture[15].

Research related to diesel dual fuel engine in improving combustion performances and emissions at high load includes Singh, et al. [17] investigate the effect of the pilot injection timing, pilot quantity and intake temperature on performances and emissions on diesel/natural gas engine at medium and high loads, and they found that by advancing the injection timing of pilot increased fuel conversion efficiency and decreased NO_x, CO and HC emissions at high load. Kerdsurwan, et al. [18] investigate the effect of advancing the pilot injection timing on engine performance on diesel dual fuel engine using diesel fuel and producer gas at low, medium and high loads, and they found that by advancing the pilot injection timing can improve BTE, fuel substitution, SFC and decreased HC, CO and soot emissions except for NO_x emission for all engine loads.

Zhou, et al.[19] investigate the effect of the pilot injection timing and pilot quantity on combustion and particle emission characteristics on diesel/natural gas engine under low, medium and high loads, and they conclude that by advancing the pilot injection timing improves combustion duration, start of combustion (SOC) and decreases HC and CO emissions significantly, but NO_x emissions slightly increase under high load. Liu, et al.[20] investigate the effect of pilot diesel quantity on emission characteristics on diesel dual fuel engine by optimizing the injection timing of pilot under medium and high loads, and they found that by advancing the pilot injection timing can reduction of CO and PM emission under high load.

Ryu,[21] , investigated the effect of diesel pilot injection timing on combustion characteristics and emissions on dual fuel diesel engines with biodiesel and CNG under high loads, He found that b,y advancing the pilot injection timing can increase cylinder pressure, heat release rate and improve start of combustion. Likewise, HC and PM emissions are decreasing a significant, but CO and NOx emissions increased under high load. Sun, et al.[22] conducted a study on the analysis of combustion performance and stability of dual fuel gas engine with variations in pilot injection timing and pilot quantity under low, medium and high loads, and they found an increase in cylinder pressure, heat release rate, specific energy consumption and reduce HC emission, but NO_x emission slightly rise under high load. Likewise, the stability of a dual diesel engine by advancing the pilot injection timing under high load can be accepted; this is marked by the coefficient of variation (CoV_{IMEP}) value under 4%.

Papagiannakis, et al. [23] investigate the effect of pilot injection timing and the mass ratio of diesel/gas to performance and emissions on dual fuel diesel engines with experimental and theoretical methods under load variations, and they found an increase in cylinder pressure, heat release rate and decrease in CO and smoke emissions despite an increase in NO_x emissions under high load. Likewise, the duration of combustion becomes shorter when the pilot injection timing was advanced at high load.

Based on the research that has been done above, it can be concluded that advancing the pilot injection timing can improve engine performance and combustion as well as emissions on diesel dual-fuel engines under high load. This study aims to optimize the pilot injection timing to improve engine performance and combustion and reduce emissions in diesel dual fuel engine under high load.

2. Methodology of research

2.1. Test engine

This research use Diamond DI 800 diesel engine, a single-cylinder using a water cooling system cylinder that was modified into a diesel dual fuel engine with diesel and CNG fuel as shown in figure 1. The engine characteristics used in this study are shown in table 1.

Table 1 The engine specification.

No	Item	Specification
1	Merk	Diamond diesel

2	Type	DI 800	
3	Model	1 cylinder, 4 stroke	
4	Bore x Stroke	82 mm x 78 mm	
5	Displacement	411 cc	
6	Max. Power	8 HP (6 KW) / 2400 rpm	
7	Continous Power	7 HP (5.22 KW)/2200 rpm	
8	Compression Ratio	18:1	
9	Cooling sistem	Hopper/ Condenser	
10	Lube capacity	1,8 liter	
11	Pilot injection timing	13^0 BTDC	
12	Valve timing	**Opening**	**Closing**
	Intake	30^0 BTDC	50^0 ABDC
	Exhaust	55^0 BBDC	35^0 ATDC

CNG with low pressure is injected through the intake, which is controlled using an electrical control unit (ECU) while diesel fuel uses a conventional system. CNG is injected at 45^0 ATDC at the suction stroke, while the injection timing of the standard pilot is 13^0 BTDC at the end of compression stroke.

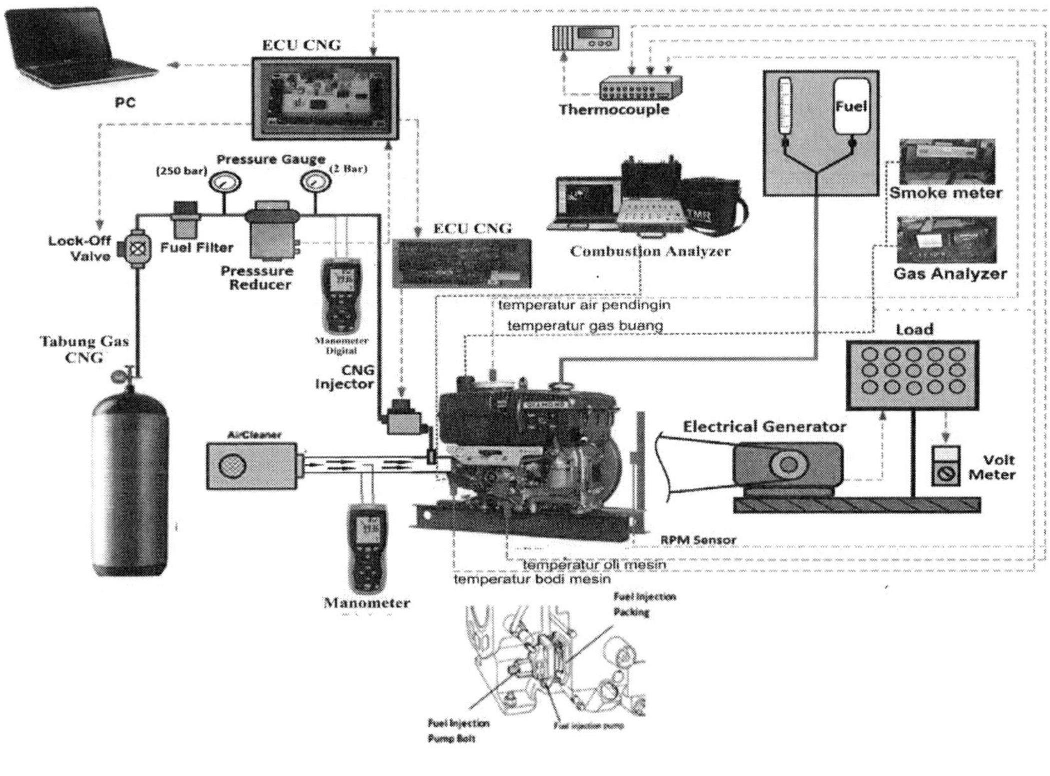

Figure 1. Experimental setup

2.2. The fuel of dual fuel system

This research uses diesel fuel as a pilot and CNG as the main fuel in diesel dual fuel engine. The diesel fuel and CNG properties used in this study are shown in tables 2 and composition of CNG at tables 3 below. The research method uses experimental methods on diesel dual fuel engine that are tested at high load. The converter kit consists of several components that work integrated to reduce the high pressure

IOP Conf. Series: Materials Science and Engineering **588** (2019) 012020 doi:10.1088/1757-899X/588/1/012020

in a CNG tank by 250 bar to a working pressure of 2 bars through the inlet port [24]. CNG is injected into the intake port using a CNG injector controlled by the ECU.

Table 2 The fuel properties[25].

Fuel Properties	Diesel	CNG
Low heating value (MJ/kg)	42,8	48,6
Cetane number	52,5	-
Octane number	-	130
Self ignition temperature (^{0}C)	316	650
A/F_{stoic} (kg/kg)	14,69	17,2
Carbon content (%)	87	75

Table 3 The composition of CNG[25].

Fuel Poperties	Volumetric concentration (%)
Methane	96.160
Ethane	1.096
Butane	0.136
Iso-Butane,n-butane	0.021
Iso-Pentane, n-pentane	0.006
N2	0.001
H2S	0.0002
H2O	0.006

2.3. *Instrument arrangement and data acquisition*

Diesel dual fuel engine is connected to an electric generator as a load. The loading of the engine is carried out with a high load using a lamp with a capacity of 4000 watts. Kistler piezoelectric pressure sensors are installed in the cylinder head which is used to measure cylinder pressure and heat release rate with a 0.20 crank angle (CA) resolution. CNG and air mass flow rate are measured using a digital manometer with an accuracy of ± 0.001. Diesel fuel consumption using a burette meter with 1 ml accuracy. The star gas 898 emission analyzer to measure HC and CO emissions, while PM emissions use a smoke tester.

2.4. *Test procedure and conditions*

The engine is heated to a working temperature by ± 80. Engine testing is carried out under high load with a constant engine speed of 1500 rpm. The test is divided into two stages: the first stage, testing the engine with standard fuel as a control variable. Second stages, testing the engine using dual fuel by advancing the pilot injection timing from 11^0 BTDC to 19^0 BTDC. Advance the injection timing of the pilot with reducing the shim plate at the injection pump holder. Each reduction of shim plate of 0.1 mm can advance injection timing of 1^0 CA.

3. Results and discussions

This discussion analyzes the effect of advancing the pilot injection timing on engine performance and combustion and emissions on a diesel dual fuel engine. Engine performance parameters include power, SFC and thermal efficiency and combustion performance including cylinder pressure, heat release rate, ignition delay, combustion duration and emissions including CO, HC, and PM. Ignition delay is defined at the start of pilot diesel to be injected to the start of combustion (SOC), while the duration of combustion is the start of combustion (SOC) to the end of combustion (EOC).

3.1 The engine performances with variations in pilot injection timing

In figure 2, shows the effect of pilot injection timing on engine performance including engine power, SFC and thermal efficiency on diesel dual fuel engine under high load. Figure 2, the thermal efficiency and power of the engine showing the same graph trend as the different pilot injection timing. The effect of advancing the pilot injection timing is more significant on thermal efficiency, while the engine power is not significant.

Increased maximum thermal efficiency occurs at pilot injection timing of 17^0 BTDC of 25.17% at constant engine speed. This is due to advancing the pilot injection timing can extend the period of ignition delay, so that the gas-air mixture can mix with the pilot fuel more homogeneously, resulting in more fire points. Thus the premixed combustion of diesel fuel is faster and triggers the diffusion of gas and diesel fuels so that thermal efficiency also increases under high load.

Figure 2, also shows that the SFC value is optimal with the pilot injection timing of 17^0 BTDC by 0.8063 kg/HP.h. This is due to greater power so that the engine is more effective at consuming fuel, characterized by increased thermal efficiency.

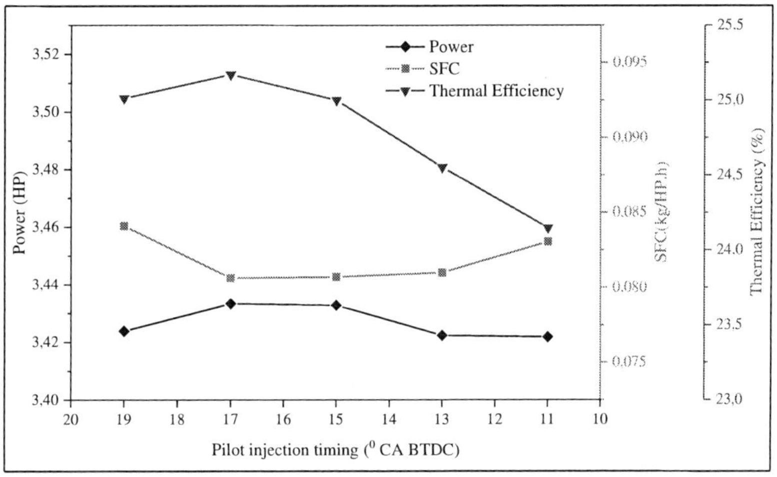

Figure 2. Engine performance with different pilot injection timing

3.2 The Cylinder pressure and heat release rate with variations in pilot injection timing

In Figure 3, shows the effect of pilot injection timing on cylinder pressure and heat release rate (HRR) under high loads. Figure 3, explains that there is a significant increase in cylinder pressure by advancing the pilot injection timing of 17^0 BTDC to 69.52 bars. This is because injection of diesel fuel is slightly accelerated so that the period of ignition delay is advanced, so the initial combustion is faster, and results in cylinder pressure peaks leading to the top dead center (TDC).

Besides that, the cylinder pressure graph at pilot injection timing of 11^0 CA BTDC, where the maximum cylinder pressure away from TDC. This is due to the longer ignition delay period in dual fuel diesel so that the pilot injection timing that is too late causes a delay in the initial ignition of combustion so that the maximum pressure occurs in the expansion stroke, so resulting in a pressure drop under high load. However, if the pilot injection timing is too early (PIT 19^0 BTDC), there is a decrease in cylinder pressure. This is because the pressure and temperature of the cylinder are low.

IOP Conf. Series: Materials Science and Engineering **588** (2019) 012020 doi:10.1088/1757-899X/588/1/012020

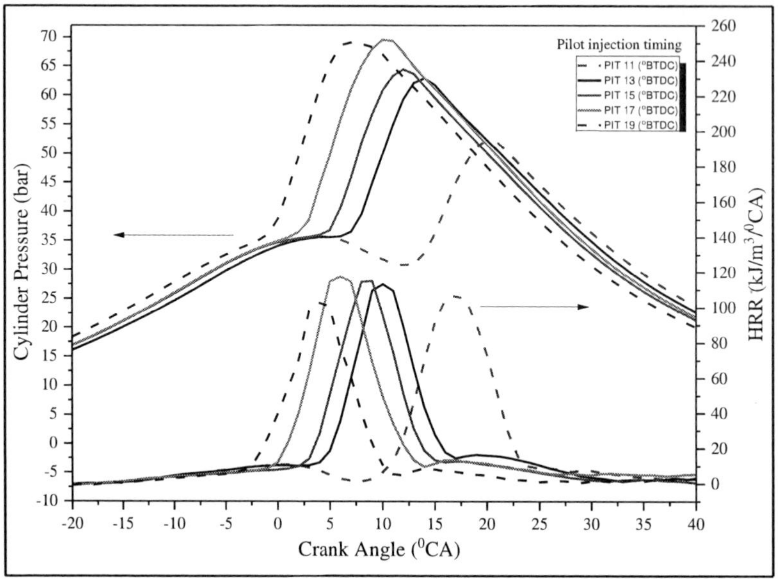

Figure 3. Cylinder pressure and HRR with different pilot injection timing

Figure 3, also shows the heat release rate (HRR) increases the significance of the pilot injection timing of 17^0 BTDC by 117.79 kJ/m^3/deg with a percentage increase of 12.41%. This proves that the increase in cylinder pressure and temperature can cause greater combustion so that the energy released becomes more abundant. The heat release rate of maximum occurs in diffusion combustion from diesel and CNG fuels, where both fuels burn optimally.

However, when the injection timing of the pilot is forwarded too early or rewind too slowly, the heat release rate (HRR) drops. Due to pressure and temperature, the cylinder is still low when the pilot injection timing is too early, while the diesel injection is too slow the pressure and cylinder temperature are too high, allowing knocking of CNG fuel and more heat being wasted through the exhaust manifold because combustion occurs at the expansion stroke.

3.3 The combustion performance with variations in pilot injection timing

Figure 4, shows the ignition delay period and duration of combustion under variations in diesel pilot injection timing on diesel dual fuel under high load. The period of ignition delay increases along with advancing the pilot injection timing. The pilot injection timing of 19^0 BTDC has a period of ignition delay that is too long at 20^0 CA. The reason is the pressure and low cylinder temperature so that the chemical delay process is longer. The advantage of a slightly longer ignition delay is to provide the opportunity for air and CNG to mix more homogeneously. However, when injection timing of pilot is slowed down, there is a significant decrease in the ignition delay period, due to the high cylinder pressure and temperature which shorten the chemical delay process, so that firing point is formed faster.

Figure 4, also shows the duration of combustion under variations in the pilot injection timing on diesel dual fuel under high load. The duration of combustion is shorter with the injection time of 15^0 BTDC pilots of 33^0 CA. This is influenced by faster combustion speeds so that combustion is completed more quickly. When the pilot injection is advanced too early the duration of combustion increases significantly because the cylinder pressure and temperature are still low.

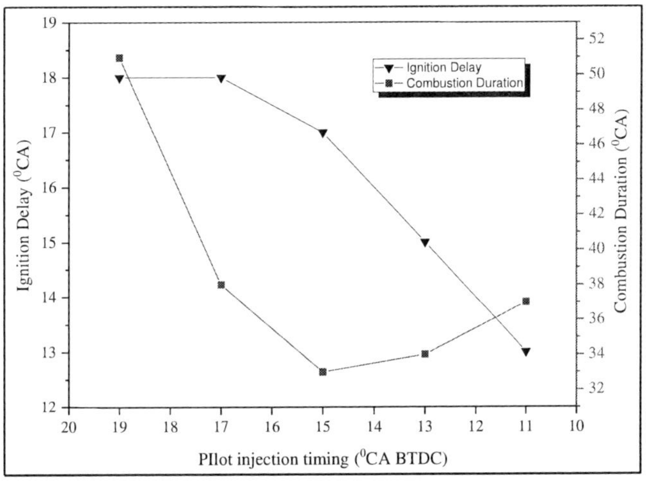

Figure 4. Combustion performance with different pilot injection timing

3.4 The emissions with variations in pilot injection timing
Figure 5, shows the effect of pilot injection timing on HC, CO and PM emissions on diesel dual fuel engine under high load. Figure 5, shows the pilot injection timing of 17^0 BTDC obtaining value optimal of CO, HC and PM emissions. A decrease in CO emission of up to 0.026%, marked by an increase in cylinder pressure and heat release rate, resulting in high temperatures so that it can help oxidized CO emissions to form CO_2 [25]. Besides, with a peak of combustion approaching TDC so that the time available for the oxidation process is longer.

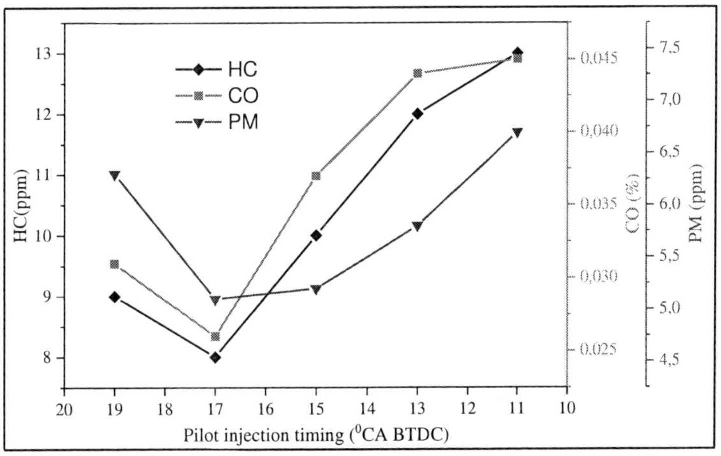

Figure 5. Emission with different pilot injection timing

Figure 5, also shows a decrease in HC emission by 8 ppm with the pilot injection timing of 17^0 BTDC, this is due to earlier burning so that the pilot fuel burns more. Likewise, advancing the injection timing of pilot increases the cylinder pressure and heat release rate so that the remaining diesel fuel can be burned in diffusion combustion. Figure 5, also shows a reduction in PM emission by 5.1 ppm with the pilot injection timing of 17^0 BTDC. Two factors, namely cause the reduction in PM emission: first, diesel fuel replaced by CNG does not have soot content, so the possibility of lower PM emissions is formed. Second, by advancing the pilot injection timing of the initial combustion initiation so that diesel fuel burns more because it mixes more with air during the ignition delay period.

4. Conclusion

In general, diesel dual fuel engine the ignition delay period is longer than that of a single fuel diesel engine, thus affecting the engine performance, combustion process, and emissions under high load. The effort taken is to set pilot injection timing so that it can control the ignition delay period that occurs. Advancing the pilot injection timing affects the maximum pressure cylinder pressure up to 69.52 bar, the maximum heat release rate up to 117.79 kJ/m^3/^{0}CA under high load. Increased combustion performance is also characterized by a reduction in the duration of combustion despite the addition of an ignition delay period by advancing the pilot injection timing under high load. Likewise, advancing the pilot injection timing can reduce emissions of HC by 8 ppm, CO by 0.026 % and PM by 5.1 ppm on diesel dual fuel engine under high load.

Acknowledgements
Thank you to the ITS DDF team who have helped in retrieving data and data processing, so that this research is well implemented.

References

[1] M E N Paloboran, I N Sutantra, B Sudarmanta, M E N Paloboran, I N Sutantra, and B Sudarmanta 2016 Performances and Emissions Characteristics of Three Main Types Composition of Gasoline-Ethanol Blended in Spark Ignition Engines *Int. Rev. Mech. Eng* **vol. 10** no. 7, p. 552
[2] M Paloboran, I N Sutantra, B Sudarmanta, and R F D Dharmawan 2017 Suitable injection duration of pure ethanol fuel for motorcyle at a high compression ratio *DYNA-Ingeniería e Ind.*, **vol. 92** no. 5
[3] M Paloboran, I N Sutantra, and B Sudarmanta *A Strategy in Adjustment of Combustion Parameters of SI-PFI Engine with Pure Bioethanol Fuelled for a High Performance and Low Emission*
[4] B Sudarmanta, B Junipitoyo, A B K Putra, and I N Sutantra 2016 Influence of the compression ratio and ignition timing on Sinjai Engine performance with 50% bioethanol-gasoline blended fuel *ARPN J. Eng. Appl. Sci.*, **vol. 11** no. 4, pp. 2768–2774
[5] M Octaviani, S Nilam, Semin, Sudarmanta, Bambang, Zaman 2018 The Implementation of CNG as Analternative Fuel For Marine Diesel Engine *Int. J. Mech. Eng. Technol* **vol. 9** no. 13, pp. 25–33
[6] B Sudarmanta *Dual Fuel Engine Performance Using Biodiesel And Syn- Gas From Rice Husk* pp. 1–8.
[7] M Y E Selim 2004 Sensitivity of dual fuel engine combustion and knocking limits to gaseous fuel composition *Energy Convers. Manag* **vol. 45** no. 3, pp. 411–425
[8] H Ambarita and E Y Setyawan *Performance and Emissions of a Small Compression Ignition Engine Run on Dual-fuel Mode (Diesel-Raw biogas) Performance and Emissions of a Small Compression Ignition Engine Run on Dual-fuel Mode (Diesel-Raw biogas)*
[9] R S Hosmath, N R Banapurmath, S V Khandal, V N Gaitonde, Y H Basavarajappa, and V S Yaliwal 2016 Effect of compression ratio, CNG flow rate and injection timing on the performance of dual fuel engine operated on honge oil methyl ester (HOME) and compressed natural gas (CNG) *Renew. Energy* **vol. 93** pp. 579–590
[10] M Y E Selim 2001 Pressure-time characteristics in diesel engine fueled with natural gas *Renew. energy* **vol. 22** no. 4, pp. 473–489
[11] D Yuvenda 2015 *Karakterisasi Performa Mesin Sistem Dual Fuel Menggunakan Pressure Reducer Adaptive Dengan Variasi Konstanta (K) Pegas Helix Tekan Dan Tekanan* no. 2014 pp. 1–8
[12] R G Papagiannakis and D T Hountalas 2003 Experimental Investigation Concerning The Effect of Natural Gas Percentage on Performance And Emissions of a DI Dual Fuel Diesel Engine

vol. 23, pp. 353–365

[13] R G Papagiannakis and D T Hountalas 2004 Combustion and exhaust emission characteristics of a dual fuel compression ignition engine operated with pilot Diesel fuel and natural gas **vol. 45** pp. 2971–2987

[14] B Yang, X Wei, C Xi, Y Liu, K Zeng, and M C Lai 2014 Experimental study of the effects of natural gas injection timing on the combustion performance and emissions of a turbocharged common rail dual-fuel engine *Energy Convers. Manag.* **vol. 87** pp. 297–304

[15] M M Abdelaal and A H Hegab 2012 Combustion and emission characteristics of a natural gas-fueled diesel engine with EGR *Energy Convers. Manag.* **vol. 64** pp. 301–312

[16] K Wannatong, N Akarapanyavit, and S Siengsanorh 2017 *Combustion and Knock Characteristics of Natural Gas Diesel Dual Fuel Engine Combustion and Knock Characteristics of Natural Gas Diesel Dual Fuel Engine.*

[17] S Singh, S R Krishnan, K K Srinivasan, K C Midkiff, and S R Bell 2004 Effect of pilot injection timing, pilot quantity and intake charge conditions on performance and emissions for an advanced low-pilot-ignited natural gas engine *Int. J. Engine Res.* **vol. 5** no. 4 pp. 329–348

18] S Kerdsuan, T Lekpradit, S Tongorn, and N Nipattummakul 2008 Effect of Advanced Injection Timing on the Performance Improvement of a Dual-Fuel Diesel Engine with Producer Gas from a Down-Draft Gasifier for Power Generation *Second Int. Energy 2030 Conf.* pp. 313– 318

[19] L Zhou *et al.* 2013 Effect of the diesel injection timing and the pilot quantity on the combustion characteristics and the fine-particle emissions in a micro-diesel pilot-ignited natural-gas engine *Proc. Inst. Mech. Eng. Part D J. Automob. Eng.* **vol. 227** no. 8 pp. 1142–1152

[20] J Liu, F Yang, H Wang, M Ouyang, and S Hao 2013 "Effects of pilot fuel quantity on the emissionsccharacteristics of a CNG/diesel dual fuel engine with optimized pilot injection timing *Appl. Energy* **vol. 110** no. x pp. 201–206

[21] K Ryu, 2013 Effects of pilot injection timing on the combustion and emissions characteristics in a diesel engine using biodiesel-CNG dual fuel *Appl. Energy* **vol. 111** pp. 721–730

[22] L Sun, Y Liu, K Zeng, R Yang, and Z Hang 2015 Combustion performance and stability of a dual-fuel diesel-natural-gas engine *Proc. Inst. Mech. Eng. Part D J. Automob. Eng.* **vol. 229** no. 2 pp. 235–246

[23] R G Papagiannakis, S R Krishnan, D C Rakopoulos, K K Srinivasan, and C D Rakopoulos, 2017 A combined experimental and theoretical study of diesel fuel injection timing and gaseous fuel/diesel mass ratio effects on the performance and emissions of natural gas-diesel HDDI engine operating at various loads *Fuel* **vol. 202** pp. 675–687

[24] D Yuvenda, B Sudarmanta, and E Alwi 2017 Analisis Kekuatan Pegas Pressure Reducer Sebagai Penurunan Tekanan Pada Mesin Duel Fuel **vol. 17** no. 2

[25] B Yang, L Wang, L Ning, and K Zeng 2016 Effects of pilot injection timing on the combustion noise and particle emissions of a diesel/natural gas dual-fuel engine at low load *Appl. Therm. Eng.* **vol. 102** no. x pp. 822–828

IOP Conf. Series: Materials Science and Engineering **588** (2019) 012021 doi:10.1088/1757-899X/588/1/012021

Symbiotic organism search for sizing and optimal location of distributed generation using novel sensitivity factor

G Setyawan[1], Umar[1,2], A Soeprijanto[1] and O Penangsang[1]

[1]Department of Electrical Engineering, Institut Teknologi Sepuluh Nopember, Surabaya, Indonesia
[2]Department of Electrical Engineering, Universitas Khairun, Ternate, Indonesia

gema.setyawan1@gmail.com

Abstract. Electric power systems consist of many interconnected buses. The most widely used network topology is Radial Distribution Network (RDN), the bus located furthest from generation has a significant voltage drop and the number of buses without any support from Distributed Generation (DG) inflicting in increased power losses and increased generation cost. DG is advantageous to reduce power losses and increasing voltage on the power system, but the location and sizing of DG in the distribution network requires great consideration. In this paper the method used to determine the location is Loss Reduction Sensitivity Factor (LRSF) and Voltage Improvement Performance Index (VIPI) and for sizing DG uses artificial intelligence Symbiotic Organism Search (SOS). This DG locating and sizing problem is associated with real power loss minimization and voltage profile improvement objectives. The proposed method has been demonstrated on IEEE 33-bus radial distribution system. Based on this research the identification of using LRSF location is better than VIPI.

1. Introduction

In the electrical system, supply of electrical energy to the load with sufficient quantities and good quality is a priority. The largest reserves of fossil energy and wisdom to reduce the effects of greenhouses gases, which promote to renewable energy. America, China and Brazil area the leading countries in the development of renewable energy potential. The number of plants in America 2010-2040 is estimated to increase by around 93 % where renewable energy sources reach 24 % with an annual increase of 2,8 % [1]. Demand for cheap electricity and quality electricity encourage application of electricity based on Distributed Generation (DG) or On Site Generation (OSG), a small scale electricity network and as a distributed energy resources [2]. Various types of DG technology are now available both renewable and non-renewable. DG which is non-renewable include Fuel Cell (FC), Reciprocating Engine (RE), GAS Turbine (GT), Micro Turbine (MT) etc. Renewable DGs include Photovoltaic (PV), Wind Turbine (WT), wave energy, geothermal energy, biomass etc. [3].

Identifying the DG concept can involve technology, location capacity and its impact on system operations [4]. In extensive area power system network, the identification of most compatible area for DG penetration is very significant. Two different sensitivity factor, based on loss reduction and voltage improvement have been proposed for selection of load buses to place the DG [5] and for size will be calculated by symbiotic organism search.

2. Mathematical Problem Formulation

The most important objective in this research to discover the optimal location and size of real power DG in RDS with a least power loss and voltage profile improvement. In this research, location is selected using LRSF and VIPI. The objective function can be separated into two parts [6], and are described as the inequality and quality constraints of the present optimization subject.

2.1. Power Losses Minimization

The major objective function of this research is to decrease the real power loss in the distribution system. It can be mathematically expressed as below:

$$P_{loss} = \sum_{i=1}^{n}\sum_{j=2}^{n} I_{ij}^2 R_{ij} \tag{1}$$

$$OF_1 = Minimize\left(P_{loss}\right) \tag{2}$$

Where P_{loss} is the real power loss of the RDS, n is the total number of buses in distribution system, I_{ij} is current branch between i and j, and R_{ij} is the resistance of the ij branch

2.2. Voltage Profile Improvement

The bus voltage of characteristic distribution network often experiences variations and even voltage may downfall under specific critical loading conditions with increasing load demand. DG are connected tightly to the load centers to avert voltage collapse and improve overall voltage profile of the distribution network. The objective function of voltage profile improvement may be expressed (3-4).

$$Voltage_{profile} = i = 1\sum_{i=1}^{n}\left(V_i - V_{rated}\right) \tag{3}$$

$$OF_2 = Minimize\left(Voltage_{profile}\right) \tag{4}$$

Where V_i is the voltage of the ith bus, V_{rated} is the rated bus voltage of the network, which is equal to 1.0 pu. All these two single objective functions are combine to form a multi-objective optimization problem, whose fitness function (FF). Where $w_1$and w_2 are weighting factors

$$FF = \min\{(w_1 \times OF_1) + (w_2 \times OF_2)\} \tag{5}$$

2.3. Constraint

The set of constraints is formed by (6-7)

$$V_{\min}^{\lim it} \le V_i \le V_{\min}^{\lim it} \tag{6}$$

$$\sum_{i=1}^{n} P_{load} \times 30\% \le \sum_{i=1}^{5} DG_i \le 50\% \times \sum_{i=1}^{n} P_{load} \tag{7}$$

Where V_i is the voltage of the ith bus, P_iDG is DG generation and P_{loss} total load in RDS. Based on location of DG, middle node and endmost node the total generation of DG cannot exceed 50 % because it will considerably reduce the system static voltage stability. However, if DG is located spread the maximum allowable penetration level is increased from 50 to 70 % [7].

3. Novel Sensitivity Factor

Novel sensitivity factor (NSF) consists of two indexes, loss reduction sensitivity factor (LRSF) and voltage improvement performance index (VIPI) [5]. In this paper, these indexes will be compared based on the effectiveness of DG location which will be placed in accordance with the objective function.

3.1. *Loss Reduction Sensitivity Factor*
LRSF is identified based on the reduction losses that develop because injection by DG, it can be defines as

$$LRSF_i = \frac{\Delta P_{loss}}{\Delta P_i} = \frac{P^i_{loss} - P^b_{loss}}{P_{DG}i^{inj}} \tag{8}$$

Where $P^i loss$ is the system loss with injection DG capacity at ith load bus $P^b loss$ is the total system loss for base case, and $P_{DG}i^{inj}$ is the increment in DG capacity at ith bus. Based on above expression, the following calculation have been observed. LRSF > 0, LRSF = 0 and LRSF < 0 indicate that injection by DG has the impact of increased.

3.2. *Loss Reduction Sensitivity Factor*
Comparable expression equivalent to the voltage improvement performance index (VIPI) for the voltage security based ranking has been developed her for the selection of load buses [8].

$$VIPI_i = \sum_{j=1}^{N} \frac{wj}{2n}\left(\frac{\Delta V^i_j}{\Delta V^{\lim}_j}\right)^{2n} \tag{9}$$

$$\Delta V^i_j = V^i_j - V^{\lim}_j \tag{10}$$

$$V^{\lim}_j = \begin{Bmatrix} V^{maks}_j, \text{if } V^i_j \geq 1.0 \\ V^{\min}_j, \text{if } V^i_j \leq 1.0 \end{Bmatrix} \tag{11}$$

$$V^i_j = \begin{Bmatrix} V^{maks}_j, \text{if } V^i_j \geq V_{\max} \\ V^{\min}_j, \text{if } V^i_j \leq V_{\min} \end{Bmatrix} \tag{12}$$

$$\Delta V^i_j = \frac{V^{maks}_j - V^{\min}_j}{2} \tag{13}$$

Where w_j is the weighting factor of bus j that is determined by the system operator depending upon the operating condition the system (=1). N is the total number of buses in the system, n the exponent of penalty function (=1 preferred), V_j^i is the voltage level at bus j.

4. Symbiotic Organism Search
SOS is a new metaheuristic developed for civil engineering application [9] [10]. This algorithm is built to resemble the symbiotic pattern of organism in an ecosystem. SOS has benefit factor (BF) and mutual vector (MV) parameters, it is consisting of mutualism, commensalism and parasitism phase are performed to implemented to main population (ecosystem) of candidate (organism) toward the global optimal region in the research space, In the mutualism phase, both organism benefit each other, in the commensalism phase, only one organism has an benefit while others are not impacted, in the parasitism phase, one organism must harm the other in order to gain the advantage. The process of three phase until stopping criteria are met as abridged in the following outline:

a. Initialize ecosystem.
b. WHILE
   Simulate ecosystem through mutualism phase
   Simulate ecosystem through commensalism phase
   Simulate ecosystem through parasitism phase
c. END

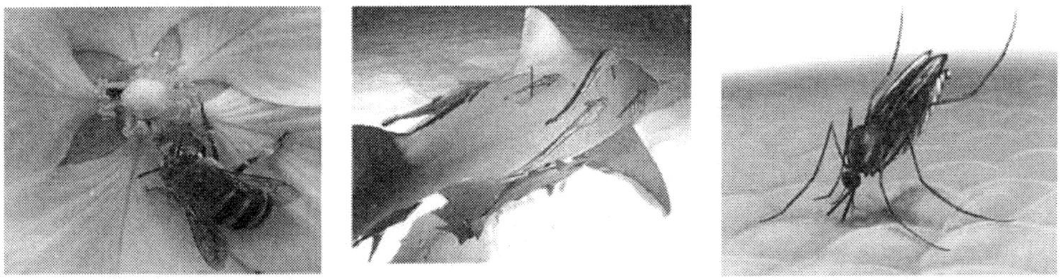

Mutualism **Commensalism** **Parasitism**

Figure 1. Symbiotic relations in ecosystem

4.1. Mutualism Phase

The mutualism phase counterfeits the mutual relations between two organisms in the ecosystem. In this phase, x_i is deliberated as the ith member of the ecosystem. Another organism x_j is then selected randomly from the ecosystem to interrelate with x_i.

$$X_i^{new} = x_i + rand(0,1) \text{ x} (X_{best} - MV \text{ x } BF_1) \quad (14)$$

$$X_j^{new} = x_j + rand(0,1) \text{ x} (X_{best} - MV \text{ x } BF_2) \quad (15)$$

$$\text{MV} = \frac{(x_i + x_j)}{2} \quad (16)$$

MV is mutual vector (that represent the mutual relationship between x_i and x_j), rand (0,1) is vector of random BF1 and BF2 are defines as benefits factors.

4.2. Commensalism Phase

New candidate solution X_i^{new} is calculating using (14) and compare with older x_i, the fittest organism as the solution to carry forward into the next iteration:

$$X_i^{new} = x_i + rand(-1,1) \text{ x} (X_{best} - X_j) \quad (17)$$

Where x_j is selected randomly from the ecosystem to interrelate with x_i, The equation $(X_{best} - x_j)$, denotes the advantage provided by x_j to help x_i to improve its degree of adaptation to the ecosystem.

4.3. Parasitism Phase

Parasite Vector is generated in the search space by duplicating organism x_i. Another x_j randomly selected dimensions using a random number. Parasite vector has a better fitness value, it will kill organism x_j and assume its position in the ecosystem. If the fitness value if x_j is better, x_j will have immunity from the parasite vector will no longer be able to live in that ecosystem.

5. Computational Procedure

The step-by-step procedure of the proposed SOS combine with NSF for the solution of the optimal placement and sizing DGs is given below:

Step 1: Identifying location for DG based on NSF equation (8-12)
Step 2: The capacity for DG as an organism based on location of DG
Step 3: Set maximum ecosize to determine the population on SOS
Step 4: Run power flow and find out real power loss and voltage profile
Step 5: Calculate the Fitness Function and recognize the best solution of organism in the ecosystem

6. Test System and Result

The step-by-step procedure of the proposed SOS combine with NSF for the solution of the optimal placement and sizing DGs is given below:

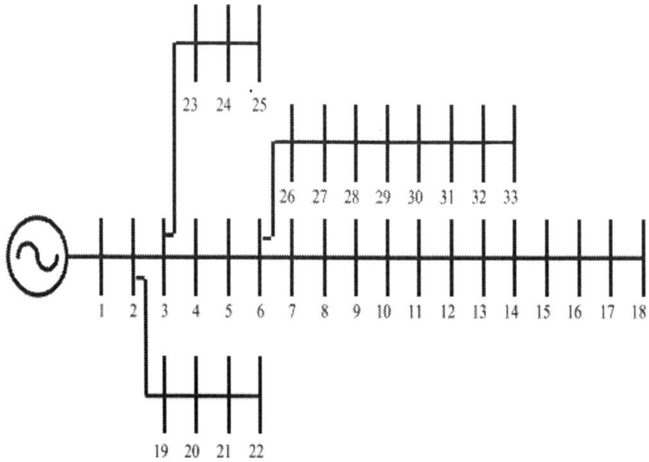

Figure 2. Single line diagram of IEEE 33 bus

6.1. Identify Location of DG

Those DGs produce only active power, for identification using LRSF, all buses on the system are tested with DG injection, equation (8) is used to get LRSF rank. The best location for DG is the minimum value LRSF. Fig 3 describes the comparison of LRSF.

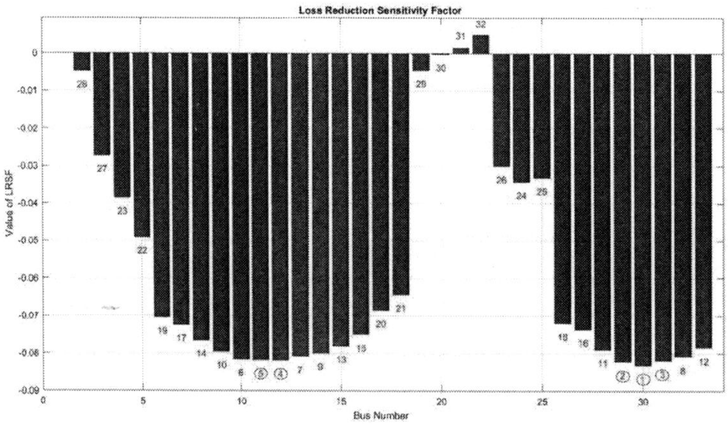

Figure 3. Comparison value of LRSF

Identify location with VIPI using equation (9-13), the bus noticed based on voltage deviation. The best location for DG is the maximum value of VIPI. Fig 4 describes the comparison of VIPI

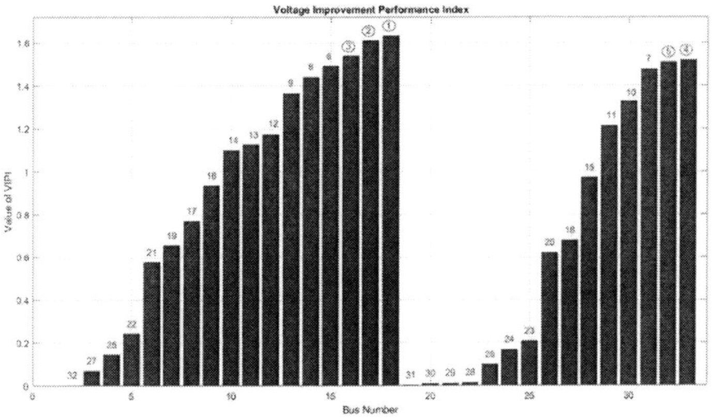

Figure 4. Comparison value of VIPI

6.2. SOS for Sizing Capacity of DG

The best location using LRSF method is 30, 29, 31, 11, and 12, while the best location using VIPI method is 16, 17 18, 33, and 32. The SOS program is built based on [9][10] using equations (14-17) and consists of 3 phases, mutualism, commensalism, and parasitism. Initialization is the number of ecosystems and the power flow backward forward as a function, equation (5) is a multi-objective function. Table 1 is a comparison when using LRSF and VIPI. Fig 5 is the voltage change before and after DG injection.

Table 1. Optimal location and sizing of DGs

Novel Sensitivity Factor + Symbiotic Organism Search						**Power Loss**	**Total Reduction Power Loss**	**Voltage Profile**
	Without DG					224,02 kW	0%	0,1284
LRSF+ SOS	11	12	29	30	31			
	520 kW	510 kW	540 kW	300 kW	380 kW	88,22 kW	60,70%	0,016
	16	17	18	32	33			
VIPI+ SOS	300 kW	300 kW	300 kW	750 kW	300 kW	101,71 kW	54,90%	0,0165

7. Summary

Placement of different method for DG has a different impact in terms of power loss reduction and voltage profiles. Using LRSF and VIPI method combined with metaheuristic SOS will give different fitness value due to differences in the location of variables in a function affecting the fitness value which in turn affect DG capacity, LRSF+SOS can reduce power losses 60,70 % while VIPI+SOS reduce power losses 54,90 % Based on this paper the identification of locations using LRSF Is better than VIPI.

Acknowledgement

This paper is published with a full support from Power System Simulation Laboratory – Institut Teknologi Sepuluh November and as contribution to Department of Electrical Engineering by awarded a master degree to the first author an enhance the ability in power system engineering.

IOP Conf. Series: Materials Science and Engineering **588** (2019) 012021 doi:10.1088/1757-899X/588/1/012021

References

[1] D Deflanti, M Falabretti and D Merlo 2003 Dispersed Generation Impact on Distribution Network Losses Electr *Power Syst. Res* **97** pp 10–18
[2] A M Borbely and J F Kreder 2001 *Distributed Generation The Power Paradigm for the New Millenium* (Florida: CRC Press LLC)
[3] E Akorede, M Funsho, Hizam, Hashim and Pouresmaeil 2010 Distributed Energy Resources and Benefits to the Environment *Renewable and Sustainable Energy Reviews* **14**, pp. 724–734
[4] W El-Khattam and M M A Salama 2004 Distributed generation technologies, definitions and benefits *Electr. Power Syst. Res* **71** no. 2 pp 119–128
[5] A K Singh and S K Parida 2016 Novel sensitivity factors for DG placement based on loss reduction and voltage improvement *Int. J. Electr. Power Energy Syst* **74** pp. 453–456
[6] P K Sultana and S Roy 2014 *Multi-objective particle swarm optimization based on fuzzy-pareto-dominance for possibilistic planning of electrical distribution systems incorporating distributed generation fuzzy set Sys* pp. 534–545
[7] L Hu, K Liu, W Sheng, Y Diao and D Jia 2017 *Research on maximum allowable capacity of distributed generation in distributed network under global energy internet considering static voltage stability* pp. 2276–2280
[8] K S S Ã and G Sudhakar 2006 *Neural network approach to contingency screening and ranking in power systems* **70** pp. 105–118
[9] M Y Cheng and D Prayogo 2014 Symbiotic Organisms Search: A new metaheuristic optimization algorithm *Comput. Struct.* **139** pp. 98–112
[10] D Prayogo 2017 Symbiotic Organisms Search with the Feasibility- Based Rules for Constrained Engineering Design Optimization *Int. Conf. Adv. Mechatronics, Intell. Manuf. Ind. Autom.*, pp. 13–18

Analysis of Factors Affecting Continuance Intention of E-Learning Adoption in Lecturers' Perspectives

Feby Artwodini Muqtadiroh[1, a], Anisah Herdiyanti[1], Indra Wicaksono[1] and Tsuyoshi Usagawa[2]

[1] Department of Information Systems – ITS, Indonesia
[2] Graduate School of Science and Technology, Faculty of Engineering, Faculty of Advanced Science and Technologoy, Kumamoto University, Japan

[a] feby.artwodini@gmail.com, feby@is.its.ac.id

Abstract. E-learning is an internet-based learning program. It has been adopted in Institut Teknologi Sepuluh Nopember (ITS) since 2006, and called SAHRE-ITS. It is managed by Directorate of Academic Affairs in ITS. Despite quite long adoption in ITS, there are still a number of problems to confront. One of them is lack of intent of the lecturers as the main and vital users to adopt e-learning continuously. The research was to identify factors affecting the continuance intention of lecturers to adopt SHARE-ITS using some model: expectation–confirmation model (ECM), technology acceptance model (TAM), theory of planned behaviour (TPB), and FLOW theory. The variables used comprised 10 variables, i.e.: perceived usefulness, perceived ease of use, attitude, perceived enjoyment, concentration, subjective norm, perceived behavioural control, confirmation, satisfaction, and continuance intention. The results of the research indicate some factors affecting the continuance intention of the lecturers to use SHARE-ITS and suggest some recommendations for strengthening those factors. This research showed the factors of lecture's satisfaction and perceived usefulness to SHAREITS positively and significantly affected the continuance intention to use SHARE-ITS. The factor of perceived ease of use among the lecturers positively and significantly affected the perceived usefulness among the lecturer towards SHARE-ITS. Those three factors were essential to consider as they affected the continued intention the lecturers to use SHARE-ITS. In addition, it is identified that the factor of perceived enjoyment in using SHARE-ITS by the lecturers positively and significantly affected their attitude in using SHARE-ITS. Accordingly, the researchers addressed recommendation within the contexts of service rendered by the administrator of SHARE-ITS, data integration, SHARE-ITS menu and navigation designs.

1. Introduction

The information technology advancement indisputably contributes positive impacts to the world in various spectrums, specifically in education sectors adopting information technology in their business processes. One of a great number of information technology contributions in education sectors is Elearning.

E-learning is an internet-based learning program functioning as pedagogical tools to address information pupils through interactions using computers [1]. Amer et al. [1], suggested that there are five benefits for lecturers from the e-learning: (1) Ease for lecturers to update instructional materials (2) personal development or research for broadening insight (3) control over the study activities of the students and more flexible lecturing processes. Due to such various positive impacts it offers, many entities are fond of implementing e-learning. Unfortunately, there are many failures in implementing it due to low level of users' continuance intention [2]. User continuance intention represents the users' intention to continuously use products or services after the users have passed system acceptance phase [3]. There are cases where the users discontinued adopting e-learning after they had initially used it. [4]. The User continuance intention in e–learning is the prime determinant and phase in the course of acceptance of a system adoption [5]. The stronger the user continuance intention is a system is, the more actual usage of the system will be [6].

There are a number of models adoptable for analyzing the User continuance intention, namely: Expectation– Confirmation Model (ECM), Technology Acceptance Model (TAM), Theory of Planned Behavior (TPB), and FLOW theory. Expectation–Confirmation Model (ECM) is a model used for analyzing the continuance of IT adoption [7]. It comprises 3 variables for explaining the continuance of usage intention: satisfaction, confirmation, and pre adoption expectation [7]. TAM is a model adopted from Theory of Reasoned Action (TRA) consisting of two important elements: Perceived ease of use and perceived usefulness [1]. TPB is a model developed from TRA explaining the intention of one to a behavior due to 2 main factors: attitude toward the behavior and subjective norms [8]. FLOW is described as a condition where those involved in an activity find no troubles [9].

At present, the adoption of e-learning is getting increased. Yet the degree of user continuance intention is still low [10]. The same happens to an e-learning system managed by Institut Teknologi Sepuluh Nopember (ITS) and called SHARE-ITS. It is administered by Directorate of Academic Affairs in ITS and was firstly launched in 2006 with 3 departments actively accepting SHARE-ITS: Information Systems Department, Informatics Department and Electrical Engineering Department. Unfortunately, despite long history of adoption in those departments, not all lecturers in those departments continuously adopt it. In addition, those three departments also operate other e-learning systems independently. In fact, the Directorate of Academic Affairs of ITS awaits that all departments can integrate and manage their subjects in SHARE-ITS in order that the Directorate of Academic Affairs can monitor, control and evaluate to maintain quality on-line lecturing.

Accordingly, the research was to analyze the factors affecting the user continuance intention in the perspective of lecturers as users of Share ITS. The perspective of lecturers was adopted as the on-line lecturing through SHARE-ITS as opened and initiated by the lecturers, instead of by the students. The research was to identify the factors affecting the intention of the lecturers to adopt SHARE-ITS sustainably and the bring positive contribution to the ITS Quality Assurance Bureau and Directorate of Academic Affairs in order that they can enhance the continuous adoption of SHARE-ITS by the lecturers for attaining the objectives of quality lecturing processes.

2. Literature Review

2.1. Previous Researches

A research on user continuance intention to e-learning by mixing the models of ECM, TAM, TPB, and FLOW was conducted by Ming-Chi Lee (2010). It was to identify the factors affecting the intention of students in the National Pingtung University in Taiwan to continuously adopt e-learning. The results of the research showed that: the variables of *satisfaction, perceived usefulness, attitude, concentration, and perceived behaviour control positively and significantly affected the continued intention* [6].

2.2. Conceptual Model

The conceptual model in the research was formulated based on some models, namely: ECM, TAM, TPB, and FLOW theories with variables of *confirmation, satisfaction, perceived usefulness, perceived ease of use, attitude, continued intention, subjective norm, perceived behavior control, subjective norm, perceived enjoyment, and concentration* [6].

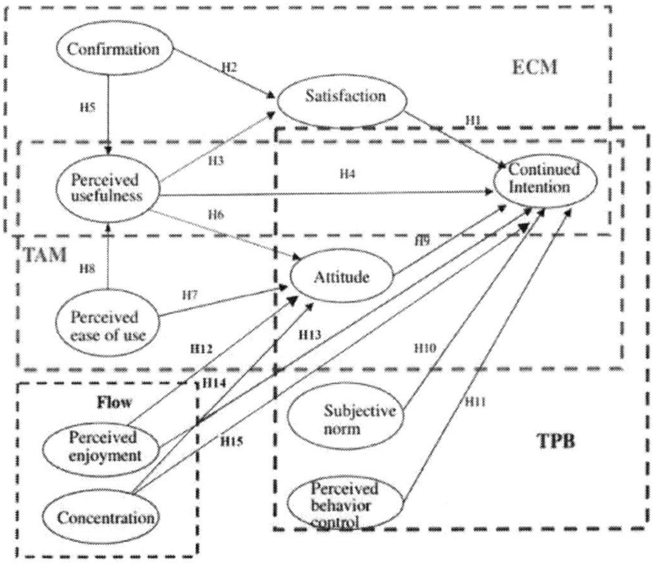

Figure 1. Conceptual Model Proposed by Ming-Chi Lee (2010)

Based on the above Figure 1, there are 4 dependent variables, comprising: *Continued Intention, Satisfaction, Perceived Usefulness, and Attitude*. In addition, there are 6 independent variables, i.e.: *Confirmation, Perceived Ease of Use, Perceived Enjoyment, Concentration, Subjective norm, and Perceived Behaviour Control.*

Confirmation is perception on suitability between compliance expectation and actual performance in a system [7]. The confirmation on this is the match between lecturers' expectation and actual performance of the SHARE-ITS.

Perceived ease of use is a perception about ease of use of technology defined as a parameter where one perceived that computers are easy to understand and use [9]. The perceived ease of use in this research is a state of condition where lecturers perceive that it is easy to use SHARE-ITS.

Perceived enjoyment is perceived as fun when doing activities by using the system [9]. The perception on enjoyment in this research is a state of condition where lecturers perceive that it is fun when using the SHARE-ITS.

Concentration is a state of condition to focus concentration on a certain activity [11]. The concentration in this research is a state of condition where lecturers can focus their attentions when using the SHARE-ITS.

Subjective norm is a social norm from other people that can influence one to behave [8]. The subjective norm in this research is a state of condition from others driving the lecturers to use the SHARE-ITS. Perceived behavior control is a perception to ease or difficulty in doing activities affected by supports, resources and challenges [12].

The perceived behavior control in this research is a state of condition where it is easy or difficult to use SHARE-ITS as affected by the supports, resources and challenges.

Continued intention is the intention of users to continuously use a system [7]. The intention in this research is the intention of lecturers as users to continuously use SHARE-ITS.

Satisfaction is a state of condition where users are satisfied with expected performance of a system [7]. The satisfaction in this research is the contentment of lecturers in using SHARE-ITS. Perceived usefulness is a perception to usefulness of a technology [7].

The perceived usefulness in this research is the usefulness perceived by the lecturers in using SHARE-ITS.

Attitude is an outlook to the use of the system in form of an acceptance or rejection as an impact when users use a technology in undertaking

IOP Conf. Series: Materials Science and Engineering **588** (2019) 012022 doi:10.1088/1757-899X/588/1/012022

3. Research Methodology

3.1. Planning Phase

It is the first phase in the research methodology. It comprises 4 main processes, i.e.: conceptual model formulation, hypothesis formulation, population and sample determination, and questionnaire building and testing.

The conceptual model formulation is supposed to identify how the method will be adopted in the research. The conceptual model is used as a basis for analyzing each data testing and model testing. The researchers adopted quantitative approach. The data were explored through survey by distributing questionnaires to ITS lecturers of SHARE-ITS and as respondents. The data were analyzed by means of quantitative and objective statistical method through scientific calculations with reference to the responses given by the respondents completing the questionnaires.

The conceptual model in the research was formulated [6] without modification. The hypotheses were formulated to identify the relationship between dependent variables and independent variables, strengthened with moderate ones. The hypotheses were also formulated based on the formulated conceptual model [6]. The arrangement of the items of statements were cited based on the researches of Bhattacheje (2001), Taylor & Tood (1995), Moon & Kim (2001) and Davis (1989).

The research subjects were ITS lecturers having used SHARE-ITS in the Information System Department, Informatics Department and Electronic Engineering Department, with minimum limit of respondents by 34% as suggested in the theory of Slovin. The prepared questionnaires were distributed to respondents both online and offline. The testing involved 20 respondents (minimum limit) completing the questionnaires. The results of the test suggested that 40 of the 44 items were state reliably and validly.

3.2. Implementation Phase

Descriptive statistical analysis was adopted to represent the questionnaire data processing results descriptively to identify the demography of respondents' data. Having completed the descriptive statistical analysis, the researchers proceeded with instrument tests.

The instrument test was to identify the reliability of the measurement test and the extent of its reliability and validity. It was conducted by testing the reliability and validity of a data. The results of the test by means of SPSS Software showed that all items were reliable and valid. Next, it was followed with classical assumption test.

The classical assumption test was adopted to assure that the resulted regression equation was accurate and consistent in estimation and not bias [13]. It comprised normality, multicollinearity and heteroscedasticity tests. The results of the tests suggested that all data are reliable. Then, it was followed up with inferential analysis.

The inferential analysis was conducted using SmartPLS Software. The model was tested and analyzed in term of outer model and inner one. The outer model test was to identify the reliability and validity of the items of indicator statement in a variable by observing the values of convergent validity, discriminant validity, composite validity, and average variance extracted (AVE). The results of the convergent validity test was valid as the loading factor of all indicator items was greater than 0.5, meaning that it was good, as the loading factor of each variable had the strongest correlation when compared to the one of other variables. The results of the composite reliability on each variable was good as the value of all composite reliability was greater than 0.7. Besides, the value of cronbach's alpha was more than 0.6. Accordingly, it was conclusive that the research variables were reliable to represent the original data scores.

Having completed the outer model test, the researchers proceeded with inner model test. The results of the inner model test showed that the regression coefficient was of all variable relationship was positive and its level of significance was more 1.68. There were 4 variable relationships having positive regression coefficient with level of significance (T-Statistic) more than 1.68, meaning that the relationships were positive and significant. The following are the results of inner model test using SmartPLS:

Table 1. Value of Path Coefficients

Variable Relationship	Name of Relationship	Parameter Coefficient	T-Table	T-Statistic
AT -> CI	Attitude -> Continued Intention	0.139	1.697	0.587
CC -> AT	Concentration -> Attitude	-0.099	1.697	0.516
CC -> CI	Concentration -> Continued Intention	-0.049	1.697	0.225
CF -> PU	Confirmation -> Perceived Usefulness	0.256	1.697	1.424
CF -> SF	Confirmation -> Satisfaction	0.279	1.697	1.081
PBC -> CI	Perceived Behaviour Control -> Continued Intention	-0.075	1.697	0.277
PE -> AT	Perceived Enjoyment -> Attitude	0.622	1.697	4.054
PE -> CI	Perceived Enjoyment -> Continued Intention	0.035	1.697	0.152
PEU -> AT	Perceived Ease of Use -> Attitude	0.219	1.697	0.855
PEU -> PU	Perceived Ease of Use -> Perceived Usefulness	0.460	1.697	2.819
PU -> AT	Perceived Usefulness -> Attitude	0.200	1.697	1.282
PU -> CI	Perceived Usefulness -> Continued Intention	0.361	1.697	1.842
PU -> SF	Perceived Usefulness -> Satisfaction	0.001	1.697	0.003
SF -> CI	Satisfaction -> Continued Intention	0.667	1.697	2.900
SN -> CI	Subjective Norm -> Continued Intention	0.018	1.697	0.102

Based on the above table, it was found that the variable relationships positively and significantly affecting were : (1) Perceived enjoyment to attitude. (2) Perceived ease of use to perceived usefulness. (3) Perceived usefulness to continued intention. (4) satisfaction to continued intention.

Next to the calculation of R-Square. R-Square is a statistical measure to know how close the data are to the fitted regression. R-Square is also known as the coefficient of determination, or the coefficient of multiple determination for multiple regression. 0% indicates that the model explains none of the variability of the response data around its mean. The following were the values of R-Square resulted by SmartPLS:

Table 2. Value of R Square

R Square of Latent Variable	
Variable	**Variable**
AT	0.622
CI	0.716
PU	0.322
SF	0.078

Based on the above Table, each of the variables was analyzed:

- The R2 of Attitude was 0.622, suggesting that Attitude could be explained by the variability of concentration, perceived enjoyment, perceived ease of use, and perceived usefulness by 62.2%.
- The R2 of continued intention was 0.716, suggesting that continued intention could be explained by the variability of attitude, concentration, perceived behaviour control, perceived enjoyment, perceived usefulness, satisfaction, and subjective norm by 71.6%.
- The R2 of perceived usefulness was 0.322, suggesting that perceived usefulness could be explained by the variability of confirmation, and perceived ease of use by 32.2%.
- The R2 of satisfaction was 0.078, suggesting that satisfaction could be explained by the variability of perceived usefulness by 7.8%

Having identified the path coefficient, the hypotheses proposed based on the conceptual model of the research through bootstrapping on the structural model were tested to identify whether they were accepted or declined. The following were the research hypotheses:

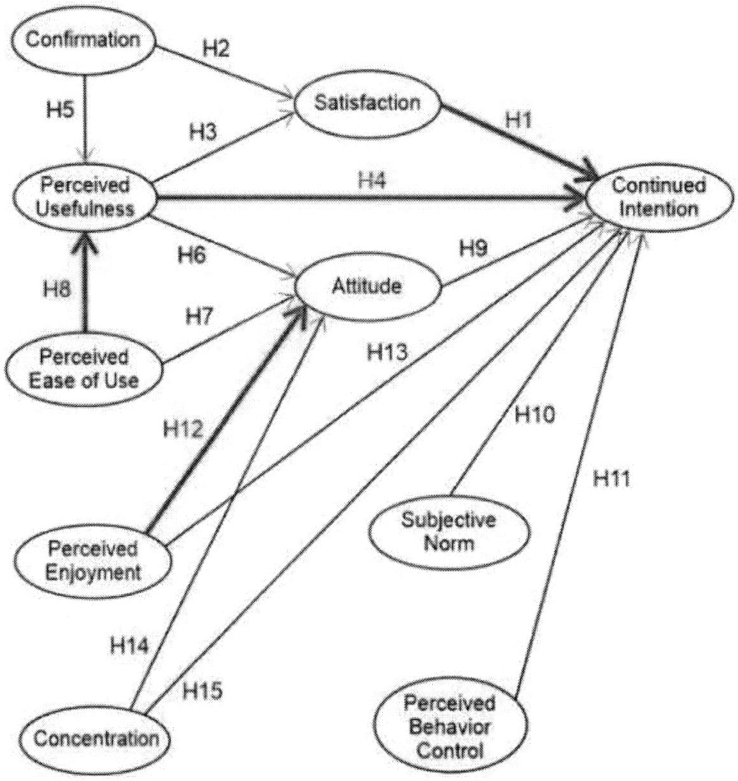

Figure 2. Research Hypotheses (Researchers, 2016)

A hypothesis is accepted when having positive loading factor and value of T-statistic greater than the one of T-Table. The following were the results of the hypothesis test:

Table 3. Results of Hypothesis Test

Hypotheses	Parameter Coefficient	T-Statistic	Remarks	Result
SF -> CI	0.667	2.900	Affecting & Significant	Accepted
CF -> SF	0.279	1.081	Affecting & Not Significant	Rejected
PU -> SF	0.001	0.003	Affecting & Not Significant	Rejected
PU -> CI	0.361	1.842	Affecting & Significant	Accepted
CF -> PU	0.256	1.424	Affecting & Not Significant	Rejected
PU -> AT	0.200	1.282	Affecting & Not Significant	Rejected
PEU -> AT	0.219	0.855	Affecting & Not Significant	Rejected
PEU -> PU	0.460	2.819	Affecting & Significant	Accepted
AT -> CI	0.139	0.587	Affecting & Not Significant	Rejected
SN -> CI	0.018	0.102	Affecting & Not Significant	Rejected
PBC -> CI	-0.075	0.277	Not Affecting & Not Significant	Rejected
PE -> AT	0.622	4.054	Affecting & Significant	Accepted
PE -> CI	0.035	0.152	Affecting & Not Significant	Rejected
CC -> AT	-0.099	0.516	Not Affecting & Not Significant	Rejected
CC -> CI	-0.049	0.225	Not Affecting & Not Significant	Rejected

With reference to the above Table 3, it suggested that there were 4 hypotheses accepted, i.e.: the hypotheses explaining the relationships between perceived enjoyment to attitude, Perceived Ease of Use to perceived usefulness, perceived usefulness to continued intention and satisfaction to continued Intention.

4. Research Findings and Discussion

4.1. Effect of Satisfaction to Continued Intention

Based on the results of inferential analysis it was found that satisfaction variable positive and significantly affected continued intention, as the value of the parameter coefficient was 0.667 and the one of t-statistic was 2.900. The value of parameter coefficient was used to explain that the positive relationship between satisfaction and continued intention. The value of T-Statistic that was greater than the one of T-Table was adopted to explain that there was significant relationship between satisfaction and continued intention. Accordingly, it was conclusive that there was positive and significant relationship between satisfaction and continued intention.
The following are recommendations for better relationship in between those two variables:

- Enhancing service deliveries through SHARE-ITS helpdesk; and
- Coordinating better with the Directorate of Information Technology and Communication in ITS

4.2. Effect of Perceived Usefulness to Continued Intention

Based on the results of inferential analysis it was found that Perceived Usefulness variable positive and significantly affected continued intention, as the value of the parameter coefficient was 0.361 and the one of t-statistic was 1,842. The value of parameter coefficient was used to explain that the

positive relationship between Perceived Usefulness and continued intention. Therefore, it was conclusive that there was positive and significant relationship between Perceived Usefulness and continued intention.
The following are recommendations for better relationship in between those two variables:

- Making enhancement and exploration of features complained by the lecturers; and
- Integrating data in SHARE-ITS with the ITS Academic Information System as these two factors are related with academic process.

4.3. Effect of Perceived Ease of Use to Perceived Usefulness
Based on the results of inferential analysis it was found that Perceived Ease of Use variable positive and significantly affected Perceived Usefulness, as the value of the parameter coefficient was 0.460 and the one of t-statistic was 2.819. The value of parameter coefficient was used to explain that the positive relationship between Perceived Ease of Use and Perceived Usefulness. The value of T-Statistic that was greater than the one of T-Table was adopted to explain that there was significant relationship between Perceived Ease of Use and Perceived Usefulness. Consequently, it was conclusive that there was positive and significant relationship between Perceived Usefulness and continued intention.
The following is a recommendation for better relationship in between those two variables: Simplifying navigation through deeper explorations (by trainings or continuous usage to get accustomed).

4.4. Effect of Perceived Enjoyment to Attitude
Based on the results of inferential analysis it was found that Perceived Enjoyment variable positive and significantly affected Attitude, as the value of the parameter coefficient was 0.460 and the one of t-statistic was 2.819. The value of parameter coefficient was used to explain that the positive relationship between Perceived Enjoyment and Attitude. The value of T-Statistic that was greater than the one of TTable was adopted to explain that there was significant relationship between Perceived Enjoyment and attitude.
Thus, it was conclusive that there was positive and significant relationship between Percevied Enjoyment and Attitude.
The following is a recommendation for better relationship in between those two variables: Enhancement of user interface by SHARE-ITS Administrator.

5. Conclusion
Based on the results of the research to identify the factors affecting the continuous intention of lecturers to use SHARE-ITS, it was concluded as follows:

- Satisfaction variable had positively and significantly affected the continued intention variable;
- Perceived usefulness variable had positively and significantly affected the continued intention variable;
- Perceived ease of use variable had positively and significantly affected the perceived usefulness variable; and
- Perceived enjoyment variable had positively and significantly affected the attitude variable.

And, it was found that there were factors directly affecting the continued use of SHARE-ITS by the lecturers, namely: satisfaction and perceived usefulness, while the one indirectly affecting the continued use of SHARE-ITS by the lecturers was the perceived ease of use. In fact, not all lecturers perceived that it was easy to use SHARE-ITS. In addition, there was a factor not related with continued intention to use SHARE-ITS by lecturers, i.e.: perceived enjoyment when using SHARE-ITS.

References
[1] Amer Al- Adwan, Ahmad Al- Adwan, Jo Smedley 2013 *Exploring students acceptance of elearning using Technology Acceptance Model in Jordanian universities*

IOP Conf. Series: Materials Science and Engineering **588** (2019) 012022 doi:10.1088/1757-899X/588/1/012022

[2] C M Chiu, S Y Sun, et al 2007 An empirical analysis of the antecedents of web-based learning continuance *Computers & Education* **49**(4) 1224–1245
[3] Hsu, Chang a Lin 2013 *an empirical study of users' continuance intention and word of mouth toward sna (social network app)*
[4] J C Roca, C M Chiu, et al 2006 Understanding e-learning continuance intention: An extension of the technology acceptance model *Human Computer Studies* **64**(6) 683–696.
[5] Chao-Min Chiu a, Meng-Hsiang Hsu b, Szu-Yuan Sun b, Tung-Ching Lin c, Pei-Chen Sun d. 2005 *Usability, quality, value and e-learning continuance decisions*
[6] Ming-Chi Lee 2010 *Explaining and predicting users continuance intention toward e-learning. National Pingtung Institute of Commerce* (Taiwan : ROC)
[7] A Bhattacherjee 2001 An empirical analysis of the antecedents of electronic commerce service continuance *Decision Support Systems* **32**(2), 201–214
[8] M Fishbein, and I Ajzen (Eds.) 1975 *Belief, attitude, intention, and behavior: An introduction to theory and research* (MA: Addison-Wesley)
[9] M Csikszentmihalyi 1997 *Finding flow: The psychology of engagement with everyday life* (New York: Basic Books)
[10] Chiu, C.-M., Sun, S.-Y., et al. 2007 An empirical analysis of the antecedents of web-based learning continuance *Computers & Education* **49**(4) 1224–1245
[11] Nugraha 2008 *Pengertian konsentrasi.* http://www.psychologymania.com/2013/04/pengertian-konsentrasi.html
[12] Lo Choi Tung 2011 *The impact of entrepreneurship education on entrepreneurial intention of engineering students* cityu university of Hong Kong

IOP Conf. Series: Materials Science and Engineering **588** (2019) 012023 doi:10.1088/1757-899X/588/1/012023

Design of ironing dies with CAD/CAM topsolid7 assistance for making 20 milimeter caliber bullet case

I M L Batan and I G A N W Baskara

Mechanical Engineering Departement, Institut Teknologi Sepuluh Nopember, Surabaya 60111, Indonesia

londbatan@me.its.ac.id , baskarawiana@gmail.com

Abstract. One of the stages to make a bullet case is ironing process, which is the process to thinning the wall part of a cup that is place between die and punch in a press tool. Die has an inner diameter difference from a small diameter to bigger diameter along the inner hole. In the other words, the inner hole of the die is conical with a slope angle is 10°. To help design ironing dies, used a Computer Aided Design (CAD) and Computer Aided Manufacturing (CAM) software TopSolid7. In this research was designed and made to make a 20 mm bullet case, from brass material CuZn30. Design process started from dimension and material strength calculation. Then the die was drawed in 3-D with CAD TopSolid 7 software. The design of die machining process is helped with TopSolid 7 CAM software. In that process design, is adding a raw material (stock), defining a reference, making toolpath, verification and G-Code making. The machining process on CNC machine become the end of design process that completed with measurement of a machining result. For example design process is designing and making an ironing die for making 20 millimeter bullet case. Ironing dies machining process is using a CNC lathe machine. The ironing die are made, measured to find out the product result spesification. From the dimension measurement of ironing die can be conclude that TopSolid 7 CAD/CAM can help the manufacture of ironing die well and can produce a precision ironing die.

1. Introduction

The manufacture of a bullet case is start with cutting a sheet material, the process is called piercing and blanking. Then, the process continue with deep drawing, which is a forming process to make cup from cutting a sheet metal (blank plate). After the cup is formed ironing process is carried out, which is the process to thinning a wall of the raw material (cup), by pressing cup with a punch when it is right in the middle of the ironing die. As a result, the cup height will be increase and the cup wall thinner equally. From several stages of ironing process on the inside diameter of dies that shrink or change gradually, will cause the height of the cup continue to increase, and at certain stages of the ironing process, a certain height of case is formed. Then the bullet case is cut off according to plan, after that one of the cup point was close, and the other point is neck case formed. According to planned of bullet caliber, then, the neck of the bullet case is cut off, therefore the final height of bullet case is obtained. The bullet case making process steps, started from forming a cup with deep drawing process, until the neck of the bullet case process (necking) and last process is cutting a case neck, as shown in Figure 1.

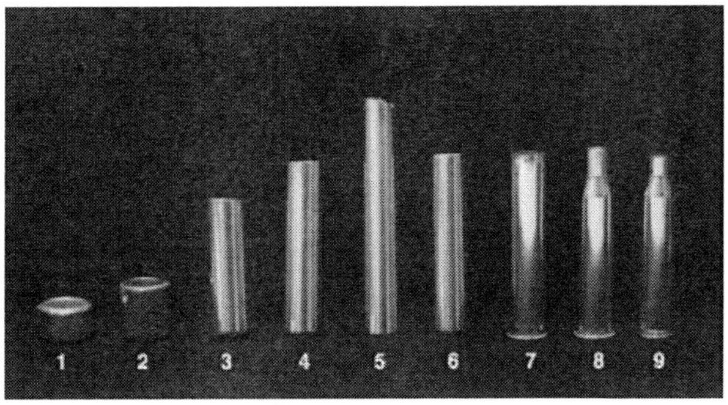

Figure 1. The Steps Of Bullet Case Making Process

It From several observations that have been made on the bullet case making process, problems arise in geometry uniformity and product dimensions that are not achieved. One reason is the hole geometry and diameter dimension on the die is rough and not precision, as a result of difficult hole making process with the size of inner diameter changing continuously make a tapered hole form (deep cone). The manufacturing experts look for solutions to make the manufacturing process with the help of CAD/CAM software. In addition to being able to compete in the market, the procurement of software is a necessity for industry to get precision geometry and product dimensions. Various brands of CAD/CAM software are produced by software developers to help the industry design and manufacturing products. One software that is currently being introduced in the Mechanical Engineering Department, FTI ITS is the TopSolid7 CAD/CAM software. The software is one of the results of the collaboration between KODAMA Japan and the Mechanical Engineering Department FTI ITS. Therefore, this research will try out to use of this software for design and manufacture of ironing dies to make 20 mm caliber case.

2. Research Method

Design and manufacture die is done by designing dimension and calculating the ironing force first, then drawing the ironing die with TopSolid 7 CAD. From the CAD image continue to design the machining process with TopSolid 7 CAM. The machining process design on TopSolid 7 software then continue to verification. The verification result become a reference to component dimension that will be produce, then the G-Code is built with CAM software assistance. The G- Code result will be entered to CNC machine for running test. To evaluate the design result, geometry measurement and workpieces dimension were carried out (ironing die) with proper measurement tool. Systematically the steps for designing 20 mm ironing die are as follow: 1) Designing ironing die, 2) Drawing ironing die, 3) Designing manufacture process of ironing die, and 4) Measuring dimension of manufactured ironing die.

3. Cup Dimension and Material

Before the die design is done, first material specification and dimension of the workpieces are determined, especially for 20 mm bullet case. For making a 20 mm bullet case, the shapes of half-finished material is cup, made from brass (CuZn30), with 40,20 mm outer diameter, and 31,20 mm inner diameter, and 26,61 mm height, as shown as Figure 2. While the expected final height of bullet case is 110 mm. The principle, to form a expected height of bullet case is need to ironing, 4 steps ironing process is needed. This means 4 type of ironing die are needed with different dimension of die inner hole. However in this research only discussed only one ironing die, which is the first stage. Whereas die design and manufacture for the next stages is no different.

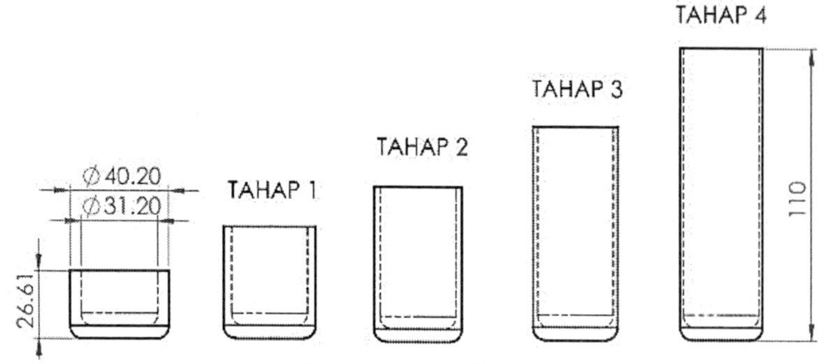

Figure 2. The Stages Of Ironing Process.

4. Ironing Dies Design

The design of ironing dies is done by calculating the thickness of cup and height of cup wall that can be achieved after ironing process. Then continue with calculating the dimension of ironing dies and ironing force is needed to carried out the ironing process. Each stages of ironing design process is desicribed as follows. Using thickness reduction ratio (Rt) 26%, which is the cup has initial outer diameter (Dl0) 40,2 mm, inner diameter (Dd0) 31,2 mm, initial thickness (t0)= 4,5 mm, and height of the cup 26,61 mm, then the thickness of cup wall from the ironing process can be calculate by calculation (1), as follow:

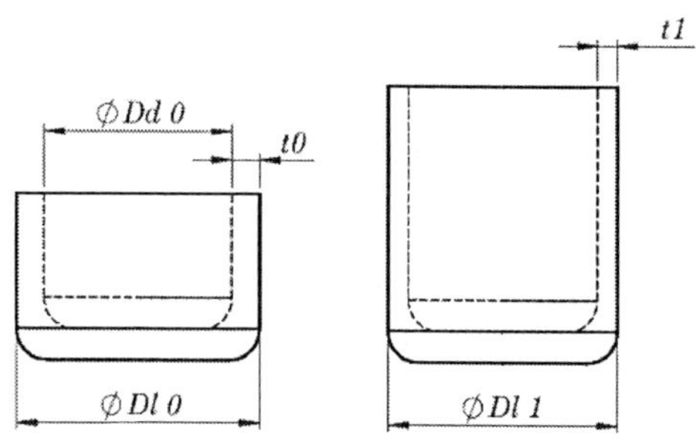

Figure 3. Ironing Process with thedimension

The thickness cup wall result from ironing process (t_1):

$$R_{t\%} = \frac{100\,(t_0 - t_1)}{t_0} \tag{1}$$

$$26\ \% = \frac{100\,(4{,}5 - t_1)}{4{,}5}$$

$$t_1 = 3,\ 33 \text{ mm}$$

From calculation result of the wall cup thickness in ironing process, the thickness of cup wall after the ironing process (t1) is 3,33 mm. Therefore, the final cup outer diameter after the ironing process is 37,86 mm (Dl1). To find out the height through the process, it is carried out with a constant volume. In constant volume method, the initial volume of the cup before ironing process considered equal to final cup volume after the ironing process. From the cup dimension, the total cup volume (Vtot) is obtained 19766,13 mm3. Based on volume constant method, the height of ironing process cup result can be calculated with equation (2):

$$Vtot = \left(\frac{\pi}{4} \times 40{,}2^2 \times L\right) - \left(\frac{\pi}{4} \times 31{,}2^2 \times (L - 5)\right) \tag{2}$$

$$19766{,}13 = \left(\frac{\pi}{4} \times 40{,}2^2 \times L\right) - \left(\frac{\pi}{4} \times 31{,}2^2 \times (L - 5)\right)$$

$$19766{,}13 = (1125{,}20L) - (764{,}15L - 3820{,}75)$$

$$19766{,}13 = 1125{,}20L - 764{,}15L + 3820{,}75$$
$$L = 44{,}16\ mm$$

From the calculation result, the height of cup is 44,16 mm. To get the geometry and cup dimension which is desired before, next calculate the die dimension that will be used for ironing process. The material that used for ironing die is SKD 11 steel with modulus elasticity 200000 N/mm^2 , ultimate tensile strength 650 MPa, yield strength 350 MPa, and density 7700 kg/ m^3. To get a proper diameter, height and thickness of the ironing die can be calculate with equation 3,4,5 as follow:
Hole diameter of the ironing die :

$$D\ dies = D\ punch + 2t + 0{,}05mm \quad (3)$$
$$D\ dies = 31{,}24\ mm + (2\ x\ 4{,}5\ mm) + 0{,}05\ mm$$
$$D\ dies = 37{,}91\ mm$$

Height of the ironing die :

$$H = \left(10 + 5T + 0{,}7\sqrt{a+b} \right) c \quad (4)$$
$$H = (10 + (5x\ 4{,}5mm) + 0{,}7\sqrt{37{,}91\ mm + 37{,}91\ mm}\ .\ 0{,}8)$$
$$H = 30{,}88\ mm$$

Thickness of ironing die :

$$T = (10 \div 20) + 0{,}8\ H \quad (5)$$
$$T = \left(\frac{10}{20}\right) + (0{,}8\ x\ 30{,}8\ mm)$$
$$T = 25{,}20\ mm$$

The hole dimension of ironing die is 37,91 mm, the minimal height of the ironing die is 30,88 mm and the minimal thickness of ironing dies is 25,20 mm. Furthermore, the ironing force was calculated and the strength of the die material was analysed, due to the ironing force. The ironing force can be calculated with equation 6, 7, 8, 9, 10 11, and 12 as follows :

$$A_0 = \frac{\pi}{4} \times (40{,}2^2 - 31{,}2^2) = 504{,}44\ mm^2 \quad (6)$$

$$A_1 = \frac{\pi}{4} \times (37{,}86^2 - 31{,}2^2) = 361{,}05\ mm^2 \quad (7)$$

$$\varphi_{max} = ln\frac{504{,}44}{361{,}05} = 0{,}334 \quad (8)$$

$$K_{str0} = 250\frac{N}{mm^2}$$

$$K_{str1} = 880 \times 0{,}334^{0.433} = 547{,}652\frac{N}{mm^2} \quad (9)$$

$$\sigma_{f,m} = \frac{250+547{,}652}{2} = 398{,}82\frac{N}{mm^2} \quad (10)$$

Therefore :

$$F_{id} = 361{,}05 \times 398{,}82 \times 0{,}334 = 48156{,}62\ N \quad (11)$$

$$F_{shear} = \frac{1}{2}\tan 10 \times 361{,}05 \times 398{,}82 = 12671{,}76\ N \quad (12)$$

$$F_{Fr,S} = \frac{2\times398{,}82\times0{,}334\times0{,}07\times361{,}05}{\sin(2\times10)} = 19713{,}24\ N \quad (13)$$

$$F_{tot} = 48156{,}62 + 12671{,}76 + 19713{,}24 = 80541{,}62\ N \approx 8{,}2\ Tonf$$

IOP Conf. Series: Materials Science and Engineering **588** (2019) 012023 doi:10.1088/1757-899X/588/1/012023

From calculation, the ironing force is 80541,62 N. While the extent of the results of the ironing process is 361,05 mm2. Furthermore, the stress received by the cup material in the ironing process can be calculated by the equation (7) as follow:

$$\sigma = \frac{F}{A} \tag{14}$$

$$\sigma = \frac{80541{,}62\ \text{N}}{361{,}05\ \text{mm2}} = 223{,}07\ \text{N/mm}^2 = 223\ \text{MPa}$$

From the calculation, the stress that cup received is 223 MPa, smaller than the ultimate tensile strength CuZn30 material 339 MPa. It can be concluded that CuZn30 workpieces material is safe for ironing process

5. Ironing Die Drawing

To realize a design become a workpiece, the draw of die is done with help of TopSolid7 CAD software. The drawing process of ironing die is helped with TopSolid 7 software start from draw a die skecth in 2 dimension (2D). The result of die drawing can be shown in Figure 4. From 2D sketch that has been draw, then its give thickness to the skecth, so the 2D skecth become 3 dimension (3D). After that, the directional hole is added to the image, finally the raw material (stock) is added for machining process. The raw material which added based from actually raw material that available.

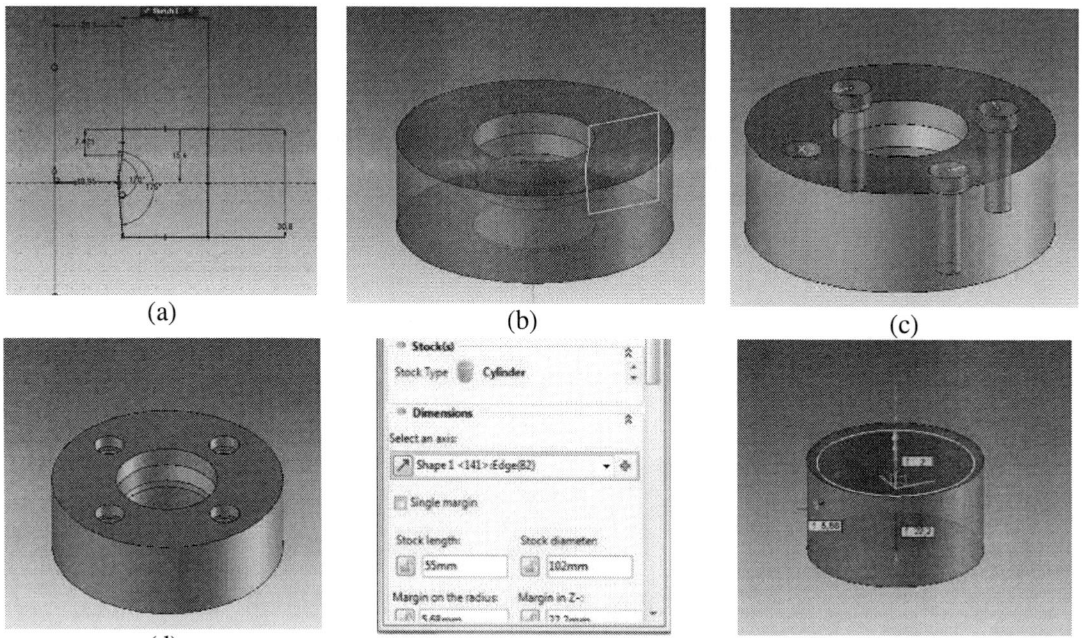

Figure 4. (a) Skecth Process (b) Giving Thickness (c) Hole Adding (d) Final 3D Design (e) Stock Dimension (f) Stock Adding

6. Design of Manufacturing Process

Toolpath is create, before design die manufacture process (machining). Toolpath making is based on the machining process planned, the toolpath result then verified to produce colours as shown in Figure 5 (b). Green colour show that the result of machining process has ± 0,1 mm dimension from planned size. while blue colour show that the result of machining process is greater than planned dimensions, therefore it means need more machining process. The stages of toolpath making process and verification process shown in Figure 5.

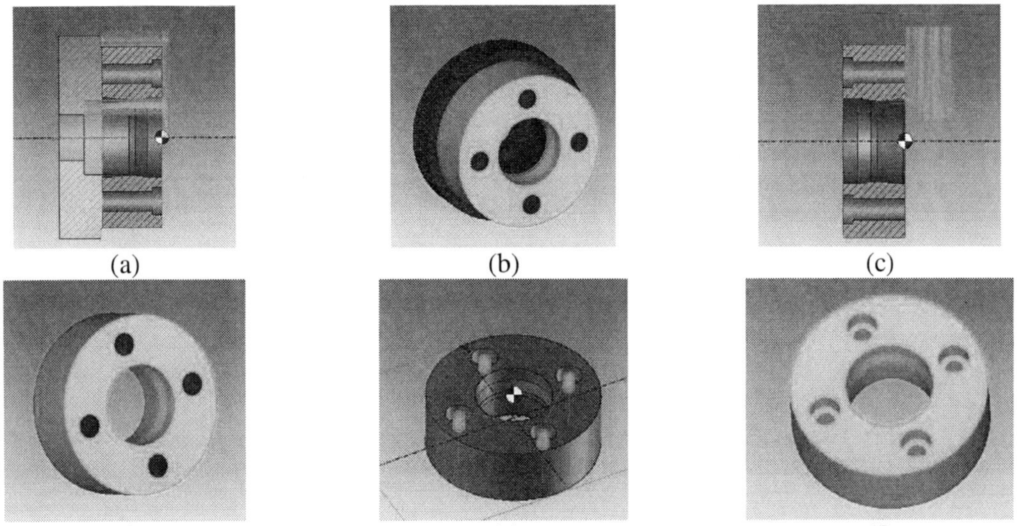

(a) (b) (c)

(d) (e) (f)

Figure 5. (a) Front Side Toolpath (b) Verification Result Of The Front Side (c) Back Side Toolpath (d) Verification Result Of The Back Side (e) Hole Making Toolpath (f) Verification Result

7. Design of Manufacturing Process

The design of manufacturing process is adjusted according to the toolpath design on TopSolid 7 software. CNC turning is used for manufacturing outer and inner diameters. Then, CNC milling is used to making holder holes. The used of cutting parameters are shown in Table 1.

Table 1. The Used of Cutting Prameters On Manufacturing Ironing Die

No	Process	Machine	Vc (m/min)	N (rpm)	F (mm/rev)	*Depth of cut* (mm)
1	Facing	CNC Turning	100	-	0,3	0,5
2	Finishing	CNC Turning	200	-	0,1	0,3
3	Eksternal Turning	CNC Turning	100	-	0,3	0,5
4	Finishing	CNC Turning	200	-	0,1	0,3
5	Center Drill	CNC Turning	20	1273	0,08	3
6	Drill Ø10	CNC Turning	20	636	0,10	5
7	Boring Ø14	CNC Turning	20	454	0,15	5
8	Boring Ø23	CNC Turning	20	276	0,15	5
9	Internal Turning	CNC Turning	100	-	0,3	0,5
10	Finishing	CNC Turning	200	-	0,1	0,3
11	Facing	CNC Turning	100	-	0,3	0,5
12	Finishing	CNC Turning	200	-	0,1	0,3
13	Internal Turning	CNC Turning	100	-	0,3	0,5
14	Finishing	CNC Turning	200	-	0,1	0,3
15	Center Drill	CNC Milling	20	1273	0,08	3
16	Drilling Ø8,5	CNC Milling	20	749	0,10	5
17	End Mill Ø14	CNC Milling	100	1500	0,5	0,02

IOP Conf. Series: Materials Science and Engineering **588** (2019) 012023 doi:10.1088/1757-899X/588/1/012023

8. Inspection of Manufactured Workpieces

The component manufacturing is done by entering G-Code result form TopSolid 7 software to used CNC machine. CNC Turning is used to manufacturing ironing die and CNC milling is used to making holder holes. Ironing die that already manufactured then be measured. Figure 6 shown the dimensions that will be measured. The calipers is used for measuring process to know the compatibility of design dimension and manufacture dimension. Dimension measurement result form ironing die shown in Table 2.

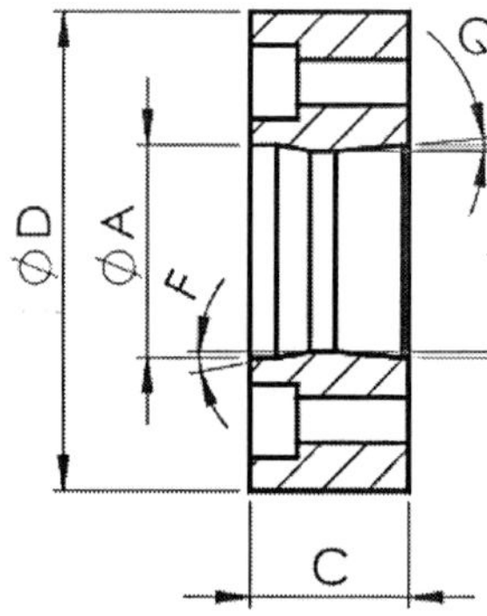

Figure 6. Ironing Die Dimensions That Will Be Measured

Table 2. Measurement Result Of Ironing Die

Part	Dimension	Measurement				
		1	2	3	4	5
ØA	40,20 mm	40,17	40,20	40,19	40,20	40,18
ØB	37,91 mm	37,91	37,95	37,92	37,92	37,93
C	30,88 mm	30,79	30,88	30,79	30,88	30,83
ØD	90,60 mm	90,59	90,60	90,60	90,58	90,59
ØE	40,20 mm	40,20	40,18	40,20	40,20	40,20
F	10°	9,83°	9,78°	9,87°	9,91°	9,97°
G	5°	4,91°	4,89°	4,93°	4,95°	4,89°

9. Conclusion

From this research can be conclude that ironing dies for 20 mm caliber bullet case making is design with die hole diameter is Ø37,91 mm with die minimal height is 30, 88 mm die minimal thickness is 25,20 mm. Ironing process for 20 mm caliber bullet case making is used to reducing an outer diameter of cup from Ø40,2 mm to Ø37,86 with height of cup result is 44,16 mm. With SKD 11 material, then ironing die was designed safe towards ironing force that will happen, and the cup material CuZn 30 can be processed properly. TopSolid 7 software can be used for help the design process and manufacturing process of 20 mm calibre bullet case. With machining process on CNC turning, the designed ironing die can be made well, and from the dimensions measurement results obtained the ironing die specifications specified in the design can be complete.

IOP Conf. Series: Materials Science and Engineering **588** (2019) 012023 doi:10.1088/1757-899X/588/1/012023

Acknowledgement
Thanks to KODAMA Corporation, Japan for granting permission to use and cooperate with the Mechanical Engineering Department, FTI ITS by providing assistance in using CAD / CAM software TopSolid7.

References
[1] Gandhi Widhi Artha 2015 *Perancangan Dies Set untuk Proses Ironing Selongsong Peluru Kaliber 20 milimeter* (Surabaya: Institut Teknologi Sepuluh Nopember)
[2] Vukota Boljanovi 2004 *Sheet Metal Forming Processes and Dies Design* (Tennessee : Industrial Press Inc)
[3] Ulrich Fischer 2010 *Mechanical and Metal Trades Handbook* (Haan-Gruiten: Verlag Europa Lehrmittel)
[4] Henkel. *LCF Behavior and Microsturcture of Alpha-Brass CuZn 30 Under Unaxial, Planar-Biaxial and Tension-Torsion Loading Conditions.* Institue of Material Engineering. Technisce Universitat Bergakademie Freiberg.
[5] Idiar 2016 *Rancang Bangun Cetakan Deep Drawing Cup Silindris untuk Selongsong Peluru caliber 20 milimeter* (Surabaya: Institut Teknologi Sepuluh Nopember)
[6] Lange and Kurt 1985 *Handbook of Metal Forming* (New York : McGraw Hill Company)
[7] M Nusron Ali *Simulation Of Ironing Process For Bullet Case To Get Minimum Forming Force With Variation Of Dies Angle And Reduction Wall Thickness* (Surabaya : Institut Teknologi Sepuluh Nopember
[8] Ivana Suchy 2006 *Handbook of Dies Design* (United States America : The McGraw Hill Companies)

IOP Conf. Series: Materials Science and Engineering **588** (2019) 012024 doi:10.1088/1757-899X/588/1/012024

Estimation of Hourly Solar Radiation on Horizontal Surface Using GAMF (Genetic Algorithm Modified Fuzzy) (Case Study in Surabaya)

Erma Hakim Setyawan[1, 2,a], Imam Abadi[1], and Sartika Arie Kusumawarni[1]
[1] Engineering Physics Departement, Faculty of Industrial Technology, Sepuluh Nopember Institute of Technology, Surabaya 60111, Indonesia
[2] SMAN 1 Porong, Sidoarjo

[a] setyawan5105@gmail.com

Abstract. The biggest renewable energy sources in Indonesia is solar energy. The installed capacity of Solar Power System in 2017 is still very far from the target. Solar radiation is very infulental on the photovoltaic performance in generating energy. The need for solar radiation estimation has become important in the design of photovoltaic system. Previous research has been done using Extreme Learning Machine (ELM) and Adaptive Neuro Fuzzy (ANFIS) methods. In this research, Genetic Algorithm Modified Fuzzy (GAMF) method used to estimate the solar radiation. Meteorogical dates used in this research are temperature, humidity, wind velocity and wind direction. There are 2 types of datas that are BMKG data and measurement data. Three experimental variations of input variation were performed for each data. For BMKG data the best estimation result is obtained when using humidity, temperature and wind velocity as variation inputs with RMSE and MAE of 145.19 and 72. While for the best result estimation result data obtained when applying humidity, temperature, wind speed and wind direction as variation inputs with values of RMSE and MAE were 1.44 and 0.65.

1. Introduction

The position of Indonesia is located in the equator, making Indonesia as a country with abundant solar energy resources, where the average solar radiation intensity is around 4.8 kWh / m^2 per day in all parts of Indonesia [1]. The data released by the EMR Ministry states that the installed capacity of the 2017 Solar Power System (PLTS) is 80 MW of the government target of 5,000 MW in 2019 and 6,400 MW in 2025 [2] Based on the data above installed capicity at this time still very far from the target of 2019 and 2025, whereas the potential for the development of PLTS in Indonesia itself is very large.

Low amount of energy generated by a PLTS can be caused by various factors. One of them is the Photovoltaic placement used for energy conversion. To obtain optimum output, Photovoltaic needs to be placed in such a way as to obtain maximum solar radiation. Photovoltaic is generally placed in a permanent position without change, so it will get maximum solar radiation only when the sun is perpendicular to the field. As the sun and Photovoltaic fields form an angle θ, there is a decrease in solar radiation. To know the amount of solar radiation received by Photovoltaic required a measuring instrument of solar radiation or the commonly used is phyranometer. This study is expected to produce solar radiation estimators at a cost that is more efficient than the previous tool. Modeling of solar radiation received at the surface of the earth, especially in the Surabaya area, further research is needed to obtain an estimator according to the conditions in Surabaya.

Previous researchs have been conducted on estimating solar radiation with various methods in [3] [4] [5]. In the research [3] modeling for hourly solar radiation using artificial neural network feedforward backpropagation. This research reach accuracy of 93.09% in estimating solar radiation per hour. It shows better accuracy than using mathematical modeling. While in research [4] esplained estimation of solar radiation per hour with method of ELM (Extreme Learning Machine), both method that used for feedforward backpropagation method have value of MSE (Mean Square Error) equal to 0,3378 and method ELM obtained MSE of 5.88E-14. In addition, for the learning speed in the testing process for feedforward backpropagation method has a duration of learning speed of 0.2171 seconds and for the ELM method has a duration of learning speed of 0.0156 seconds from the estimated solar radiation. It shows that ELM method is better in terms of accuracy and learning speed compared with feedforward backpropagation method. Antother study [5], estimating solar radiation using ANFIS (adaptive Neuro Fuzzy) method. In this research we got comparison between estimation using ELM and ANFIS method, ANFIS has better RMSE and MAE compared to ELM. The smallest RMSE value is 0.87 and the smallest MAE is 0.22. This shows that in this case ANFIS method has better estimation ability than using ELM.

In the next research will be developed a new method in determining the estimated solar radiation per hour on horizontal surface (case study in Surabaya) using method of GAMF (Genetic Algorithm Modified Fuzzy).

2. Methodology

Estimation to hourly solar radiation on horizontal suface using GAMF method in Surabaya developed due to research method shown in figure below:

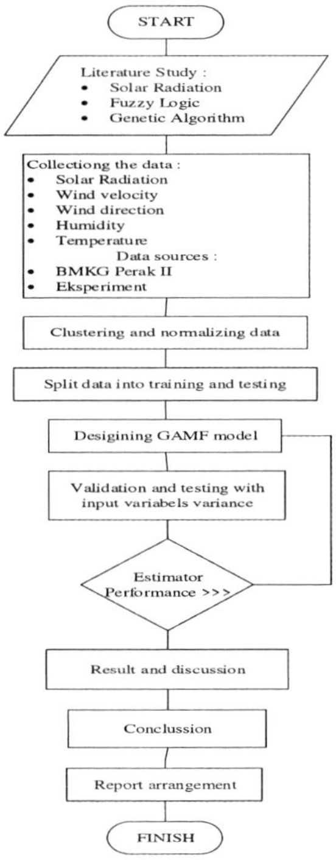

Figure 1. Methological Flowchart

2.1 Collectiong the Data

The research using two kind of data from The Meteorology, Climatology, and Geophysics Body (BMKG) Perak II Surabaya and experiment. The research uses hourly data. These are solar radiation on a horizontal surface (W/m^2), air temperature (^{o} C), relative humidity (%), wind speed (m/s) and directions. That data were taken during January 2015 until December 2017 for BMKG data and February 2018 until April 2018 for experiment data.

Training and testing data classification

Both of data divided into two two parts they are training and testing with ratio 80 :20. The training and testing process will be divided into 3 experiments with variation of input variables. The input variables used are: humidity, temperature, wind speed and wind direction. While the output variables are solar radiation. This is done to find the input variation that produces the best estimation output. The input variations scheme are explain in figure below.

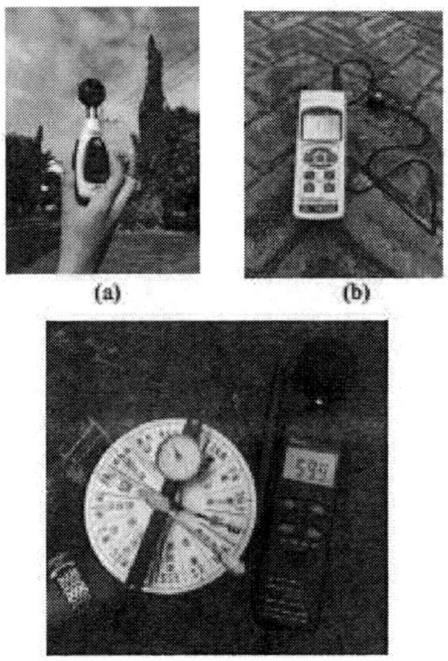

Figure 2. Collecting the data (a) Wind velocity (b) Solar radiation (c) Wind direction, Humidity and Temperature

Table 1 : The input variations scheme

Experiment	Input variations
P1	Temperature, Humidity
P2	Temperature, Humidity, Wind velocity
P3	Temperature, Humidity, Wind velocity, Wind direction

Determining number and type of membership function

To get a good estimation result required the right type of membership function. In this research used membership type function gauss with membership membership function 3. Membership function used is gauss type because it is considered the most optimal compared to other membership function types that is trim, trap and gbell. The membership function obtained from the training results for the four input variables used are as follows:

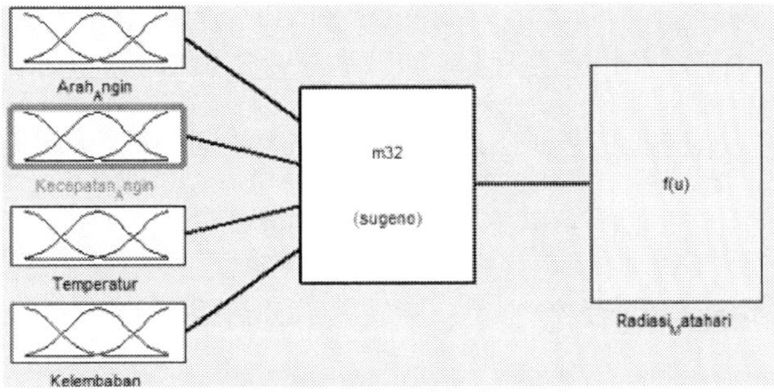

Figure 3. Membership Functions

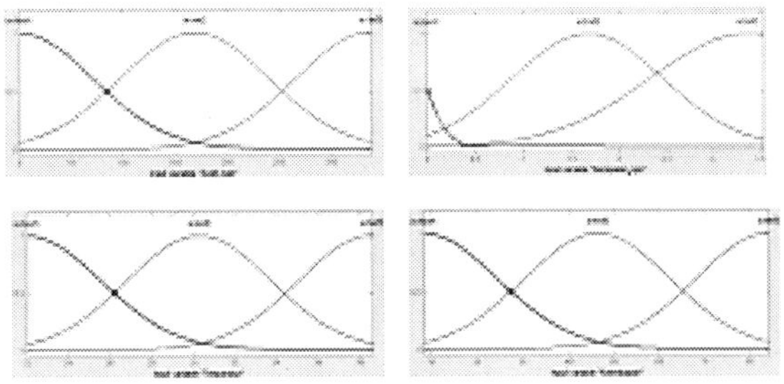

Figure 4. Membership Function Curve

Number of membership functions used for this research are 3, from the training process generates a rule based for output and input pair with the number of rule based 9 for 2 variations of input variables, 27 for 3 input variations and 81 for 4 variations of input variables.

Analyze the Estimation Result

After getting the estimated output obtained from MATLab with the "evalfis" command then calculation done with Microsoft Excel to get the error value. The type of calculation performed are Root Mean Square Error (RMSE) and Mean Absolute Error (MAE). The calculation formula for determining RMSE and MAE is as follows.

$$\text{RMSE} = \sqrt{\frac{1}{N}\sum_{t=1}^{N}(y_t - t_\tau)^2} \quad (1)$$

$$\text{MAE} = \frac{1}{N}\sum_{t=1}^{N}|(y_t - t_\tau)| \quad (2)$$

GA Optimization

The estimation results obtained from the fuzzy method given as the initial population then optimized to obtain more optimum results with the genetic algorithm method. The optimization process is carried out for the three experiments of input variation. The objective function used in the optimization process using genetic algorithm is error. The magnitude of the error can be calculated from the difference in target solar radiation with the estimated solar radiation obtained from the fuzzy method.

3. Result and Discussion

3.1. BMKG Data Analysis

Data analysis has been done on secondary data obtained from BMKG Maritim Perak II Surabaya within 3 years period from January 2015 until December 2017. The data obtained consist of wind speed, wind direction, temperature, humidity, and solar radiation with the amount of 10542 data pairs. The secondary data characteristics BMKG Maritime Perak II Surabaya is as follows:

Table 2. Characteristic of Data BMKG

	Temperature (°C)	Humidity (%)	Wind speed (m/s)	Wind direction (°)	Solar radiation (W/m^2)
Range	22,4-37,7	30,94-100	0-8,75	0-359	0-1207
Span	15,3	69,06	8,75	359	1207
Mean	28,94	76,33	1,87	157,3	400,96
Std. Dev	2,45	11,48	1,17	96,04	295,33
Max	37,7	100	8,75	359	1207
Min	22,4	30,94	0	0	0

There are two types of classification data, training data and data testing. Classification is done with the ratio of 80:20 so that obtained training data amounted to 8434 pairs of data and data testing amounted to 2108 pairs of data.

3.2. Experiment Data Analysis

The experiment data is obtained from measurement with measuring instrument. The amount of experiment data are 685 pairs of data and classified into data training and data testing with a ratio of 80:20. So that obtained training data amounted to 548 pairs of data and data testing amounted to 137 pairs of data. The characteristics of the measurement data are as follows.

Table 3. Characteristic of Experiment Data

	Temperature (°C)	Humidity (%)	Wind velocity (m/s)	Wind direction (°)	Solar radiation (W/m^2)
Range	22,6-38,6	8,5-84,2	0-3,53	0-337,5	21,8-1073
Span	16	75,7	3,53	337,5	1051,2
Mean	30,62	35,35	0,79	127,77	534.15
Std. Dev	4,77	17,4	0,573	96,04	313,704
Max	38,6	84,2	0	359	1073
Min	22,6	8,5	0-3,53	0	21,8

3.3 Modeling Analysis (BMKG Data)

Based on the results of the computations performed using MATLab software using the Fuzzy-Sugeno method, the estimation stage is divided into 2, namely training and testing. Here is the result of training from simulation on experiment P1, P2, P3 with Fuzzy method and GAMF.

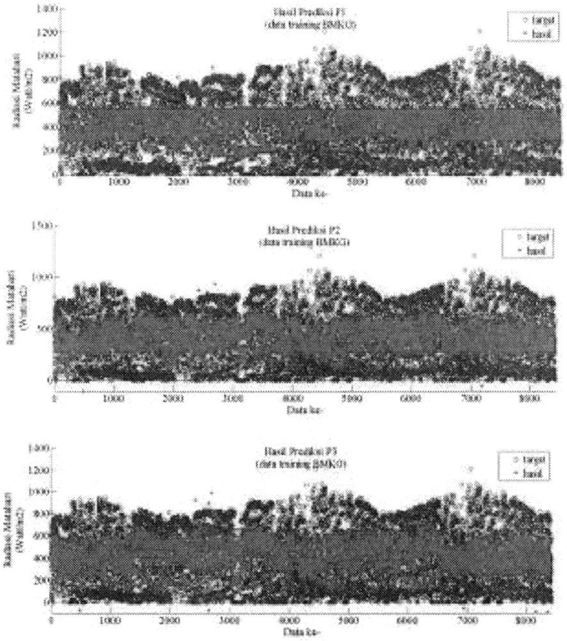

Figure 5. Estimation Result from Fuzzy Merhod (BMKG training data) (a) P1 (b) P2 (c) P3

Figure 5 shows the distribution of target solar radiation and estimation results for each experiment using the Fuzzy method for BMKG training data. In Figure 5 (a) a maximum error of 2119435,15, a minimum error of 0.01, and a MAE of 232. In Figure 5 (b) a maximum error of 1879834,90, a minimum error of 0, and a MAE of 219. 5 (c) maximal error of 1493389,78, minimum error of 0.01, and MAE of 209. Thus, the best estimation results obtained when using variations of input variables as many as 4 pieces

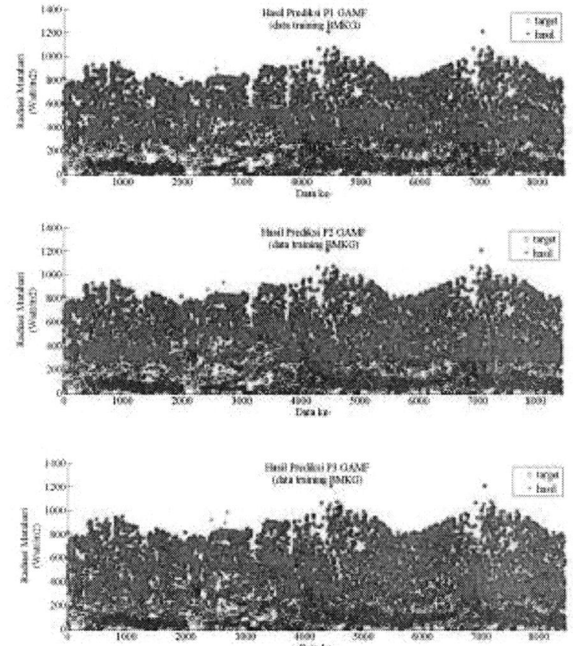

Figure 6. Estimation Result from GAMF method (BMKG training data) (a) P1 (b) P2 (c) P3

In Figure 6 we can see the spread of target solar radiation and estimation results for each experiment using the GAMF method. In this Figure it can be seen that the spread of the estimation results is more evenly distributed than the previous figure, indicating that the estimation is close to the target. Based on data calculation performed, GAMF method has a smaller MAE value when compared to Fuzzy method that is 115; 109 and 104 for each input variation.
Based on the results of the training, we have obtained the comparison of Root Mean Square Error (RMSE) and Mean Absolute Error (MAE) results for the Fuzzy and GAMF methods as follows:

Table 4. RMSE and MAE Analysis from BMKG data training

P-	Structure	RMSE *(training)*		MAE *(training)*	
		Fuzzy	**GAMF**	**Fuzzy**	**GAMF**
P1	3, Gauss, 100	271,22	191,95	232	115
P2	3, Gauss, 100	262,15	186,35	219	109
P3	3, Gauss, 100	253,56	181,25	209	104

Based on Table 4 we can see significant decrease in RMSE and MAE in GAMF method when compared with Fuzzy method. The smallest RMSE and MAE, 181,25 and 104, were obtained when the input variation was 4, so the most optimum estimation result for BMKG data training was using P3 experiment with GAMF method.
As for testing phase BMKG data for experiments P1, P2 and P3 with Fuzzy and GAMF method can be seen in the picture below.

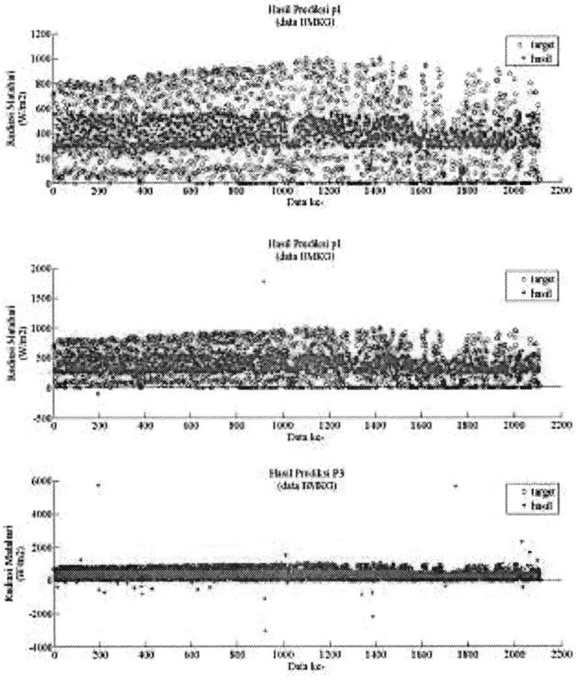

Figure 7. Estimation Result from Fuzzy method (BMKG testing data) (a) P1 (b) P2 (c) P3

Figure 7 shows the distribution of target solar radiation and estimation results for each experiment using the Fuzzy method. The data used is BMKG training data. In Fig. 7 (a) the maximal error is 577830.89, the minimum error is 0.0044, and the MAE is 265. In Figure 7 (b) the maximal error is 1775652,14, the minimum error is 0.1542, and the MAE is 257. In fig. 7(c) maximum error of 5616792,55, minimum error of 0.0630, and MAE of 289. Thus, for training data using BMKG data, the best estimation results obtained when using 3 variation of input variables.

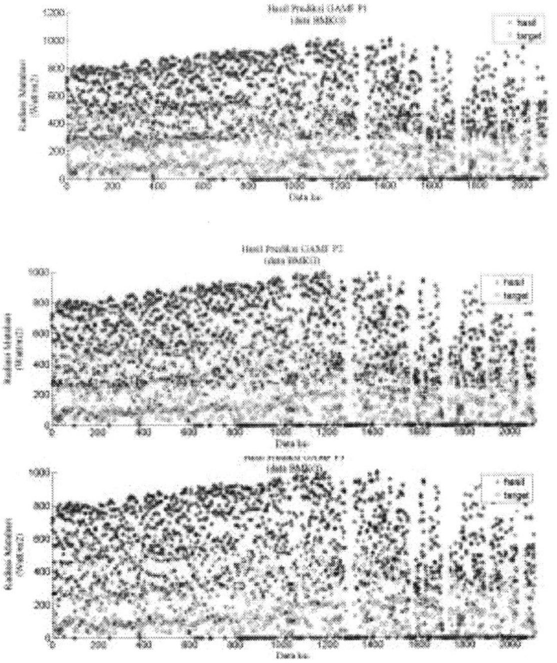

Figure 8. Estimation Result from GAMF method (BMKG testing data) (a) P1 (b) P2 (c) P3

Figure 8 shows the distribution of target solar radiation and estimation results for each experiment using the GAMF method. In this Figure it can be seen that the spread of the estimation results is more evenly distributed than the previous figure, indicating that the estimation is close to the target. Based on data processing done, GAMF method has a smaller MAE value when compared to Fuzzy method that is equal to 74; 65 and 72 for each input variation.

The three experiments that have been done using BMKG data testing obtained comparison of Root Mean Square Error (RMSE) and Mean Absolute Error (MAE) for Fuzzy and GAMF method are as follows:

Table 5. RMSE and MAE Analysis from BMKG data testing

P-	**Structure**	**RMSE** *(testing)*		**MAE** *(testing)*	
		Fuzzy	**GAMF**	**Fuzzy**	**GAMF**
P1	3, Gauss, 100	301,19	138,67	265	74
P2	3, Gauss, 100	297,89	126,61	257	65
P3	3, Gauss, 100	388,46	145,19	289	72

Based on Table 5 we can see significant decrease in RMSE and MAE on GAMF method when compared with Fuzzy method. For BMKG data, the smallest RMSE and MAE values of 126.61 and 65 are obtained when the input variation is 3, so the most optimum estimation result for BMKG data training is using P2 experiment with GAMF method.

3.4.Modeling Analysis (Experiment Data)

Based on the results of the computations performed using MATLAB software using the Fuzzy-Sugeno method, the estimation stage is divided into 2, namely training and testing. The following is the result

IOP Conf. Series: Materials Science and Engineering **588** (2019) 012024 doi:10.1088/1757-899X/588/1/012024

of the training of measurement data from the simulation on experiment P1, P2, P3 with Fuzzy method and GAMF:

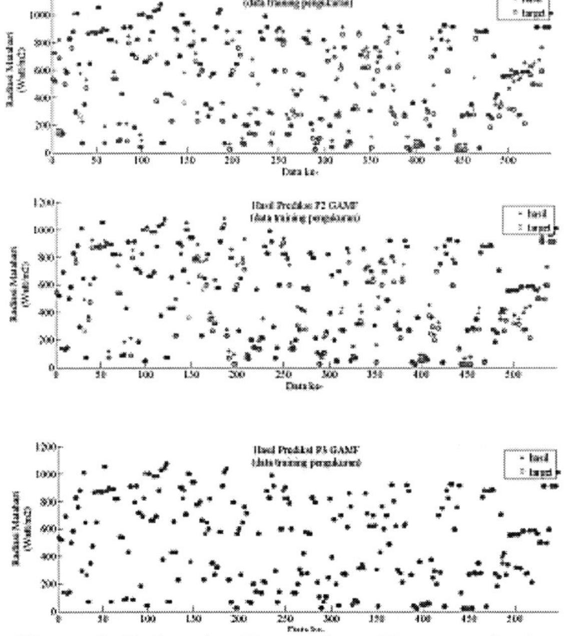

Figure 9. Estimation Result from Fuzzy method (Experiment training data) (a) P1 (b) P2 (c) P3

Figure 9 shows the distribution of target solar radiation and estimation results for each experiment using the Fuzzy method. The data used is BMKG training data. In Figure 9 (a) a maximum error of 226.23, a minimum error of 0.3642, and a MAE of 140.24. In Figure 9 (b) the maximal error is 349.39, the error is at least 0.0213, and the MAE is 94.45. there is Figure 9 (c) maximal error of 10.62, minimum error of 0.0025, and MAE of 2.94. Thus, for training data using BMKG data, the best estimation results obtained when using 4 variations of input variables.

In Figure 10 we can see the distribution of target solar radiation and estimation results for each experiment using the GAMF method. In this Figure it can be seen that the spread of the estimation results is more evenly distributed than the previous figure, indicating that the estimation is close to the target. Based on data processing performed, GAMF method has a smaller MAE value than Fuzzy method that is equal to 56; 35 and 0 for each input variation.

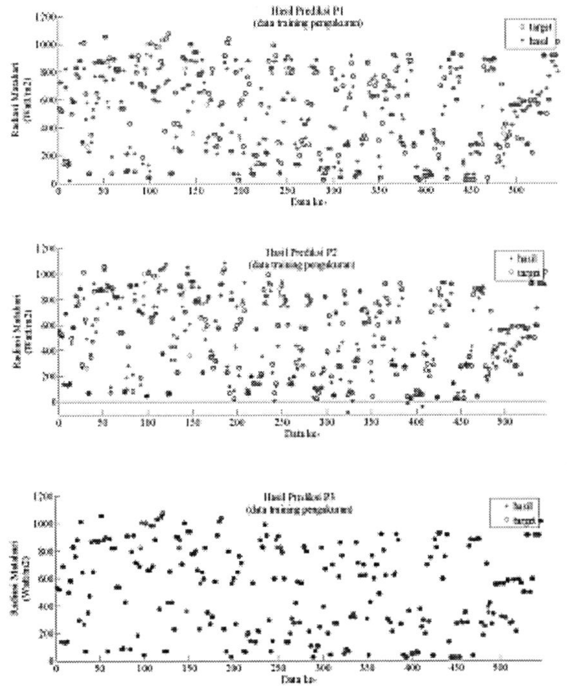

Figure 10. Estimation Result from GAMF method (Experiment training data) (a) P1 (b) P2 (c) P3

Based on the results of the training, we have obtained the comparison of Root Mean Square Error (RMSE) and Mean Absolute Error (MAE) results for the Fuzzy and GAMF methods as follows:

Table 6. RMSE and MAE Analysis from experiment data training

P-	**Structure**	**RMSE *(training)***		**MAE *(training)***	
		Fuzzy	**GAMF**	**Fuzzy**	**GAMF**
P1	3, Gauss, 100	140,24	97,43	112	56
P2	3, Gauss, 100	94,45	69,34	70	35
P3	3, Gauss, 100	2,94	0,54	2	0

Table 6 shows a significant decrease in RMSE and MAE in the GAMF method when compared to the Fuzzy method. The smallest RMSE and MAE are 0,54 and 0, obtained when the input variation is 4, so the most optimum estimation result for the training of measurement data is using P3 experiment with GAMF method.

As for testing phase BMKG data for experiments P1, P2 and P3 with Fuzzy and GAMF method can be seen in the picture below:

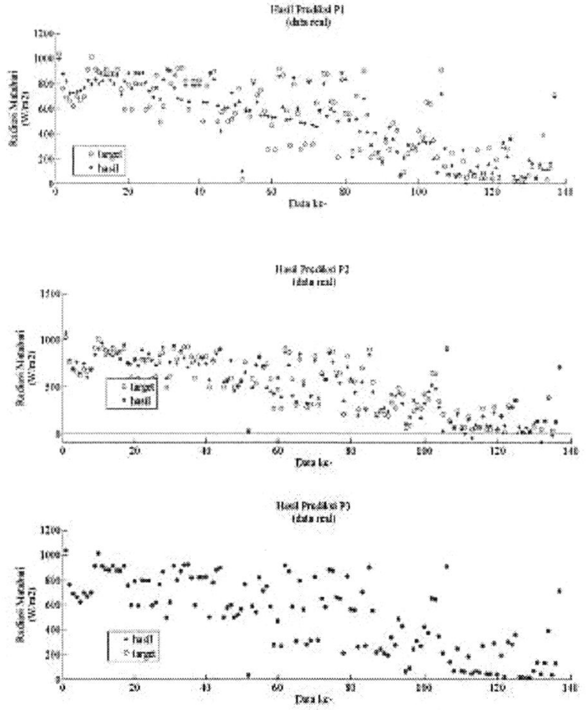

Figure 11. *Estimation Result from Fuzzy method (Experiment testing data) (a) P1 (b) P2 (c) P3*

Figure 11 shows the distribution of target solar radiation and estimation results for each experiment using the Fuzzy method. The data used is BMKG training data. In Figure 11 (a) a maximum error of 226.23, a minimum error of 0.36, and a MAE of 126.44. In Figure 11 (b) a maximum error of 303.0732, a minimum error of 0.0213, and a MAE of 8.68. There is Figure 11 (c) maximal error of 9.38, minimum error of 0.0035, and MAE of 2.1. Thus, for training data using BMKG data, the best estimation results obtained when using 4 variations of input variables.

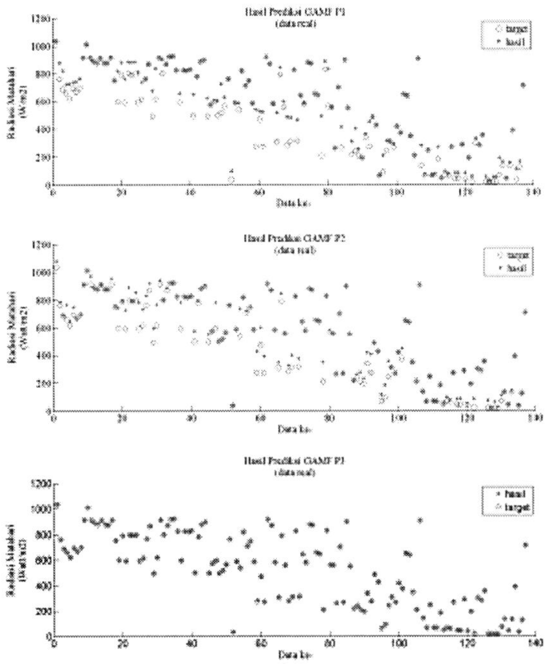

Figure 12. *Estimation Result from Fuzzy method (Experiment testing data) (a) P1 (b) P2 (c) P3*

In Figure 12, we can see the spread of target solar radiation and estimation results for each experiment using the GAMF method. In this Figure it can be seen that the spread of the estimation results is more evenly distributed than the previous figure, indicating that the estimation is close to the target. Based on data processing done, GAMF method has a smaller MAE value than Fuzzy method is 49.85; 30.31 and 0.65 for each input variation.

Based on the result of radiation estimation by using MATLab software on data testing result of measurement then continued with data calculation using microsoft excel to get comparison of result of Root Mean Square Error (RMSE) and Mean Absolute Error (MAE) for Fuzzy and GAMF method as follows:

Table 7. RMSE and MAE Analysis from experiment data testing

P-	**Struktur**	**RMSE *(testing)***		**MAE *(testing)***	
		Fuzzy	**GAMF**	**Fuzzy**	**GAMF**
P1	3, Gauss, 100	126,44	87,21	99,31	49,85
P2	3, Gauss, 100	84,68	57,34	65,78	30,31
P3	3, Gauss, 100	2,10	1,44	1,26	0,65

Based on Table 7 we can see significant decrease in RMSE and MAE in GAMF method when compared with Fuzzy method for all varieties of input variables. In the table, the smallest RMSE and MAE values of 1.44 and 0.65 are obtained when the input variation is given 4 with the GAMF method, so the most optimum estimation result for the measurement data testing is using P3 experiment with the moisture input variable, temperature, wind speed and wind direction.

4. Conclusion

Based on data analysis and discussion about solar radiation estimation using GAMF method, it can be drawn the conclusion as follows:

- The best estimation parameter for BMKG data is using 3 variation of input variables they are humidity, temperature and wind speed. As for the data measurement results, using variations of 4 inputs they are humidity, temperature, wind speed and wind direction. It can be seen based on the smallest RMSE and MAE values.
- The design of the GAMF method for estimation of solar radiation using the Sugeno-type Fuzzy, and the estimates produced by Fuzzy are optimized using genetic algorithm methods. In Fuzzy model type of membership function used is Gaussian with number of membership function as much 3. On GA optimization objective function used is error, with amount of iteration counted 100.
- Accuracy of solar radiation estimation results by using GAMF method can be seen from result of RMSE and MAE produced. For data BMKG, RMSE and MAE most optimum that is equal to 126,61 and 65. As for experiment data are 1,44 and 0,65.

References

[1] I. Abadi, A. Musyafa and A. Soeprijanto 2014 Extreme Learning Machine Approach to Estimate Hourly Solar Radiation On Horizontal Surface (PV) in Surabaya -East Java *International Conference on Information Technology, Computer and Electrical Engineering (ICITACEE)* **vol. 1** pp. 370-374

[2] R. Wardani, *"Direktorat Jendral Energi Baru Terbarukan dan Konservasi Energi,"* 14 September 2017. [Online]. Available: http://ebtke.esdm.go.id. [Accessed 8 Desember 2017].

[3] A. Kassem 2009 Development of Neural Network Model to Estimate Hourly and Diffuse Solar Radiation on Horizontal Surface at Alexandria City (Egypt) *Applied Science Research* **vol. 5**

[4] N. U. Hidayatullah 2014 *Estimasi Radiasi Matahari Perjam Pada Permukaan Horizontal Dengan Extreme Learning Machine (Studi Kasus Di Surabaya)* (Surabaya : Teknik Fisika ITS)

[5] I. Abdurrahman 2017 *Estimasi Radiasi Matahari Perjam Pada Permukaan Horizontal Dengan Adaptive Neuro Fuzzy Inference System (Anfis) (Studi Kasus Di Surabaya),"* (Surabaya : Teknik Fisika ITS)

Numerical Analysis of Heterogeneous Oxidation Reaction on Multi-stage Air Inlet Downdraft Gasifier

Arif Rahman Saleh[1, 2,a], Bambang Sudarmanta[1,] and Depi Rustam Ependi[1]

[1]Department of Mechanical Engineering, Faculty of Industrial and Technology, Institut Teknologi Sepuluh Nopember.
[2]Department of Mechanical Engineering, Faculty of Engineering, Universitas Pasir Pengaraian.

[a] ariefrahmansaleh@upp.ac.id

Abstract. Gasification technology is one of the thermochemical conversion methods that can be a solution to overcome the negative municipal solid waste (MSW) impact. Syngas application from gasification on the internal combustion engine has problems with tar content. one method for reducing tar content is by means of a multi-stage air inlet in the reduction zone to initiate heterogeneous oxidation reactions. In this study a three-dimensional computational fluid dynamic model was used to simulate gasification reactions. The simulation was carried out using a one-phase model with k epsilon turbulence model, P1 radiation model, devolatilization model using kobayashi model, MSW fuel modeled as a discrete phase that moves in the continuous phase, and the reactions that occur are volumetric and particle surface reaction in which the interaction of turbulence-chemistry using finite-rate/eddy-dissipation. The ratio of air used is 10%, 20%, 30%, and 40% between oxidation and reduction. Based on the simulation results, there is an increase in syngas composition, especially CO because there is a heterogeneous oxidation reaction between carbon in the reduction zone with oxygen which also triggers a rise in temperature in the oxidation zone. CO2 concentration decreases due to an increase in the rate of boudouard reaction. but when the amount of air continues to increase there is a decrease in temperature in the oxidation zone because the amount of air for the oxidation reaction decreases, the simulation results obtained can be used to optimize the downdraft gasifier by adding air input to the reduction zone

1. Introduction

An increase in the population in the world has a negative impact on the amount of municipal solid waste that continues to grow. Based on data on world population growth, an increase in the world's population, especially in urban areas, in 2020 reaches 66% [1]. This number will increase the population of the world's population which in 2015 7 million people. The handling of municipal solid waste currently varies from accumulation in final landfills, incinerator combustion, methane gas harvesting, and using thermochemical processes namely gasification technology. when compared to direct combustion,

IOP Conf. Series: Materials Science and Engineering **588** (2019) 012025 doi:10.1088/1757-899X/588/1/012025

gasification has advantages because in addition to producing heat that can be converted into energy, this technology also produces gas (syngas). Direct MSW combustion has a negative impact on the environment because it causes CO and CO2 emissions, while landfill accumulation also causes CH4 emissions and affects the quality of groundwater due to liquid waste deposits (leachate) [2, 3].

Therefore, from the various methods of handling MSW above, the most environmentally friendly is Gasification. with gasification technology, solid waste is converted into energy in the form of gas (syngas) [4]. The gasification reaction occurs through several stages, namely drying, pyrolysis, partial oxidation, and reduction. The resulting syngas consists of combustible gas components (CO, H2, CH4) and non-combustible (CO2, O2, N2, H2O). besides, gasification also produces residues in the form of ash and tar. This syngas can be used for both thermal and power applications. for the power application itself, including being used as fuel on diesel fuel motors with dual fuel mode. However, to be able to be used on the syngas diesel engine that is produced, it must meet the minimum standards of tar 100 mg / Nm3 [4, 5]. Several studies have been carried out to reduce the tar content in syngas by using primary and secondary methods. The primary method reduces the tar content in situ in the gasifier. while the secondary method decreases the tar content after leaving the gasifier using various methods including cyclone, water scrubber and dry filter [6].

The primary method that has been carried out by several researchers is to modify the structure of the gasifier with multilevel air input. In the condition of one level, the air for gasification enters the partial oxidation zone while in the multilevel condition the air enters the zone of pyrolysis, oxidation, and reduction. To input air in the pyrolysis zone, a study was conducted by Sudarmanta and arif [7-10] which modifies the gasification of an airway into two airways. Bentzen [11] air entering the pyrolysis zone and gave bulkhead using the grate to separate pyrolysis and partial oxidation zone. Bui [12] adding air to the pyrolysis zone to determine the effect of air input to the performance of gasification. Martinez dan galindo [13, 14] vary the percentage of air input in the pyrolysis and oxidation zone to obtain the optimum ratio. From the various studies reported that there was an increase in gasification performance in the form of an increase in the volumetric gas percentage especially CO and also a pyrolysis zone temperature increase, besides that the tar content could also be reduced to 80% compared to a single air input. In addition to input pyrolysis air, shi and nerijus [15, 16] conduct experiments with air input in all three zones namely pyrolysis, oxidation, and reduction. The results of his research report that when air is put into the reduction zone, the air will tend to react with carbon compared to gas. Therefore, heterogeneous reactions in the reduction zone occur in an oxidative environment. In addition, because the air in the reduction zone oxidizes a small amount of carbon, an increase in temperature increases the CO2 and H2O reduction reaction rate to be more intensive. In addition, the amount of unoxidized residual charcoal is also reduced because it has reacted with oxygen from the air. However, from various studies there has been no numerical research for multilevel air input using a computational fluid dynamic (CFD) model.

Numerical modeling is needed because not all phenomena in the gasification process can be measured through experiments. other than that, for the purpose of numerical modeling optimization is needed before physical modification or optimization is carried out. Various CFD gasification modeling studies have been conducted by several researchers including Gupta and Simone [17, 18] modeling the gasification reaction to find out the temperature distribution along the gasifier. therefore, the purpose of this study is to analyze the effect of air input in the reduction zone on operating conditions (temperature distribution, pressure, and speed) and gasification performance (syngas composition, low heating value, and cold gas efficiency).

2. Material and Method

2.1. Computation model

The type of gasifier used in this numerical analysis is the downdraft gasifier. This gasifier design uses two air inputs, namely the oxidation zone and the reduction zone. 3-dimensional modeling is done is to simplify the form of the gasifier by making a fluid domain only on the inside of the gasifier. Figure 1 shows a picture showing the domain of this simulation where air enters the oxidation and reduction zone through five 10° slope nozzles. The fuel used is MSW pellets with a composition of 60% organic waste and 40% non-organic. The diameter is 6 mm with an average length of 10-15 mm. fuel characterization is done using the ultimate test, proximate, and bomb calorimeter to determine the composition of the fuel. The 3D model is meshed by using a mixture of tetrahedral and hexahedral models. The number of elements is 839,159. To obtain accurate simulation results, a grid independency analysis was conducted.

2.2. *Numerical Model*

Numerical simulation analysis was carried out using FLUENT 15 version of student software. The equations involved in this simulation are flow equations, turbulence, species transport, energy, and radiation. The turbulence model used is realizable k-ε. The transport equation for the species involved in the simulation used the species transport equation model so that the mass fraction and mole fraction for species can be calculated. The chemistry interactions between species are modeled with the finite-rate/eddy-dissipation model. The fuel is modeled into a particle in the form of spherical and combusting particle. The fuel particle is modeled as a discrete phase that moves in the continuous phase and is injected into the surface area above the gasifier. fuel decomposition in the pyrolysis process is modeled with Kobayashi (two competing rates). Radiation heat transfer that occurs in the gasifier is modeled with the P-1 model. Because the gasifier used is downdraft and syngas move down the gasifier, the gravitational force is activated in the direction of the Y axis. The initiation of combustion is done by patching a temperature of 1000°C in the oxidation region.

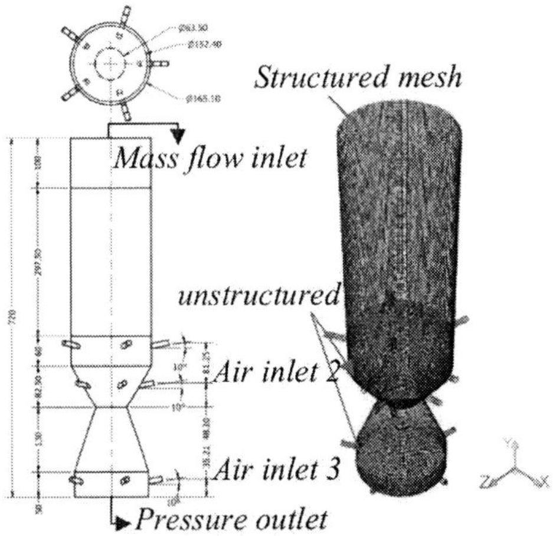

Figure 1. Geometry and Meshing model

Table 1. Properties of municipal solid waste pellet

Analysis	Contents	% mass
	Fixed carbon	20.19
Proximate	Volatile	65.78
	Moisture	9.82

	Ash	4.21
	Carbon	47.265
Ultimate	Hydrogen	6.7
	Oxygen	45.545
	Nitrogen	0.49
Calorific Value (J/kg)	1.348 x 10^8	

2.3. Reaction Model

The stages of the gasification reaction simulation are modeled according to the sequence of stages of gasification starting from drying, pyrolysis, oxidation, and reduction. The first pellet fuel through drying process is modeled with material droplets. Furthermore, it will undergo an evaporation process into hot steam because the temperature of the particle is lower than its intake temperature (law 1 and law 6 thermodynamics) [19].

$$T_p < T_{vap} T_p \quad (1)$$

$$m_p \leq (1 - f_{v,0})\, m_{p,0} \quad (2)$$

Decomposition modeling uses the Kobayashi model where the decomposition process is controlled by two devolatilization kinetic parameters (R1 and R2) with two different temperatures. Decomposition occurs when the fuel particle temperature reaches the evaporation temperature so that the particle mass does not exceed its non-volatile mass [19].

$$\mathcal{R}_1 = A_1 e^{-(E_1/RT_p)} \quad (3)$$

$$\mathcal{R}_2 = A_2 e^{-(E_2/RT_p)} \quad (4)$$

$$\frac{m_v(t)}{(1-f_{w,0})m_{p,0} - m_a} = \int_0^t (\alpha_1 \mathcal{R}_1 + \alpha_2 \mathcal{R}_2) \exp\left(-\int_0^t (\mathcal{R}_1 + \mathcal{R}_2)\, dt\right) dt \quad (5)$$

When the decomposition process has occurred and produces a solid product in the form of charcoal, then there will be particle surface reaction which consumes the fraction fb [19].

$$m_p < (1 - f_{v,0})(1 - f_{w,0}) m_{p,0} \quad (6)$$

$$m_p > (1 - f_{v,0} - f_{comb})(1 - f_{w,0}) m_{p,0} \quad (7)$$

The surface reaction rate is determined by kinetic and diffusion rate based on a kinetic/diffusion-limited theory by Baum and Street [19].

$$D_0 = C_1 \frac{[(T_p + T_\infty)/2]^{0.75}}{d_p} \quad (8)$$

With kinetic rate:

$$\mathcal{R} = C_2 e^{-(E/RT_p)} \quad (9)$$

Are weighted to yield a char combustion rate of

$$\frac{dm_p}{dt} = -A_p p_{ox} \frac{D_0 \mathcal{R}}{D_0 + \mathcal{R}} \quad (10)$$

The rate of reaction of partial fuel combustion occurs very quickly and is influenced by temperature and turbulence. The rate of chemical reactions that occur in the gasification stage is calculated using an Arrhenius rate and turbulence mixing rate using the finite-rate/eddy-dissipation model. The kinetic parameters and reactions involved in the gasification stage are presented in table 2.

Table 2. kinetic parameter for homogeneous and heterogeneous reactions [17]

Heterogeneous reactions	Kinetic parameter, A ($kg/m^2/s/Pa^{0.5}$), E (J/kmol)
$C(s) + 0.5O_2 \rightarrow CO$ (R1)	A1 = 0.052 E1 = 6.1 x 10^7
$C(s) + CO_2 \rightarrow 2CO$ (R2)	A2 = 0.0732 E2 = 1.125 x 108
$C(s) + H_2O \rightarrow CO + H_2$ (R3)	A3 = 0.0782 E3 = 1.15 x 108
Homogeneous reactions	
$CO + 0.5O_2 \rightarrow CO_2$ (R4)	A4 = 2.239 x 10^{12} E4 = 1.674 x 10^8
$H2 + 0.5O_2 \rightarrow H_2O$ (R5)	A5 = 6.8 x 10^{15} E5 = 1.67 x 10^8
$CH_4 + 0.5O_2 \rightarrow CO + 2H_2$ (R6)	A6 = 4.4 x 10^{11} E6 = 1.25 x 10^8
$CO + H_2O \leftrightarrow CO_2 + H_2$ (R7)	A7 = 2.34 x 10^{10} E7 = 2.883 x 10^8
$CH_4 + H_2O \rightarrow CO + 3H_2$ (R8)	A8 = 8.7 x 10^7 E8 = 2.51 x 10^8

2.4. Simulation process

The computational process is carried out using a 3-dimensional model which is analyzed using the finite volume method. The air ratio or equivalent ratio used is 0.4. while the comparison made is for the single stage condition or input air in the oxidation zone and multi stage or input air in the oxidation and reduction zone. The ratio used in this study are 10%, 20%, 30%, and 40%. The boundary conditions used in this numerical analysis are presented in table 3.

Table 3. Boundary condition [20]

Boudary condition	value
MSW particle inlet temperature (°C)	30
MSW flow rate (kg/s)	0.00374
Air inlet flow rate stage 1 (kg/s)	0.003366
Air inlet flow rate stage 2 (kg/s)	0.000374
MSW particle desity (kg/m^3)	370
MSW HHV (KJ/kg)	13415

3. Result and Discussion

In Figure 2 and Figure 3 shows the simulation results of temperature distribution along the gasifier with the difference in air ratio between reduction and oxidation zones from 10% to 40%. When the air ratio at the right level from a single state becomes a level starting from 10% to 40% what happens is an increase in temperature in the reduction zone. While the temperature in the oxidation, pyrolysis, and drying zones decreases. The increase in temperature in the reduction zone occurs because of a partial oxidation reaction between carbon and oxygen contained in the air. This oxidation reaction produces a certain amount of heat resulting in an increase in temperature. While the decrease in temperature in the oxidation zone occurs due to a decrease in the amount of air. As a result, the amount of oxidized charcoal decreases and the rate of oxidation reaction decreases. Therefore, there is a decrease in temperature in the pyrolysis and drying zone because both of them require heat. However, when viewed by the overall temperature distribution, the decrease in temperature in the oxidation, pyrolysis and drying zones does not exceed the minimum temperature limit in the zone. Instead the reduction zone to get additional heat energy can affect the reactivity of the gasification reaction (R2, R7, and R8).

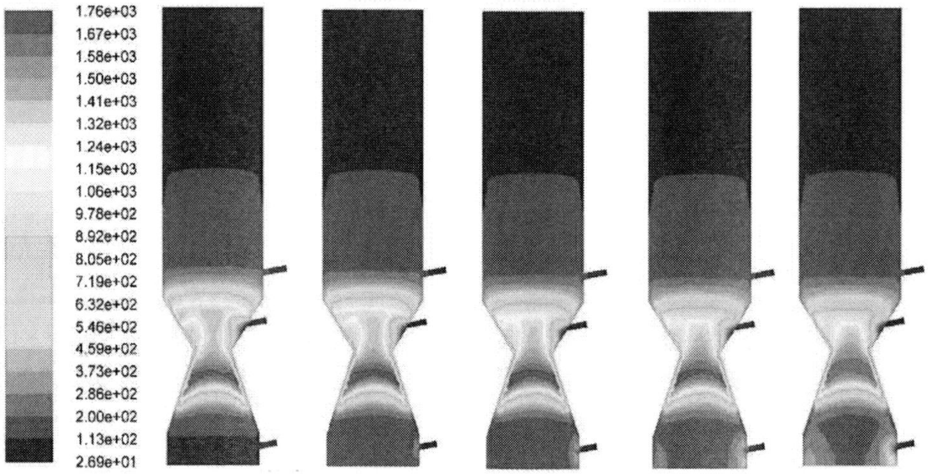

Figure 2. Temperature profile visualization from left to right air ratio (0, 10, 20, 30, 40%)

In addition to affecting change in the temperature distribution, inclusion of air in the reduction zone also leads to changes in volumetric percentage of flammable gas. when the temperature in the reduction zone rises, there is an increase in the endothermal reaction rate R2 therefore in figure 3 when the percentage of air by 10% is added to the reduction zone, the percentage of volumetric CO increases. Besides that, H_2 also increases because there is an increase in the reaction rate of R7 due to the increase in temperature.

Increase in CO, H_2, and CH_4 respectively from 24.3% to 24.82%, 13.04% to 14.2%, and 1.61% to 1.76%. while CO_2 decreased from 20.23% to 1.82%. However, when the amount of air continues, the composition of CO, H_2, CH_4 and CO_2 decreases. This happens because more carbon is consumed so the number of reactants for the reaction R2 and R3 decreases.

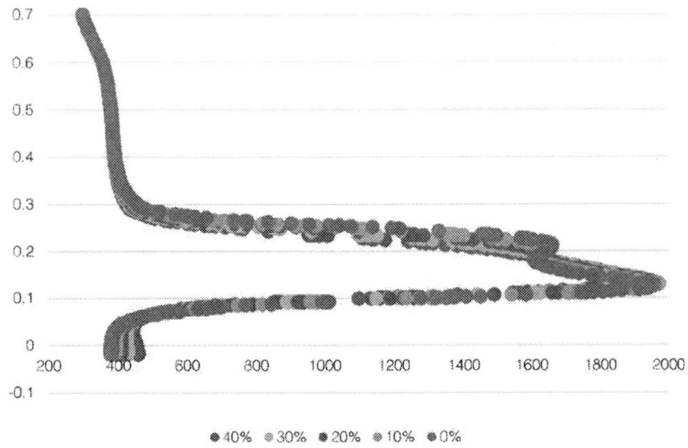

Figure 3. Temperature profile

When compared with the experimental results, the temperature distribution in this numerical simulation is greater. if in the experiment the highest temperature is 1,300°C, in numerical simulations it reaches 1,700°C. this is because in the numerical simulation the gasifier wall is assumed to be adiabatic while in the experiment there is a lot of heat loss. the same thing with composition and LHV is also greater than numerical simulation results compared to experiments. but overall both temperature and syngas composition have the same trend of increase and decrease as the experiment. however, further experimentation needs to be done to reduce the difference to smaller ones.

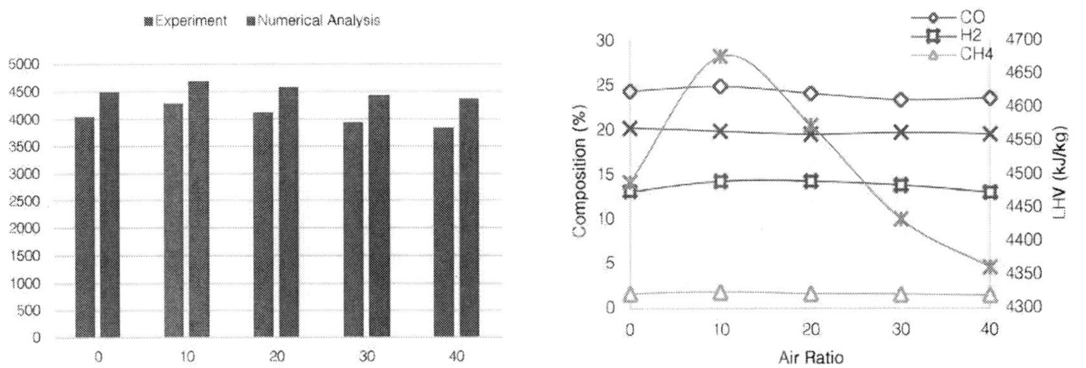

(a) (b)

Figure 4. a) Comparation of Low Heating Value from Numerical and Experimental Analysis, b) Syngas Composition and LHV

Changes in the low heating value (LHV) as an indicator of the performance of gasification is affected by the volumetric percentage of flammable gas. when viewed from the pattern of increase and decrease, the optimum LHV is obtained when the amount of air input in the reduction zone is 10%. Although there was an increase in the rate of carbon consumption which affected the rate of fuel consumption, but because the percentage of volumetric flammable gas decreased, LHV also decreased Comparison between the results of experiments with numerical analysis is presented in Figure 4.

4. Conclusion

The 3D model was created to simulate the gasification of MSW pellets by adding air inlet on the pyrolysis zone ranging from 10% to 40%. The simulation results revealed that the addition of air in the reduction zone was able to improve gasification performance in the form of an increase in temperature in the reduction zone and a decrease in temperature in the oxidation, pyrolysis and drying zones which are still above the minimum temperature value. In addition, there is also an increase in the volumetric gas percentage, especially CO. low heating value and cold gas efficiency also increased. Henceforth, with this numerical model it can be used as a reference to optimize gasification performance by adding air to the reduction zone.

References

[1] E Damanhuri 2005 *Some principal issues on municipal solid waste management in Indonesia.* Expert Meeting on Waste Management ..., 1-10.

[2] U Arena 2012 Process and technological aspects of municipal solid waste gasification. *A review. Waste Management* **32** (4) 625-639.

[3] S Thakare and S Nandi 2015 Study on Potential of Gasification Technology for Municipal Solid Waste (MSW) in Pune City *Energy Procedia* **90** 509-517

[4] P Basu 2010 *Biomass Gasification and Pyrolysis Handbook* 3-21

[5] T A Milne, R J Evans and N Abatzaglou 1998 *Biomass Gasifier "Tars": Their Nature, Formation, and Conversion*

[6] L Devi, K J Ptasinski and F J J G Janssen 2002 A review of the primary measures for tar elimination in biomass gasification processes *Biomass and Bioenergy* **24** (2) 125-140.

[7] Bambang Sudarmanta D B M D F W J *Karakterisasi Gasifikasi Biomassa Sekam Padi Menggunakan Reaktor Downdraft dengan Dua Tingkat Laluan Udara.*

[8] Agus Wijianto B S 2002 *Karakterisasi Gasifikasi Downdraft Berbahan Baku Cangkang Kelapa Sawit Dengan Varasi Gasifying Agent*

[9] B Sudarmanta et al. *The effect of two stage gasifying agent on biomass downdraft gasification to the gasifier performance.*

[10] A R Saleh and B Sudarmanta *Experimental investigation on multi-stage downdraft gasification: Influence of air ratio and equivalent ratio to the gasifier performance.*

[11] J D Bentzen and R M Hummelshøj 2000 Low Tar and High Efficient Gasification Concept *ECOS 2000 Conference* University of Twente, Enschede, The Netherlands

[12] T Bui, R Loof and S C Bhattacharya 1994 *Multi-stage reactor for thermal gasification of wood.* Energy **19** (4) 397-404.

[13] J D Martínez et al 2011 Experimental study on biomass gasification in a double air stage downdraft reactor *Biomass and Bioenergy* **35** (8) 3465-3480

[14] A L Galindo et al 2014 Biomass gasification in a downdraft gasifier with a two-stage air supply: Effect of operating conditions on gas quality *Biomass and Bioenergy* **61** 236-244.

[15] H Shi, W Si and X Li 2016 The concept, design and performance of a novel rotary kiln type air-staged biomass gasifier *Energies* **9** (2) 1-18.

[16] N Striugas et al 2014 An evaluation of performance of automatically operated multi-fuel downdraft gasifier for energy production *Applied Thermal Engineering* **73** (1) 1149-1157

[17] R Gupta, P Jain and S Vyas 2017 *CFD Modeling and Simulation of 10KWE Biomass Downdraft Gasifier* **7** (4) 0-2.

[18] M Simone, C Nicolella and L Tognotti 2013 Numerical and experimental investigation of downdraft gasification of woody residues *Bioresource Technology* **133** 92-101

[19] A Inc 2017 *ANSYS FLUENT Theory Guide* Release 18.2 **15317** 724-746

[20] A R Saleh, B Sudarmanta and D R Ependi 2018 Numerical study of gasification pelletized municipal solid waste using multi-stage air inlet downdraft gasifier In *Materials Science and Engineering Conference Series* **Vol. 434** No. 1 012177

IOP Conf. Series: Materials Science and Engineering **588** (2019) 012026 doi:10.1088/1757-899X/588/1/012026

A Study on conceptual design of mini FSRU as LNG receiving facility

I W G K D D Putra[1], K B Artana[1,2], I M Ariana[1,2] and L G M P Sudiasih[1,a]

[1] Department of Marine Engineering, Faculty of Marine Engineering, Institut Teknologi Sepuluh Nopember, Surabaya 60111, Indonesia
[2] Center of Excellence in Maritime Safety and Marine Installation (PUI-PT Keselamatan Kapal dan Instalasi Laut)

mirahpardani@gmail.com

Abstract. One of prioritizes of the government of Indonesia is to provide electricity in Papua regions to reduce electricity disparity in Indonesia. According to recent Power Supply Business Plan of PT Perusahaan Listrik Negara (Persero), there are more than 50 power plants are planned to build in Papua and Maluku regions. One of power plants which has considerable high capacity is located in Jayapura, Papua Province. This study addresses the conceptual design for three power plants in this area with capacities of 50 MW, 50 MW and 40 MW. The power plants are assumed to be supplied by natural gas from Floating Storage and Regasification Unit (FSRU). This concept was chosen because of the sloping contour of the seabed and to reduce the use of land for the receiving terminal. To supply the FSRU, a 30,000 m^3 small scale LNG vessel is used for a shuttle vessel with estimated roundtrip is 15 days. The conceptual design consists of determination of cargo tank capacity, designing lines plan and general arrangement of FSRU, stability calculation, Process Flow Diagram (PFD), Process and Instrumentation Diagram (P&ID) for loading and regasification processes at the FSRU.

1. Introduction

The electricity demand in Indonesia has increased every year. According to power supply business plan of PT. PLN (persero), the increase in electricity demand reaches 8.1% per year [1]. For this reason, the Indonesian government planned to provide 35,000 MW of electricity, mainly in areas that are still shortage of electricity. To fulfill this plan, the government purposed to build a power plant and upgrade the existing power plant. As in Papua, it has been planned to make several power plants in Jayapura. The power plant in Jayapura region consists of three power plants with the capacities of Jayapura Main Power Plant (MPP) 50 MW, Jayapura Peaker Power Plant 50 MW, and Jayapura Load Follower Power Plant 40 MW. These three power plants are designed to be supplied by LNG as their main fuel. An LNG shuttle vessel with a size of 30,000 m^3 is used to supply the receiving terminal with estimated roundtrip is 15 days. To be able to receive LNG, it requires a facility that can store, process and distribute LNG to the plant.

The LNG receiving terminal is a facility required before supplying gas to the power plant. In general, this receiving terminal consists of a storage unit, a processing unit for LNG regasification and an LNG distribution unit to the plant. There are several alternatives of LNG receiving terminals that can be used such as offshore based, onshore based or a combination of both. Onshore based is a land receiving terminal where all storage tanks and regasification units are located on land. Offshore based when the floating structure is used while all the storage, regasification and other systems is inside the floating structure. Some types of offshore receiving terminals are Floating Storage Regasification Unit (FSRU), Floating Storage Unit (FSU) and Floating Regasification Unit (FRU). FSRU is a unit that can work alone, while FSU and FRU can be combined with onshore receiving terminals.

The selection of this terminal cannot be done arbitrarily. Terminal selection refers to the advantages in both technical and economic aspects. In this study the type of receiving terminal in the form of FSRU was chosen to supply power plants in Jayapura. The FSRU was chosen because of the sloping contour of the seabed and the area is not available to build a land terminal. In determining the proper FSRU design, it requires various considerations, both technical and economic aspect. This paper discusses the technical considerations of building an LNG terminal in supplying LNG to the Jayapura power plant. Conceptual design consists of general arrangement design, system design, 3D design, and stability analysis in accordance with IMO regulations. General Arrangement is the determination of the main dimensions of the ship and the arrangement of components and equipments on the ship that refers to DNV GL. The systems designed are in the form of Process Flow Diagrams (PFD), and Process and instrumentation Diagrams (P&ID) of the loading system and regasification system. To achieve this, a special methodology and calculation are needed to determine the specific specifications of the entire equipment. Determining the LNG needs of each plant to know the total needs of the LNG for 15 days, so that the specifications of the tank can be determined.

2. Literature Study

LNG is a clean energy and relatively safe compared to other fuels such as gasoline and diesel oil because the flammable temperature of the LNG is relatively higher. When the LNG tank is leaked, LNG will instantly vaporize and move upwards due to the mass of natural gas is lighter than the surrounding air [2]. Basic information about LNG and its properties are needed to be considered in order to determine how LNG processing equipment can work.

FSRU is one option that can be considered as a receiving terminal when the onshore based receiving terminal more difficult to build in certain area due to lack of are onshore to install the facilities. The study revealed that the LNG facilities in form of FSRU has price around 50-60% of the onshore terminal price [3]. Capital expenditures for FSRU may be lower than onshore, due to the potential efficiency of construction costs, through repeated acquisition (iteration of the design) where there is a generic design concept, with minor adaptation for each particular project. [4]

The construction times of FSRU about half than that of the onshore terminal LNG and other advantage of FSRU is that this facility can be moved to other location when necessarily required. The selection process of the type of LNG receiving terminals can be found in the study entitled "Study of Selection Technology LNG Receiving Terminal for Bali Island (case study of LNG supply from Tangguh LNG Plant to Bali)"[5]. There are several criteria used in the selection process such as: land area, safety aspect to surrounding residential and working areas, and implementation period. In their study, combination of Floating Storage Unit (FSU) and Floating Regasification Unit (FRU) were selected as LNG receiving terminals in Benoa Bali over other two alternatives namely Gravity Based Structure and land based receiving terminal.

Previous study on LNG distribution in Papua region can be found in Antara, G.B [6]. In the study, the LNG distribution is done using two LNG ships with the capacity of 10,000 m3 and 23,000 m3 and onshore based terminal LNG receiving facilities is considered to be used in Jayapura. However, this study considers an offshore based terminal by means of FSRU while FSRU is supplied by LNG shuttle carrier which has 30,000 m3 in capacity.

3. Several Consideration on Designing FSRU Facility in Jayapura

Mobile Power Plant (MPP) Jayapura is located in the Holtekamp, North Jayapura, as well as two other power plants that have been planned to be construct and operated near to MPP. These power plant are located quite close to the coastal area, but in a position on the edge of a bay where the sea level is very shallow. The position of the power plant can be seen in Figure 1.

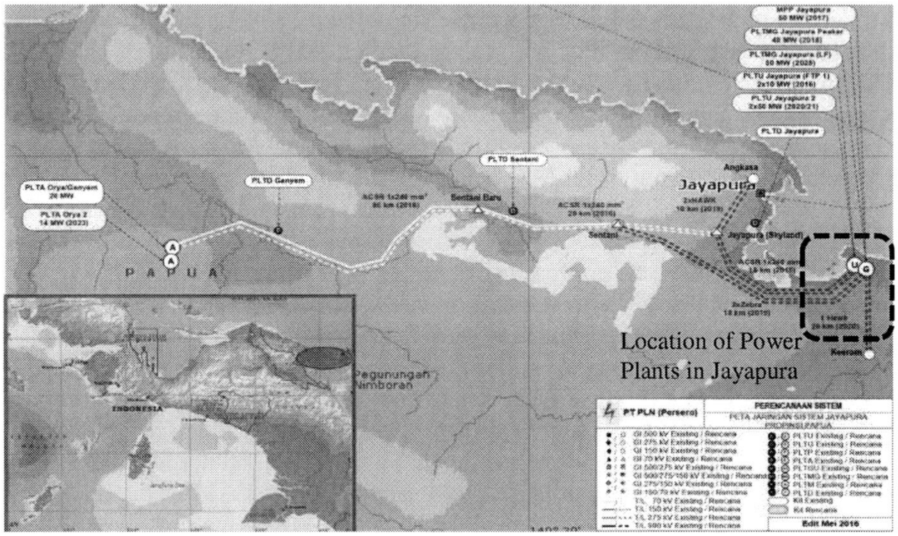

Figure 1. Location of Power Plants in Jayapura Region [1]

The selection of dimension of FSRU highly depends on demand as well as the location including the water depth in the location. Other consideration of locating FSRU is dimension of the available shuttle vessel in such away the shuttle vessel can be moored directly to the FSRU or by means of intermediate jetty facility. The shuttle vessel must be able to supply the FSRU in close proximity so that the condition remains safe. The shuttle vessel that consider in this study has a draft of 7.4 m, hence the location to allow certain draft to be served when the jetty is located approximately 1500 m away from shoreline. Total length of trestle including in the shore approach is 1,910 m, where the distribution pipeline will be stretched along this length. The location and length of jetty as location of FSRU can be seen in Figure 2. This figure also give some information regarding to condition of water area i.e. the bathymetry conditions.

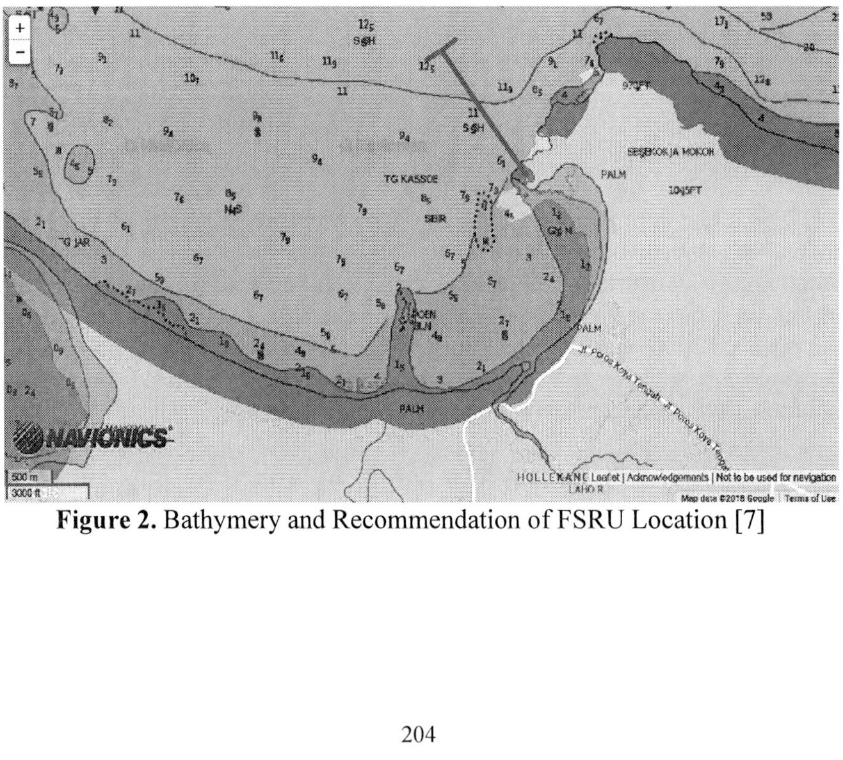

Figure 2. Bathymery and Recommendation of FSRU Location [7]

According to the duration of roundtrip of shuttle vessel to deliver LNG to Jayapura from supply location, the minimum capacity of LNG containments in FSRU must able to save an LNG for 15 days or about 10,365 m^3. Four type C tanks with each capacity of 3,750 m^3 are installed in the FSRU. Type C tank is an independent tank that can withstand greater pressure, so the formation time of the boil-off gas (BOG) is longer. This tank is also considered to be more efficient in overcoming the sloshing effect of liquid loads. Local stress also does not occur in some parts, when it is given more force, so that the tank is safer even though the condition is not full.

In selecting the volume of the tank also through several considerations that the tank cannot be fully filled 100% or emptied until 0%. Referring to DNV Part 5 [8], the maximum volume can be filled at 98% of the total tank volume. Thas is done to prepare the space when BOG is formed, then there is no overpressure inside the tank. While the minimum volume that must be contained in the tank is 5% of total tank volume. This is done to keep the low temperature inside the tank and prevent thermal shock due to extreme temperature differences during tank filling.

4. Conceptual Design of FSRU Facility in Jayapura

Basically, the barge type of FSRU is considered in this study. The principal dimension of the FSRU is determined through various stages of configuration so the result is an appropriate ratio between length, draft and breadth. This configuration is done iteratively so that the dimensions of the FSRU and area for putting all of required components on the barge can be optimized. The most suitable configuration is obtained by adopting and referring to Damen Stan-Pontoon B32 SPo9832 [9]. The principal dimension of FRSU barge to be built as receiving terminal in Jayapura as follows: length overall (LOA) 97.5 m, Breadth 32.2 m and Height 8.1 m.

To design and determining number of required compartment and room space in the FSRU, it is necessary to calculate frame spacing and determine the location of watertight bulkhead on the ship. Referring to DNV GL [10], the double bottom height is 1.6 m. In accordance with the length of the ship and the location of the engine room at the aft side of the ship, the FSRU must have at least four watertight bulkheads. The bulkheads include collision bulkhead, after peak bulkhead, engine room bulkhead, and bulkhead in the cargo area. Hence, the arrangement of the tanks on the FSRU is given in Figure 3.

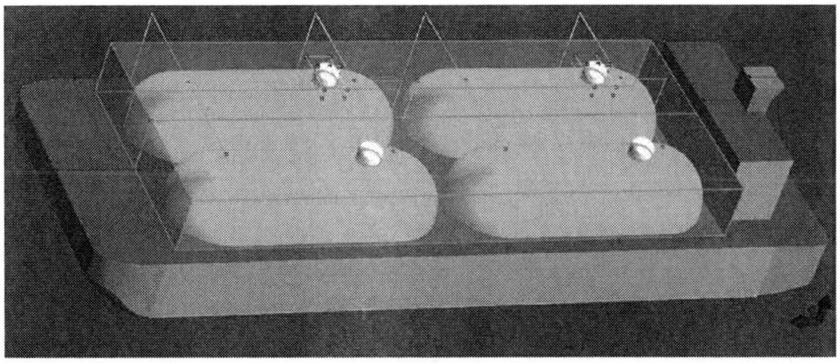

Figure 3. Configuration of the LNG Tank, Compartment, Superstructure and Deck Houses

LNG loading header is a pipe used to drain LNG from the manifold to the LNG tank. LNG loading process uses a pump from shuttle vessel with capacity of 550 m^3/h. Hence, the transfer process can be done in at least 5 hours. Condition inside the pipe must be kept at an allowable speed to avoid electron displacement. According to DNV [8], the thickness of pipe is 4.19 mm with internal diameter of 264.62 mm.

The regasification process starts from pumping the LNG to the LNG header that will be distributed through two different lines. High pressure line which serves to supply natural gas demand at MPP Jayapura, while low pressure line supplies the needs of Jayapura peaker and Jayapura load follower power plant. The spesification of high pressure and low pressure lines is the same with 5.54 mm in thickness and internal diameter of 598.52 mm. Figure 4 shows that after the LNG pump output there is a flow meter that will automatically open the valve when the LNG flow is considerably very low. It will connect to the return pipe that make LNG flow back to the tank, this happens at the time of starting the pumps. LNG flows toward two lines which is on the high pressure line is given the high pressure pump to increase the pressure. High pressure pump is chosen instead because its dimensions, weight and power tends to be smaller than putting a high pressure compressor before the metering unit.

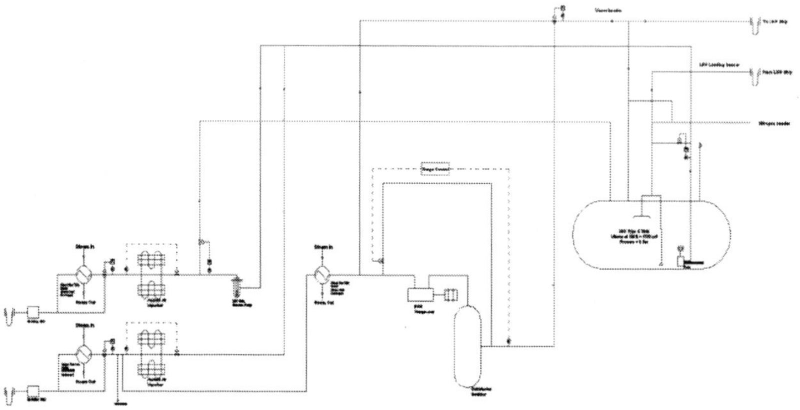

Figure 4. Process Flow Diagram of FSRU

LNG in the tank will certainly change to gas phase due to receiving heat from its surroundings. Therefore, the gas formed is allocated to the needs of generator and boiler at FSRU. If the gas formed exceeds its need, then the gas would be channeled toward the low pressure line by BOG compressor. The BOG compressor equipped with supporting facilities such as anti-surging and scrubber. That equipment aims to keep the BOG compressor operate in secure conditions. If the gas flow is reduced then the BOG compressor will receive back pressure which is very dangerous for the compressor's life time. To prevent this condition, anti-surging is installed and serves to drain the compressed gas to the suction line of compressor. In the other hand, pressure differences will forming droplets and when these droplets flowing directly towards the BOG compressor, this condition will impact to the life time of compressor. A gas scrubber is installed to help separation of the droplets with the gas. BOG heater is installed as it is needed to increase the temperature when the temperature of the gas coming out of the compressor is low. The BOG heater used two pieces of Thermax TVN Automatic Direct Steam Vaporizer T25S which having maximum capacity 25 SCFH [11].

Figure 5 shows an arrangement of regasification facilities that is located on the fore side of FSRU. The regasification facilities designed at FSRU are Ambient Air Vaporizers (AAV), heater and pumps. This designed was referred to other study found in [12] that the AAV is ranked number two, one level above an open rack vaporizer (ORV). This AAV also appropriate to be operated at ambient temperature above 18°C. The same specification of AAV is fitted for the low pressure and high pressure lines. The AAV installed is Thermafin Supergap with having maximum capacity of LNG is 77 m^3/h and temperature output is 10°C at ambient temperatur of 20°C [11].

Figure 5. The Arrangement of Regasification Facilities in FSRU Jayapura

Stability analysis is considered for two conditions, at empty load condition (about 5% load and holds ballast) and full load condition (minimum 76% load). While in both condition, it is assumed that fuel oil, lubricating oil, and fresh water are in full condition. The Jayapura FSRU's stability booklet is calculated based on the Internasional Maritime Organization (IMO) regulation that should be applied for all ships [13]. The stability booklet according to "Trim & Stability Booklet" [14] at the Jayapura FSRU is as follows:

Table 1. Stability Booklet

Condition		Unit	1 Empty Load Condition	2 Full Load Condition
Light Weight		(t)	5805.05	5805.05
Fuel Oil		(t)	97.68	97.68
Lube Oil		(t)	1.259	1.259
Ballast Water		(t)	3882.78	0
Fresh Water		(t)	6.183	6.183
Displacement		(t)	9791	14262
Draft	fore	(m)	3,.336	5.037
	aft	(m)	3.11	4.207
	mean	(m)	3.223	4.622
Trim		(m)	0.83	0.48
T.P.C		(t)	32.081	32.081
M.T.C		(m)	254.37	248.92
K.G		(m)	5.645	4.378
G.M		(m)	16.05	25.199
Stability	Judgment		Good	Good
	Area 0-30	(m-deg)	159	131
	Area 30-θ u	(m-deg)	76,4	58,6
	Area 0-40 θ u	(m-deg)	235	189.9
	Max GZ	(m)	7.704	5.9

Table 1 shows that stability condition of FSRU can be considered good or acceptable that all condition comply with the IMO regulations. It can be seen that the righting arm (GZ) in the condition of empty load is greater than the condition of full load, this means that maximum heeling occurs in an empty condition is greater than the full condition.

5. Conclusion

The conceptual design of FSRU that will be used to serve power plants in Jayapuran has been shown. Several aspect in design also discussed such as determining principal dimension, general arrangement of FSRU, arrangement of loading of LNG as well as regasification unit and unloading of gas to power plants. To assure the stability of the FSRU, the stability analysis also conducted and discussed in this paper.

6. Future Direction of Research

Future research has been prepared to complete this existing research. Further research aims to measure the safety level of the system that has been made. The research is in the form of risk assessment to the loading system and LNG regasification system. This risk assessment is carried out considering the operation of the FSRU in an open space using equipment that generates heat. In addition, with a working fluid is kind of flammable gases, it is very possible if a fire can occur during the operations. In general following steps will be conducted for the risk assessment: (1) hazard identifications, (2) frequency analysis, (3) consequence analysis, (4) risk representation and (5) propose mitigation if necessary.

References

[1] Ministry of Energy and Natural Resources 2016 RUPTL (General Electricity Supply Plan).
[2] Soegiono and K B Artana 2006 *Indonesian LNG Transportation* (Surabaya: Airlangga University Press)
[3] J Henderson 2017 *The Outlook for Floating Storage and Regasification Unit (FSRUs)*
[4] M Giranza and A Bergmann 2017 *An Economic Evaluation of Onshore and Floating Liquefied Natural Gas Receiving Terminals: The Case Study of Indonesia*
[5] M Maria (n.d.) *Study of Selection Technology LNG Receiving Terminal for Bali Island (case study of LNG supply from Tangguh LNG Plant to Bali)*
[6] G B Antara 2014 *Economic Optimization and Analysis of Distribution of LNG to Power Plant in the Papua Region*
[7] Port Maps 2018 Live AIS and meetingpoint for yachties https://www.portmaps.com/eng/Map
[8] DNV-GL 2011 Part 5 Chapter 5 *In Liquefied Natural Gas*
[9] Damen Shipyard (n.d.) Damen. Retrieved from https://www.damen.com/en
[10] DNV-GL. (n.d.) Part 3 Chapter 1. In Hull Equipment, Supporting Structure and Appendages
[11] PaxOcean. (n.d.) Product-Brosure-25000m3-FSRU
[12] Fluor John Mark 2013 *LNG Vaporizer Selection Based on Site Ambient Temperature*
[13] IMO 2008 Review of MEPC.1/CIRC.511 and Relevant MARPOL ANNEX 1 and ANNEX VI Requirements.
[14] Trim & Stability Booklet (n.d.) Retrieved from http://vessel-sea.com/outline/pdf/2-Trim_Stability_Booklet.

IOP Conf. Series: Materials Science and Engineering **588** (2019) 012027 doi:10.1088/1757-899X/588/1/012027

Hybrid Saliency-SVM Method Implementation for Automatic Data Training Selection in Image Segmentation

Rully Soelaiman[1,a], Chastine Fatichah[1] and Aisha Yuliandari[1]

[1] Department of Informatics Engineering, Faculty of Information Technology, Institut Teknologi Sepuluh Nopember, Surabaya, Indonesia 60111

[a] rully@if.its.ac.id

Abstract. Image segmentation is one of the most important step in computer vision and image processing, which later will be used in image retrieval, object identifying and data classification. Image segmentation can be seen as classification problem, namely by marking each pixels according to certain characteristics. Support Vector Machine (SVM) is a classification method included in supervised learning. Supervised learning is a method which requires training and testing. Training sample used in training process isn't always exist in few cases, especially in the image segmentation case. This particular research implemented SVM-based method which is Saliency-SVM for automatic data training selection in image segmentation. This method generates data training using SVM-based visual saliency detection where there is pre-segmentation step and trimap formation based on saliency information from visual saliency detection, HSV color space quantitation, histogram analysis and local homogeneity threshold. Data training produced is pixel belonging to positive (object) and negative (background). The step before segmentation done with SVM is feature extraction to create input vector in SVM. Object segmentation on the image is done by SVM based on SVM Trained Model. Test result from Saliency-SVM for this image segmentation has average accuracy value up to 94,84 % compared to the image ground truth

1. Introduction

Image segmentation is one of the most important step in computer vision and image processing, which later will be used in image retrieval, object identifying and data classification. Image segmentation can be seen as classification problem, namely by marking each pixels according to certain characteristics. Several classification methods have already successfully performed image processing, Support Vector Machine (SVM) is one of the method among others.

Support Vector Machine (SVM) is a classification method included in supervised learning. Supervised learning is a method which requires training and testing. [1]. SVM used a few training sample which are marked for classification learning, meanwhile marked training sample doesn't always exist in few cases, especially in image segmentation case [2]. In that case, this research will implement new approach which combine Visual Saliency Detection and SVM classifier for image segmentation. This method produced training data using SVM-based visual saliency where there are pre-segmentation and trimap formation steps based on saliency from visual saliency detection, HSV color space quantitation, histogram analysis and local homogeneity threshold. After training data formed, we extract the feature and classification with SVM.

The next part of this paper composed of Research Methodology in part 2, performance discussion from method for few scenarios in Testing and Analysis in part 3, and Conclusion in part 4.

2. Research Methods

This part will discuss about methods used in this research. This research composed of 3 main parts; training data selection which covers pre-segmentation process and trimap formation, HSV color space quantitation, histogram analysis and pixel training selection with local homogeneity threshold. The next part is features extraction which are color, saliency, spatial and texture features. The last part is pixel classification using SVM.

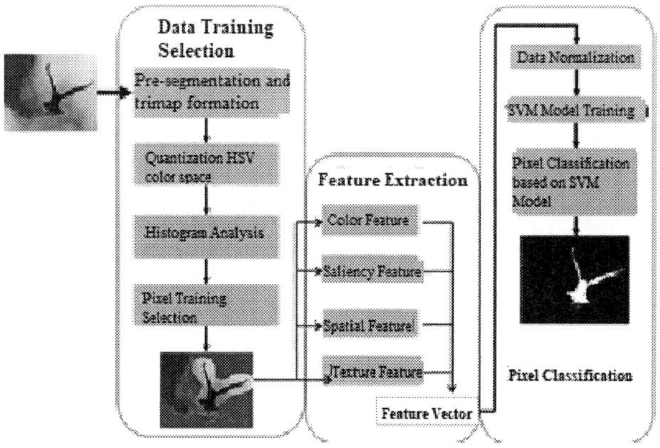

Figure 1. General illustration of image segmentation using Saliency-SVM

2.1. Training Data Selection

This particular pixel training selection with SVM-based visual saliency covers pre-segmentation process and trimap formation, where estimating object location and background based on saliency map resulted from visual saliency detection on input image. Trimap produced will be further identified with histogram analysis for dominant color selection which have been quantitatized before in HSV color space, and later selected in pixel training with local homogeneity threshold. Those identifications were done to determine whether those pixels are within positive training set or negative training set.

2.1.1. Pre--segmentation and trimap formation

Pre-segmentation and trimap formation were done based on saliency information generated from visual saliency detection with spectral residual approach [3]. Visual saliency detection were used for generating saliency map, which then used for approximate object from an image. Spectral residual analyzes spectrum log from image, saliency map formed based on spectral residual from amplitude spectrum from fourier transform on the image. Spectrum fase on fourier transform is the key to calculate location from the most prominent part of it. Use the following equation to form saliency map with spectral residual:

$$A(f) = absolute(F(I(x)) \tag{1}$$

$$P(f) = angle(F(I(x)) \tag{2}$$

$$L(f) = log(A(f)) \tag{3}$$

$$R(f) = L(f) - h_n(f) * L(f) \tag{4}$$

$$S(f) = g(x) * ||F^{-1}[exp\, R(f) + i * P(f)]||^2 \tag{5}$$

Where F^(-1) is Inverse Fourier transform, P(f) is phase spectrum and g(x) is 2D Gaussian Filter with σ = 8 for a better visual effect. To identify location of an object, do binary image with otsu method :

$$OM(x,y) = \begin{cases} 0 & if\ SM(x,y) < t \\ 1 & if\ SM(x,y) \geq t \end{cases} \quad (6)$$

Threshold from this binerization is determined by discrimination criteria (σ_B^2, σ_W^2) from 2 class (salient object and background). For white part (object) valued 1 is R_o whereas black pixel valued 0 is R_b.
To eliminate unnecessary pixel around boundary R_o, do morphology operation with:

$$M_o = R_o \ominus E_{r_e} \quad (7)$$

Boundary R_o depreciated with erosion operation. M_o is Mask Object and $\ominus E_{re}$ is erosion operator which shows depreciation R_o on pixel r_e.
R_o expanded with dilated operation. M_b is Mask Background and $\oplus D_{r_d}$ dilation operator which shows R_o expansion towards pixel r_d. Dilation result from R_o minus by R_o and combine the result with R_b.

$$M_b = ((R_o \oplus D_{r_d}) - R_o) \cup R_b \quad (8)$$

Structure element used in this morphology operation is square, with width of 70 pixels.
M_o and M_b resulted from morphology operation serve as standard location in trimap formation. Where Trimap (T_o, T_r dan T_b) inside RGB color space no longer in binary. T_o is a prominent part in the pre-segmentation process, T_b is background and T_r are vague pixels from remaining area of T_o and T_b. Figure 2 shows illustration of pre-segmentation process and trimap formation.

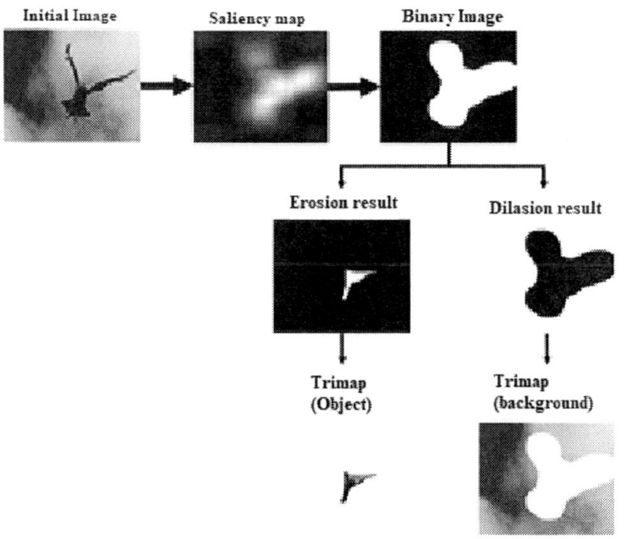

Figure 2. Pre--segmentation and trimap formation illustration

2.1.2. HSV color space quantitation

T_o and T_b in RGB color space converted to HSV color space and quantitated to generate 1-dimention vector. HSV (*Hue*:[0,360], Saturation:[0,1], and Value:[0,1] is able to emphasize on human visual perception, which proved in having better result for image segmentation from RGB color space. To decrease computation complexity in dominant color determination T_o and T_b in histogram analysis, do HSV color space quantitation on T_o and T_b.

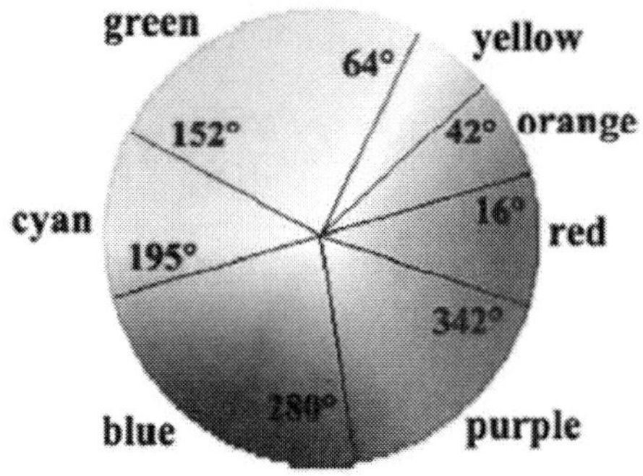

Figure 3. Hue Channel Quantitation Scheme [2]

Figure 3 shows scheme result quantitated into 7 non-uniform bins represented from 0-6 and each shows main color with Equation 9.

$$H = \begin{cases} 0 & if\ h \in (342,16] \\ 1 & if\ h \in (16,42] \\ 2 & if\ h \in (42,64] \\ 3 & if\ h \in (64,152] \\ 4 & if\ h \in (152,195] \\ 5 & if\ h \in (195,280] \\ 6 & if\ h \in (280,342] \end{cases} \quad (9)$$

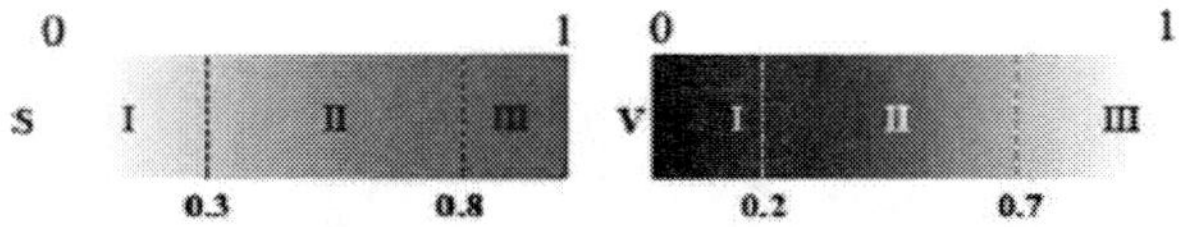

Figure 4. Quantitation scheme S&V channel [2]

$$S = \begin{cases} 0 & if\ s \in [0,0.3) \\ 1 & if\ s \in [0.3,0.8) \\ 2 & if\ s \in [0.8,1) \end{cases} \quad (10)$$

$$V = \begin{cases} 0 & if\ v \in [0,0.2) \\ 1 & if\ v \in [0.2,0.7) \\ 2 & if\ v \in [0.7,1) \end{cases} \quad (11)$$

When the value of S is large enough, for instance S better than 0.8, part III from figure 4 can be considered as pure red. When the value of V is small enough, for instance V smaller than 0.2 part I from figure 4 can be considered as pure black area. So, 3 non-uniform bins expressed as 0 until 2 are sufficient to represent value and saturation information.

According to the scheme above, vector 1-dimension feature built by 3 channel values as in Equation 12.

$$L = Q_s Q_v H + Q_v S + V \quad (12)$$

As the often used quantization method, quantization coefficient determined as Q_s=Q_v=3 then L can be calculated with Equation 13.

$$L = 9H + 3S + V \tag{13}$$

Thus, the 3 channels (hue, saturation and value) can be distributed in 1-dimention vector L and L$\in \{0,1,\ldots,62\}$. Since quantization resulted in only 63 bins, then computation complexity will decrease significantly. Considering non-uniform characters on those 3 channels, quantization result is more similar to human visual method [2].

2.1.3. Histogram Analysis

Histogram analysis is the step for choosing histogram peaks in T_o and T_b that has been quantized in HSV color space.

The peak in histogram can be information from image's color. For image with color, dominant color can be identified with peaks in global histogram. Peak selection in histogram was done to decrease color amount in an image which has been quantized in HSV color space and ignoring colors which appear not as often. Those colors are then dominant in T_o and T_b. Assumption from [2] which has been tested in 1000 image, that no more than 3 or 4 significant dominant colors describe part T_o and T_b. Steps in histogram analysis shown below:

1. Calculate global histogram from T_o that has been quantized in HSV color space.

$$H^o = \frac{Num(f(x,y)=l_i}{Num(T_o)}, (x,y) \in T_o, l_i \in \{0,1,\ldots,62\} \tag{14}$$

   Where $Num(T_o)$ = pixel count in T_o. $Num(f(x,y)$ = Pixel count with color level l_i in T_o.
2. Identify all peaks. $P_{ko} : P_{l_1}, P_{l_2}, \ldots, P_{l_k}$, l_i is index from peak i, and $l_1<l_2<\ldots<l_k$.
3. Calculate maximum and minimum peak from H^o. Erase lowest peak based on T_{ho}. Calculate T_{ho} with Equation 15 and a new peak will be formed $P_{ho} = P_{l_1}, P_{l_2}, \ldots, P_{l_h}$.

$$\begin{aligned} \mu_m &= \frac{P_{max}+P_{min}}{2} \\ \sigma_m &= \sqrt{\frac{\sum_{i=1}^{k} =(P_{l_i}-\mu_m)^2}{k}} \\ T_{ho} &= \mu_m - \sigma_m \end{aligned} \tag{15}$$

4. Erase few peaks according to threshold width T_{wo}, T_{wo}= 20, assumed there are no more than 3 or 4 dominant colors on object T_o. For two adjacent peaks P_{l_i} *and* P_{l_j}, if $(l_j-l_i)<T_{wo}$ then save highest valued peak and remove other peaks from P_{ho}.
5. The result is peak sequence P_{no} and dominant color from T_o, determined as $C_o : l_1, l_2, \ldots, l_n$.
   Do T_b as well with steps above. After dominant color chosen, then SVM training dataset consists of TS_p (positive training set) and TS_n (negative training set) can be formed with Equation 16.

$$\begin{aligned} TS_p &= \{(x,y)|f(x,y) = i, (x,y) \in T_o, \in C_o \\ TS_n &= \{(x,y)|f(x,y) = i, (x,y) \in T_b, \in C_b \end{aligned} \tag{16}$$

2.1.4. Local Homogenity Threshold

Generally, the amount of SVM training dataset is too large to be used as training directly. Neighborhood homogeneity threshold is used to choose pixels on TS_p and TS_n as training sample in SVM training. For pixel $p(i,j)$ in training set TS_p , TS_n local homogeneity on $n \times n$ neighborhood calculated with Equation 17. M_p is local homogeneity with $n \times n$ = 3, $d(p,q)$ is Euclidean color

IOP Conf. Series: Materials Science and Engineering **588** (2019) 012027 doi:10.1088/1757-899X/588/1/012027

distance between pixels p and q on quantized HSV color space. If $M_p \leq T_{lh}$ then that pixel belongs to TS_p and TS_n. The value of T_{lh} is assigned by user, the better value of T_{lh} inputted, the value of TS_p and TS_n generated will be more precise.

$$M_p = D_p^{n\times n} = \sum\nolimits_{q \in N_p^{n\times n}} d(p,q) \tag{17}$$

2.2. Feature Extraction

The feature extraction is done by taking color value of r, g, and b on image in RGB color space, intensity (i) on an image quantized in HSV color space, saliency value on saliency map, spatial value (x,y) and texture value which is local energy and local gradient value [4] from an image with steerable filter applied.

2.3. Pixel Classification with Support Vector Machine

After feature value from each pixel extracted, to generate image segmentation, those values will be classified using Support Vector Machine. Data normalization, SVM training and SVM testing will be done in this stage.

2.3.1. Data Normalization

Normalization is a preprocessing technique to do data scaling and standardization. Normalization done here is by scale normalization, which scale data in a certain range where the common range is 0-1. This is done to avoid data range from being too distant.

2.3.2. Classification with SVM

SVM falls under supervised learning. Supervised learning is a method requiring training and testing. SVM method attempt to seek a separator function with decent generalization. Generalization is an ability of hypothesis to classify a group of training in the middle of two classes. Searching equivalent optimal hyperplane by maximizing margin or distance between boundary field on both class [5].

In the plane formed from data pattern, SVM will form hyperplane separating 2 classes. The simplest form of SVM is linear SVM. As a reference, training set D can be denoted with Equation 18.

$$D = \{(x_i, y_i) | x_i \in R^p, y_i \in \{-1,1\}\}_{i=1}^{n} \tag{18}$$

The value of $y_i \in \{-1,1\}$ represents class index or label valued 1 or -1 from dataset x_i. The best hyperplane separates data with $y_i = 1$ and $y_i = -1$. The hyperplane can be denoted in Equation 19.

$$w \cdot x - b = 0 \tag{19}$$

Data point x_i included in class 1 can be denoted as point which fulfills Equation 20 and data point x_i included in class -1 can be denoted with Equation 21.

$$w \cdot x - b \geq 1, y_i = 1 \tag{20}$$

$$w \cdot x - b \leq 1, y_i = -1 \tag{21}$$

Ideal hyperplane has maximum margin in separating data classes. In accordance with application of distance calculation formula between point and line, distance between nearest data and central boundary of hyperplane is $\frac{1}{||w||}$. To maximize margin, value of $||w||$ needs to be minimized. Minimization of $||w||$

value means completing quadratic programming problem. This problem will search for minimum point from Equation 22 with regard to boundary from Equation 23.

$$f: \frac{1}{2}||w||^2 \tag{22}$$

$$y_i(x_i \cdot w + b) - 1 \geq 0, \forall_i \tag{23}$$

This quadratic programming problem can be solved with various computation techniques. Lagrange Multiplier is one of them. Lagrange Multiplier used variable α to simplify calculation with the value of it being 0 or positive for Equation 24.

$$L(w, b, \alpha) = \frac{1}{2}||w||^2 - \sum_{i=1}^{l} \alpha_i(y_i(x_i \cdot w + b) - 1) \tag{24}$$

Optimal value from Equation 26 can be calculated by minimizing L towards w and b, and maximizing L towards α_i. Equation 24 can also be modified so it will only contain α_i by noticing the value of L = 0 on the optimum gradient point into Equation 25 with boundary from Equation 26. The data correlated with positive α_i is the support vector.

$$\sum_{i=1}^{l} \alpha_i - \frac{1}{2}\sum_{i,j=1}^{l} \alpha_i \alpha_j y_i y_j x_i x_j \tag{25}$$

$$\alpha_i \geq 0 \ dan \ \sum_{i=1}^{l} \alpha_i y_i = 0 \tag{26}$$

Basically SVM is a linear classification model, but it can be used to classify non-linear using kernel trick as in Equation 27.

$$K(x_i x_j) = \emptyset(x_i) \cdot \emptyset(x_j) \tag{27}$$

Kernel trick in SVM is a method to classify non-linear data by transforming real data space into data space with higher dimension. Most used Kernels in SVM are shown in Table 1

Table 1. Commonly used Kernel functions in SVM

Kernel Name	Inner Product Kernel
Linear	$x^T x$
Polynomial	$(x^T x_i + 1)^p$ with unrestricted p value, determined by user.
Radial-basis function (RBF/Gaussian)	$exp\left(-\frac{\|\|x-x'\|\|_2^2}{2\sigma^2}\right)$ with unrestricted σ value, determined by user.

3. Testing and Result Analysis

The testing in this research compares segmentation yielded by application with ground truth, by calculating accuracy using Equation 28. Testing is done on 50 images taken from benchmark dataset in [6].

$$ER = \frac{N_f + N_m}{N_t} \times 100\%$$
$$Accuracy = 100 - ER \tag{28}$$

Where variable N_f shows the amount of false-segmented on image pixels, N_m shows the amount of miss-segmented on image pixels and N_t is the total image pixels.

First test was done by comparing segmentation result with ground truth using initial parameter. Accuracy result for 50 images using initial parameter reached 93.69%, which means the segmentation outcome has been pretty good.

Tabel 2. Parameter and Initial Methods

No.	Parameter/ Method	Value/ method name
1.	Resize method on visual saliency detection	*'bilinear'*
2.	Value of *n* in average filter convolution matrix in visual saliency detection	3
3.	Pixel width on morphology operation in the pre-segmentation process and trimap formation	70
4.	Threshold on local homogeneity	0
5.	Local window on local homogeneity	3 x 3
6.	SVM kernel function on pixel classification process	Linear
7.	Value of parameter C SVM on pixel classification process	1

Figure 5. Examples of Segmentations Result

The second test compares segmentation outcome with ground truth based on differences in resize method and *n* value on different average filter convolution matrix in visual saliency detection process. Resize methods comparison used in visual saliency detection are bilinear and bicubic. While *n* value in average filter convolution matrix is 3, as referenced in reference literature, 5 and 7. Table 3 shows the average accuracy for 50 images.

Tabel 3. Test results with different parameters in visual saliency detection

Resize Methods	Average Filter Convolution Matrix 3	5	7
Bilinear	**93.69**	93.61	93.54
Bicubic	93.65	93.53	93.52

Third test was done by comparing segmentation outcome with ground truth based on pixel width differences on morphology operation in pre-segmentation process and different trimap formation. This was done on pixel width on morphology operation with the values of 70 and 75. Best accuracy resulted by pixel width of 75 with the average accuracy of 93.82%.

The next test was done by comparing segmentation outcome with ground truth based on threshold value differences in local homogeneity threshold process, where this threshold value can affect data training amount generated in pixel classification to produce image segmentation. This test will use threshold values of 0, 50 and 100. Table 4 shows average accuracy result from 50 images.

Tabel 4. Average accuracy results with different thresholds

Threshold value	0	50	100
Accuracy	93.82	94.23	**94.26**

The test continues by comparing segmentation outcome with ground truth based on kernel function differences and parameter C value in pixel classification with SVM to create image segmentation. This test uses linear, polynomial and rbf kernel functions with parameter C value of 0.1 and 1. Average accuracy result from 50 images is shown in Table 5. Rbf kernel function has the best accuracy result with C value of 1 and average accuracy result of 94.84%.

Tabel 5. Accuracy result with different kernel functions and C parameters

		C value	
Kernel Function	**Parameter**	**0.1**	**1**
Linear		94.45	94.26
Polynomial	$p = 3$	94.10	94.23
	$p = 4$	94.15	94.20
Rbf	$\sigma = 0.25$	93.45	93.35
	$\sigma = 0.5$	94.71	**94.84**

4. Conclusion

From previously done tests and test result analysis towards this system, it can be concluded that:

1) Saliency-SVM method can generate training data and execute color image segmentation well with average accuracy reaching 94.84%.
2) Image segmentation result with Saliency-SVM is influenced by resize method and n value in average filter convolution matrix in visual saliency detection process, albeit with insignificant difference of accuracy. Resize bilinear method has the best accuracy result with the n value of 3.
3) Image segmentation result with Saliency-SVM is influenced by pixel width on morphology operation in pre-segmentation proceess and trimap formation, which has the best accuracy with pixel width of 75.
4) Image segmentation result with Saliency-SVM is influenced by threshold value in pixel training selection with local homogeneity threshold, where data training amount generated will affect image segmentation result. Best accuracy was produced by threshold value of 100, which means the more pixel training formed, the better image segmentation generated.
5) Image segmentation result with Saliency-SVM is also affected by kernel function and parameter C on pixel classification process with SVM. Rbf kernel function has the best accuracy result with σ value of 0.5 and C parameter of 1.

5. Acknowledgments

Author A.Y expresses the highest gratitude towards Allah SWT for all the grace and mercy, enabling author to finish this research swiftly. Author also expresses gratitude to Mr. Rully Soelaiman and Mrs. Chastine Fatichah who helped a lot in finishing this research

References

[1] A Y W S M D A N S M Dinda Firly 2014 *Impelementasi Support Vector Machine dengan Optimasi Pruning pada Persoalan Klasifikasi Pola* (Surabaya: Institut Teknologi Sepuluh Nopember)

[2] W W Xuafei Bai 2014 Saliency-SVM: An automatic approach for image segmentation *Neurocomputing* **vol. 136** pp. 243-255

[3] X H a L Zhang 2007 *Saliency Detection: A Spectral Residual Approach*

[4] Q-Y W H-Y Y J B Xiang-Yang Wang 2011 Color image segmentation using automatic pixel classification with support vector machine *Neurocomputing* **vol. 74** pp 3898-3911

[5] D F S M R S S M Astris Dyah Perwita 2014 *Uji Kinerja Penggabungan Algoritma Support Vector Machine dan Simulated Annealing pada Permasalahan Klasifikasi Pola* (Surabaya : Institut Teknologi Sepuluh Nopember)

[6] J S N-N Z X T a H-Y Tie Liu [Online]. Available: http://research.microsoft.com/en-us/um/people/jiansun/SalientObject/salient_object.htm. [Accessed 03 June 2015].

IOP Conf. Series: Materials Science and Engineering **588** (2019) 012028 doi:10.1088/1757-899X/588/1/012028

Exterior Electric Hatchback Crossover Car Design with Strong, Vicious and Sharp Concept

Baroto Tavip Indrojarwo[1,a], Andhika Estiyono[1] and Abdur Rosyid Pratama[1]

[1] Department of Industrial Design, Faculty of Architecture Design and Planning, Institut Teknologi Sepuluh Nopember, Surabaya, Indonesia 60111

[a] tavip.baroto@gmail.com

Abstract. Identity of design of an electric car is necessary to show electrical technology and added values for consumers. Design of electic cars today, in the meaning of appearance and image, are judged by consumers as wierd and ugly compared to design of internal combustion engine (ICE) cars. The distinguished factor of electric car compared to ICE car is the technology that provides environment friendly and fuel energy crisis solution. Symbols of electric car can actually be build to be an interesting form and shape, such as smart, lightweight, efficient and clean. This research aims to design a small-medium electric car with 4-passenger car for young urban family consumer. The methods are consecutively: kind of car defining, 2D sketching, ergonomic or anthopometrical study, engineering and passenger package study, 3D styling and part design. The concept of form has adapted by keywords: strong, vicious, and sharp. Definition of ergonomic study based on Indonesian people with percentile of 5% women and 95% of men. Definition of engineering package based on Li-ion battery with size of 24 kWh placed under floor and single electric motor package placed on the front with front wheel system. The design result is a small medium crossover car with dimensions of wheelbase of 2500 mm, height of 1534 mm, front overhang of 858 mm and rear overhang of 723 mm. Design character of strong, vicious and sharp are transformed by small sharp headlamp and rear lamp; small grille with LED lines; smooth sculpting line on entire body; deep sculpting line on side body, firm and bold form of front and rear fender. The design outputs are 3D model of exterior, 3D mode; of body parts, 3D animation and physical scale model

1. Introduction

This research aims to support development of the national electric car program in the year of 2013, whereas Institut Teknologi Sepuluh Nopember (ITS) is one of the five universities that taken part to it. The basic platform is small medium car that accommodates 4 passengers in urban area. The possibile kind of this platform are sedan, hatchback, SUV(sport utility vehicle) and minibus (MPV-multi purpose vehicle) in the size of small-medium car. Sedan is a three-box passenger car consists of bonnet, passenger cabin and trunk with low-seat position (about 300 mm). Hatchback is a sedanbased two-box car, consists of bonnet and cabin with no box of trunk, because it is integrated with cabin. The trunk consists of a single door that are lift-up opened. SUV is a two-box car with high ground clearance (about 200 mm) with high-seat position. MPV is a single box or two-box car with with highseat position (about 500 mm). Sedan and hatchback excel in comfort and appearance, whilst SUV and MPV excel in offroad capability and load capacity.

Specific design identity of electric cars in order to make differentiation to ICE car can enhance the needs and wants for consumer. By considering aspects of universal design so that electric car design can be compete in the market. The purpose of this research is to design a 4-seat passenger electric car that suites to urban living for young family. According to that needs and wants, this research hypothesis is a stylish crossover hatchback that fits to accomodate activity of daily businees and weekend holiday with semi-offroad ability to answer general infrastructure condition of Indonesia city.

Figure 1. Basic Figure of Hatchback Car

2. Literature Review

2.1. The Best Research and Product

Design or styling of Range Rover Evoque is “a tasteful”, with excellent proportions without use of many lines and curves or folds on the body. The design makes Evoque seen as new image of urban crossover cars in the cities. Sporty mesh grill, futuristic layout and lamp shape, side air vents and floating roof. “Simplicity Design” used by the Range Rover is actually good proportion of basic form that can form a new image of SUV. Evoque design becomes inspiration of this research in styling concept that be build by good proportion from basic forms.

2.2. Proportion of Hatchback and Crossover

The body side proportion of hatchback and crossover car can be described belows.

Figure 2. Car Types Classification

In H-Point, the 4-passenger electric car will be included in the Economic Car category, the largest group in the passengercar market. Hatchback Crossover is a car used in urban areas with average long size of 3400 mm.

2.3. Interior Environment Study

Interior environment dimension of car bases on passenger seat position (H-Point). The study of seat position can be described as follows.

Determining size of car must consider size of the passengers inside. Design experts stated that the formation of the car starts from the inside out. It aims to make passengers comfort and safe in the cabin. For most types of passenger cars, the H-Point is relatively low to facilitate access in and out of the car and provide a low center of gravitation. In addition to human anthropometry, there are several important points in determining hardpoint because interior is the most important factor in determining the exterior dimensions of the car as it is directly related to the user's comfort factor.

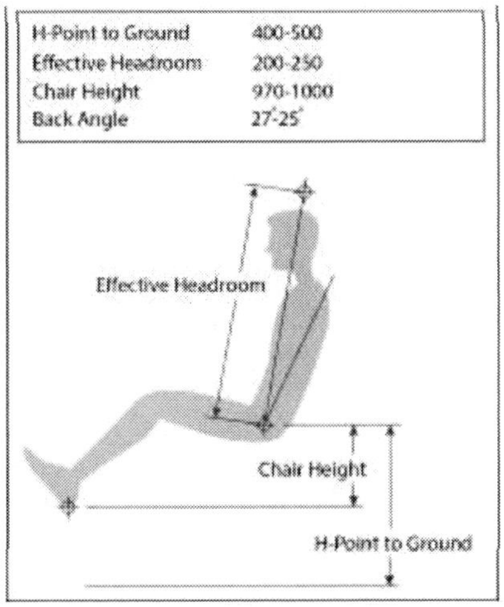

Figure 3. Various Driver height from ground and postures

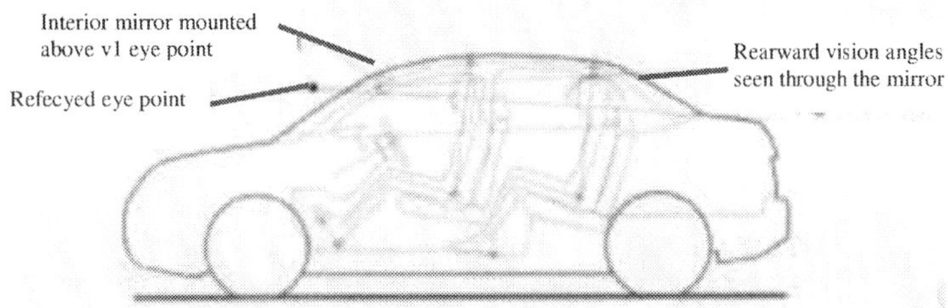

Figure 4. Visibility through rear window

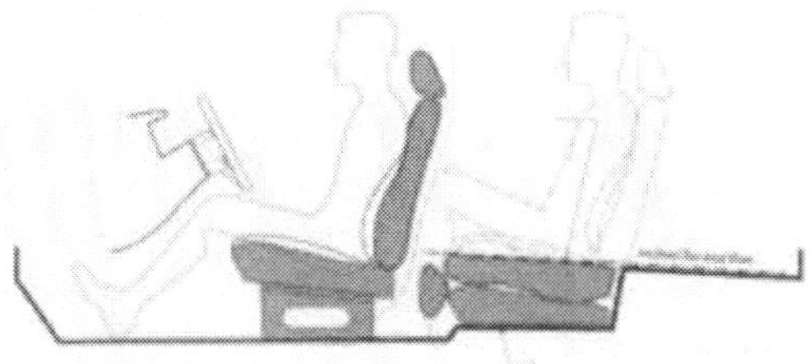

Figure 5. Rear Seat folding mechanism

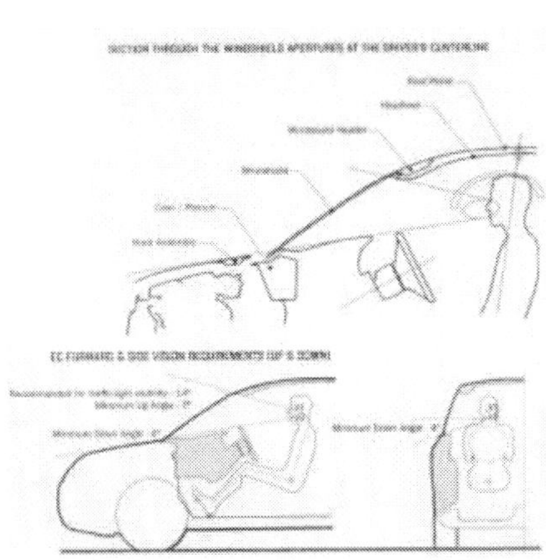

Figure 6. Terms of installation of a car windshield

2.4. Engineering Packaging Study

In determining electrical components and layouts for electric drive, different treatment with the machine, in general, is required. Supporting components are also different from ICE cars.

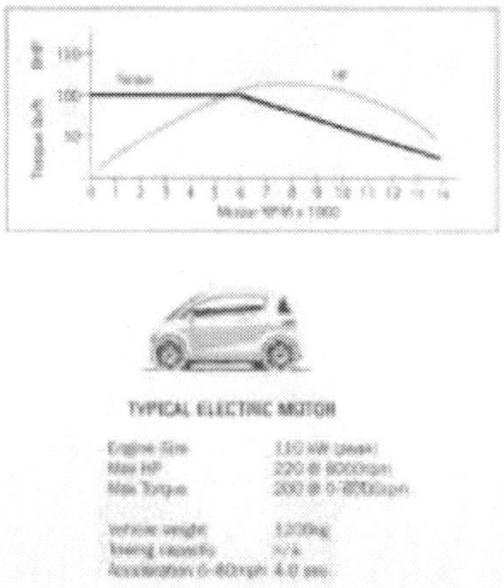

Figure 7. Specification of the electric motor

The motor used is usually smaller in size but the fuel storage or motor battery is larger so that the shape of the car becomes longer. Therefore, the battery is usually placed under the floor.

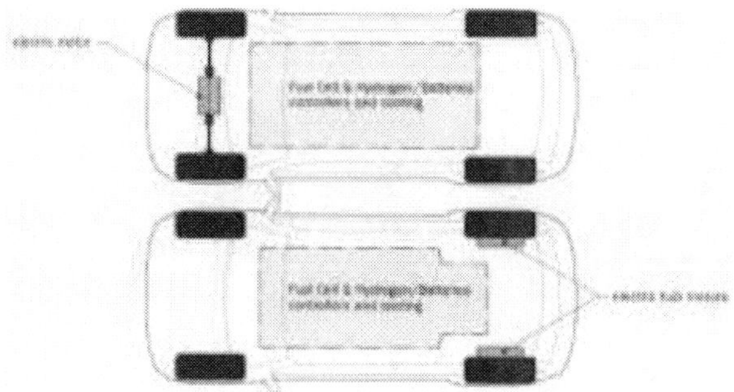

Figure 8. Layout type of propulsion motors in electric cars.

3. Research Methodology

The research process can be described belows:

a. Product planning: analyzing the problems of the product on the market, including the needs and wants of consumers, competitors, as well as ongoing trends. This 3 study is definition of car kind by MSCA (market survey and competitor analysis) that obtained a car that suited to the variable of: comfort, young look/appearance and semi-offroad capability. MSCA result is a combination of hacthback and SUV with small-medium size. This car is then defined as Hatchback Crossover Car. Hatchback represents of good apearance and comfort, whilst SUV represents of semi-offroad capability and load capacity. The seat position is defined as high-low seat position (about 400 mm).

b. Benchmarking of size and proportion to existing products with similar attributes to define golden proportions.

c. Ergonomic study: determining consumers or user needs by analizing their activity and dimension in order to realize comfort and safety. The result is configuration of passenger and engineering package with each dimension.

d. Design ideation: developing shape and form of overall body. The result is facade design and 3D appearance that transforms the concept of strong, vicious and sharp.

e. Final design: developing design into 3D digital geometry for overall exterior body and main single parts. The output is 3D digital geometry which can run CNC machne to make positive model and mold for production process

3.1. Brenhcmarking Study

- Objectives: determine base and dimensions of car that will be used as platform of the electric car design
- Variable: interior space, number of seats, layout, volume
- Sample: Mini Countryman, Nissan Juke and Suzuki SX-4
- Method: observation and experiment
- Result:
  - Dimension (L/H/W in mm): 3840-3860/1680- 1720/1480-1520.
  - Best proportion: Countryman; best strong image: Nissan Juke; best sharp image: Juke; best vicious image: SX-4

3.2. Consumers Behaviour Analysis

- The 25-30 years old, with energic and attractive soul
- Young family (3-4 people), a new family with 1- 2 kids.
- Living in urban areas.
- High mobility in daily activity.
- Fun behavior in business and family communication.
- Smart attitude in technology usage.
- Friendly spirit to the environment.

3.3. Ergonomic Study of Passenger Package

By combining data used in midsize car and electric car, interior dimensions which have been applied to human anthropometry of 95 percentile of men and 5 percentile of women can be determined as follows.

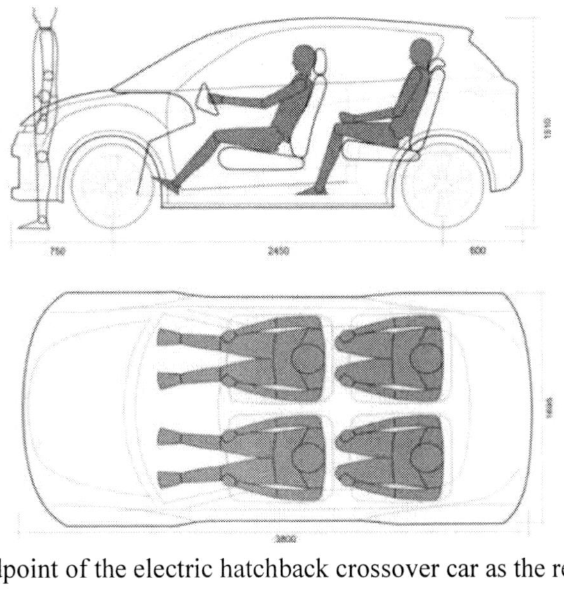

Figure 9. Hardpoint of the electric hatchback crossover car as the research result

Passenger height of the 95 percentile men is similar to187.5 cm and the 5 percentile women is similar to 163.6 cm. The men have upward vision angle of 18° and downward vision angle of 11° , whereas the women have upward vision angle of 32° and downward vision angle of 7° .

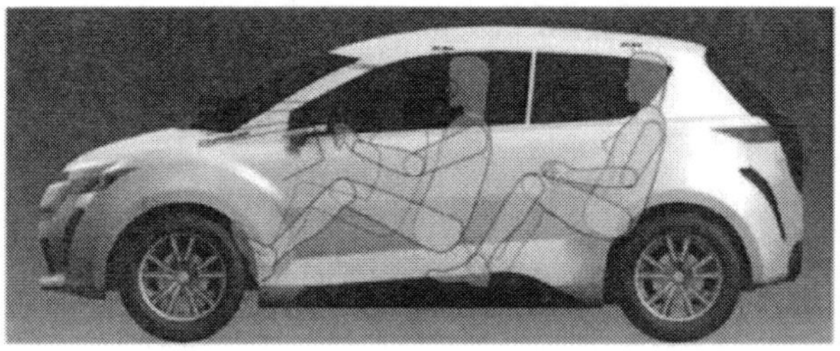

Figure 10. Anthropometric study result of 95 percentile-men and 5 percentile-women

3.4. Ergonomic Study of Engineering Package

Main engineering elements that must be considered in design are battery, motor, converter and drive train components. The result of this study can be desribed as follows.

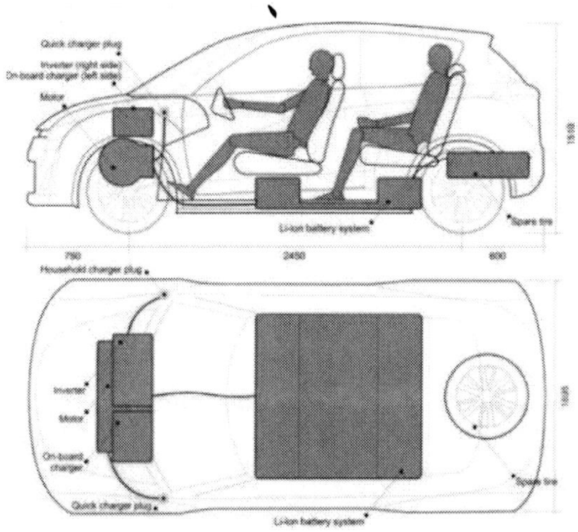

Figure 11. Engineering package result of the electric crossover-hatchback car

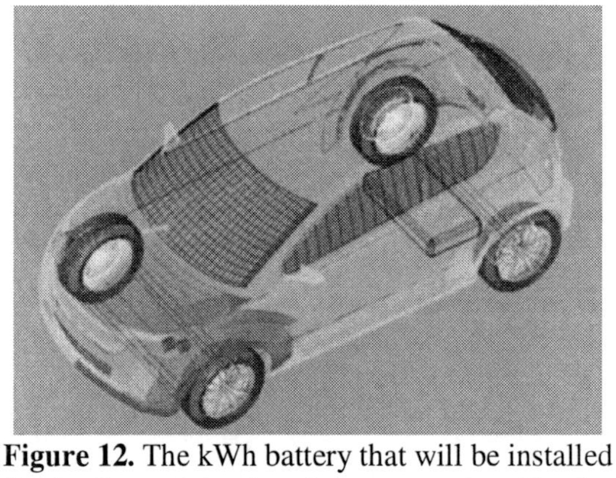

Figure 12. The kWh battery that will be installed under the floor of the electric crossover-hatchback car.

Battery and electric motors that applied in this research are adapted from Nissan Leaf electric car, i.e: lithium-ion battery 24kWh, with 48 lithium-ion per module and electric motor of 80 kw AC motor, 187 lb-ft and 107 horsepower. The result researches above then developed into basic dimension approaches that can be described at table belows.

Table 1. Basic Dimension

Element	Dimension
Overall L/W/H	3850 / 1700 / 1500 mm
Wheelbase	2450 mm
Wheel Dimension	185 mm , 55 mm , 16' up to 215 mm, 55 mm, 18'
Front/rear tread	1480 / 1485 mm
Ground Clearance	190 mm

Engineering aspects that can be implemented to the design, research by electric and mechanic team can be described at table belows.

Table 2. General Engineering Specification

Element	Specification
Motor	80 kW AC synchronous electric motor, 107 hp
Battery	24 kWh lithium-ion (Li-ion)
System Charge	6.6 kW Onboard Charger
Distance Cruise	160 km
Structure System	Tubular frame
Body Structure	Monocoque
Suspension System	Front: McPherson Strut + Coil Spring Rear: Torsion Beam + Coil Spring
Gross Weight	1480 kg
Curb Weight	980 kg

Integration of ergonomics study and engineering package assumption produces general lay-out of electric car design. This lay-out provides ergonomic space for 4 passengers with optimum space at human body, leg and head room at any gestures. This lay-out also provides suitable arrangement of electric equipment. Design of body structure can also be developed clearly caused of this definitive lay-out.

4. Design

4.1. Styling Concept
The styling concept is determined from the needs and wants of strong, vicious and sharp character. These keywords than develop into the image board that represents to it, as described as follows.

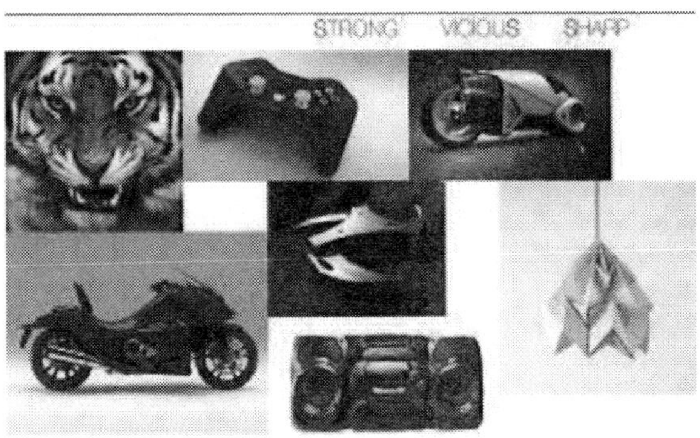

Figure 13. Image board of products that express strong, vicious and sharp images that represent the target consumer

The image board then develop into hand sketches of perspectives below.

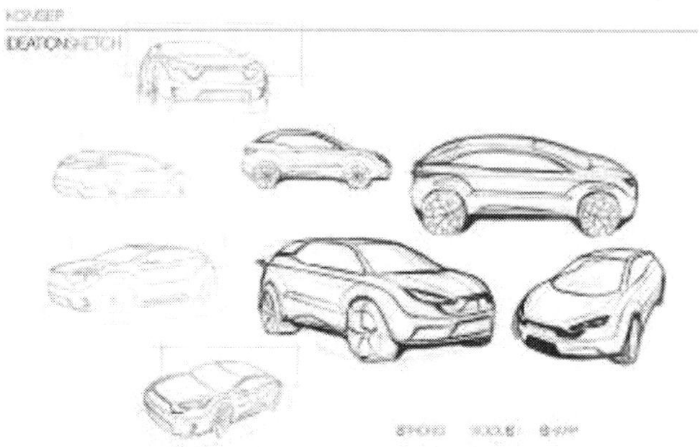

Figure 14. Hand Sketch of The Overall Car Exterior

From the image board above, sketches of form and line represent strong by firm constant line harmony, while vicious is obtained from rising line from one side and sharp is shown by lines that has pointed to form angle.

4.2. The 3D Digital Modelling

The 3D digital modeling process is developed from hand sketches by graphic computer. The result can be described as follows.

Figure 15. Surfacing Process in 3D Digital Modeling

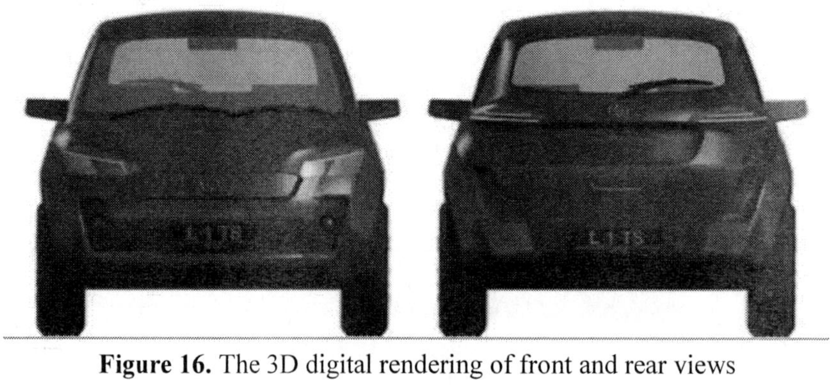

Figure 16. The 3D digital rendering of front and rear views

Figure 17. The 3D digital rendering of side and top views.

Figure 18. The 3D digital rendering of ¾ front-side and ¾ rear-side perspective view

4.3. The 3D Digital Modelling of Body Parts

The next step of this research is to explode the entire exterior model into body parts. The main body parts are sturcture elements, i.e.: side panels (left and right), roof and floor. The second main ones are filler elements, i.e: doors (sides and rear), bumpers (front and rear), engine hood, widshield and all glass windows. All design of body parts can be figured as follows.

In order into production phase, each body parts must be run into single part production required, such as: mold, press machine, cutting machine, finishing machine and mechanicalelectrical support element. The main body parts (side panels, floor and roof) must be arranged in order to form basic shape. After that, each secondary main body part and other mechanical-elecrical supported part can be assembled following the production line and procedure. This research results.

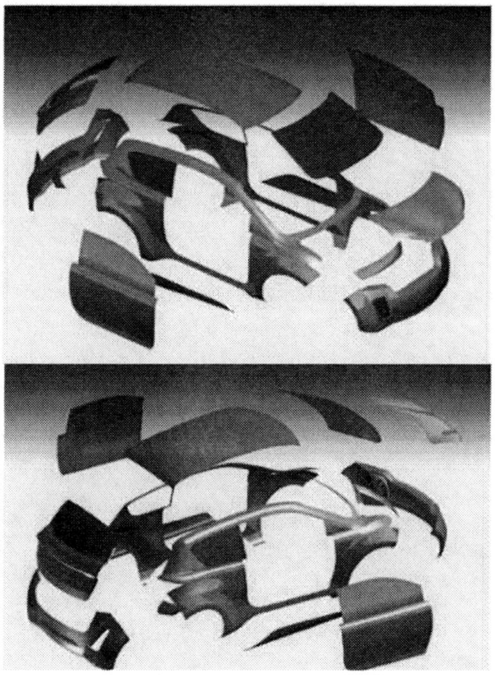

Figure 19. The 3D digital geometry of main single parts of exterior

5. Conclusion

From previously design, it can be concluded that:

1) The style of strong, vicious and sharp can be obtained by :
   - Strong: proportion consists of wheelbase of 2500 mm, height of 1534 mm, front overhang of 858 mm and back overhang of 723. This proportion figures sporty and muscle image.
   - Vicious: short bonnet, bigger wndshield and sharp headlamp from front to side with LED figure as 6 vicious factor . Small and lines of grill form also enhance the image of vicious.
   - Sharp: combination of smooth and sharp body sculpting and carving figure sharp concept. Form of head lamp and rear lamp that adapted from blade shape also sthengthen the sharp image
2) Styling character that has been obtained from determining line characteristics:
   - Strong: line arrangement that forms firm, dynamic and bold .
   - Vicious: curve line that raising from bottom to top .
   - Sharp: line s that form small pointed angle .
3) Dimensions of the battery under the passenger seat can provide a good ergonomics level for the interior package for 2 adults (height 187.5 cm) in front and 2 children in rear seat .
4) The 3D digital geometry of overall exterior body roles as a visualization of real appearance of design of hatchback crossover electric car that shows its styling, (form, dimension, line, surface and color).

5) The 3D digital geometry of single part body roles as a platform for manufacturing production process that are ready to apply to make positive model and mold by CNC machin e .
6) This reasearch results 16 body parts in 3D digital geometry, i.e: 2 side panels (left and right), roof, floor, 3 doors (L, R and rear), 2 bumpers (front and rear), engine hood, grill, 2 front fenders (L and R), 2 head lamps (L and R) and 2 rear lamps (L and R).

References

[1] Ghassan, Aysar 2009 *Blue in A Green World* Northumbria University
[2] Macey, Stuart and Wardle, Geoff 2008 H -Point *The Fundamentals of Car Design & Packaging* (C.A : Design Studio Press -Art Center Collage of Design Pasadena)
[3] Panero, Julius and Zelnik, Martin 1979 *Human Dimension and Interior Space* (Whitney Library of Design Wayson – New York : Guptill Publication)
[4] Samboro and M Y Alief 2013 *Styling of Exterior of 4 - Seater Electric Car based on City Car* (Surabaya : Department of Industrial Design - ITS)
[5] The BMW Group *Design that Moves* The BMW Design
[6] Volkswagen Tiguan 2011 *Top Rated Compact Crossover Suv Autobytel Autopacific Consumer Award* www.autobytel.com 15 November 2013
[7] Jendro Trimaryanto 2011 *Design Development of the Micro Car Gea as Stylish City Car that Suites to Indonesia People* (Surabaya : Department of Industrial Design - ITS)
[8] Wiel, Jam Willem Vander 2012 *Future of Automotive Design & Material, ATC* Available: http//acemr.eu, date accessed: 17 November 2013

IOP Conf. Series: Materials Science and Engineering **588** (2019) 012029 doi:10.1088/1757-899X/588/1/012029

The Vibration Reduction in Multilevel Structure Model with the addition of Tuned Liquid Column Damper (TLCD):Study of the Influence of variations in TLCD's mass ratio & orifice cross-sectional area

H L Guntur[1,a], A A A Daman[1] and W Hendrowati[1]

[1] Department of Mechanical Engineering, Institut Teknologi Sepuluh Nopember, Surabaya, Indonesia 60111

[a] haruslg@me.its.ac.id

Abstract. The multilevel buildings are vulnerable to the excitation forces such as winds and earthquakes. The analysis of vibration reduction in multilevel building is established in this study. The experimental and simulation method are conducted in laboratory scale. The multilevel structure model is design to be two-floor building model. The Tuned Liquid Column Damper (TLCD) is placed on first-floor of the building model and then it is placed on second-floor of the building model. The TLCD's mass ratio and its orifice cross-sectional area are varied. The experimental results correspond well to the simulation results. The TLCD's mass ratio does not provide significant difference, while the variation of orifice cross-sectional area provides significant differences.

1. Introduction

Nowadays, the residential buildings were constructed upwards in order to resolve the limited area problem. However, the increasing of building's height causes the increasing of its vulnerability to the excitation forces which are from winds and earthquakes. The excitation forces from winds and earthquakes can produce excessive vibration to buildings and it can generate the massive failure to the building. Therefore, the buildings require the additional system which is able to reduce the vibration caused by the winds or earthquakes. Dynamic Vibration Absorber (DVA) which is consist of a mass-spring system can be used to reduce the vibration of system in translational and rotational motions [1]. Furthermore, the Tuned Mass Damper (TMD) is developed to reduce the vibration of structure [2,3]. However, the utilization of DVA and TMD are not quite necessary in a building. Thus, the Tuned Liquid Column Damper (TLCD) which utilizes motion of liquid to reduce the structural vibration is established [4]-[12]. The validation of proposed formula and experimental results of a SDOF shear-type model utilized by three TLCD variations has been obtained [4]. Zhu [5] researched the multi-tuned liquid column dampers (MTLCD) by using multiple shaking table. Then, the effects of nonlinear air spring in TLCD were investigated [6]. A study of an SDOF structure equipped with TLCD which with the blocking ratio variation of orifice has been conducted [7]. The verification using finite element analyses was conducted to investigate the performance of TLCDs placed on a prototype structure [8]. The performance of TLCD in reducing the vibration of wind turbine has been researched [9].

The aim of this paper is to study the influence of TLCD's position to structural vibration reduction. Aside, the variation of TLCD's orifice cross-sectional area and the mass of TLCD are analysed to identify the optimum cross-sectional area and the mass of TLCD in vibration reduction.

2. Method

Both of simulation method and experimental method were conducted in laboratory scale. The main system consists of two plates which represent the floor of buildings and is supported by four rods as shown in fig. 1. The 2-Degree of Freedom (2-dof) system is placed on translational shaking table (TST) that will be the source of excitation force to the 2-dof system. Figure 1(a) represents the main system without TLCD which consists of the TST system (1), the first floor (2) and the second floor (3). While fig. 1(b) shows the main system with TLCD (4) laid on the first floor. The first variation in this study is the position of TLCD. The TLCD is laid on the first plate and then it is placed on the second plate after the data of the first floor are done. Aside of the position of TLCD, the variation of TLCD cross sectional area (A) is conducted. Then, the mass ratio (Mf/Ms) is varied by 5% and 7%. The first method of simulation is modelling the main system and translational shaking table as shown in fig. 2, where:

M_S : the mass of translational shaking table
M_1 : the mass of the first floor
M_2 : the mass of the second floor
M_f : the mass of fluid
K_S : the stiffness constant of TST
K_1 : the stiffness constant of the first-floor rod
K_2 : the stiffness constant of the second-floor rod
C_S : the damping constant of the TST
C_1 : the damping constant of the first-floor rod
C_2 : the damping constant of the second-floor rod
B : the width of TLCD
h : the height of liquid
X_S : the displacement of TST
X_1 : the displacement of the first-floor
X_2 : the displacement of the second-floor
Y : the displacement of liquid

IOP Conf. Series: Materials Science and Engineering **588** (2019) 012029 doi:10.1088/1757-899X/588/1/012029

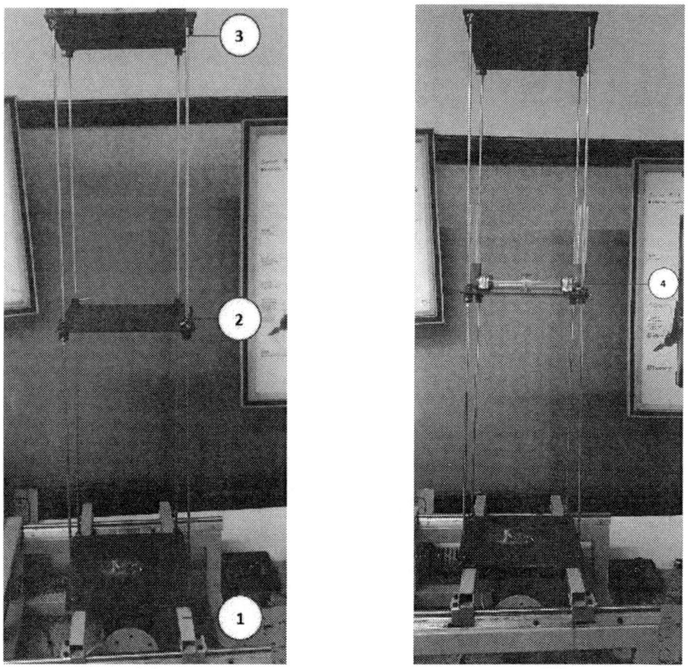

Figure 1. The experimental set-up of two-floor building without TLCD (a) and with TLCD placed on first floor (b)

The motion equation of each mass is derived from *free-body diagram.* The motion equation of TST system is expressed as:

$$M_s \ddot{X}_s + (C_s + C_1) \dot{X}_s + (K_s + K_1) X_s = C_1 \dot{X}_1 + K_1 X_1 + m\omega_e{}^2 R \cos \omega_e t \tag{1}$$

While the motion equation of mass 1 (first floor) and mass 2 (second floor) are written as:

$$M_1 \ddot{X}_1 + (C_1 + C_2) \dot{X}_1 + (K_1 + K_2) X_1 = C_1 \dot{X}_s + C_2 \dot{X}_2 + K_1 X_s + K_2 X_2 \tag{2}$$

$$M_2 \ddot{X}_2 + C_2 \dot{X}_2 + K_2 X_2 = C_2 \dot{X}_1 + K_2 X_1 \tag{3}$$

The equation (1)-(3) can be written as matrices form and become the equation of main system without TLCD:

$$\begin{bmatrix} M_s & 0 & 0 \\ 0 & M_1 & 0 \\ 0 & 0 & M_2 \end{bmatrix} \begin{Bmatrix} \ddot{X}_s \\ \ddot{X}_1 \\ \ddot{X}_2 \end{Bmatrix} + \begin{bmatrix} C_s + C_1 & -C_1 & 0 \\ -C_1 & (C_1 + C_2) & -C_2 \\ 0 & -C_2 & C_2 \end{bmatrix} \begin{Bmatrix} \dot{X}_s \\ \dot{X}_1 \\ \dot{X}_2 \end{Bmatrix} +$$

$$\begin{bmatrix} (K_s + K_1) & -K_1 & 0 \\ -K_1 & (K_1 + K_2) & -K_2 \\ 0 & -K_2 & K_2 \end{bmatrix} \begin{Bmatrix} X_s \\ X_1 \\ X_2 \end{Bmatrix} = \begin{Bmatrix} m\omega_e{}^2 R \cos \omega_e t \\ 0 \\ 0 \end{Bmatrix} \tag{4}$$

While, the matrices of main system with TLCD laid on first floor are define by:

IOP Conf. Series: Materials Science and Engineering **588** (2019) 012029 doi:10.1088/1757-899X/588/1/012029

$$\begin{bmatrix} M_s & 0 & 0 & 0 \\ 0 & M_1 + M_f & 0 & \rho AB \\ 0 & 0 & M_2 & 0 \\ 0 & \rho AB & 0 & M_f \end{bmatrix} \begin{Bmatrix} \ddot{X}_s \\ \ddot{X}_1 \\ \ddot{X}_2 \\ \ddot{Y} \end{Bmatrix} + \begin{bmatrix} C_s + C_1 & -C_1 & 0 & 0 \\ -C_1 & (C_1 + C_2) & -C_2 & 0 \\ 0 & -C_2 & C_2 & 0 \\ 0 & 0 & 0 & C_f \end{bmatrix} \begin{Bmatrix} \dot{X}_s \\ \dot{X}_1 \\ \dot{X}_2 \\ \dot{Y} \end{Bmatrix} + \begin{bmatrix} (K_s + K_1) & -K_1 & 0 & 0 \\ -K_1 & (K_1 + K_2) & -K_2 & 0 \\ 0 & -K_2 & K_2 & 0 \\ 0 & 0 & 0 & K_f \end{bmatrix} \begin{Bmatrix} X_s \\ X_1 \\ X_2 \\ Y \end{Bmatrix} = \begin{Bmatrix} m\omega_e{}^2 R \cos \omega_e t \\ 0 \\ 0 \\ 0 \end{Bmatrix} \quad (5)$$

$$\begin{bmatrix} M_s & 0 & 0 & 0 \\ 0 & M_1 & 0 & 0 \\ 0 & 0 & M_2 + M_f & \rho AB \\ 0 & 0 & \rho AB & M_f \end{bmatrix} \begin{Bmatrix} \ddot{X}_s \\ \ddot{X}_1 \\ \ddot{X}_2 \\ \ddot{Y} \end{Bmatrix} + \begin{bmatrix} C_s + C_1 & -C_1 & 0 & 0 \\ -C_1 & (C_1 + C_2) & -C_2 & 0 \\ 0 & -C_2 & C_2 & 0 \\ 0 & 0 & 0 & C_f \end{bmatrix} \begin{Bmatrix} \dot{X}_s \\ \dot{X}_1 \\ \dot{X}_2 \\ \dot{Y} \end{Bmatrix} + \begin{bmatrix} (K_s + K_1) & -K_1 & 0 & 0 \\ -K_1 & (K_1 + K_2) & -K_2 & 0 \\ 0 & -K_2 & K_2 & 0 \\ 0 & 0 & 0 & K_f \end{bmatrix} \begin{Bmatrix} X_s \\ X_1 \\ X_2 \\ Y \end{Bmatrix} = \begin{Bmatrix} m\omega_e{}^2 R \cos \omega_e t \\ 0 \\ 0 \\ 0 \end{Bmatrix} \quad (6)$$

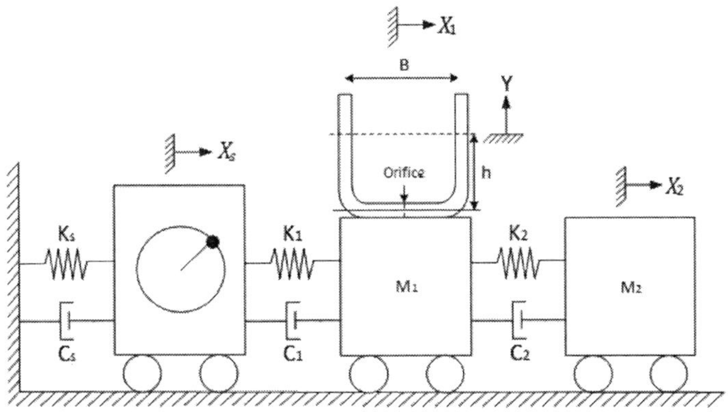

Figure 2. The dynamics model of main system with TLCD laid on the first floor

Then, the block diagram is built from the equations above. The simulation results are the displacement and the acceleration of the main system without TLCD and with TLCD in variation of mass ratio, cross sectional area, and TLCD's position. The results that are obtained from experimental method are the acceleration of main system without TLCD as well as the acceleration of main system with TLCD in variation of mass ratio and TLCD's position only. The acceleration responses in experimental method are collected by using accelerometer and oscilloscope.

3. Results and Discussion

The discussion will be divided into three concerns which are the effect of the mass ratio, the effect of TLCD's position, and the effect of TLCD's cross sectional area variation to the vibration responses. The bode diagram (fig. 3) is obtained from simulation method to acknowledge the natural frequency of system without TLCD, system with TLCD placed on first-floor, and system with TLCD placed on second-floor. The information that can be obtained from figure 3 is that the second and third natural frequency of system with TLCD are lower than the natural frequency of system without TLCD. It means that the utilization of TLCD is convenient to reduce the vibration in second and third natural frequency of system without TLCD. Therefore, the simulation method and the experimental method are conducted in input frequency of 8 Hz or about 50,26 rad/s.

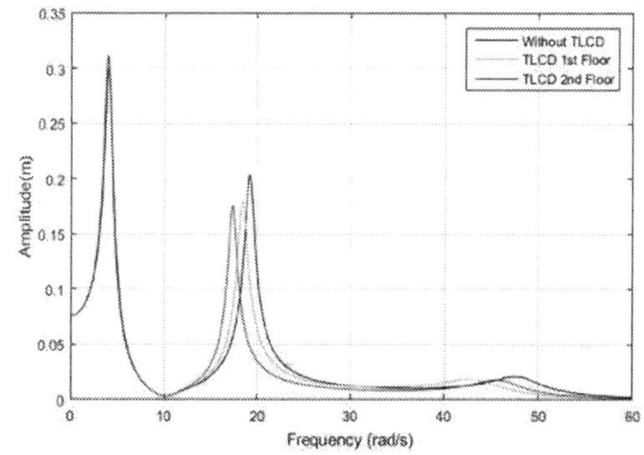

Figure 3. The bode diagram of system without TLCD and with TLCD

The simulation results are shown in fig.4-7. The acceleration responses of first-floor without TLCD and with TLCD in mass ratio variation are shown in fig. 4, while the acceleration responses of second-floor without TLCD and with TLCD in mass ratio variation are shown in fig. 5. Figure 4 shows that the difference of the maximum acceleration of first-floor without TLCD and the maximum acceleration of first-floor with TLCD placed on second-floor is relatively small. However, the additional of TLCD laid on first floor as well as laid on second floor can reduce the maximum acceleration response of first floor. The lowest maximum accceleration response occurs when model is utilized by TLCD laid on first-floor.

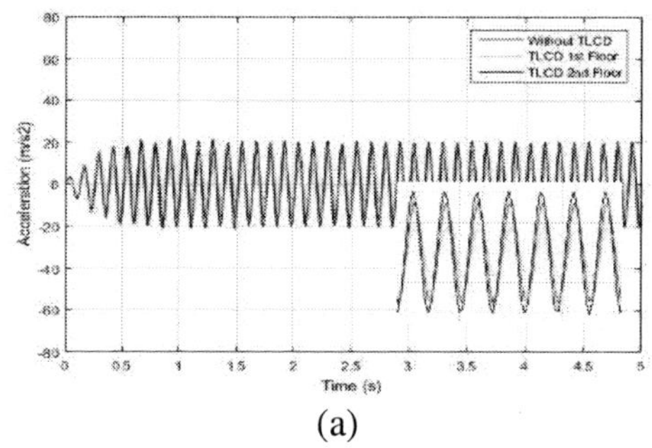

(a)

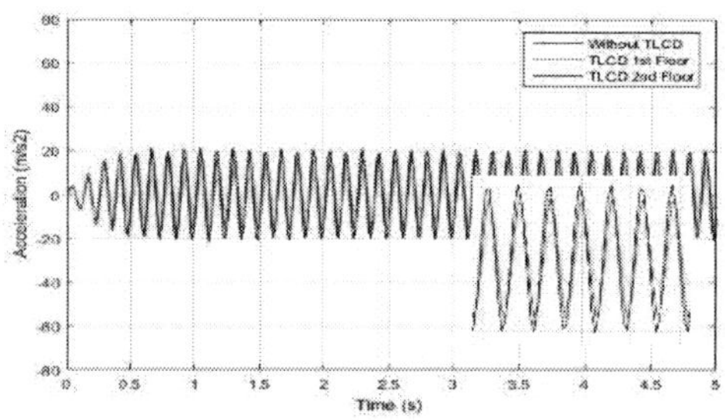

(b)

Figure 4. The acceleration response of first-floor with mass ratio (Mf/Ms) 5% (a), (Mf/Ms) 7% (b) obtained by simulation method

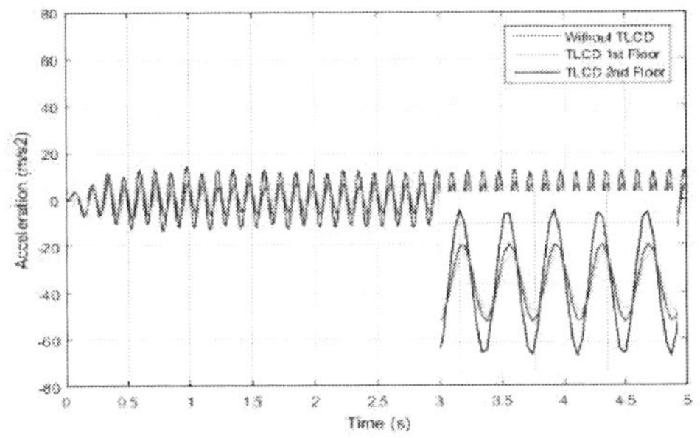

(a)

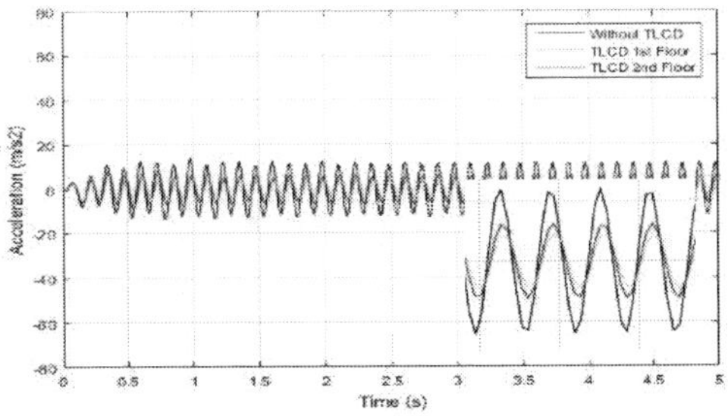

(b)

Figure 5. The acceleration response of second-floor with mass ratio (Mf/Ms) 5% (a), (Mf/Ms) 7% (b) obtained by simulation method

The difference of the maximum acceleration response of system with TLCD and without TLCD is significant when it is inspected from second floor's response (fig.5). The lowest maximum acceleration still occurs in utilization of TLCD on first-floor. The response of second-floor (fig.4) is lower than the response of first floor (fig.5). The effect of mass ratio variation to the acceleration response of both floor is not significant. The root mean square (RMS) of the acceleration shown in table 1 presents the detail of the acceleration magnitude of first-floor and second-floor in variation of mass ratio and TLCD's position. Table 1 shows that the acceleration RMS of first-floor in any arrangement is higher than the acceleration RMS of second-floor. The effect of mass ratio does not provide significant differences than the effect of TLCD's position to the acceleration RMS of first-floor and the acceleration RMS of second-floor.

Table 1. The acceleration RMS obtained from simulation method

Mass ratio	**TLCD's position**	**The acceleration RMS of 1st-floor (m/s^2)**	**The acceleration RMS of 2nd-floor (m/s^2)**
5%	Without TLCD	14,49	8,69
	With TLCD laid on 1st-floor	5,7	3,49
	With TLCD laid on 2nd-floor	11,89	4,71
7%	Without TLCD	14,49	8,69
	With TLCD laid on 1st-floor	5,39	3,29
	With TLCD laid on 2nd-floor	11,79	4,48

IOP Conf. Series: Materials Science and Engineering **588** (2019) 012029 doi:10.1088/1757-899X/588/1/012029

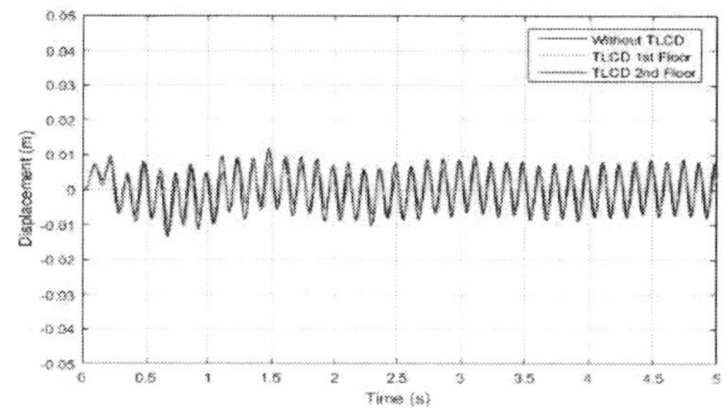

(a)

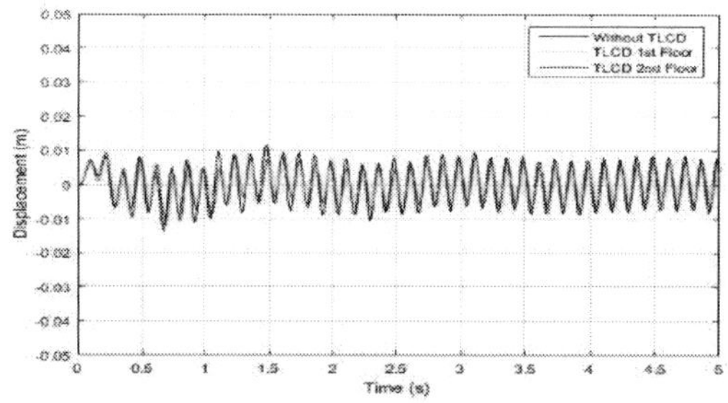

(b)

Figure 6. The displacement response of first-floor 5% (a) and the 7% (b)

Figure 6 represents the displacement response of first-floor in variation of mass ratio and in variation of TLCD's position. While, figure 7 shows the displacement response of second-floor in variation of mass ratio and in variation of TLCD's position. The displacement response of second-floor is smaller than the displacement response of first-floor. It corresponds well with the acceleration response discussed above. The utilization of TLCD in any position can reduce the displacement of first-floor as well as the displacement of second-floor. On the other hand, the mass ratio variation of TLCD still does not provide significant difference to the displacement response.

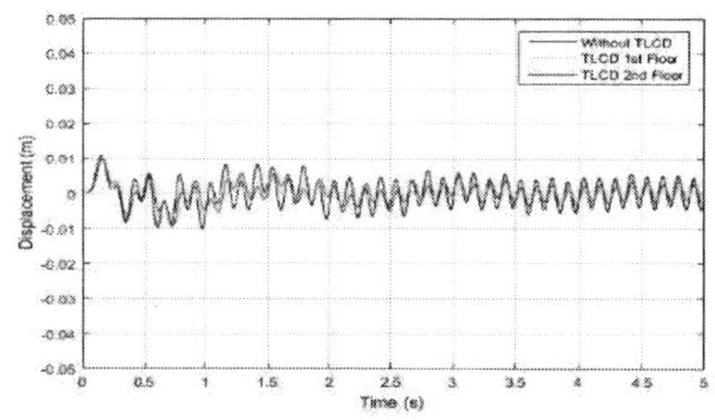

(a)

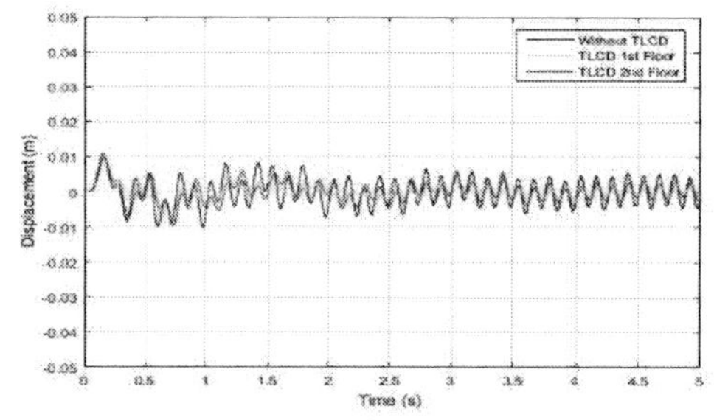

(b)

Figure 7. The displacement response of second-floor 5% (a) and the 7% (b)

Figure 8(a) represents the effect of orifice cross-sectional area variation to the acceleration of first floor when TLCD is placed on first floor. The variation of cross-sectional area affects the acceleration of first floor significantly. The acceleration of first floor is reduced significantly with the addition of TLCD laid on first-floor. The increasing of cross-sectional area causes the decreasing of the acceleration of first floor. While the position of TLCD on second floor (fig. 8(b)) insignificantly affects the acceleration of first floor. However, the utilization of TLCD placed on second floor is still reducing the acceleration of first floor. The acceleration responses of second-floor in three arrangements are shown in fig.9. Compare to the response of first-floor, the response of second-floor is smaller. The effect of TLCD's position to the acceleration of second-floor provides the same results as to the acceleration of first-floor. The system with TLCD laid on first-floor generates smaller magnitude of the acceleration than the system with TLCD laid on second-floor.

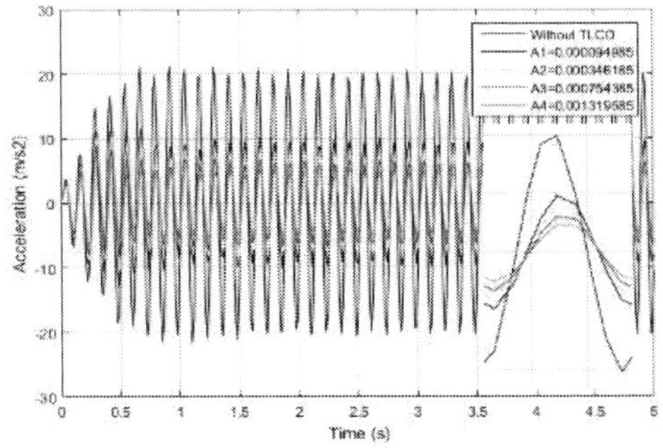

(a)

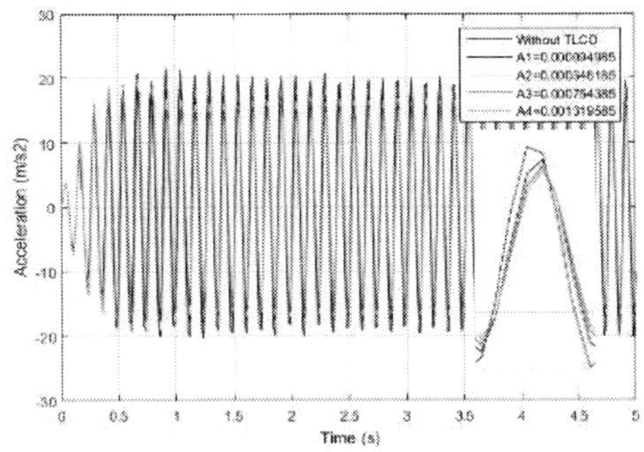

(b)

Figure 8. The acceleration response of first floor in variation of TLCD cross-sectional area when TLCD placed on first floor (a) and when TLCD placed on second floor (b)

IOP Conf. Series: Materials Science and Engineering **588** (2019) 012029 doi:10.1088/1757-899X/588/1/012029

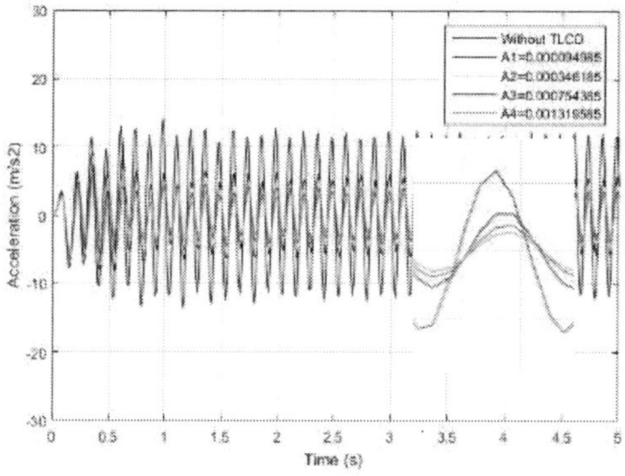

(a)

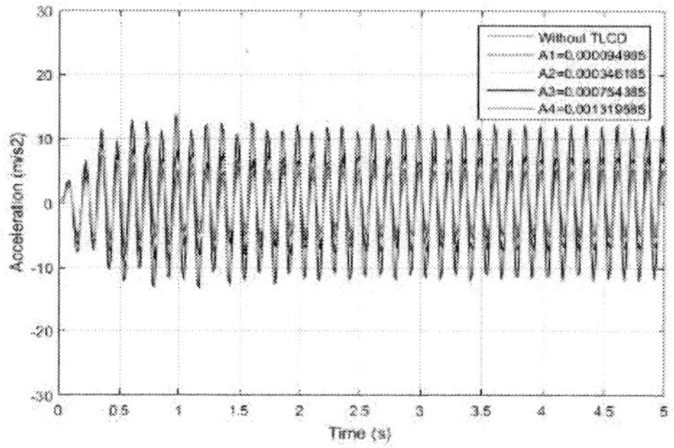

(b)

Figure 9. The acceleration response of second floor in variation of TLCD cross-sectional area when TLCD placed on first floor (a) and when TLCD placed on second floor (b)

The acceleration responses of system obtained from experimental method are shown in fig.10-11. Both figures show that the maximum acceleration responses of system with TLCD are lower than the maximum acceleration of system without TLCD. The detail information of the acceleration responses is shown in table 2. Table 2 represents the acceleration RMS of system in three-arrangements. The experimental results show the same trend as the simulation results. The acceleration response of system with TLCD's mass ratio 5% is almost the same as the acceleration response of system with TLCD's mass ratio 7%. The acceleration response of system with TLCD laid on first-floor is lower than the acceleration response of system with TLCD laid on second-floor.

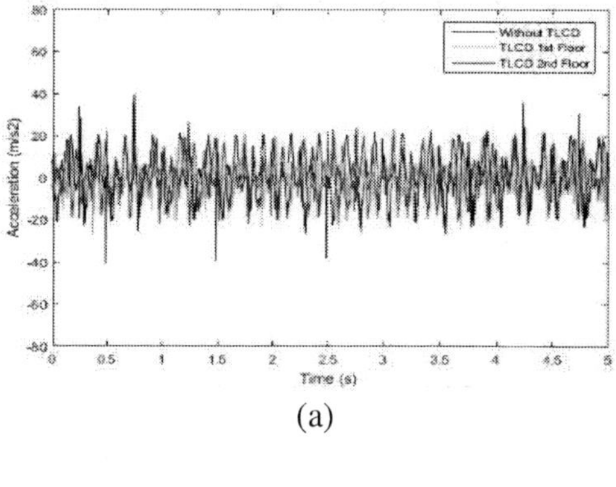

(a)

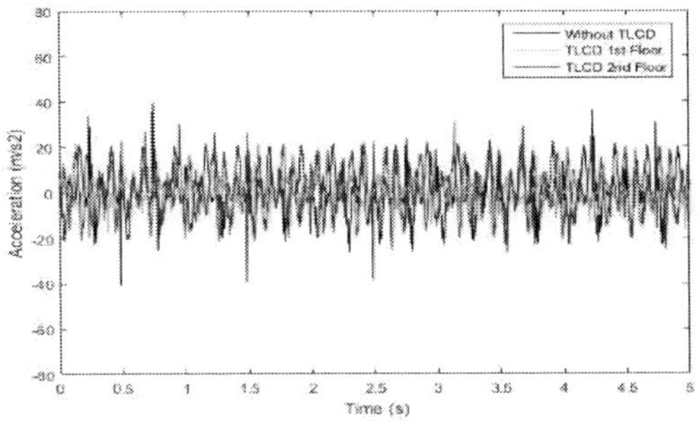

(b)

Figure 10. The experimental result of the acceleration response of first-floor with mass ratio (Mf/Ms) 5% (a), (Mf/Ms) 7% (b)

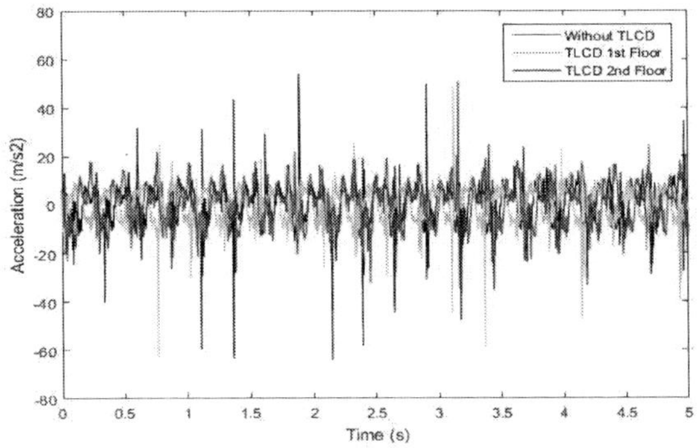

(a)

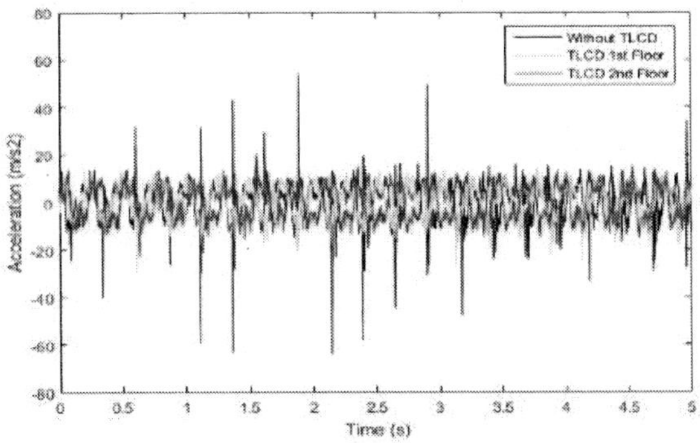

(b)

Figure 11. The experimental result of the acceleration response of second-floor with mass ratio (Mf/Ms) 5% (a), (Mf/Ms) 7% (b)

Table 2. The acceleration RMS obtained from experimental method

Mass ratio	TLCD's position	The acceleration RMS of 1st-floor (m/s^2)	The acceleration RMS of 2nd-floor (m/s^2)
5%	Without TLCD	10,38	8,33
	With TLCD laid on 1st-floor	5	7,02
	With TLCD laid on 2nd-floor	9,85	7,05
7%	Without TLCD	10,38	8,33
	With TLCD laid on 1st-floor	4,23	6,09
	With TLCD laid on 2nd-floor	9,99	6,75

4. Conclusion

From the results and discussions above, there are several points that can be concluded in this study, such as:

a. The optimum position of TLCD in reducing vibration is on the first-floor. It means the position of TLCD should be closed to the source of excitation.
b. The increasing of TLCD's orifice cross-sectional area results in the decreasing of the acceleration of main system.
c. The mass ratio variation does not provide significant differences to the system.
d. The maximum vibration reduction is about 62,6%.

References

[1] Daman A A A, Guntur H L, Susastro 2016 *The influence of dynamic vibration absorber to reduce the vibration of main system with 2-DoF* (AIP Conference Proceedings 1778, 030028 (2016); doi: 10.1063/1.4965762

[2] Varadarajan N and Nagarajaiah S 2004 Wind Response Control of Building with Variable Stiffness Tuned Mass Damper Using Empirical Mode Decomposition/Hilbert Transform *Journal of Engineering Mechanics* **vol 130**

[3] Kwok KCS and B Samali 1995 Performance of tuned mass dampers under wind loads *Engineering Structures* **vol 17** p 655-667

[4] Matteo A D, Iacono F L, Navarra G and Pirrotta A 2014 *Experimental validation of a direct pre-design formula for TLCD Engineering Structures* **vol 75** p 528-538

[5] Zhu F, Wang JT, Feng J, and Lu LQ 2017 Real time hybrid simulation of full-scale tuned liquid column dampers to control multi-order modal responses of structures *Engineering Structures* **vol 138** p 74-90

[6] Bhattacharyya S, Ghosh AD and Basu B 2017 Nonlinear modelling and validation of air spring effects in a sealed tuned liwuid column damper for structural control *Journal of Sound and Vibation* **vol 410** p 269-286

[7] Min KW, Kim YW and Kim J 2015 Analytical and experimental investigations on performance of tuned liquid column dampers with various orifices to wind-excited structural vibration *J. Wind Eng. Ind. Aerodyn* **vol 139** p 62-69

[8] Altunisik AC, Yetisken A and Volkan K 2018 Experimental study on control performance of tuned liquid column dampers considering different excitation directions *Mechanical Systems and Signal Processing* **vol 102** p 59-71

[9] Buckley T, Watson P, Cahill P, Jaksic V and Pakrashi V 2018 Mitigating the structural vibrations of wind turbines using tuned liquid column damper considering soil-structure interaction *Renewable Energy* **vol 120** p 322-341

[10] Kim H and Adeli H 2005 Wind-Induced Motion Control of 76-Story Benchmark Building Using the Hybrid Damper-TLCD System *Journal of Structural Engineering* **131** (12) pp 1794-1802

IOP Conf. Series: Materials Science and Engineering **588** (2019) 012029 doi:10.1088/1757-899X/588/1/012029

[11] Chakraborty S, Debbarma R and Marano G C 2012 Performance of Tuned Liquid Column Dampers Considering Maximum Liquid Motion in Seismic Vibration Control of Structures *Journal of Sound and Vibration* **331** pp 1519-1531
[12] Shum K M 2009 Closed Form Optimal Solution of a Tuned Liquid Column Damper for Suppressing Harmonic Vibration of Structures *Engineering Structures* **31** pp 84-92

Acknowledgments
Authors wishing to acknowledge support from Penelitian Terapan Unggulan Perguruan Tinggi (PTUPT) that has been funding this study.

IOP Conf. Series: Materials Science and Engineering **588** (2019) 012030 doi:10.1088/1757-899X/588/1/012030

Physical and Optical Characterization of Mn-doped ZnS Nanoparticles Prepared via Reverse Micelle Method Assisted by Compressed CO2

Rahizana Mohd Ibrahim[1,4], Masturah Markom[1,2,a], Huda Abdullah[3], Muneer M. Ba-abbad[1] and Kamal Firdausi Abd Razak[4]

[1] Chemical Engineering Programme, Faculty of Engineering and Built Environment, Universiti Kebangsaan Malaysia,43600 Bangi, Selangor, Malaysia.
[2] Centre for Sustainable Processing Technology (CESPRO), Faculty of Engineering and Built Environment, Universiti Kebangsaan Malaysia,43600 Bangi, Selangor, Malaysia.
[3] Centre of Advanced Electronic and Communication Engineering (PAKET), Faculty of Engineering and Built Environment, Universiti Kebangsaan Malaysia, 43600 Bangi, Selangor, Malaysia
[4] Department of Petrochemical Engineering, Politeknik Tun Syed Nasir Syed Ismail, Hab Pendidikan Tinggi Pagoh, 84600 Muar, Johor, Malaysia

[a] Corresponding author: masturahmarkom@ukm.edu.my

Abstract. This paper presents the study on the Mn-doped ZnS (ZnS:Mn) nanoparticles synthesized using reverse micelle method assisted by compressed CO2 at 60 bar and 40oC. Effects of retention time (30-120 min) on the particle formation were studied. The optical properties, size, morphology, and structure of the resulting particles were investigated. The results showed that ZnS:Mn nanoparticles were successfully formed with particle size ranging from 2 to 3 nm, which showed quantum size effects. The highest photoluminescence (PL) intensity ratio was found at 60 min retention time. The increase in photoluminescence (PL) intensity ratio indicated an increased homogeneity of nanoparticle growth with decreased surface defects.

1. Introduction

Zinc sulfate (ZnS) is a nontoxic II–VI semiconductor material with direct and large bandgap that is widely studied because of its numerous potential applications, including solar cells, bio-imaging, wavelength-tunable lasers, and electronic and optoelectronic nanodevices [1–3]. The presence of optical impurities affects the physical and chemical characterization of ZnS. Incorporating suitable dopants is an approach to enhancing the potential of ZnS semiconductor as a phosphor.

Issues commonly encountered in the synthesis of nano-particles are size control and agglomeration of the particles. In the case of doped ZnS nanoparticles, a suitable synthesis method is not only required to prevent agglomeration but also to improve the optical properties. Among various techniques for producing nanoparticles, reverse micelle method is considered a promising technique to prepare less agglomerated and more monodispersed nanoparticles [4-6]. The advantage of this method is the ability

to control excess particle growth when particle size approaches that of a H2O nanodroplet. However, difficulty is usually encountered in post process recovery step that affect particle size and dispersion [7-8].
In recent years, some researchers have carried out combination of antisolvent CO2 and reverse micelle method for the formation of nanoparticles [9-11]. It is expected that the properties of the water-in-oil in reverse micelle system can be easily tuned by the pressure and temperature of CO2. Nanoparticle synthesis using reverse micelle assisted by CO2 as antisolvent is an interesting phenomena and needs further study. Therefore, the aim of this work is to produce ZnS:Mn nanoparticles with better particle characteristics and optical properties.

2. Materials and Methods

2.1. ZnS:Mn Particle Synthesis
The ZnS:Mn solution was prepared using reverse micelle method reported previously [12]. In this study, the desired amount of ZnS:Mn2+ (4%) sample solutions were then loaded into a 1 L standard steel cylinder with a sight glass. The cell was then sealed and pressurized with CO2 antisolvent with four different retention times (30 – 120 min). The conditions in the vessel was kept at relatively low pressure of 60 bar and temperature of 40oC. The precipitated ZnS:Mn particles were collected and then washed by aqueous ethanol solution, centrifuged and dried in an oven

2.2. Particle Characterization
The X-ray diffraction (XRD) patterns of the powdered samples were investigated using a D8 Advance AXS X-ray diffractometer at diffraction patterns ranging from 20° to 80°. The crystal size was estimated at full width half maximum (FWHM) of the XRD peaks. The fluorescence measurements were performed on a photoluminescence (PL) SP920 spectrophotometer. Physical characterizations of the particles were then performed using transmission electron microscopy (TEM).

3. Results and Discussion

3.1. Physical Characterization of ZnS:Mn
The XRD patterns (Figure 1) of ZnS:Mn nanoparticles at different retention times exhibit three diffraction peaks at 2θ = 28.50°, 48.30o, and 56.50°, which corresponded to the (111), (220), and (311) planes, respectively. These peaks were indexed as a zinc blende structure corresponding to JCPDS file no. 05-0566 with no impurity peak observed. The average of crystal size as calculated according to the Debye-Scherrer formula [13] ranging between 2.0 to 3.0 nm (Table 1), which fell within the quantum confinement region. However, additional peak appeared at 32° when retention time was set at 120 min. It shows that the dissolution of CO2 at long retention time in the micelle core leads to micellar instability and gives excess energy to segregate the Mn2+ atom out of the ZnS lattice. Generally, the shape of the particles for the samples are spherical and they are homogeneously dispersed as characterized by TEM (Figure 2). These results provide indirect evidence of the role of CO2 to effectively improve surfactant capping on the surface of ZnS:Mn nanoparticles.

Table 1. Characterization of ZnS:Mn Particles

Retention Time (min)	Particle Size (nm)	Emission (nm)	PL Intensity Ratio
30	2.53	602	12.02
60	2.21	608	32.42
90	2.39	608	23.31
120	2.48	608	19.28

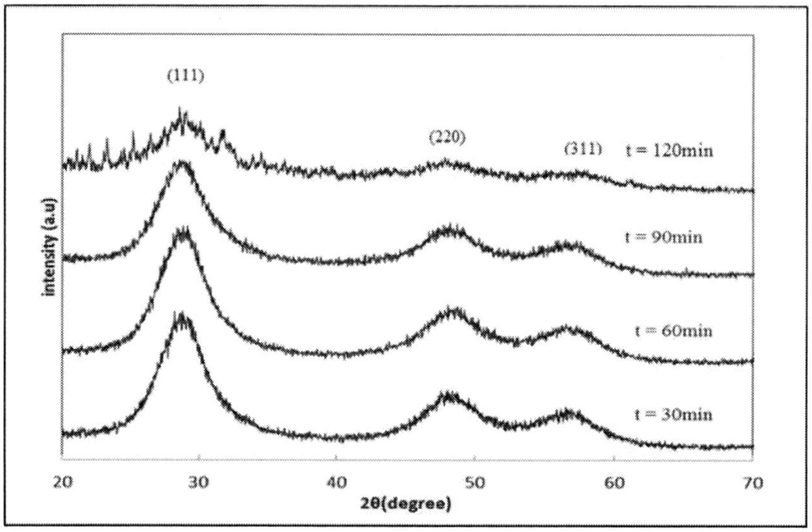

Figure 1. XRD patterns of ZnS:Mn at different retention times

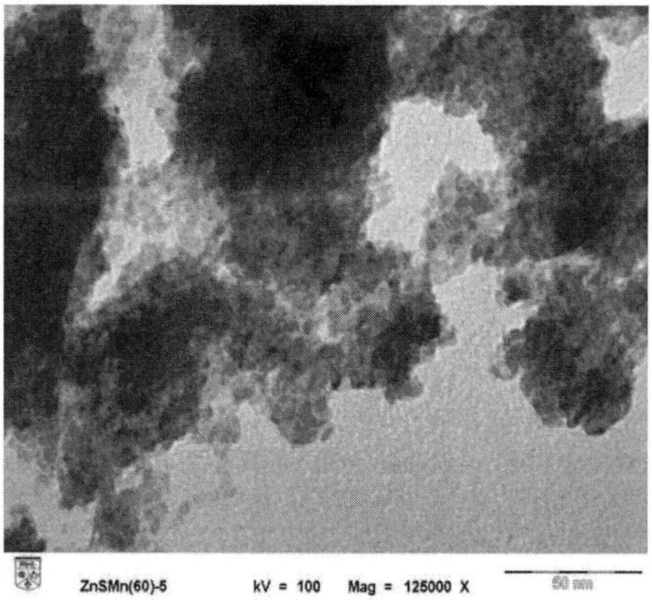

Figure 2. Transmission electron microscope (TEM) image of ZnS:Mn at 60 minutes

3.2. Characterization of ZnS:Mn

Figure 3 shows that even though the ZnS:Mn particles showed small size differences at different retention times, the relative intensities of PL varied significantly. This might be due to the control of the droplet size by surfactant concentration in the water–heptane matrix. The orange emission intensity increased linearly with the quenching of the blue emission and started to decrease when using the samples with 90 min retention time. Table 1 and Figure 3 show high value of PL intensity ratio at 60 min exposure to compressed CO2, thus indicating the formation of monodisperse nanoparticles.

This phenomenon could be discussed based on the nanoparticles-surfactant-solvent interaction. Upon pressurization with CO2, the gas continues to dissolve into the micelle core and decreases the solvation of the AOTcap nanoparticles, thus causing the removal of water in the reverse micelle droplet. Therefore, trap sites on the nanoparticle surface were decreased and crystallinity was promoted, which dramatically improved the PL intensity.

Previously, several researchers studied the modification of ZnS:Mn nanocrystal by using different surfactants[14- 15], W ([H2O]/[CTAB]) values [16] and types of mixing method in a reverse micelle of H2O/AOT/n-heptane [17]. However, tuning the properties of surfactant solution using CO2 is a new progress in modification of nanoparticles. The mechanism for CO2 to stabilize reverse micelle and improve the surface defect of nanoparticles has never been discussed before. Therefore, this research will greatly contribute to the understanding of nanoparticles formation using compressed CO2 as the antisolvent.

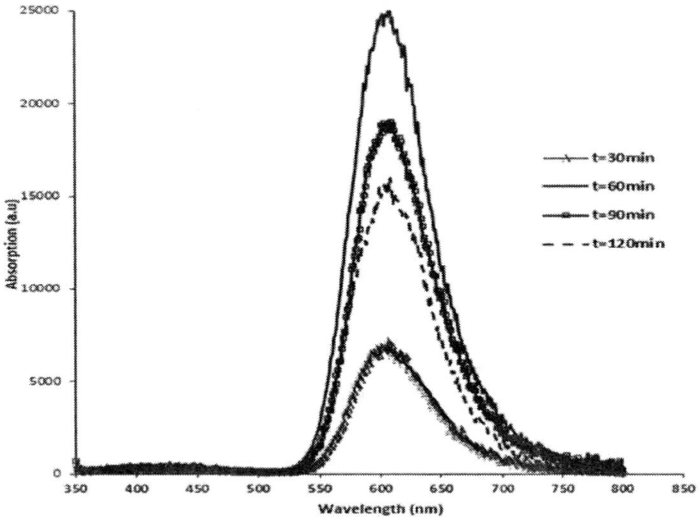

Figure 3. Photoluminescence spectrum of ZnS:Mn at different retention times

4. Conclusion

A quaternary AOT/CO2/n-heptane/H2O reverse micelle system was found to be effective in producing uniform, monodisperse and nanosize Mn-doped ZnS particles. The ZnS:Mn nanoparticles exhibited strong quantum-size effect due to dopant incorporation into the ZnS host structure, which reduced surface defects and enhanced optical properties.

5. Acknowledgements

The authors are grateful to the financial support by the Ministry of Higher Education Malaysia under grant FRGS/1/2017/TK02/UKM/02/4.

References

[1] S Sen, P Sharma, C S Solanki and R Bandyopadhay J 2010 *Trends in Chem* **1** 14-17
[2] M Sharma, T J, Sukhvir Singh and O P Pandey 2012 *Solar Energy* **86**(1) 626-633
[3] Y Liu, J C Hua, C L Ngob, S Prikhodkob, S Kodambakab, J L Li and R Richards 2011 A*ppl. Catal. B: Environ.* **106**: 212–219
[4] S N Khadzhiev, K M Kadiev, G P Yampolskaya and M K Kadieva 2013 *Adv in Colloid and Interface Sci* **197-198** 132-145
[5] X Fang, T Zhai, U K Gautam, L L L Wu, Y Bando and D Golberg 2011 *Prog. in Mat. Sci*. **56** 175-287
[6] H Hu and W Zhang 2006 *Opt. Mat.* **28** 536-550
[7] R N Bhargavaa, D Gallaghera and T 1994 Welker J *Luminescence* **60&61** 275-280
[8] S K Mehta, S Kumar and M Gradzielski J 2011 *Colloid and Interface Sci* **360** 497-507

[9] P Brown, M J Wasbrough, B E Gurkan and T A Hatton 2014 *Langmuir* **30** 4267-4272
[10] M A Malik, M Y Wani and M A Hashim 2012 *Arab. J. Chem.* **5** 397-417
[11] I R Kamrupi and S K Dolui 2012 *J. Exp. Nanosci* **7** 243-253
[12] R M Ibrahim, M Markom, H Abdullah and K Firdausi 2016 Size control and photoluminescence enhancement of Mn2+-doped ZnS nanoparticles synthesized by AOT-reverse micelle modified by compressed CO2 in Materials Today: Proceedings 3, 2 ed. Elsevier Ltd pp. 479 – 484
[13] R M Ibrahim, M Markom and H Abdullah 2015 ECS J *Solid State Sci. Tech.* **4**(2) R31-R37
[14] D Singha, N Barman and K Sahu 2014 *J. Colloid and Interface Sci.* **413** 37-42
[15] R M K Whiffen, D J Jovanovi, Ž Anti, B Bártová, D Mi-livojevi, M D Dramićanin and M G Brik 2014 *J. Luminescence* **146** 133
[16] V V Tatarchuk, A P Sergievskaya, T M Korda, I A Druzhinina and V I Zaikovsky 2013 *Chem. Mat.* **25**(18), 3570-3579
[17] R Vishwakarma 2015 *J. Theo. App. Phys.* **9** 185-192

IOP Conf. Series: Materials Science and Engineering **588** (2019) 012031 doi:10.1088/1757-899X/588/1/012031

Determination of Shelf Life and Proximate Analysis of Orumy Beverage Produced by Micro, Small and Medium Enterprise INOKAM in Dolly Distric

Yatim Lailun Ni'mah[1,a] and Vennycha Dwi Ameswari[1]

[1] Department of Chemistry, Faculty of Natural Science, Institut Teknologi Sepuluh Nopember, Sukolilo,Surabaya, Indonesia, 60111

[a] yatimnikmah@gmail.com

Abstract. In this research, Orumy beverage have been made with the main ingredient of the Eucheuma cottoni seaweed spesies. For labeling and commercialization purposes, it is necessary to determine the shelf life and nutritional contents of seaweed beverages. The shelf life investigation in this study was conducted by Accelerated Shelf Life Testing (ASLT) method. While on the determination of nutritional content, the proximate analysis method was used. The Orumy beverage used in this research consists of three flavors: strawberry, lemon, and lychee. This study results that the shelf life of Orumy beverages with strawberry, lemon, and lychee flavors are 44 days, 45 days and 38 days at 30°C storage temperature respectively. The nutrients available in Orumy beverages with strawberry, lemon, and lychee flavors are ash content by 0.11 ± 0.01%, 0.11 ± 0.00%, 0.12±0.02%, moisture content by 89.54±0.01%, 88.53±0.01%, 88.52±0.01%, protein content by 0.65±0.22%, 0.33±0.23%, 0.33±0.23%, crude fiber content by 3.34±0.01%, 3.49±0.01%, 3.39 ± 0.01%, fat content by 2.71±0.24%, 2.75±0.28%, 2.87±0.21%, and carbohydrate by 2.71±0.24%, 4.78±0.55%, 4.78±0.05%.

1. Introduction

Surabaya is the capital of East Java Province, the center of regional government, politics, trade, industry, education and culture. Surabaya is one of the metropolis that has a variety of socio-cultural, Surabaya is a city with a variety of faces. Surabaya is also a city that has the six largest localization points in Southeast Asia including Gang Dolly, Bangunsari, Kermil-Tambak Asri, Klakah Rejo, Moroseneng and Putat Jaya [1]. Most people who live, depend on the localization area, it is a reason to maintain prostitution activities. In addition to supporting survival, the absence of special abilities and low education is also the background of the development in this prostitution activity. The development process of prostitution activities is also supported by a market mechanism that makes prostitution a profitable sex business that is badly needed [2]. Prostitution activities do not always run smoothly without problems because they not only cause economic turmoil but also present social problems and are considered to be contrary to social and religious norms that apply in the pattern of life arrangements of a community group.

Finally on June 19, 2014, the localization of "Dolly" located in the village of Putat Jaya was declared closed by Mrs. Dr. Ir. Tri Rismaharini, M.T as Mayor of Surabaya. However, many polemics occurred, especially the problem of employment for the people affected by the closure. So before this politician was made, she also made politicians that every prostitution activity in the Dolly area would be given Rp 3.000.000 for capital to make new businesses and social ministers also give Rp 4.200.000. However, this effort did not run successfully and only reduced prostitution activities temporarily. To overcome this, some community organizations have provided assistance, both in the socio-cultural field such as assistance in improving the quality of education in localization areas, conducting religious activities, community services. In addition, assistance is provided in the economic field such as the development of the Dolly area as a tourist center and building mass media that focuses on children and women [1]. However, no one has conducted food related training. While there are many people in the MSMEs and made Dolly set up a business in the food sector, for example is a beverage product based on "Orumy" seaweed made by MSMEs INOKAM. (Inovasi Kampung Mandiri) Jalan Putat Jaya Gang III-A RT. 03 RW. III Kecamatan Sawahan, Surabaya.

Currently practical packaging drinks are quite popular in the public. So that it becomes very dangerous if there is no spesific understanding for the community, because a lot of food and drinks are dangerous to consume, especially the problem of food and beverage expiration date. Besides the expiration date, nutritional information from food or drinks that will be consumed is also something that needs attention. The expiration date on the food or beverage label is the time limit for which the food has passed from the date of consumption [3]. Foods or drinks that have passed the expiration date are harmful to health if they are still consumed. So the shelf life information is one of the information that must be entered by the producer on the packaging of food products.

Shelf life assessment can be carried out in accelerated conditions (Accelerated Shelf Life Test) which is able to predict the shelf life of the product. The ASLT method uses conditions in the environment that can accelerate the reaction to a decrease in the quality of food products [4]. This ASLT method is very good to use because the testing time is relatively short with high accuracy. Furthermore, the Arrhenius approach is used as an equation to account for changes in activation energy [5]. So, in this research ASLT method was used in determining the shelf life of seaweed drinks.

Another important thing that must be considered besides the expiration date label is the label of nutritional information on food products. Because nutritional information shows the main components of an ingredient and the nutritional content of food. For foodstuffs, the main components consist of water content, ash content, carbohydrates, protein, and fat [6]. Proximate analysis is generally not expensive and relatively easy to do [7]. The survey results of the products of MSMEs INOKAM Orumy were not found on the expiration date and the nutritional content of the packaging so it was necessary to pay attention the importance of information about the expiration date and information on the nutritional content of seaweed produced by the Dolly people.

2. Materials and Methods

Orumy beverage were used as ingredients in this research. This beverage made by INOKAM UMKM (Kampung Mandiri Innovation) Jalan Putat Jaya Gang III-A RT. 03 RW. III Sawahan, Surabaya. The reagents used in this research were 50% NaOH, NaOH 0.02 N, $H_2C_2O_4$ 0.02 N, HCl 0.02 N, NaOH 3.25%, 1.25% H_2SO_4 , anhydrous Na_2SO_4, n-hexane, ethanol 96%, phenolphthalein indicator, methyl red indicator, and methyl blue indicator.

2.1. Production of Orumy Beverage

Dried Eucheuma cottonii seaweed spesies were soaked in water for 24 hours. Then seaweed is washed using running water until it is clean and drained. After that, 225 grams of seaweed were weighed, then the seaweed was roughly cut and blended with 1 liter of water. Blended seaweed is put into 2 liters of boiling water. Then add 400 grams of sugar, 2 tablespoons of salt, and 1 gram of sodium benzoate which has been dissolved in 0,5 liters of water, then stirred until boiling. Added 2 grams of CMC-Na which has been dissolved in 0,5 liters of water, then stirred until boiling. Then add 100 mL of fructose, then stir and wait for it to boil.

After boiling, seaweed is immediately removed and filtered using a stainless steel filter. After that drink seaweed is added flavor and citric acid warm. Then seaweed beverage are packed in plastic bottles that have been sterilized first with warm water and labeled. Seaweed drinks in plastic bottles are awaited until cool, after that the bottled bottoms are closed with a seal. Seaweed beverage used in this research were produced by MSMEs INOKAM (Inovasi Kampung Mandiri) Jalan Putat Jaya Gang III-A RT. 03 RW III Kecamatan Sawahan, Surabaya.

2.2. Shelf-life Determination of Orumy Beverage

Packaged seaweed drinks have a net weight of 350 mL. The drinks are stored in crisis temperature variations of 10, 20 and 30°C. Observations are carried out periodically every 7 days from day 0 to day 42, so there are 7 observation points. In this research two parameters were used to determine shelf life, namely water content and pH value.

The results of data analysis for each parameter are plotted against time (days) and obtained linear equations. Three equations are obtained for three storage temperature conditions with the following equation:

$$y = ax + b \qquad (1)$$

Order reaction for a parameter is chosen by comparing the regression value (R^2) of each linear equation at the same temperature. The order reaction with a greater R^2 value is the reaction order used as a parameter for determining shelf life.

After obtaining a linear equation for each storage temperature, the slope value (equation 1) which shows the change in product characteristics is calculated as ln a. Then the value of ln a is plotted on the graph of the Arrhenius equation. In the graph of the Arrhenius equation, ln a is the ln k value plotted against 1/T (K^{-1}). From the graph of the Arrhenius equation, the slope and intercept values of the linear regression equation 2.

$$\ln k = \ln k_0 \frac{Ea/_R}{1/T} \qquad (2)$$

From the equation, the value of k_0 is obtained which is an exponential factor and shows a decrease in the quality of the product that stored at certain temperature. From the equation, the activation energy value (Ea) is also obtained which is a reaction to changes in the characteristics of the stored product. Next, the model of the reaction rate equation for temperature is determined, where the k value indicates a decrease in product quality which can be calculated using the following equation 3.

$$k = k_0.e^{-Ea/RT} \qquad (3)$$

Based on the Arrhenius equation (equation 3) and the calculation of k values, the shelf life of seaweed drinks can be calculated using the order of zero (equation 4) or the first order (equation 5).

$$t \text{ zero order} = \frac{\Delta A}{k} \qquad (4)$$

$$t \text{ first order} = \frac{\ln\left(A_0/_A\right)}{k} \qquad (5)$$

2.3. Deternination of Orumy Beverage Nutrition Content

Determination of seaweed content was carried out using proximate analysis based on SNI 01-2891-1992. Proximate analysis consists of ash content, water content, protein, crude fiber, fat, and carbohydrates.

2.3.1. Ash Content Analysis

Analysis of the contents of ash using the ignition method. 5 grams of wet sample is weighed in a porcelain cup that has been dried and weighed. After that, the sample is heated until it turns into charcoal using a bunsen heater until the sample produces no more smoke. Then the sample is put into a furnace with a temperature of 600°C until the ignition process is complete. After the ignition process is complete, the cup containing ash is weighed to reach a constant mass.

2.3.2. Water Content Analysis
Water content analysis was carried out using the gravimetric method by drying in an oven to obtain a constant mass. The wet sample was weighed 2 grams in a porcelain cup which was dried and weighed. Then the sample is put in the oven at 105°C for 3-5 hours. After drying, the sample is cooled in the desiccator for 15 minutes, then the mass of the cup containing the dry sample is weighed to reach a constant mass.

2.3.3. Protein Content Analysis
Rough protein content analysis using the Kjeldahl method is based on the determination of nitrogen content to calculate the ptorein content that contained in the sample. 0.1 gram of wet sample is weighed and put into a test tube, after which 2 mL of concentrated H_2SO_4 is added and 1 gram of anhydrous Na_2SO_4 catalyst. Then the mixture is destructed for 2 hours at 85°C, then cooled. After the destruction process, the mixture was transferred to the distillate flask and added 50 mL aqua DM and 10 mL of 50% w/v NaOH solution then the distillation process was carried out. The results of the distillation process are accommodated in an erlenmeyer containing 10 mL 0.02 N HCl, each of 5 drops of methyl blue and methyl red indicator. The distillation process is carried out until the volume of the solution on erlenmeyer reaches 2 times the initial volume. After that, the distillate solution was titrated using standardized NaOH with 0.02 N $H_2C_2O_4$ solution. The end point of the titration process is marked by the color change from purple to green. Then take note the volume of NaOH used and made blank as a comparison

2.3.4. Crude Fiber Content Analysis
Crude fiber analysis using hydrolysis method. Weighed as much as 2 grams of dried sample, then put into a Three-neck Rounded Flask. Then added 50 mL of H_2SO_4 solution and the mixture was refluxed for 30 minutes while stirring using a magnetic stirrer with a speed of 400 rpm. After 30 minutes, 50 mL of 3.25% NaOH solution was added, then reflux process was carried out for 30 minutes while stirring using a magnetic stirrer at a speed of 400 rpm. Then the sample in a hot state is immediately filtered using filter paper that has been dried and weighed. The precipitate contained in the filter paper was washed successively with 1.25% H_2SO_4, hot aquademin, and 96% ethanol each as much as 10 mL. Filter paper is removed and placed in a watch glass, then dried using an oven at 105°C for 1 hour. After the oven, the filter paper is cooled in a desiccator and weighed until a constant mass is obtained.

2.3.5. Fat Content Analysis
Fat content analysis was carried out using the Soxhlet method. Dry samples weighed 5 grams, then extracted using n-hexane organic solvents in soxhlet for 6 hours. Extraction results are collected in a round flask, evaporated with an oven at 50°C until the fat deposits are left on the bottom of the round pumpkin. After that, the extraction results are cooled in a desiccator and weighed until a constant mass is obtained.

2.3.6. Carbohydrate Content Analysis
The total carbohydrate content in the sample was determined using the method by different.

3. Results and Discussion

3.1. Production of Orumy Beverage
Seaweed used for this research is Eucheuma cottoni seaweed spesies. In the seaweed, the immersion process is needed with clean water for 24 hours to remove the dirt. After that, the seaweed is washed

again using running water until clean. The seaweed is drained to reduce the water content. Then seaweed is cut into small pieces and mashed using a blender.

Seaweed that has been smooth then filtered first because it wants a low viscosity drink. Furthermore, the smooth seaweed is processed by heating until boiling by adding other food additives such as, salt, sugar, fructose, Carboxyl Methyl Celulose (CMC), sodium benzoate, flavoring, and citric acid. After the processing is complete, seaweed drinks are immediately removed and filtered with a stainless steel filter to separate the seaweed that still has a rough texture. Then added flavor to the seaweed beverage. In this research, 3 seaweed flavors were used to compare and analyze, namely seaweed beverage with the addition of lemon, lychee, and strawberry flavoring. After adding flavoring, seaweed beverage were also added to citric acid in each taste. Orumy beverage in a bottle are waited until cold, then the bottle is sealed with a seal.

3.2. Shelf-life Determination of Orumy Beverage

The shelf life of Orumy beverage produced by MSMEs INOKAM is estimated by the Arrhenius model of the Accelerated Shelf Life Testing (ASLT) model which is widely used to estimate the shelf life of food products that are easily damaged by chemical reactions. In the ASLT method, temperature is a key parameter for determining damage because of the temperature increases, the damage reaction will be faster [5]. Storage temperatures used in this study are temperatures of 10, 20, and 30°C. Storage temperature is chosen because seaweed is a product that must be stored in cold temperatures. To determine the decline in the quality of seaweed beverage, the main parameters used are the pH value and water content that has been determined as a parameter to determine the decline in the quality of seaweed beverage due to the influence of temperature and storage time. The values of the two parameters are then plotted on the Arrhenius model.

3.2.1. The Water Content Parameter

Water Content is an important component in food ingredients because water content can affect the texture, appearance and taste of food products in general. The water content in a food product can determine the freshness and durability of the food product. If the water content is relatively high, it can be estimated that the material will be damaged faster by microbes that are easily breed in food products or by other reactions. The higher of water content in a food product will affect the freshness of the food [8][9].

Measurement of water content using a drying method (gravimetric). In the drying method (gravimetric) is based on measuring the reduction in the weight of the dried material. Selection of drying temperature in this method is very influential on the results of moisture content. This method is chosen because it is accurate, relatively easy, inexpensive, and can do sample analysis simultaneously [10]. Measurement of water content was carried out on Orumy beverage stored at 10, 20, and 30°C for 42 days. Water content is measured by removing the water in a sample of Orumy beverage through evaporation. Evaporation of water content in Orumy beverage is done by oven at 105°C for 3-5 hours. Based on the comparison of linear regression results, the water content parameter follows a zero order reaction. The results of determining the water content of Orumy beverage with strawberry, lemon, and lychee flavors stored at different storage temperatures, each taste plotted against storage time at different storage temperatures can be seen in Figure 1.

Water content of Orumy beverage changes differently at each storage temperature condition. The changes of water content in strawberry, lemon and lychee flavored Orumy beverage at different temperature storage are shown in Figure 1. Based on Figure 1, the linear regression equation in sample 283, 293 and 303 can be seen in Table 1, 2 and 3.

Table 1. The Equations of Linear Regression for Sample of Strawberry Flavors

Sample	Equations	R^2 Value
283	$y = -0.0051x + 89.036$	0.050
293	$y = 0.0051x + 88.536$	0.041
303	$y = -0.0102x + 88.929$	0.333

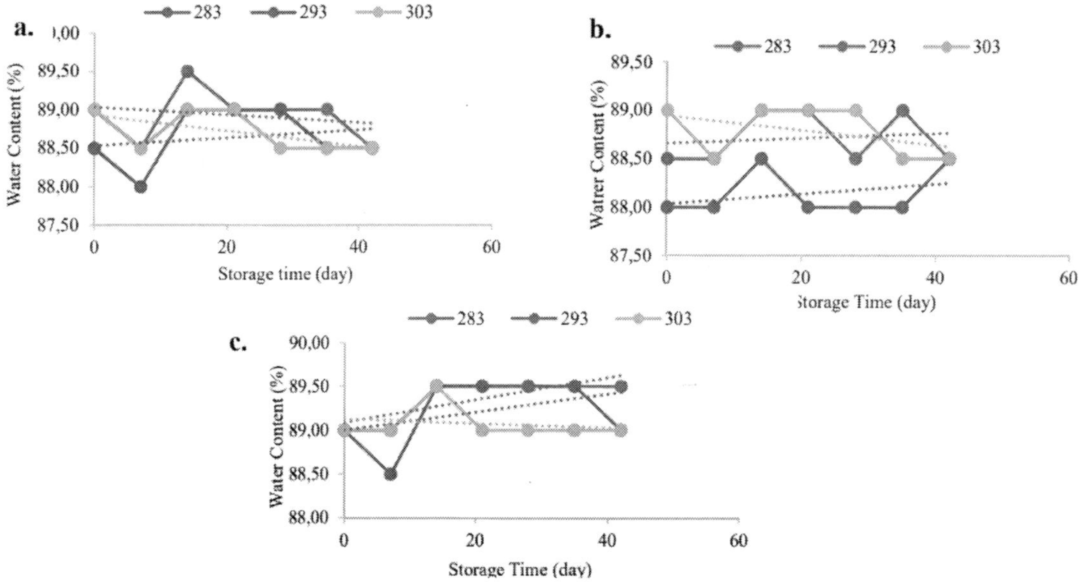

Figure 1. The Change of Water Content in (a) Strawberry Flavor, (b) Lemon Flavor, (c) Lychee Flavor

Table 2. The Equations of Linear Regression for Sample of Lemon Flavors

Sample	Equations	R^2 Value
283	y = 0.0026x + 88.661	0.020
293	y = 0.0051x + 88.036	0.100
303	y = -0.0077x + 88.946	0.187

Table 3. The Equations of Linear Regression for Sample of Lychee Flavors

Sample	Equations	R^2 Value
283	y = 0.0128x + 89.089	0.625
293	y = 0.0102x + 89.000	0.153
303	y = -0.0026x + 89.125	0.041

From the linear equation obtained the slope value at each storage temperature is plotted against 1/T in the Arrhenius equation.

Figure 1 showed the effect of storage time on changes in water content in samples of Orumy beverage stored at different temperatures. The changes of water content in products are factors that greatly affect the decline in the quality of food products. Water content is closely related to water activity, where changes in water content are possible due to water activity (a_w) which is also closely related to the growth of bacteria, fungi, and other microbes. The higher of water activity, shows that the more bacteria that grow [9].

3.2.2. The pH Parameter

PH value is one of the parameters to determine the change in acidity of a product. If the level of acidity (pH) in a food ingredient is lower, then the food is not suitable for consumption because indirectly a decrease in pH indicates internal changes in products such as microbial activation [11][12].

IOP Conf. Series: Materials Science and Engineering **588** (2019) 012031 doi:10.1088/1757-899X/588/1/012031

The pH value of Orumy drinks was measured using a pH meter which was calibrated first with a phosphate buffer pH 4, 7, and 10. pH measurements were made on Orumy beverage stored at 10, 20, and 30°C for 42. Based on the results of the comparison of linear regression values, the pH parameter follows a zero-order reaction. PH value of Orumy beverage strawberry, lemon, and lychees flavor stored at different storage temperatures, each taste plotted against storage time at different storage temperatures can be seen in Figure 2

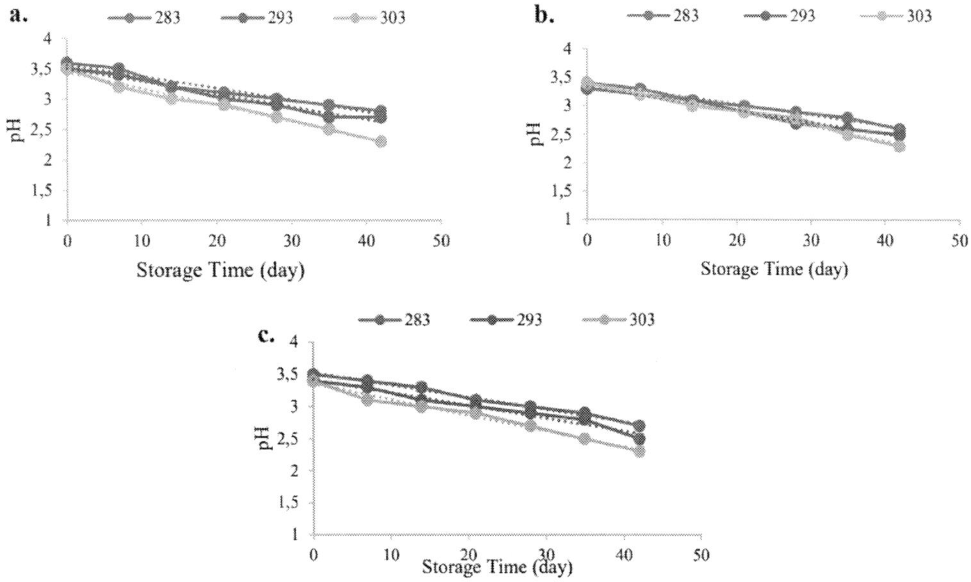

Figure 2. The Change of pH in (a) Strawberry Flavor, (b) Lemon Flavor (c) Lychee Flavor

The pH value of Orumy beverage changes differently in each storage temperature condition. The effect of storage time on the pH value of strawberry and lemon flavored orumy beverage can be seen in sample 303 which is stored at 30°C and has the highest change in moisture content of samples stored at other temperatures (Figure 2). Based on Figure 2, the linear regression equations in sample 283, 293, and 303 respectively, can be seen in Table 4, 5 and 6.

Table 4. The Equations of Linear Regression for Sample of Strawberry Flavors

Sample	Equations	R^2 Value
283	y = -0.0184x + 3.400	0.987
293	y = -0.0204x + 3.328	0.985
303	y = -0.0250x + 3.396	0.980

Table 5. The Equations of Linear Regression for Sample of Lemon Flavors

Sample	Equations	R^2 Value
283	y = -0.0194x + 3.564	0.960
293	y = -0.0209x + 3.496	0.972
303	y = -0.0270x + 3.439	0.989

Table 6. The Equations of Linear Regression for Sample of Lychee Flavors

Sample	Equations	R^2 Value
283	y = -0.0189x + 3.525	0.989
293	y = -0.0199x + 3.417	0.970
303	y = -0.0245x + 3.357	0.982

From the linear equation obtained the slope value at each storage temperature is plotted against 1/T in the Arrhenius equation.

The changes in pH values indicate the presence of decay due to microbes contained in drinks stored for 42 days under different temperature conditions. If the level of acidity (pH) in a food ingredient is lower, so the food is not suitable for consumption because indirectly a decrease in pH indicates internal changes in products such as microbial activation [11][12].

3.2.3. Kinetics Reaction Model with Arrhenius Equations

Estimation of shelf life of Orumy beverage using the Arrhenius model's Accelerate Shelf Life Test (ASLT) method. This method uses conditions in an environment that can accelerate the reactions of food product quality degradation [4]. To determine the shelf life value with the Arrhenius equation, the values ln k and 1/T are needed, where the value of k is obtained from the slope value of the graph of water content and pH against time. The Arrhenius plot for moisture and pH parameters can be seen in Figure 3.

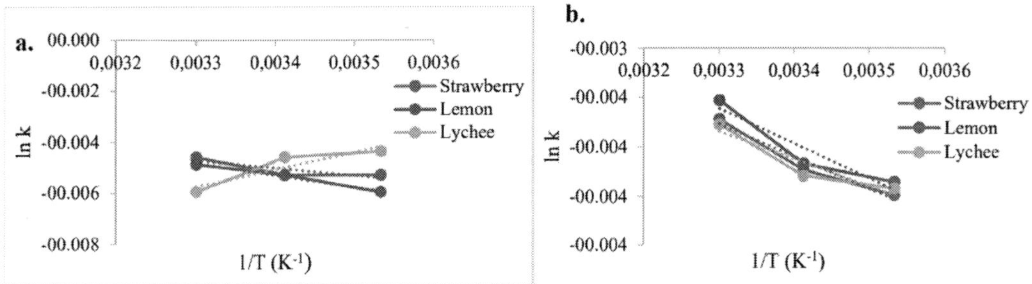

Figure 3. Arrhenius Plot for (a) The Changes of Water Content, and (b) The Changes of pH Value

Based on the Arrhenius plot (Figure 3) on the parameters of water content and pH, the equation can be seen in Table 7 and 8.

Table 7. The Equations of Linear Regression for Water Content Parameter in Orumy Beverage

Sample	Equations	R^2 Value
Strawberry	y = -2936.9x + 4.983	0.732
Lemon	y = -4665.9x + 10.571	0.986
Lychee	y = 6775.7x – 28.108	0.840

Table 8. The Equations of Linear Regression for pH Parameter in Orumy Beverage

Sample	Equations	R^2 Value
Strawberry	y = -1308.8x + 0.611	0.958
Lemon	y = -1407.9x + 1.001	0.896
Lychee	y = -1104.6x – 0.092	0.879

The equation obtained is used to determine the activation energy (Ea). Activation energy is obtained from the slope of the Linear Regression equation. Activation energy is a quantitative way to effectively compare samples. In Orumy with strawberry, lemon, and lychee flavors based on water content parameters, the activation energy value is 5.833; 9.266; and 13.457 kcal/mol. While based on pH parameters for strawberry flavor, lemon flavor, and lychee flavors, the activation energy was 2.599; 2.799; and 2.194 kcal/mol. Activation energy is the energy needed to activate the damage and reaction process faster. So that the decline in the quality of seaweed is getting faster too [13][14].

Based on the calculation of the Ea value obtained from the parameters of water content and pH value, to determine the shelf life of Orumy beverage with the ASLT method the smallest parameter is chosen, namely the pH parameter. In this study, the zero order reaction kinetics is used (equation 4). Shelf life can be determined from the value of the constant decrease in quality with the initial and final quality values of drinks stored at 30°C. Based on the results of calculations from intercept, slope, and 1/T (equation 3), the quality decrease constants (k) in the strawberry, lemon, and lychee flavored Orumy were obtained based on the pH parameter of 0.024; 0.026; and 0.028. In the previous research, the shelf life of Orumy beverage with original flavor was 17 days and the taste of green tea was 34 days at a storage temperature of 30°C [15]. The results of the determination of shelf life in this research, obtained the shelf life of strawberry flavored Orumy beverage for 44 days, lemon flavor for 45 days, and lychee flavor for 38 days at a storage temperature of 30°C. Addition of citric acid to Orumy drinks can extend the shelf life of the product. Shelf life can also be extended by lowering the product storage temperature. At a storage temperature of 10°C, the shelf life of strawberry-flavored orumy beverage becomes 60 days, lemon flavor becomes 63 days, and 49 days for lychee flavor.

3.2.4. Determination of Orumy Beverage Nutrition Content

Determination of the nutritional content of Orumy's drinks using the method of proximate analysis. Proximate analysis needs to be done because the analysis in these foods is related to the level of nutrition found in these foods. Proximate analysis carried out consisted of ash, water, protein, crude fiber, fat, and carbohydrate content by different [10]. This proximate analysis as a whole refers to SNI 01-2891-1992.

Determination of ash, water, and protein content was used as a sample of wet Orumy drinks so it was not necessary to prepare samples before testing. However, for determination of crude fiber content and fat content, free water samples are used so that the sample preparation must be done first. Dry samples are obtained from the samples of wet Orumy beverage that have been removed from moisture or free samples of water with an oven at 50-70°C and ensure that the sample is completely dry and free of water. Samples that have been dried are marked by nonstickiness and can be smoothed. Then the dry sample is mashed with mortar and pestle which aims to enlarge outside the surface of the sample so that it is easier to analyze. Proximate analysis for each sample with different flavors was carried out twice repetition.

While the determination of carbohydrate content using the method by different in the proximate analysis is strongly influenced by other nutrient content. So the method by different is also called the indirect method which is obtained from the results of 100% deducted from the content of other nutrients such as ash, water, protein, crude fiber and fat content [16]. The results of the proximate analysis can be seen in the Table 9.

Table 9. The Results of Determination Nutrient Content

Nutrient content	Strawberry flavor	Lemon flavor	Lychee flavor
Ash (%)	0.11 ± 0.01	0.11 ± 0.00	0.12 ± 0.02
Water (%)	89.54 ± 0.01	88.53 ± 0.01	88.52 ± 0.01
Protein (%)	0.65 ± 0.22	0.33 ± 0.23	0.33 ± 0.23
Crude Fiber (%)	3.34 ± 0.01	3.49 ± 0.01	3.39 ± 0.01
Fat (%)	2.71 ± 0.24	2.75 ± 0.28	2.87 ± 0.21
Carbohydrate (%)	3.65 ± 0.02	4.78 ± 0.55	4.78 ± 0.05

4. Conclusion

Based on research that has been carried out the result was two. First, the lower storage temperature will extend the shelf life of the product. The results of this reseach was obtained by using pH parameter. So, the shelf life of Orumy beverages (with strawberry, lemon, and lychee flavors) in 30°C of storage temperature is 44 days, 45 days and 38 days. In the other side, in the storage temperature of 10°C the shelf life of the strawberry flavored will last for 60 days, the lemon flavor will last for 63 days, and the lychees flavor will last for 49 days. Second, this research also determined the nutritional content of Orumy beverage. Orumy beverage strawberry, lemon, and lychees flavors containing ash 0.11±0.01%, 0.11±0.00%, 0.12±0.02%, water 89.54±0.01%, 88.53±0.01%, 88.52±0.01%, protein 0.65±0.22%, 0.33±0.23%, 0.33±0.23%, crude fiber 3.34±0.01%, 3.49±0.01%, 3.39±0.01%, fat 2.71±0.24%, 2.75±0.28%, 2.87±0.21%, and carbohydrate 2.71±0.24%, 4.78±0.55%, 4.78±0.05%.

5. Acknowledgments
The author would like to thank the Instrumentation and Analytical Science Laboratory, Fundamental Laboratory, Chemistry Department FIA ITS, and LPPM ITS for their support until the completion of this research, as well as all parties involved in research and drafting of this research paper.

References
[1] A Amirah 2015 Cultural, Social, and Economic Perspectives in Making a Criminal Policy (Dolly Prostitution) *The Journal of MacroTrends in Social Science* **1**(1) 45-53
[2] A M Moefad 2015 Komunikasi Masyarakat Eks Lokalisasi Pasca Penutupan Dolly *Jurnal Komunikasi Islam* **5** 145-184
[3] Dirjen POM 1991 Dirjen POM No. 02591/B/SK/VIII/1991 tentang Perubahan Lampiran Peraturan Menteri Kesehatan Nomor 180/Men.Kes/Per/IV/1985 tentang Makanan Kadaluwarsa. Menteri Kesehatan.
[4] T Labuza and D Riboh 1982 Theory and Aplication of Arrhenius Kinetics to The Prediction of Nutrien Losses in Food *Food Technology* **36** 66-74
[5] H Anagari, S A Mustaniroh and Wignyanto 2011 *Penentuan Umur Simpan Minuman Fungsional Sari Akar Alang-Alang Dengan*
[6] Metode Accelerated Shelf Life Testing (ASLT) (Studi Kasus di UKM "R. Rovit" Batu, Malang) *Agrotek* **5** 133-143
[7] Y H Hui 2006 *Handbook of Food Science, Technology, and Engineering Volume 1* (Boca Raton: Taylor & Francis Group)
[8] Ensminger and Audrey 1994 *Foods and Nutrition Enscyclopedia Volume 1 (2nd ed.)* (Boca Raton: CRC Press LLC)
[9] H Herawati 2008 Penentuan Umur Simpan pada Produk Pangan *Jurnal Litbang* **27** 124
[10] F G Winarno 2004 *Kimia Pangan dan Gizi* (Jakarta: Gramedia Pustaka Utama)
[11] (AOAC) Association of Official Analytical Chemist 2005 *Official Methods of Analysis (18th ed.)* (Mayland (US): Association of Official Analytic Chemist Inc)
[12] C G Perdani, K H A Wijaya and S Kumalaningsih 2017 Karakteristik Bubuk Lobak, Nanas Madu dan Kemiri dengan Metode Pengeringan Foam Mat Drying *Industria: Jurnal Teknologi dan Manajemen Agroindustri* **6** 103-111
[13] M Ganje, S M Jafari, A Dusti, D Dehnad, M Amanjani and V Ghanbari 2016 Modeling Quality Changes in Tomato Paste Containing Microencapsulated olive Leaf Extract by Accelerated Shelf Life Testing *Food and Bioproducts Processing* **97** 12
[14] E T Haryati, F Heppy and K Ahmadi 2015 Pendugaan Umur Simpan Menggunakan Metode Accelerated Shelf Life Testing (ASLT) dengan Pendekatan Arrhenius pada Produk Tape Ketan Hitam Khas Mojokerto Hasil Sterilisasi *Jurnal Pangan dan Agroindustri* **3** 156
[15] T P Labuza and M K Schmidl 1985 Accelerated Shelf Life Testing of Foods *Food Technology* **39**(9), 57-62, 64, 134
[16] N I Agustin 2017 *Penentuan Umur Simpan Dengan Metode Accelerated Shelf Life Testing (ASLT) dan Analisis Proksimat Pada Minuman Rumput Laut* (Surabaya)
[17] E Chivandi, N Mukonowenzou, and D Berliner 2016 The coastal red-milkwood (Mimusops caffra) seed: Proximate, mineral, amino acid and fatty acid composition *South African Journal of Botany* **102** 137

IOP Conf. Series: Materials Science and Engineering **588** (2019) 012032 doi:10.1088/1757-899X/588/1/012032

Study on the structural evolution of SiO_2 extracted from Sidoarjo mud

Devita Rachmat[1,a] **and Doty Dewi Risanti**[1,b]

[1] Department of Engineering Physics, Faculty of Industrial Technology, Institut Teknologi Sepuluh Nopember, Sukolilo,Surabaya, Indonesia, 60111

[a)]devitarachmat@gmail.com, [b)]d.d.risanti@gmail.com

Abstract. Sidoarjo mud is known to have a large amount of silica mineral i.e 44.8% with a purity of 88.4 to 95.7%. In this study, the annealing process in extracted silica was carried out for 5 hours to observe the structural changes at 800°C, 1000°C and 1150°C. Thermogravimetric analysis of extracted silica shows that glass temperature transition (Tg) occurs at onset temperature of 800°C and crystalline phase take places at above 1150°C. The results are confirmed by X-Ray Diffraction (XRD). It indicates that there is a transformation phase of amorphous SiO_2 to triclinic crystalline structure with increasing amount of tridymite phase as the temperature increases from 800°C up to 1150°C. The changes in SiO_2 phase structure are also confirmed by FTIR results that show a transmittance shift of Si-O bending band slightly to lower wavenumbers and increasing transmittance intensity of Si-O stretching bonds as the heating treatment temperature is increased. Based on the UV-Vis results, the optical band gap shows no dependencies on heat treatment temperature, whereas the Urbach energy is proportional to the increasing temperature, in which SiO_2 sample heated at 800°C has the lowest Urbach energy or the fewer defects. Urbach energy values of extracted silica are ranging from 7.91-17.33 eV. In accordance with Urbach energy, the DSSC with photoanode containing SiO_2 heated at 800°C shows the best performance with efficiency of 0.077 %.

1. Introduction

Silica is found in many free sources in nature, one of them is Sidoarjo mud which is known from the previous research it has silica content of 44.8% [1]. Silica extracted from Sidoarjo mud has an amorphous phase structure with a purity of 88.4 to 95.7% at pH 7 [1],[2]. The previous study has been also carried out that amorphous silica extracted from Sidoarjo mud can be used as corrosion inhibitors and also as a scattering material on Dye Sensitized Solar Cell (DSSC) photoanode [3].

Amorphous silica is known to form a network of randomly interconnected tetrahedrals [4]. These tetrahedrals are composed of one silica atom at the centre and form oxygen atoms at the corners while each oxygen atom is shared by two tetrahedral atoms such that the total number of O atoms is twice that of the number of Si atoms [4]. However, the bond angle between Si and O atoms has various values allowing it to form crystalline structures in different forms such as quartz, tridymite and cristobalite and also can easily form the amorphous structures whereas amorphous SiO_2 is known to have more dangling bonds which further causes a localized state [5]. This localized state between conduction band and valence band causes electrons unable to move freely [5]. Hence the amorphous and crystalline SiO_2 have different properties such as the energy gap value of amorphous silica is 9.3 eV [6], which is higher compared to crystalline one, i.e 5.48-6.3 eV [7]. In this study silica will be obtained by extract Sidoarjo mud using coprecipitation method. The phase transformation of the amorphous phase of silica to crystalline, was carried out through the calcination process for 5 hours at high temperature but still below its melting point of 1700°C, the corresponding properties were studied.

2. Experimental

2.1. The Preparation of silica from Sidoarjo mud

Sidoarjo mud was washed using aquadest to remove impurities. Then the mud is dried and crushed. The crushed mud is then weighed in 10 grams and washed using HCl for 12 hours. Subsequently, it was reacted with NaOH 7M and stirred for approximately 1 hour at 80°C with a speed of 2 mods. Afterwards, the slurry solution was filtered using Whattman filter paper No. 15. The filtrate solution was titrated using HCl while stirring at temperature of 40°C to obtain a pH value of 7. The extracted silica was precipitated and washed using distilled water to remove the content of acid, base and salt. Then, the silica precipitation was dried in an oven at 80°C for 12 hours. The extracted silica was heating for 5 hours at temperatures of 800, 1000 and 1150°C to ensure that the sample has fully transformed.

2.2. Characterizations

Thermogravimetric Analysis (TGA) was carried out on 14.9 mg Sidoarjo mud and 16.3 mg silica extracted from Sidoarjo mud using Linseis STA PT 1600 thermal analyzer in a temperature range of 0°C to 1200°C with a heating rate of 10°C/minute. Identification of transformation in crystalline phase was observed by X-Ray Diffractometer (Philips X'pert MPD 30 kV, 40 MA). Fourier Transform InfraRed (FTIR) test was performed using a Thermo Nicolet i510 Spectrometer. Subsequently, the absorption of silica in visible light (200 nm - 800 nm) was characterized using a UV-Vis Spectrometer (Lambda 750 UV/VIS/NIR spectrophotometer). Furthermore, to find out the effectiveness of silica extracted from Sidoarjo mud as scattering material in the DSSC photoanode, the I-V test was done in Photonics Material Laboratory, Department of Physics, ITB.

3. Results and discussion

3.1. TGA Analysis

Figure 1. shows the TGA curves of pristine Sidoarjo mud and SiO_2 extracted from Sidoarjo mud. Weight loss stage of pristine Sidoarjo mud take places in two steps. The first stage occurs at temperature below 100°C, which should be due to the evaporation of physically adsorbed water molecules and the second was between 400 – 600°C, that is attributed to dehydroxylation process or hydroxyl groups (water crystal) releases process. The two weight loss stages are also observed in SiO_2 extracted from Sidoarjo mud with its pronounced decrease up to 400°C as indicative of the presence of water evaporation and melting of some impurities.

Our silica TG curve indicates two distinct phase transformations, i.e. glass transition onset temperature is at 800°C and crystallization at about 1150°C. The larger temperature gap of the two processes may result in the more stable glass phase obtained [8]. According to the phase diagram of silica the glass transition is related to β tridymite[8]. These two transformation cannot be seen in TGA curve of pristine Sidoarjo mud, since there exists a more complex compounds in it.

3.2. XRD Analysis

X-ray powder diffraction (XRD) studies were performed on all samples studied to confirm their amorphous and crystalline states (Figure 2). SiO_2 extracted from Sidoarjo mud without heat treatment has the amorphous phase as shown from the broad peak at 2θ value of 22° [1],[2]. The transition from amorphous to crystalline silica of Sidoarjo mud takes place further as the temperature increases [9].

SiO_2 in triclinic structure is obtained for all heat treated samples. However, the tridymite structure is also observed and its fraction is increased. This result indicates that extracted silica undergoes a very slow transformation towards a stable tridymite structure starting from 1000ºC. This is also supported by the TGA results (Figure. 1) showing the larger gap between glass transition and crystalline temperatures.

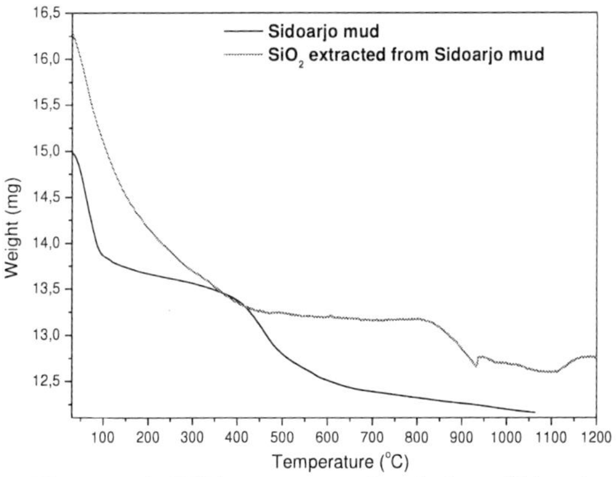

Figure. 1 TGA curves of pristine Sidoarjo mud and SiO_2 extracted from Sidoarjo mud

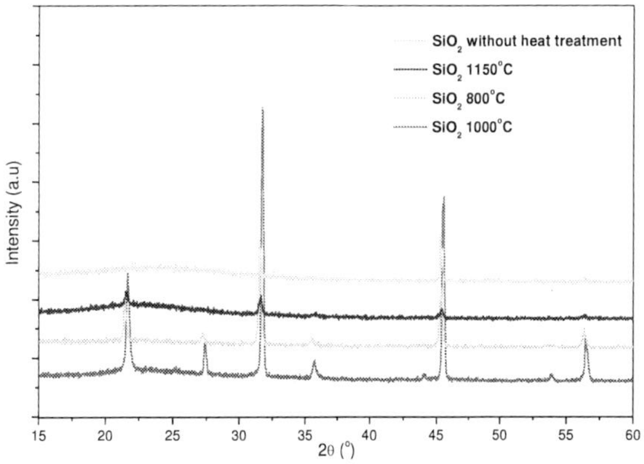

Figure. 2 XRD spectra of extracted SiO_2 heated at various temperatures

3.3. FTIR Analysis
In the IR spectra of silica extracted from Sidoarjo mud as illustrated in Figure 3. The rocking, bending, and stretching vibrational bands of Si-O can be observed at ~450 cm^{-1}, ~800 cm^{-1}, and ~1073 cm^{-1}, respectively [8]. There are two typical features for each sample, namely, a shift in the transmittance band of Si-O bending bond to a slightly lower wave numbers and an increase in transmittance intensity of Si-O stretching bond as heat treatment temperature was increased [8].
The strong absorption bands in the region of ~3400 cm^{-1} and 1650 cm^{-1} which only appears in silica without heat treatment indicates the stretching vibrations between hydrogen and some other atoms as the existence of water content [9]. The presence of water content in silica extracted from Sidoarjo mud is related to TGA curve in Figure. 1, i.e. a continuous decrease up to 400ºC, and it is confirmed by the loss of hydrogen from the surface that is observed by IR spectroscopy.

3.4. UV-Vis Analysis
UV-vis characterizations results were used to determine the optical band gap (E_g) using Tauc plot. E_g is determined by extrapolation of a linear part of the curve to intercept the energy axis ($\alpha h\upsilon = 0$)[10]. Figure. 4 shows that the increase in heat treatment temperature tends to increase the optical band gap values as written in Table 1. Moreover, Urbach energy (E_u) which can be obtained from the reciprocal of the slope of ln α against photon energy ($h\upsilon$) is shown in Figure. 5 and listed in Table 1. It is known that Urbach energy (E_u) value represents structural defect. Urbach energy of heat treated silica are getting higher as the temperature increases. Whereas E_u is constant or weakly dependent on temperature [11].

3.5. J-V Curves Analysis
In general, all photoanode samples containing SiO_2 show higher efficiency compared to that of without SiO_2 addition (see Table 2.). Since the amount of silica heated at 1150ºC is not sufficient for DSSC fabrication, therefore we cannot measure its J-V characteristics. There exists a so-called parasitic effects in DSSC that caused by internal resistances, i.e. series resistance (R_S) and shunt resistance (R_{SH}). Those resistances may lead to power loss in DSSC. As illustrated in Figure 6, SiO_2

IOP Conf. Series: Materials Science and Engineering **588** (2019) 012032 doi:10.1088/1757-899X/588/1/012032

heated at 800°C has higher series resistance compared to that of SiO_2 heated at 1000°C. All samples studied show a non ideal J-V curve, which is attributed to the presence of those resistances. Photoanode containing extracted silica can enhance the performance of TiO_2 photoanode in DSSC up to 0.055 %. Extracted silica is beneficial for improving open circuit voltage. On the other hand, its effect on the short circuit current is somewhat complex. This is because of the presence of particularly series resistance which becomes more obvious for silica heated at 1000°C. It is likely that extracted SiO_2 in the form of triclinic structure results in an almost ideal J-V curve. This is also supported by the fact that the Urbach energy of this condition is the lowest.

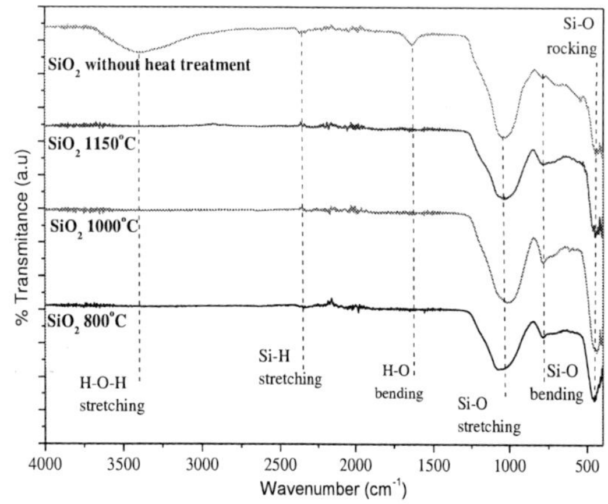

Figure. 3 IR spectra of Sidoarjo mud and extracted SiO_2 heated at various temperatures

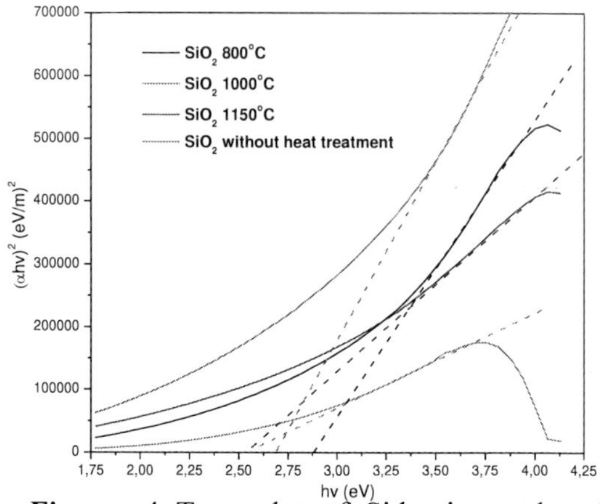

Figure. 4 Tauc plot of Sidoarjo mud and extracted SiO_2 heated at various temperatures

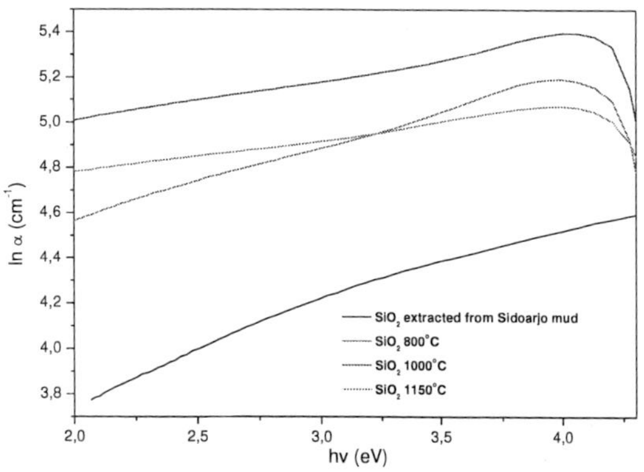

Figure. 5 Plots of ln (α) against hv of extracted SiO_2 in various temperatures

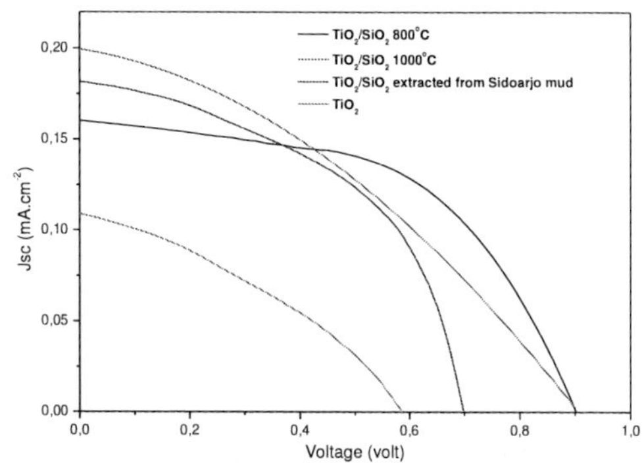

Figure. 6 J-V curves of extracted SiO_2 to modifying TiO_2 photoanode in DSSC

Table 1. Urbach energy values of SiO_2 extracted from Sidoarjo mud

Sampel	Urbach energy (eV)	*Optical gap (eV)*
SiO_2 without heat treatment	7.91	2.55
SiO_2 800°C	7.43	2.875
SiO_2 1000°C	14.39	2.7
SiO_2 1150°C	17.33	2.55

Table 2. Efficiency of DSSC

Photoanode sample	Efficiency (%)
TiO_2	0.022
TiO_2/SiO_2 without heat treatment	0.062
TiO_2/SiO_2 800°C	0.077
TiO_2/SiO_2 1000°C	0.064

4. Conclusions

SiO_2 extracted from Sidoarjo mud heated at 800 – 1150°C undergoes structural transformation : amorphous to triclinic to tridymite with hexagonal structure. Extracted silica has glass transition temperature at 870°C which corresponds to the formation of tridymite. It is known that the rise in Urbach energy (E_u) of extracted silica is proportional to the increasing temperature. DSSC photoanode containing extracted silica can enhance the performance of TiO_2 photoanode in DSSC up to 0.055%.

5. Acknowledgments

The authors gratefully thanks to the Ministry of Research Technology and Higher Education of Republic Indonesia through Penelitian Terapan Unggulan Perguruan Tinggi 2018 for supporting the financial aid during this project

References

[1] R S Mustopa *et al* 2013 Physical Properties Characterization of Porong Sidoarjo Mud and Its Potentials as CO Gas Adsorbent Materials *AIP Conference Proceedings 1554*, 75.

[2] A J H Pribadi *et al* 2013 Pengaruh pH Akhir Larutan pada Sintesis Nanosilika dari Bahan Lusi dengan Metode Kopresipitasi *Inovasi Fisika Indonesia 2* (3) 7–10.

[3] H Budiarti *et al* 2017 Structure Using Lapindo Mud Extract via Sol-Gel Method *Procedia Engineering 170* 65–71.

[4] D Ren et al 2018 The electronic and optical properties of amorphous silica with hydrogen defects by *ab initio* calculations *Journal of semiconductors* **Vol. 39** No. 4

[5] T Demuth et al 1999 Polymorphism in silica studied in the local density and generalized-gradient approximations *Journal physics Condens. Matter* 11 : 3833-3874.

[6] Z A Weinberg et al 1979 Exciton and hydrogen diffusion in SiO_2. *Journal of applied physics* 50.

[7] The quartz page. The silica group. Web. Last modified January 12nd, 2014. http://www.quartzpage.de/gen_mod.html

[8] Y H Roos 2010 Glass transition temperature and its relevance in food processing. *Article. Review of food science and technology* 469 - 496

[9] M H S Abadi et al 2015 Effects of Annealing Temperature on Infrared Spectra of SiO_2 Extracted From Rice Husk *Journal of ceramic science and technology* **Vol. 6** no. 1, 41-45.

[10] D Waroquiers 2013 *Electronic and optical properties of crystalline and amorphous silica from first principle* Universite catholique de louvain

[11] F N C Anyaegbunam et al 2018 A study of optical band gap and associated urbach energy tail of chemically deposited metal oxide binary thin films *Digest Journal of Nanomaterials and Biostructures* **vol. 3** no. 3, 847-856.

.

IOP Conf. Series: Materials Science and Engineering **588** (2019) 012033 doi:10.1088/1757-899X/588/1/012033

Enrichment Chromite Sand Grade Using Magnetic Separator

S Subandrio[1*], W Dahani[1], M Alghifar[1], and T T Purwiyono[1]

[1]Department of Mining Engineering, Trisakti University, Jakarta, Indonesia

*Corresponding author : Subandrio@trisakti.ac.id

Abstract. To fullfil requirements of mettalurgical grade 48% Cr_2O_3 with Cr:Fe ratio 3:1 need the right beneficiation, one of them is using magnetic separator. Chromite sand sample obtain from Konowe-Southeast Sulawesi have low grade, so it needs process to enrichment Cr:Fe Ratio. Cr:Fe ratio feed was 0.55, after the process for size variation, roll speed, feeder speed and electric current decline Cr:Fe Ratio around 50.5%. When size -65+100 mesh Cr:Fe ratio was 0.282. At roll speed 20 RPM, feeder speed 50 Kg/Hr and electric current 1.5 A have Cr:Fe ratio 0.3525. Therefore, it needs more beneficiation before using magnetic separator for chromite sand from Konowe-Southeast Sulawesi.

1. Introduction

Chromite ore is the only commercial source of chromium in the metallurgical industries [1]. It occurs as a chromium spinel, a complex mineral containing chromium, iron, aluminium and magnesium in changeable proportions depending upon the deposit [2]. Chromite deposits can be estimated using geostatistics and geotechnical methods [3,4].

Chromite can be classified on the Cr/Fe ratio. The highest-grade chromites are those having Cr/Fe ratio of more than 2 and containing a minimum of 46 to 48% Cr_2O_3 [5]. Chemical-grade (high-iron) chromite contain large amounts of iron, which often results in Cr/Fe ratios from 1.5 to 2.1, although the total amount of chromium ranges from 40-46% Cr_2O_3. Refractory-grade chromite contains relatively large quantities of Al_2O_3 (> 20%) and have (Cr_2O_3+Al_2O_3) levels of more than 60% [6].

One of the chromite potentials in Indonesia is located in Sulawesi. Metallurgical quality concentrate, refractories or chemicals can be increased by beneficiation through gravitational, magnetic and flotation processes on chromite ores [7,8]. Chromite sand from Konowe-Southeast Sulawesi has a low Cr/Fe ratio of 0.9 so it is melted with an arc furnace to increase the Cr/Fe ratio [9] and eliminate the surfactant templates used in the materials [10,11,12]. Increasing Cr/Fe ratio to 2.31 (Cr_2O_3 grade 48.9%) using the wet shacking table and can reprocess it to 3.35 with Cr_2O_3 levels of 52.3% using High Intensity magnetic separator (IRMS) [13].

In this study, the sample of chromite sand from Konowe-Southeast Sulawesi was only processed using magnetic separator. Cr, Fe and others are transition metal elements that have paramagnetic or ferromagnetic properties. If the material is only processed using a magnetic separator, it will simplify and accelerate the beneficiation process [14].

2. Experimental

Sample taken from Konowe-Southeast Sulawesi contain chemical composition as shown in the table 1. Sample weighed 200 gr and be examined into 4 variable. There are size variation, roll speed, feeder speed and electric current. For size variation tested is -35+48, -48+65, -65+100 and -100 mesh. While for roll speed variation that is 15,20,25,30 RPM. In feeder speed were tested 45, 50, 55 and 60 Kg/Hr.

also for electric current 1.25, 1.5, 1.75 and 2 Ampere complete as shown in the table 2. Therefore, posititon of feeder and splitter kept constant druing testing.

TABLE 1. Chemical composition of chromite sand from Wosu-Morowali

$Fe\ Cr_2\ O_4$	$Ca\ Fe\ Si_2\ O_6$	$Fe_2\ Mg\ O_4$	$Si\ O_2$	$Ca\ Ti\ O_3$	$Mg\ Al_2\ O_4$	Unidentified
26	21.6	19.3	17.2	10.3	5.4	3.4

TABLE 2. Research design

No.	Roll Speed (RPM)	Feeder Speed (Kg/Hr)	Electric Current (A)	Size Feed (*Mesh*)
1.	20	50	1.5	**–35+48**
2.	20	50	1.5	**-48 +65**
3.	20	50	1.5	**65 +100**
4.	20	50	1.5	**-100**
5.	**15**	50	1.5	-65 +100
6	**20**	50	1.5	-65 +100
7.	**25**	50	1.5	-65 +100
8.	**30**	50	1.5	-65 +100
9.	20	50	**1.25**	-65 +100
10.	20	50	**1.50**	-65 +100
11	20	50	**1.75**	-65 +100
12	20	50	**2.00**	-65 +100
13	20	**45**	1.5	-65 +100
14	20	**50**	1.5	-65 +100
15	20	**55**	1.5	-65 +100
16	20	**60**	1.5	-65 +100

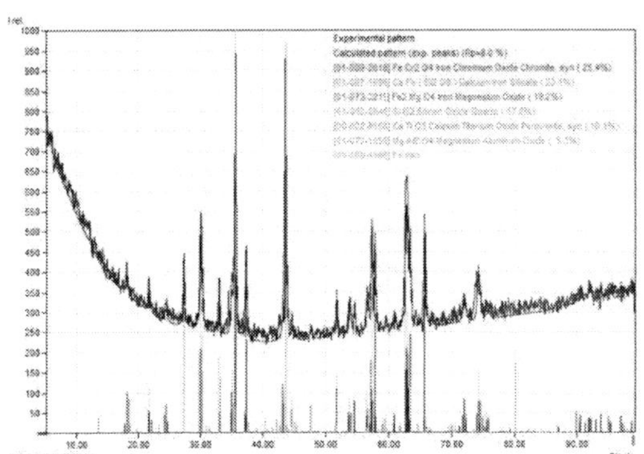

FIGURE 1. Result difaktogram XRD of chromite sand

FIGURE 2. Dry magnetic separator

TABLE 3. feed grade of chromite sand from Wosu-Morowali

Size	Feed Grade		Cr:Fe Ratio
	Cr	Fe	
-35+48#	12	44.2	0.271
-48+65#	22.9	40	0.573
-65+100#	24.3	44.2	0.550
-100#	12.2	42.6	0.286

3. Result and Discussion

As shown table 3, on each size have best Cr:Fe ratio amount 0.573 at size -48+65 mesh. However, in testing variation of roll speed, feeder speed and electric current use size -65+100 mesh because it have the best grade of Cr 24.3% nevertheless doesn't have best ratio as -48+65 mesh 0.55. in this test hope the highest grade of Cr at -65+100 mesh it can reduce silica (SiO_2) and iron (Fe) grade.

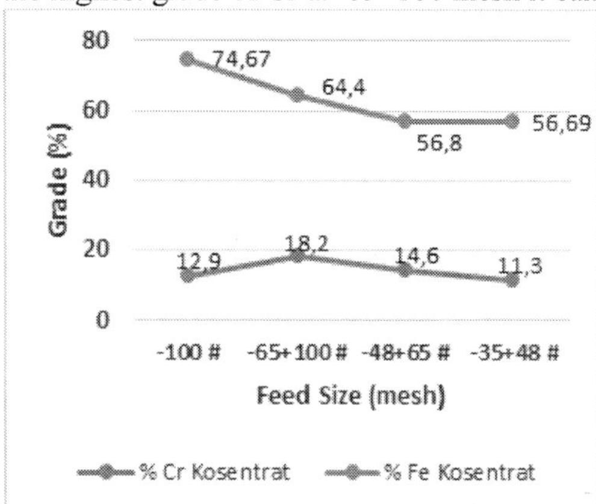

FIGURE 3. Concentrate Cr and Fe grade

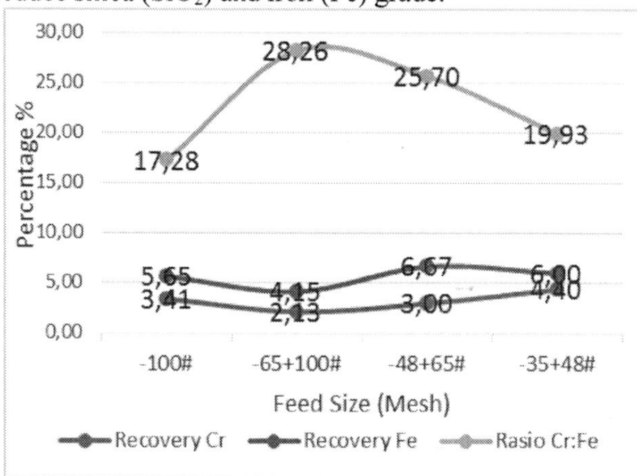

FIGURE 4. Cr:Fe ratio and Recover Cr with Fe

In this research to test influence of size variation against recovery and Cr:Fe ratio. Can be seen in figure 3, at -65+100 mesh position is the best condition of Cr grade 18.2% and the greater of sample size will make Fe grade tends to go down. Therefore, the best of Cr:Fe ratio also in the -65+100 mesh 0.282 however Cr:Fe best variation on the -35+48 mesh is 4.4%

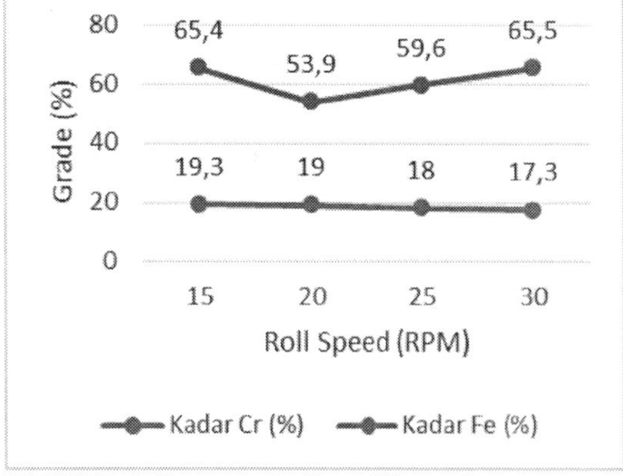

FIGURE 5. Concentrate Cr and Fe grade

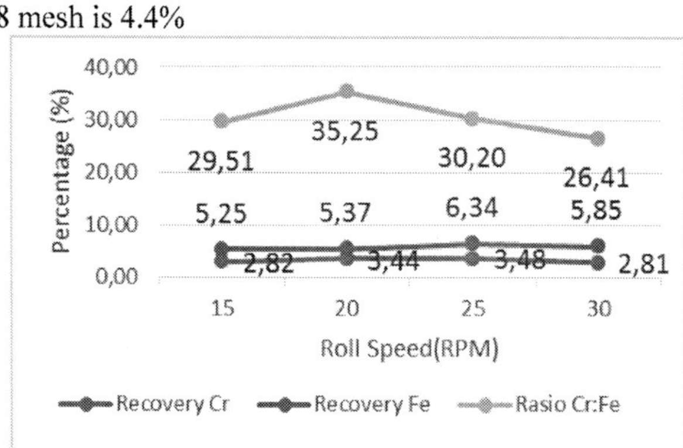

FIGURE 6. Cr:Fe ratio and Recover Cr with Fe

On roll speed it can be seen that the faster roll speed will make Cr grade decrease, however it will make lowest point of Fe grade and after that it will increase again. The best roll speed of Cr:Fe ratio is 0.35, however an increase roll speed will make Cr:Fe ratio decrease because Fe grade will increase but Cr grade decrease. The best of recovery Cr roll speed is 3.48%.

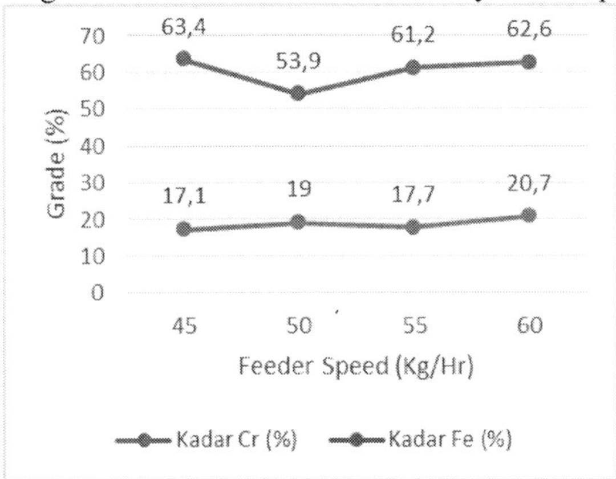

FIGURE 7. Concentrate Cr and Fe grade

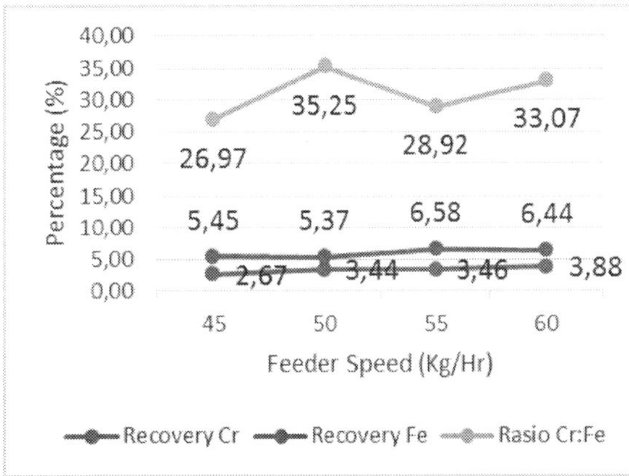

FIGURE 8. Cr:Fe ratio and Recover Cr with Fe

On feeder speed (Kg/Hr), Cr grade formed will be better if feeder speed faster although on roll speed 55 Kg/Hr have decrease. As well Fe grade will decline on roll speed 50 Kg/Hr and little increase feeder speed will increase the Fe grade. So that, Cr grade on roll speed 55 Kg/Hr have significant decline and the best Cr:Fe ratio on roll speed 50 Kg/Hr is 0.3525. the best of recovery Cr roll speed 60 Kg/Hr is 3.88%.

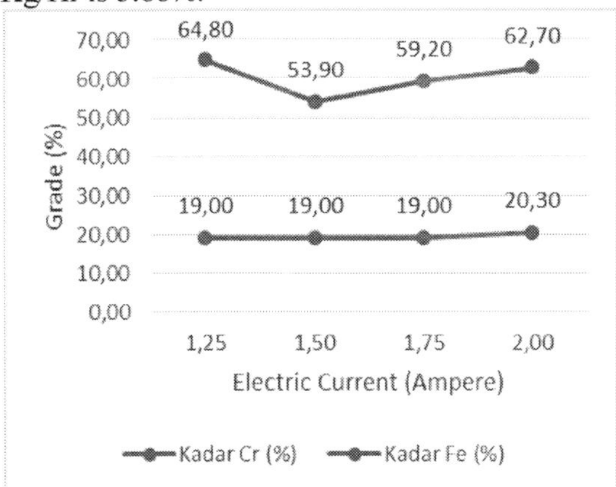

FIGURE 9. Concentrate Cr and Fe grade

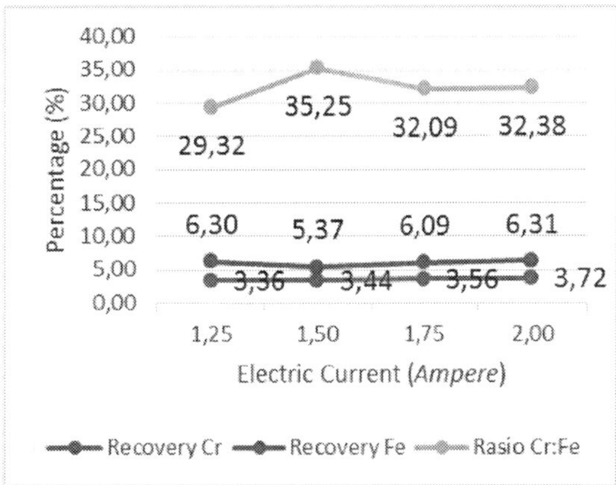

FIGURE 10. Cr:Fe ratio and Recover Cr with Fe

On electric current, an increase electric current will increase Cr grade. However on electric current 1.5 A will make lowest grade of Fe and slowly increase with additional electric current. Then, the best Cr:Fe ratio on electric current 1.5A is 0.352. an additional electric current will make recovery Cr increase and the best recovery Cr when 2A is 3.72.

4. Conclusion

If observed further, Cr:Fe ratio on feed more better than after process. Cr:Fe ratio on feed is 0.55 but after process obtained Cr:Fe ratio for size variation 0.282 and roll speed, feeder speed, electric current is 0.352. A decrease in Cr:Fe ratio because Cr grade wasted with silica (SiO_2). Beneficiation is needed to use shacking table before magnetic separator for chromite sand from Konowe-Southeast Sulawesi.

Acknowledgments
The author would like to thank the laboratory of mineral processing Bandung Institute of Technology (ITB), Dr. Ir. Edy Sanwani, M.T. who has provided support, encouragement and opportunity to do this research. For XRD and XRF analysis was supported by Institut Teknologi Sepuluh Nopember (ITS).

References

[1] Gupta A, and Yan D S 2006 An Introduction Mineral Processing Design And Operation, Amsterdam: Elsevier.
[2] Gu F and Wills B A 1988 Chromite-Mineral and Processing *Minerals Engineering* **1** 235-240.
[3] Marwanza I, Badaruddin S P and Azizi M A 2018 Copper Resource Estimation in PT X Batu Hijau, Regency of West Sumbawa, West Nusa Tenggara Province using Geostatistical Method *IOP Conference Series: Earth Environmental Science* **212**.
[4] Azizi M A, Marwanza I, Amala S A and Hartanti N A 2018 Three Dimensional Slope Stability Analysis of Open Pit Limestone Mine in Rembang District, Central Java *IOP Conference Series: Earth Environmental Science* **212**.
[5] Sen R 2010 Physical and Chemical Characterisation of Briquettes Made From Indian and South African Chromite Concentrates and Their Smelting Behaviours, *PhD thesis*, Jadavpur University.
[6] Bhandary A K, Gupta P, Mukherjee S, Chaudhuri M G, and Dey R 2016 Benefication of Low Grade Chromite Ore and Its Characterization for the Formation of Magnesia-Chromite Refractory by Economically Viable Process *International Journal of Chemical and Molecular Engineering* **10** 1096-1104.
[7] Subandrio, Dahani W, and Purwiyono T T 2017 Optimizationof Gravity Chromite Processing with Shaking Table *Journal Petro* **6** 2.
[8] Seifelnasr A, Tammam T, and Abouzeid A 2012 Gravity Concentration of Sudanese Chromite Ore using Laboratory Shaking Table *Physicochemical Problem of Mineral Processing* **48**(1) 271-280.
[9] Nurjaman F, Subandrio S, Ferdian D, and Suharno B 2018 Effect of Basicity on Beneficated Chromite Sand smelthing process using submerged arc furnace *AIP Conference Proceedings* 1964.
[10] Zi S C, Chandren S, Yuan L S, Razali R, Ho C S, Hartanto D, Mahlia T M I, and Nur H 2016 New Method to synthesize mesoporous titania by photodegradation of surfactant template *Solid State Sciences* 83-91
[11] Kurniawati R, Iryani A and Hartanto D 2018 The effect of 2-propanol on the shifting band gap of ZSM-5/TiO_2 Composite Prepared Via Sol-Gel Method *AIP Proceeding* **2049** 020089
[12] Pambudi A, Kurniawati R, Iryani A and Hartanto D 2018 Effect of calcination temperature in the synthesis of carbon doped TiO_2 without external carbon source. *AIP Conference Proceedings*
[13] Tripathy S K, Singh V and Ramamurthy Y 2012 Improvement in Cr:Fe Ratio of Indian Chromite Ore for Ferrochrome Production *International Journal of Mining Engineering and Mineral Processing* **1** 101-106.
[14] Rao R B, Dasand B, and Sastri S R S 1997 Improving Grade of Indian Chromite Ores by High–Gradient Magnetic Separation *Magnetic and Electrical Separation* **8** 175-183.

IOP Conf. Series: Materials Science and Engineering **588** (2019) 012034 doi:10.1088/1757-899X/588/1/012034

Tricycle Applications for Physical Therapy Sufferers

I Made Londen Batan[1,a], Teditya Nico Aji Sukma Lutiawan[1,b] and Lutfi Agus Salim[1,c]

[1] Mechanical Engineering Departement, Institut Teknologi Sepuluh Nopember, Surabaya 60111, Indonesia

[a]londbatan@me.its.ac.id, [b]tedityanico@gmail.com, [c]as.lutfi90@gmail.com

Abstract. Bicycles are one of the sports equipment as well as a mode of transportation that is environmentally friendly, when using a bicycle, all members of the body perform movements. In this study, a tricycle with a delta concept was made, that is 2 wheels behind and 1 wheel in front. This bike is designed to help stroke sufferers perform physical rehabilitation independently by pedaling a bicycle. To find out the benefits of bicycles to help cure strokes, especially the legs of the patient, a stroke test were carried out for 1 month (30 days) by 4 stroke patients, who were the respondents in this research. Each respondent pedaled the bike once every 3 days, so the amount of therapy became 9 times a month for each respondent. During pedaling respondents vital sign was recorded, the value of heart recovery rate (HRR) and amount of energy consumption were calculated. To find out the usefulness of bicycle therapy for stroke patients, 4 stroke sufferers were also selected who were doing stroke rehabilitation as respondents in this research, but did not pedal a bicycle. Furthermore, all respondents, whether bicycle therapy or not, performed a fitness test using the six minute walking test (6-MWT) test method. The method is used to determine the maximum oxygen volume capacity (VO2 max) of a person who walks on a certain path for 6 minutes. From the test, it can be seen that the distance taken by the respondent after walking for 6 minutes. The distance traveled is converted in the form of VO2 Max capacity. The test results shows that respondent therapy using a bicycle has a significant increase in body fitness, compared to respondent therapy without using a bicycle.

1. Introduction

Stroke makes the sufferer experience interference in the body's coordination system. One method of healing stroke is to use therapy that aims to restore the body's coordination function. Various tools were created by experts to help stroke patients perform physical therapy to speed up healing or recovery of patients from stroke. Besides the conventional treatments in the physical rehabilitation section of the hospital, equipment that's widely used to help treat stroke patients is a static bicycle, Leg and Arm / Upper Body Trainer, and a treadmill. Static bicycle therapy equipment is a common therapeutic tool in hospitals and health and sports equipment stores. Static bicycles can be used for rehabilitation of patients who are sick with calcification, knee pain, stroke, or post-accident that are recommended by doctors to train their legs, knees or joints.

Leg and Arm / Upper Body Trainer is a special rehabilitation tool developed by the MOTOmed company, as shown in figure 1. This device functions like a static bicycle, which is there is a stroke for the feet. The difference is in this tool, the patient does not need to move seats from a wheelchair or ordinary chair that has wider surface area and is comfortable than a small static bicycle seat. In addition there are also strokes for hands and monitors that can indicate feedback from the rehabilitation being performed.

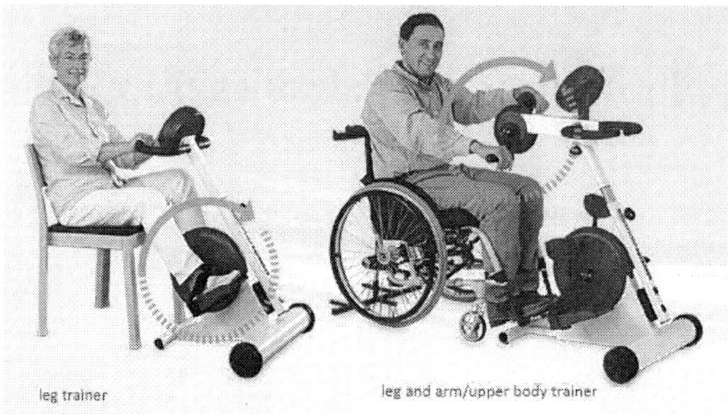

Figure 1. Leg and Arm / Upper Body Trainer along with examples of its use [MOTOmed]

The new study found that a structured and progressive walking exercise program using a treadmill, was effective in helping patients to build strength and balance after one year suffer from stroke. The study also found that 52 percent of stroke patients who participated, in physical therapy programs that included in treadmill walking exercises and home programs that focused on progressive strength and balance exercises experienced an improvement in walking ability. This trial was held by Locomotor Experience Applied Post-stroke (LEAPS) and led by physical therapist Pamela W. Duncan, and involved 408 participants who had a stroke. Researchers initially hypothesized that physical therapy with the help of a treadmill and coupled with a walking program, were more effective than exercises just performed at home. But after one year, both methods proved to have the same effectiveness that is able to increase walking speed, balance, and quality of life, "said Duncan of Duke University School of Medicine in Durham, North Carolina". Examples of the use of treadmills for stroke therapy can be seen in Figure 2.

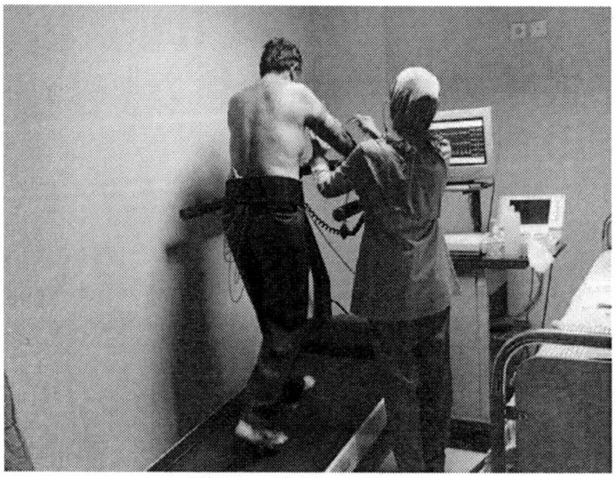

Figure 2. Treadmill

According to Duncan, research conducted for six months concluded that both forms of physical therapy programs, both treadmills and physical exercises at home are more effective and superior, when compared to the usual standard care provided.

The above equipment can only be used in a fixed place, can not help stroke patients to move places independently, such as going to the park, seeing the environment or to the market. Therefore, since 2013 at the Product Design and Development Laboratory, the Mechanical Engineering Department of

ITS, three-wheeled bicycles have been designed and developed as a physical therapy aid for stroke sufferers by several designers [1] [2] [6] . Three-wheeled bicycles developed are intended to help stroke patients carry out physical therapy as well as psychological therapy, where patients not only live inside the house, but also can see and enjoy the environment, and can even go pedaling to public places, such as parks , markets and other places [5]. The entire bicycle design is based on the tadpole concept, which is a tricycle, where 2 wheels are installed in the front and 1 wheel is mounted behind. In principle, the bicycle is very helpful for stoke sufferers to do therapy by pedaling. But the design still has problems in setting the right and left turn movements by the patient. This is because, the 2 front wheels are quite heavy to be bent, on the other hand stroke sufferers often do not have the strength to move things by hand. To overcome this problem, since 2016 a delta concept tricycle was developed, that is a tricycle with 1 wheel in front and 2 wheels behind [3] [4] . The prototype of a bicycle therapy design with the delta concept as shown in figure 3. To find out the benefits of bicycles as a stroke rehabilitation aid, stroke patients were tested with stroke requirements in accordance with the standard and procedure for stroke patients.

2. Research Method

The method used in this study is an experiment using a bicycle. Respondents in this study were stroke patients who were undergoing therapy at the Haji Surabaya General Hospital Medical Rehabilitation Installation. Respondents were divided into two groups, namely: therapy groups using tricycles (post stroke bikes) and therapy groups without using bicycles. Respondents underwent a physical fitness examination using the 6-MWT method at the beginning of therapy and one month after therapy. Therapy using a bicycle after a stroke is done dynamically through a predetermined path. Before the implementation of therapy, vital sign respondents were taken, and continued with warming up for 5 minutes. During the respondent's performs therapy, the speed of the bicycle was maintained in the range of 4-6 Km / hour, this was done to avoid the excessive workload on the respondent's body and avoid the occurrence of accidents when respondent's perform the therapy, considering that this was the first therapy to use a bicycle dynamically for patients post stroke. Vital sign measurements include measurement of heart rate, blood pressure and respiratory intensity (Respiration). The measuring instrument used to measure vital sign is a cyclometer, tens meter, stop watch. During the paddle test, respondent's pedaled two sessions with a duration of each session for 10 minutes. At the end of each session a vital sign was taken and the respondent was asked to rest for 3 minutes. Experimental results are displayed in the graph of the heart rate, blood pressure and respiratory intensity (Respiration). Besides that, there is also a graph of the development of the heart recovery rate (HRR), the VO_2 Max chart, and the graph of the energy consumption of the respondent during pedaling. At the end of the research, the fitness of each respondent will be evaluated, both respondent therapy with bicycle and respondent who do therapy manually without pedaling a bicycle. From these results it can be seen the usefulness of tricycle as a physical therapy aid for stroke patients.

3. Experimental Trials

3.1. Testing Procedures

In this study, the type of stroke therapy that will be given to post-stroke respondents is dynamic paddle therapy. Therefore, to ensure the safety of respondents and ensure each respondent gets identical treatment, it is necessary to have testing procedures.

Before paddling therapy using a post stroke bikes is started, measurement of the vital sign respondent is done first, then the measurement results are recorded in the data sheet. Then the respondent warms up for 5 minutes, after warms up the respondent will performs the first stage of paddle therapy for 10 minutes. After 10 minutes, the respondent is rested for 2 minutes and the vital sign is recorded. Then the respondent entered the second stage paddle therapy for ten minutes. After 10 minutes the respondent is rested for 2 minutes and the vital sign is recorded. The stages of therapy were repeated 9 times. The 6-MWT measurement was done before the first therapy, after the nineth.

IOP Conf. Series: Materials Science and Engineering **588** (2019) 012034 doi:10.1088/1757-899X/588/1/012034

Respondents in this research were post stroke patients who had passed the critical period and bed rest. This means that post-stroke patients who are respondents are post-stroke patients with a stable vital sign, so that it will allow them to perform maximum paddle therapy and the risk of having another stroke is small. Respondents who participated in this research amounted to 8 respondents, the 8 respondents will be divided into two groups, namely the paddle therapy group (A, B, C, and D) and the non-paddle therapy group (E, F, G, and H). Respondent data in this study can be seen in table 1.

Table 1. List of respondent to find out the tricycle usefulness

Name	Age (Yrs)	Weight (Kg)	Type of Stroke	Group
Winarno	62	68	Stroke Infark	Paddle therapy group
Suyono	55	58	Stroke Infark	
Lexi	72	65	Stroke Infark	
Ahyar	45	62	Stroke Infark	
Soekardi	66	58	Stroke Infark	Non Paddle therapy group
Gunawan	52	58	Stroke Infark	
Suyono	64	69	Stroke Infark	
Puamin	58	57	Stroke Infark	

3.2. Results Of Vital Sign Measurement and Fitness

From the research that have been carried out, there are vital sign charts which include: heart rate, blood pressure, and respiratory intensity, as well as graphs of heart recovery rate (HRR) development, VO2 max chart, and consumption graph of respondent energy paddle.

3.2.1. Heart Rate.

During therapy by pedaling a bicycle, the heart rate of paddle therapy group was measured. Measurement results are displayed in the graph and can be seen in Figure 5.

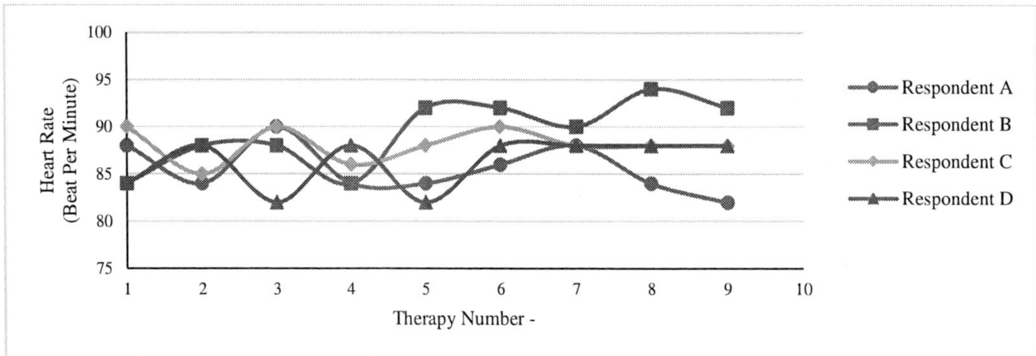

Figure 3. Development of heart rate (Heart rate) of respondents A, B, C and D

From graph in the figure 3, it can be seen that the heart rate of each respondent that is the respondents A, B, C and D are stable, during the four respondents doing therapy using post stroke bikes. The respondent's heart rate is said to be stable if there is no extreme change that exceeds the safe limit, namely (50-80)% of the maximum heart rate of each respondent.

3.2.2. *Blood Pressure*

Respondent's blood pressure was measured during a paddle therapy with tensimeter. The results of blood pressure measurement from paddle therapy group can be seen in Figure 4.

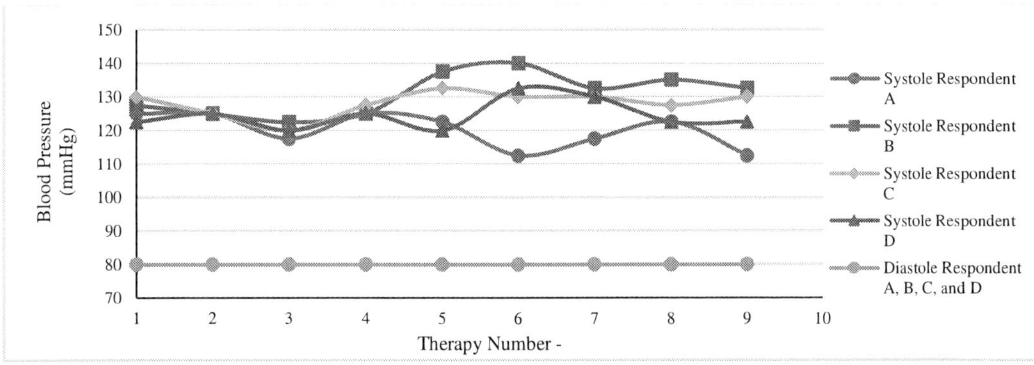

Figure 4. Development of blood pressure (TD) of respondents A, B, C and D

From graph in Figure 4, it can be seen that the blood pressure of each respondents, respondents A, B, C and D who were doing bicycle therapy after stroke is stable. The respondent's blood pressure is said to be stable if there is no extreme change that exceeds the safe limit. However, for respondent A requires special attention because blood pressure tends to decrease, a decrease in blood pressure can be caused by disruption of nerve function that regulates blood pressure, but it can also be caused by dehydration and drug use. Besides respondent A, respondent B also needs special attention, this is because the respondent's blood pressure is close to the upper limit of the blood pressure that is permitted. Stable blood pressure will reduce the risk of the respondent having a stroke again when the respondent undergoes therapy using a bicycle after a stroke.

3.2.3. *Respiratory Intensity*

The respiratory intensity in question is the amount of inhalation and exhalation during a certain period. The results of the measurement of respiration intensity for 2 x 10 minutes pedaling is poured in graphical form. The following is a graph of respiratory intensity of respondents A, B, C and D during therapy using a post stroke bicycle, can be seen in Figure 5.

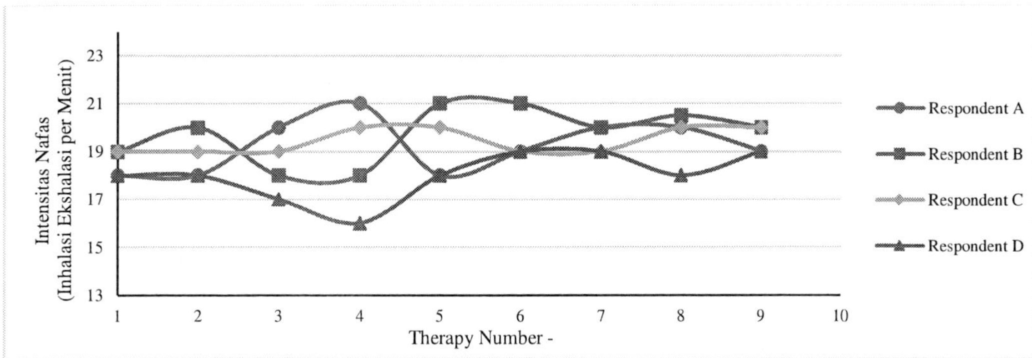

Figure 5. Development of respiration intensity of respondents A, B, C and D

From the graph above, it can be seen that the respiratory intensity of each respondent, namely respondents A, B, C and D is stable during post stroke bicycle therapy. Respondent's respiration intensity is said to be stable if there is no extreme change that exceeds the safe limit. However, for respondents A and B need to get special attention, this is because the respondent's A and B respiration intensity exceeds the allowed upper limit of 20 inhalation exhalations per minute.

3.2.4. Heart Recovery Rate (HRR)

In this sub-chapter will be discussed about the heart recovery rate (HRR) of respondents A, B, C and D during post stroke bicycle therapy. Heart recovery rate is obtained from the results of a reduction in heart rate just after physical activity with a heart rate at 2 minutes after a person has done physical activity. Using the formula of equation (1), the HRR of each respondent can be known, as follows:

$$HRR = HR_{\substack{just\ after \\ exercise}} - HR_{\substack{2\ minute\ after \\ exercise}} \quad (1)$$

After the respondent's heart rate was measured, then the HRR value of each respondent was calculated and presented in graphical form as shown in figure 6 below.

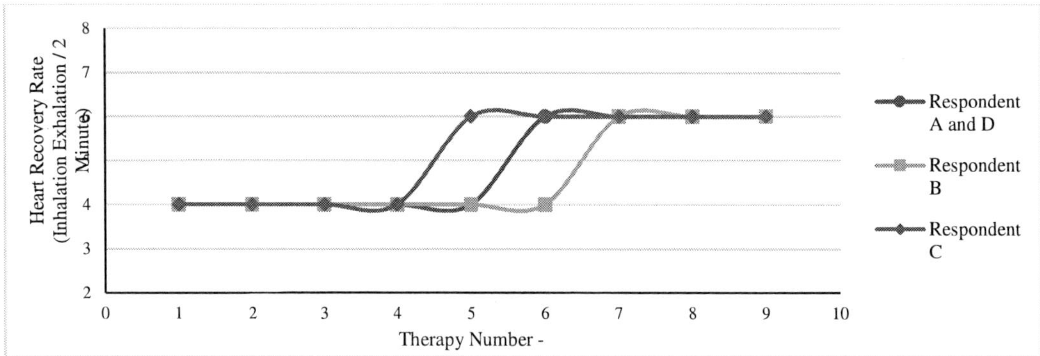

Figure 6. Development of heart recovery rate of respondents A, B, C and D

From the graph above, it can be seen that the heart recovery rate of each respondent, namely respondents A, B, C and D, began to develop better, this was indicated by an increase in heart recovery rate of each respondent. Heart recovery rate from respondents A, B, C and D is far from the normal number of 22-52 beats per 2 minutes. Heart recovery rate values that are still far below these standards indicate that the heart condition of respondents A, B, C and D is in a bad condition. An increase in the heart recovery rate of respondents means that the heart rate of respondents A, B, C and D increases with the burden of the activities carried out, but the heart of respondents A, B, C and D takes longer than normal people to return to heart rate normal (resting pulse). On the heart recovery rate chart shown in Figure 8, it is known that the increase is very extreme, which is 2 beats per 2 minutes. This extreme increase occurs due to the calculation method used, where the pulse rate calculation method used is calculating the pulse at the first 15 seconds then multiplied by 4 to get pulses per minute of the respondent's pulse. From these results obtained the pulse of the respondent for session 1 therapy and session 2 therapy, then searched for the average pulse rate from both sessions. With this measurement method, the accuracy obtained is at least 2 beats per 2 minutes.

3.2.5. VO_2 Maximum

VO_2 Max measurement, aims to get the amount of O_2 that can be processed by the respondent's body during carriying out paddling therapy using a post stroke bicycle. VO_2 Max as long as the patient performs therapy using a post stroke bicycle is obtained from an equation which is a function of the maximum heart rate (HR Max) and the resting heart rate of the respondent (HR Rest) listed in equation (2).

$$VO_2\ Max = 15{,}3\ x\ \frac{HR\ Max}{HR\ Rest} \tag{2}$$

Where :
$HRMax$ = Heart rate just after therapy
$HRRest$ = Heart rate at 2 minutes after therapy (recovery heart rate).

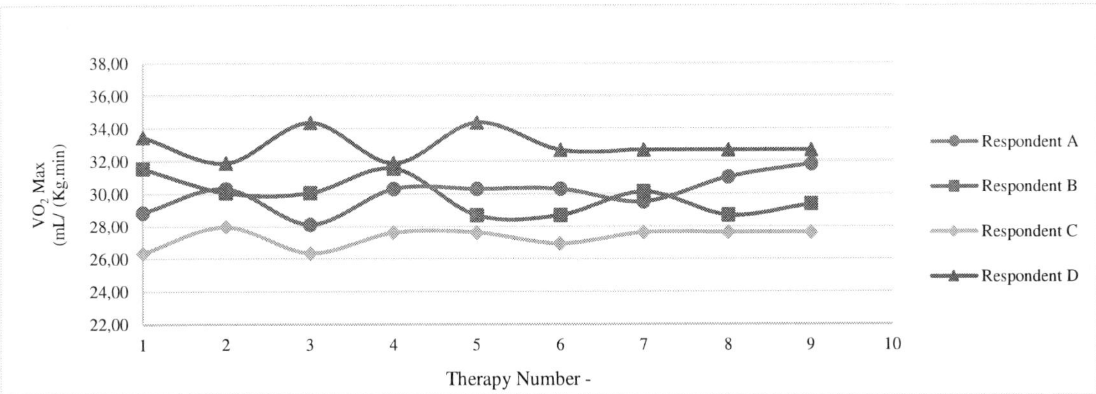

Figure 7. Development of VO2 Maximum capacity of respondents A, B, C and D

From the graph above, it can be seen that the VO2 Max capacity of each respondent is respondents A, B, C and D. For respondents A and C experienced an increase in VO2 Max, while respondents B and D experienced a decrease in VO2 Max capacity. Respondent A experienced an increase of 10.28% of VO2 Max capacity compared when the respondent A performed the first time post stroke bicycle therapy. Respondent B experienced a decrease of 6.93% from VO2 Max capacity at the first time post stroke bicycle therapy. The decrease in VO2 Max capacity from respondent B was influenced by the heart rate of respondents who tended to be high in the last few therapies without following the increase in HRR of respondent B. Respondent C experienced an increase of 4.88% of VO2 Max capacity compared when the respondent C performed the first time post stroke bicycle therapy. Respondent D experienced a decrease of 2.38% of VO2 Max capacity compared when the respondent D performed the first time post stroke bicycle therapy. The cause of the decreasing VO2 Max capacity from respondent D was the heart rate of respondent D did not experience a big change as long as respondent D underwent post stroke bicycle. This is because the height of the bicycle is not in accordance with the respondent D body posture which causes respondents feels uncomfortable and afraid to fall when use post stroke bicycle which causes the respondents' heart rate to increase.

3.2.6. Energy Consumption

Paddle energy is the amount of energy used by respondents to carry out paddling therapy activities using post stroke bicycle. The value of the energy released by the respondent was obtained by using the quadratic regression equation of the respondent's heart rate listed in equation 3. From the quadratic regression equation can be known the amount of energy released by the respondent during therapy and when the respondent rested. The results of the calculation of the respondent's energy calculation can be seen in Figure 8

$$KE = Et - Ei \tag{3}$$

Where :
KE = Energy consumption for certain activities (Kcal / minute)
Et = Expenditures of energy at a certain work time (Kcal / minute)
Ei = Energy expenditure at rest (Kcal / minute)

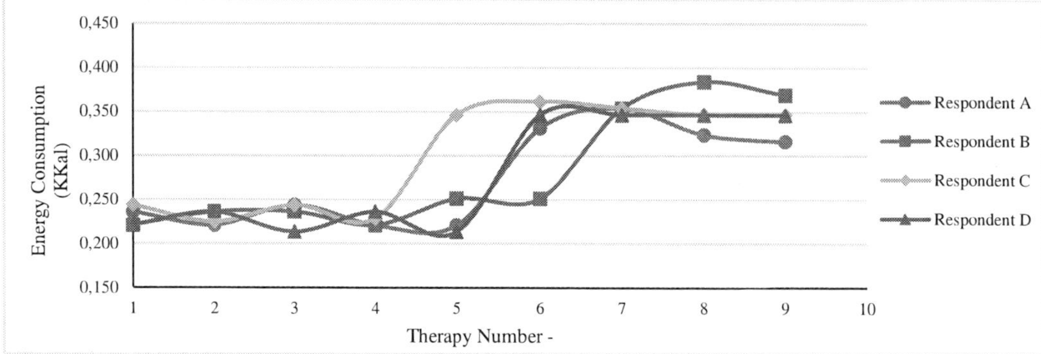

Figure 8. Development of energy consumption of respondents A, B, C and D

From the graph of energy consumption of each respondent in Figure 8, it can be seen that the energy consumption of each respondent A, B, C and D are as follows: Energy consumption of respondent A experienced an increase of 34.04% compared when the respondent A performed the first time post stroke bicycle therapy. Energy consumption of respondent B experienced an increase of 67.05% compared when the respondent B performed the first time post stroke bicycle therapy. Energy consumption of respondent C experienced an increase of 42.27% compared when the respondent C performed the first time post stroke bicycle therapy. Energy consumption of respondent D experienced an increase of 56.82% compared when the respondent D performed the first time post stroke bicycle therapy. On the graph of energy consumption of respondents A, B, C and D there is a drastic increase. Keep in mind the amount of energy consumption needed to perform activities is a function of heart rate. So that the rhythm of the heart rate of the respondent is very influential on the graph of energy consumption.

4. Usefulness Analysis of Post Stroke Bicycle

To determine the usefulness of post stroke bicycle in helping to recover the health of stroke patients. Then the VO_2 Max analysis from the two groups that became respondents in this research was carried out. The measurement of VO_2 Max in both groups used the 6 minute walking test method. So that the measurement results of VO_2 Max from both groups can be compared both descriptively. This method is used because, being able to interpret respondent VO_2 Max well, besides this method also does not cause fatigue effect on the respondent and can be done independently and does not require a lot of money. From the results of testing the 6 minute walking test on both groups of respondents, the data obtained increased VO2Max of each group as in table 2 as follows.

From the table it can be seen that, there was a difference in the increase in VO_2 Max capacity in both therapy groups. Therapy groups that use post stroke bicycle have an average increase in VO_2 Max capacity of 1,892 mL / (Kg.min). While the group of therapy without using post stroke bicycle has an average increase in VO_2 Max capacity of 1,282 mL / (Kg.min). This means that with the help of post stroke bicycle for 30 days, the VO_2 Max capacity of the post stroke bicycle therapy group is 47.58% greater than the therapy group without using post stroke bicycle.

Table 2. Enhancing Body Fitness for Each Group

Respondents		VO2 Max Capacity (Using 6 MWT Method) (mL/(Kg.minute))			
		1st Evaluation	2nd Evaluation	Δ VO_2 Max	Mean of oxygen volumen (Δ VO_2 Max) each group
Paddle therapy group	A	28,95	31,16	2,216	1,892
	B	31,68	33,32	1,646	
	C	26,93	28,69	1,761	
	D	32,90	34,84	1,945	
Non – Paddle therapy group	E	29,71	31,10	1,393	1,282
	F	31,45	32,57	1,122	
	G	33,25	34,51	1,260	
	H	27,98	29,33	1,352	

5. Conclusion

From the usefulness analysis that has been done, it was found that the body fitness of the respondent of paddling therapy using post stroke bicycle experienced an increase of 47.58% more than the respondents of therapy without using post stroke bicycle. So it can be concluded that the tricycle is very helpful for the patient to perform physical therapy, and in a certain time can recover patient health, at least in 1 month the patients more be able to move their legs.

References

[1] Anonim 2013 Treadmill Test (Exercise Stress Test), available at http://kardioipdrscm.com/en/portfolio/treadmill-test-exercise-stress-test/#sthash.ulFYJ0vx.dpbs Accessed at 04 August 2018.
[2] Rodika 2016 *Rancang Bangun Sepeda Untuk Pasien Pasca Stroke* (Surabaya : Thesis Magister Mechanical Engineering Department, Institut Teknologi Sepuluh Nopember)
[3] Chandra and Rahardian 2017 *Perancangan Rangka Sepeda Pasca Stroke Dengan Konsep Delta (1 Roda Depan dan 2 Roda Belakang) (*Surabaya : Final Project, Institut Teknologi Sepuluh Nopember)
[4] Salim and Lutfi Agus 2017 *Rancang Bangun Sepeda Pasca Stroke Konsep Delta* (Surabaya : Final Project, Institut Teknologi Sepuluh Nopember)
[5] Potter & Perry 2005 *Buku Ajar Fundamental Keperawatan Konsep, proses, dan Praktik." 4th Edition* Volume 1. (Jakarta : EGC)
[6] Luthfianto and Heri 2016 *Pengujian Sepeda Pasca Stroke Untuk Pasien Stroke di Rumsh Sakit Haji Surabaya* (Surabaya : Final Project, Institut Teknologi Sepuluh Nopember, Surabaya)
[7] Björn Ekblom *The Bicycle Account 2014* (Stockholm, Sweden : Karolinska Institute)
[8] JF Burr et al 2011 *The 6-minute walk test as a predictor of objectively measured aerobic fitness in healthy working-aged adults*" available at https://www.ncbi.nlm.nih.gov/pubmed/21673494 Accessed at 05 Agustus 2018.

IOP Conf. Series: Materials Science and Engineering **588** (2019) 012035 doi:10.1088/1757-899X/588/1/012035

The effect of rice husk ash addition as pozzolan on geopolymer binder using alkali activators

R Bayuaji , D P Dibiantara, R Ilmiah , N A Husin , MS Darmawan , R Buyung and S Subekti

[1] Department of Civil Infrastructure Engineering, Institut Teknologi Sepuluh Nopember, Sukolilo,Surabaya, Indonesia, 60111

bayuaji@ce.its.ac.id

Abstract. Geopolymer concrete is an example of innovative concrete that does not use cement as a binder and generally fly ash used in recent research. However, fly ash is categorized as a hazardous and toxic waste [8]. This research studied about another binder, besides fly ash, in this case is rice husk ash made as a geopolymer paste alkali activated by a chemical mixture of sodium silicate (Na2SiO3) and sodium hydroxide (NaOH) confronted to a fly ash mixture paste as comparative material. The compositions made were, 100% rice husk ash, a mixture of 50% rice husk ash and 50% fly ash, and 100% fly ash as control mixture. After making the mixture, the composition will be tested to setting time-test in the form of paste. The size of specimen was cylindrical 2,5 cm in diameter 5 cm in height and cube size 15 cm x 15 cm x 5 cm. Each type of specimen was projected to porosity test, compressive strength test, UPV (Ultrasonic Pulse Velocity) and permeability test on 3 days, 28 days and 56 days of sample's age. From the result of setting time, it was found that the composition of 100% rice husk ash had the longest on binding, which were 129 minutes for early binding and 170 minutes for late binding. The lowest result of porosity test was 100% of fly ash at 56 days of sample's age testing with a porosity of 20%. The lowest result of compressive strength was 100% of rice husk ash tested in 3 days of sample's age showed a compressive strength at 0.65 MPa. Based on the test results of UPV (Ultrasonic Pulse Velocity), the lowest value of wave velocity was 531.667 m/s, belong to 100% rice husk ash mixture at 3 days of sample's age testing, with very poor binder quality qualification. The test result of highest permeability kT value with very poor-quality index binder was 100% rice husk ash with period of testing time in 3 days 0,047.10-16 m^2 .

1. Introduction

Concrete as one of building structures' materials already widely known because almost all existing buildings now use concrete as basic materials. The strength of concrete is very dependent to the composition and strength of each of its constituent materials, one of those is cement. From the earth conference, which was held in Bali on December 2007 themed about climate change, it was stated that the total number of cement production in the world was recorded 1.5 billion tons. In this case, cement industries release 1.5 billion tons carbon dioxide into nature. According to The International Energy Authority: World Energy Outlook, the amount of carbon dioxide produced in 1995 was 23.8 billion tons. Another fact, based on these data, Portland cement production accounts for seven percent of the carbon dioxide produced from a variety of sources that may lead to global warming. There are several researches concern in the concept of cement replacement in concrete production, in this particular case is rice husk ash mixed with fly ash. As it is well known that fly ash has high silica content and can be used as a binder (pozzolan) for the partially replacement of cement in concrete production. In addition, since fly ash is classified as a waste, which is hazardous and toxic materials, it should be deprecated [8]. This problem becomes a factor to reduce the usage of fly ash as a geopolymer binder, substituting materials in example rice husk ash and alkali activator. Rice husk is burned at high temperatures (400-800˚C) to produce rice husk ash containing high silica (SiO_2), or known as pozzolan in rice husk ash, is 94-96% and pass a sieve no. 200 so that the rice husk ash can fill in the empty spaces between grains in order to increase the adhesion between the particle. Silica content in 10% rice husk ash usage, in the mixture, improves compressive strength. Alkaline activator used was sodium silicate (Na_2SiO_3) which serves acceleration of the polymerization reaction and alkali solution used was sodium hydroxide (NaOH), which serves to assist the process of bonding between the particles [6]. In this study, the researcher used a binder mixture of rice husk ash and fly ash with the intention of determining the effect of rice husk ash addition in binder geopolymer using comparative other binder, namely fly ash, with alkali activator sodium silicate (Na_2SiO_3) and sodium hydroxide (NaOH). The samples were then examined in setting time, porosity, UPV, permeability, and compressive strength tests during the binding process up to its paste form.

2. Composing Materials

Fly ash is one of by-product material in coal industry, which is used as a binder in geopolymer concrete. Fly ash is determined as a pozzolanic material for it is consisted of alumina-silica compounds. These compounds potentially react to calcium hydroxide of conventional cement and water and then forms a strength compound into the hydrating cement. It is indicating the ability of fly ash to chemically reacts and performs, also having properties similar to pozzolanic material, presented by nature, i.e. the volcanoes ashes or other sedimentary materials [1].

Rice husk is the result of a rice mill industry, burned at 400-800˚C temperature [7]. The highest content of silica (SiO_2), which is obtained by sun drying for 1 hour, amounted up to 88.97%, compared to the oven drying (190˚C) for 1 hour was up to 83.15%. The most common amount of the silica content of rice husk ash is up to 90-96%. Silica, contained in rice husk ash, has hydrated amorphous structure [5].

Fly ash and rice husk ash are material that would be made as a geopolymer binder. Before making a geopolymer concrete, the researcher tested chemical compound of the material. The tests conducted were XRF (X-Ray Fluorescence), XRD (X-Ray Diffraction), and SEM-EDX (Scanning Electron Microscope). The results of these tests can be seen in Table.1, Table.2, Figure.1, Figure.2, Figure.3, and Figure.4 as following:

Table 1. The Test result of XRF Fly Ash of Paiton Power Plant

Compound	Content (%)
SiO_2	47,10
Al_2O_3	24,25
Fe_2O_3	16,07
CaO	5,830
SO_3	0,206
MgO	2,620
Na_2O	0,645
K_2O	1,640

Table 2. The test result of XRF Rice Husk Ash

Compound	Content (%)
SiO_2	96,40
K_2O	2,720
CaO	0,639
MnO	0,073
Fe_2O	0,093
CuO	0,028
ZnO	0,042
Re_2O_7	0,021

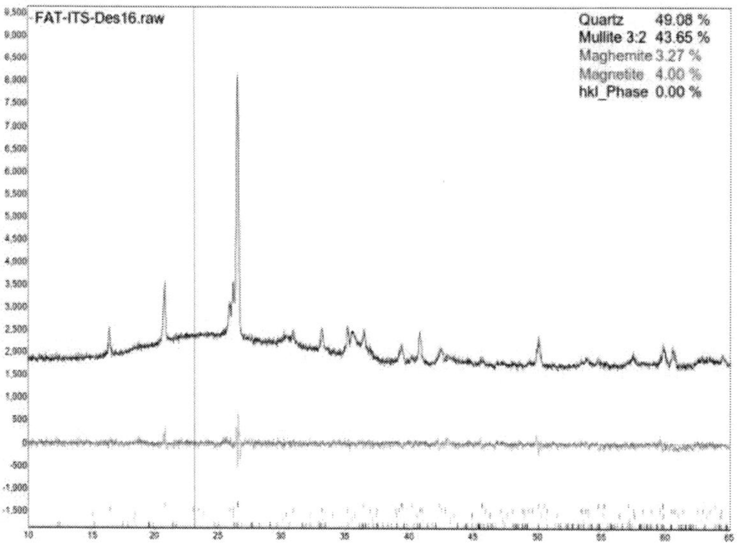

Figure 1. The Test Result of XRD Fly Ash Paiton

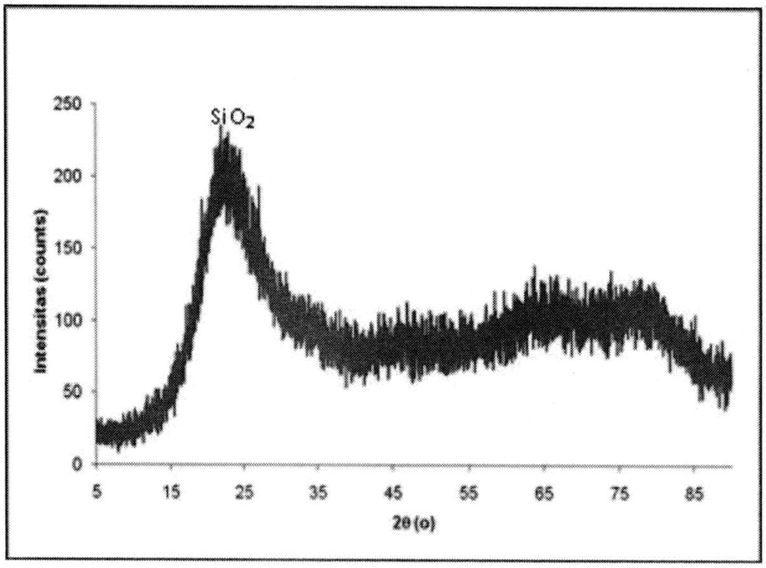

Figure 2. The Test Result of XRD Rice Husk Ash

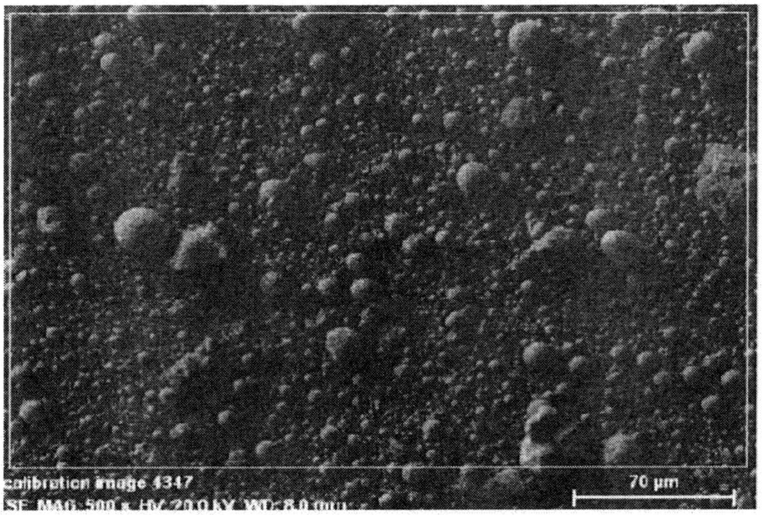

Figure 3. The Test Result of SEM-EDX Fly Ash Paiton

Figure 4. The Test Result of SEM-EDX Rice Husk Ash

Referring to the compound test, XRF (X-Ray Fluorescence), fly ash had a CaO content of less than 10% is 5.83%, in this research, the researcher used fly ash type F [1]. From the compound test XRD (X-Ray Diffraction) of fly ash, could be seen there was SiO_2 content of 49.08% and the test results of SEM-EDX (Scanning Electron Microscope) shown the form of spherical particles of fly ash that improves workability, so it could reduce the use of water, simplify the binding inter-particles and minimize the space between the mixed material [4]. From the test results of the compound, XRF (XRay Fluorescence) of rice husk ash, the SiO2 content of rice husk ash was 96.40%, the addition of rice husk ash proportion circa 10% improves the compressive strength. This was occurred as rice husk ash is a pozzolanic material. Based on XRD, crystal content of rice husk ash is shown as the highest peak on the XRD diagram, specifically at 2D°. Crystal silica could improve the compressive strength as rice husk ash is a pozzolanic material [3] and from the results of SEM-EDX (Scanning Electron Microscope) test, the form of rice husk ash particle, which resembles as a hollow sponge, requires a high-water demand. This shape of particle caused the involvement of rice husk ash affects to a very slow process in specimen hardening [2].

IOP Conf. Series: Materials Science and Engineering **588** (2019) 012035 doi:10.1088/1757-899X/588/1/012035

The combination of Sodium Silicate (Na_2SiO_3) and Sodium Hydroxide (NaOH) liquid were aimed to help prevent chemical reactions among aluminum and silica, contained in fly ash and rice husk ash. Sodium hydroxide is an alkaline compound that is very reactive when mixed with water or distilled water [9].

3. Methodology

The methodology used in this research was an experimental study with direct examination in the laboratory. This research intentioned to know the effect of rice husk ash to replace fly ash in geopolymer binder with alkali activator mixture of Sodium Silicate (Na_2SiO_3) and Sodium Hydroxide (NaOH).

In the production of this geopolymer binder, fly ash was taken from PT. PJB UP Unit 1-2 Paiton (Power Plant), Probolinggo, East Java. Rice husk was taken from the rice milling in Situbondo, East Java. Rice husk was burned in the charcoal factory in Balongbendo, Krian, Sidoarjo, East Java

Geopolymer binder compositions made were 100% rice husk ash, a mixture of 50% rice husk ash and 50% fly ash, and 100% fly ash, blended with the alkali activator molarity of 12 M, weight ratio of Na_2SiO_3 to NaOH statically maintained at 1,5. Every pastes of composition mixtures were examined into setting time. The specimen made into cylindrical shape of 2,5 cm in diameter and 5 cm in height were attempted into porosity test and compressive strength test. Meanwhile, the specimen of cubical shape of 15 cm x 15 cm x 5 cm were directed into UPV test (Ultrasonic Pulse Velocity) and permeability test, at the age of 3 days, 28 days and 56 days of samples.

The production of specimens:

1. Material preparation, consisting of rice husk, fly ash, NaOH solution and Na2SiO3
2. Rice husk drying under the sunlight for 3 hours, repeat until dry
3. Burning the rice husk at ± 700°C temperature up to 4 hours
4. Sift the rice husk ash using a sieve No. 200
5. Test the chemical properties of fly ash and rice husk ash
6. The mixture of Na2SiO3 and NaOH is made to approach 1.5 in weight ratio.
7. The composition of geopolymer binder mixture in weight ratio of fly ash and rice husk ash (materials) to alkali activator can be seen in Table.3 following:

Table 3. Comparison of Weight Mixed

Fly ash : Husk ash (%)	Material : Alkali (%)
100 : 0	45 : 55
50 : 50	74 : 26
0 : 100	65 : 35

8. Vicat test (initial setting time and final setting time) 12M ratio of 1.5. Following Table.4. lists the results:

Table 4. The Result of Setting Time

Binder Name	Binder Code	*Setting Time* (minute)	
		Initial	*Final*
100% of Rice husk ash	ASP12-1,5	129	170
100% of fly ash	FA12-1,5	18	45
Mix 50% of fly ash dan 50% of rice husk ash	ASPFA12-1,5	73	110

9. Amount of each samples' mixtures attempted into tests :

Table 5. The Variation of Object Test

Test	Ratio of Fly ash to Husk ash			Total
	100:0	50:50	100:0	
Compressive Strength	9	9	9	27
Porosity	9	9	9	27
UPV dan Permeability	9	9	9	27

10. Cylindrical sample of 2,5cm in diameter and 5cm in height, examined at the age of 3 days, 28 days and 56 days of samples. The results presented in Table.6 :

Table 6. The Result of Cylinders Testing

Binder Name	Binder Code	Binder's Age/Time	compressive strength (MPa)	Porosity (%)
100% Rice husk ash	ASP 12-1,5	3 days	0,65	FAILED
		28 days	1,62	
		56 days	2,58	
100% fly ash	FA 12-1,5	3 days	19,12	26,2
		28 days	34,90	21,6
		56 days	43,32	20,0
mix 50% fly ash and 50% Rice husk ash	ASPFA 12-1,5	3 days	1,53	58,0
		28 days	3,33	55,2
		56 days	7,65	45,1
75% fly ash and 25% Rice husk ash	ASPFA [B]	3 days	7,95	-
25% fly ash and 75% Rice husk ash	ASPFA [A]	3 days	0,95	-

11. Cubical sample of 15 cm x 15 cm x 5 cm test at the age of 3 days, 28 days and 56 days of samples. Obtained the results as Table.7. following:

Table 7. The Results of Cube Testing

Binder Name	Binder Code	Binder 's Age/ time	UPV (m/s)	*Permeability* (10^{-16} m^2)
100% rice husk ash	ASP 12-1,5	3 days	531,7	FAILED
		28 days	810,0	
		56 days	1635	
100% *fly ash*	FA 12-1,5	3 days	1711,7	0,020
		28 days	2187,5	0,004
		56 days	2777,5	0,002
Mix of 50% *fly ash* and 50% rice husk ash	ASPFA 12-1,5	3 days	679,17	0,047
		28 days	1074,2	0,016
		56 days	1140,0	0,010

12. Analyzing the result of testing
13. Making conclusion based on the analysis

4. Data Analysis

Hereby the results of examination undergone into the specimens, specifically setting time, porosity, compressive strength, permeability, and UPV tests of geopolymer binders' specimens (12M with 1,5 ratio of alkali activator) presented in Figure 5 and Figure 10. From the graphic, the longest time setting (initial setting time and final setting time) had by the mixture of 100% rice husk ash composition, the initial setting time is 129 minutes and the final setting time is 170 minutes. The fastest setting time (initial setting time and final setting time) had by 100% fly ash composition that has initial time setting time 18 minutes and the final setting time is 45 minutes. Due to the shape of rice husk ash particles, that resembles as a hollow sponge captured on Figure.4, the mixture requires a high-water demand and might need an addition of superplasticizer, a chemical functioning to reduce the use of excess water, so it's taking a longer time to set, compared to fly ash' control mixture.

Otherwise, the fly ash has a spherical particle shape, shown in Fig.3, which increases the workability and reduces the water usage, in other word, the mixture is easier to bind to each other and minimize the space between materials mixed. Highest compressive strength was tested at the age of 56-day of 100% fly ash mixture, resulted 43.32 MPa. Highest compressive strength tested at the age of 3-day held into the composition of 75% fly ash : 25% rice husk ash (ASPFA [B]) 12M-1,5 mixture and 25% fly ash : 75% rice husk ash (ASPFA [A]) 12M-1.5 is 100% fly ash mixture, which is 21.12 MPa. Moreover, the results of the lowest compressive strength tested on 3-day sample's age of 100% rice husk ash composition at 0.65 MPa. So, it could be concluded, when the matrices of materials perfectly bind each other, the compressive strength tend to be higher. The highest porosity was found on 3-day of sample's age of 50% rice husk ash and 50% fly ash composition by 58% in value, but the lowest porosity test resulted on 56 days of sample's age of 100% fly ash composition by 20% in value. The failure of permeability examination was occurred on 100% rice husk ash composition, at all age testing of 56 days, 28 days and 3 days. It was caused by the form of hollow particles that degrade the adhesion of particles. The highest UPV test results held on 56 days examination of 100% fly ash composition in the amount of 2777.5 m/s (shoddy binder quality) and for the lowest UPV test results was 3 days of 100% rice husk ash composition amounted to 531.6 m/s (binder quality was very poor). The highest permeability testing held on 3-days sample's age of 100% rice husk ash and fly ash by 100% 0,047.10-16 m² including normal binder quality and for the lowest permeability test result was the testing of 56-day 100% fly ash composition by 0,00233.10-16 m² (binders' quality is very good). There was a failure in testing permeability to 100% rice husk ash in all age testing at 56 days, 28 days, and 3 days. It was caused of the form of hollow particles results in lower adhesion between the particles.

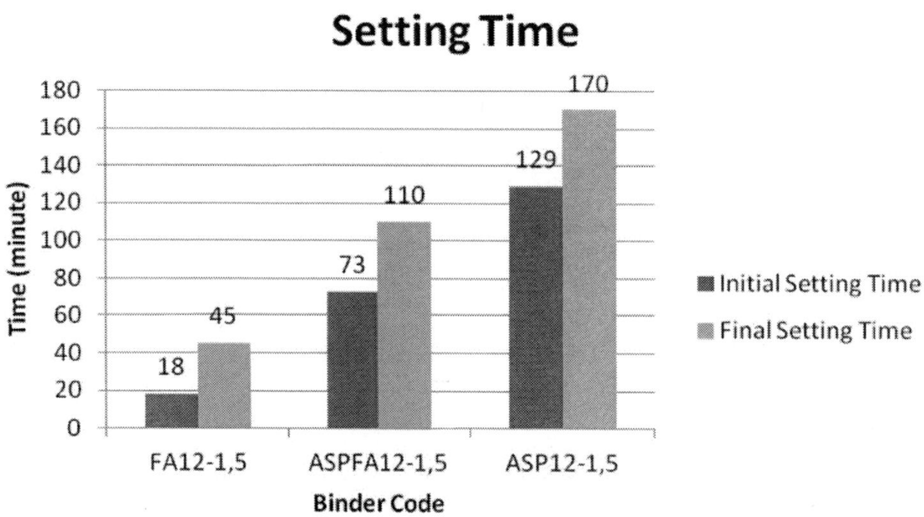

Figure 5. The Test Result of Setting Time (initial and final time)

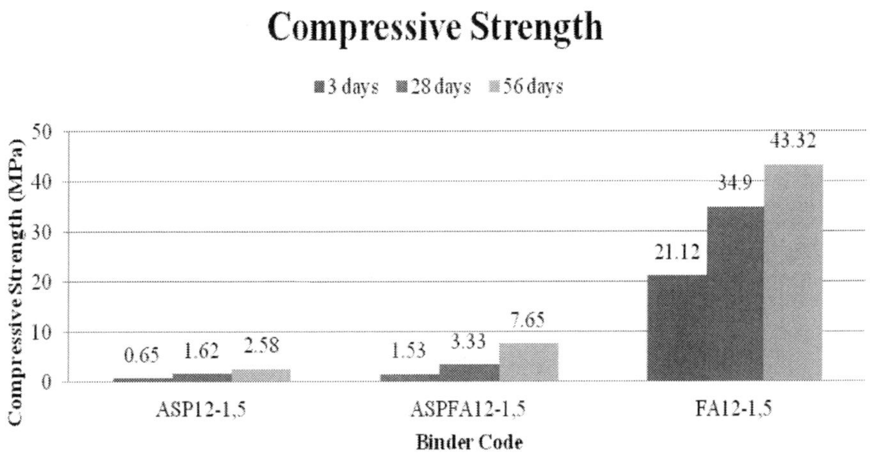

Figure 6. Compressive Strength Test Results 3 Days, 28 days, and 56 days

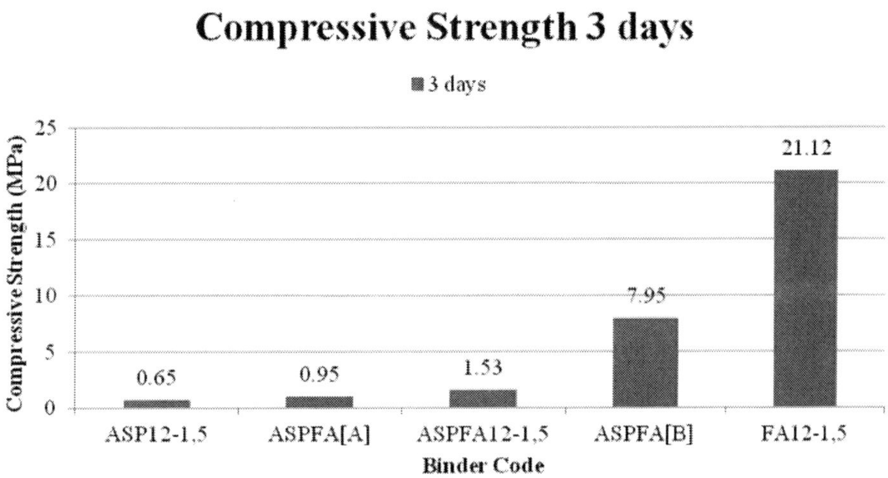

Figure 7. Compressive Strength Test Results 3 Days

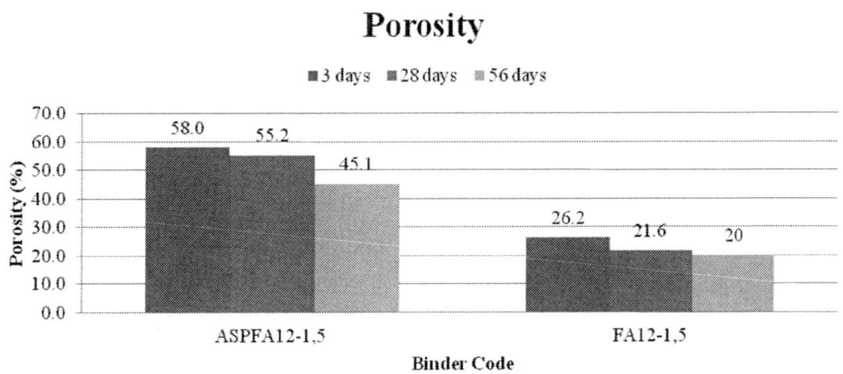

Figure 8. Porosity Test Results

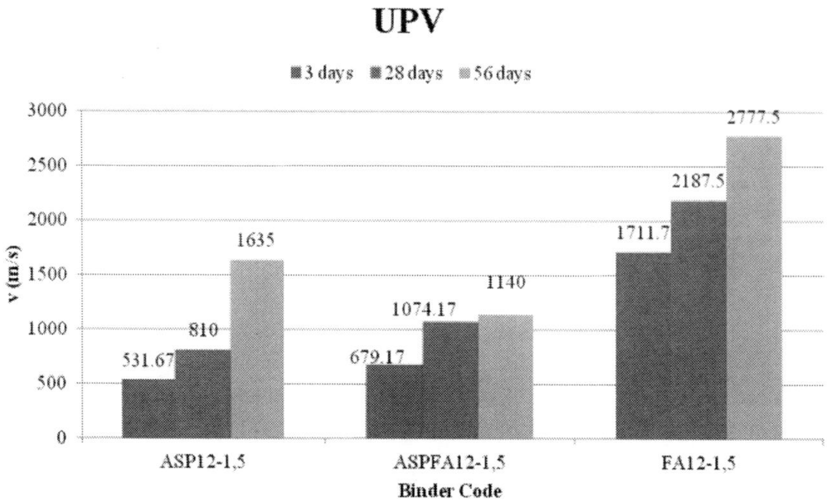

Figure 9. UPV Test Results (Ultrasonic Pulse Velocity)

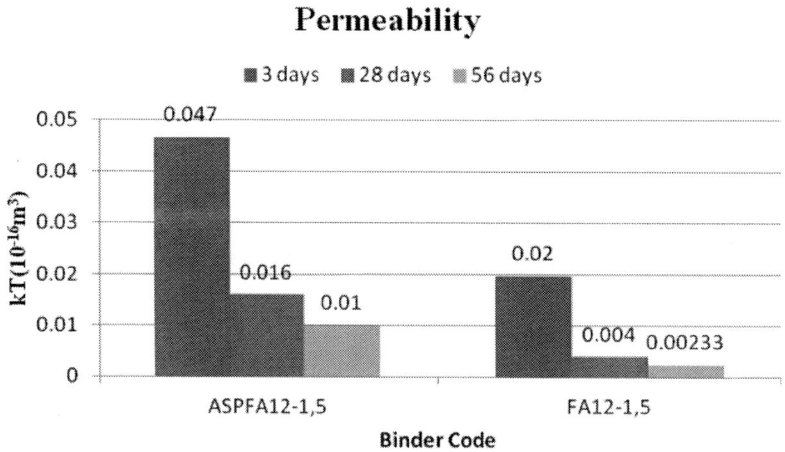

Figure 10. Permeability Test Results

5. Conclusion

The best mixture is presented in the composition of 100% fly ash with the compressive strength up to 43,32 MPa. The inclusion of ash to the geopolymer concrete in big quantities could provide high compressive strength. The use of rice husk ash in binder decreases the compressive strength and its sponge-like particles tend to absorb more water into the mixture and low the reactivity. The advantage of rice husk ash usage is seen from its slowing setting time-characteristic. As it is well known, the problem of Indonesian material geopolymer binder is its rapid setting time, which interferes the usage of geopolymer binder for industrial usage.

References

[1] ACI 226.3R-3, ACI Manual of Concrete Practice Dilaporkan oleh ACI Committee 226. American Concrete Institute, Farmington Hills, Michigan

[2] Ade Ilham 2005 Pengaruh sifat fisik dan kimia bahan pozzolan pada beton kinerja tinggi **Vol. 13** No. 3, Edisi xxxiii Oktober 2005

[3] Anam M Samsul and Trianto Wawan 2013 Pengaruh Penggunaan Bone Ash dan Rice Husk Ash Terhadap Sifat Mekanis Pasta Semen. Konferensi Nasional Teknik Sipil Universitas Sebelas Maret (Surabaya : Institut Teknologi Sepuluh Nopember)

[4] D Hardjito , Steenie E Wallah, Dody M J Sumajouw and B V Rangan 2004 Factors Influencing the Compressive Strength of Fly Ash-Based Geopolymer Concrete *Jurnal Dimensi Teknik Sipil* **6** 2:88-93

[5] H Harsono 2002 Pembuatan Silika Amorf dari Limbah Sekam Padi *Jurnal Ilmu dasar FMIPA Universitas Jember Jawa Timur* **3**(2) : 98-102

[6] D F Houston 1972 Rice Chemistry And Technology , American Association Of Cereal Chemist, Inc. Minnesota

[7] Diah Kusumantara 2009 *Pengaruh Faktor Air Semen terhadap Campuran 50% Semen Portland dan 50% Abu Sekam Padi* (Jakarta : Universitas Indonesia)

[8] PP. no. 101. 2014 Pengelolaan bahan berbahaya dan beracun. Peraturan Pemerintahan Republik Indonesia : Jakarta.

[9] P Septia 2011 *Studi Literatur Pengaruh Konsentrasi NaOH dan Rasio NaOH:Na_2SiO_3 Rasio Air/Prekursor, Suhu Curing, dan Jenis Prekursor terhadap Kuat Tekan Beton Geopolimer* (Jakarta : Universitas Indonesia)

IOP Conf. Series: Materials Science and Engineering **588** (2019) 012036 doi:10.1088/1757-899X/588/1/012036

The Chemical and Mineral Composition Effect on Rock Mass Strength in Kaliwadas Limestone Hills, Karangsambung, Kebumen, Central Java

W Dahani[1,a], I Marwanza[1], A Susilo[1], F A Pramesti[2], and R Kurniawati[2]

[1]Department of Mining Engineering, Trisakti University, Jakarta, Indonesia.
[2]Department of Chemistry, Institut Teknologi Sepuluh Nopember, Surabaya, Indonesia

[a]Corresponding author : wiwik_d@trisakti.ac.id

Abstract. The Kaliwadas Hills, Karangsambung, Kebumen, Central Java are composed of limestone lithology, which has been deformed in the form of cracks. The chemical and mineral rock composition is related to the adsorbed water content and rock strength. This paper will analyze the correlation between the chemical composition of rocks and adsorbed water content with rock strength of Kaliwadas Hills. Rock sample was soak in an acid solution that represent rainwater. It was known that if the crack exposed to rainwater (acid water) continually, it will change the chemical composition of rocks and water content adsorbed. It is resulting the decrease of rock mass strength that can lead to aftershocks landslide.
Keyword : Kaliwadas Hill, limestone, rock strength

1. Introduction

Rocks have various types of minerals, for example calcite, quartz, kaolin, chromite [1] and other minerals that can be used as natural resources for the synthesis of inorganic materials [2]. Synthesis of inorganic materials can be done by various methods, for example, sol-gel, solid state, and hydrothermal method [3]. Synthesized inorganic materials can be used as adsorbent [3] or photocatalyst [4] depends on their properties. For example, zeolite that has a high active surface area can be used as supporting material on photocatalyst TiO_2 [5] and calcite which usually found in limestone rock is used as building material because it has cementation effect [6].

Kaliwadas Hill is a type of limestone hill located in north of Karangsambung Village, Kebumen, Central Java. This limestone hill is located in a densely populated area. In 2016 this hill had already suffered a landslide. Hundreds of people were rescued. In 2017, some researchers found a fracture in this hill that is getting wider and deeper over the time [7]. This fracture if left unchecked is feared to caused another landslide. To find out the possibility of this happening, it is necessary to analize the rock strength of the hill.

This study discusses the effect of rainwater on the strength of Kaliwadas Limestone Hill. Analysis of the hill's chemical composition was carried out using X-Ray Fluorescence (XRF) and X-Ray Diffractometer (XRD). To determine the impact of rainwater on the strength of the slope, modeling was carried out by soaking rock samples from the slope with acid water representing rainwater. The strength of the hill was measured with Point Load Index.

IOP Conf. Series: Materials Science and Engineering **588** (2019) 012036 doi:10.1088/1757-899X/588/1/012036

2. Methods

Determination of rock's chemical content effect on the rock strength can be done by a simple method. Rock sampling was done in 4 different locations in Kaliwadas Limestone Hill, Karangsambung, Kebumen, Central Java that represented by point I, A, B, and C. Rock sample was soak in an acid solution (pH 6) in a various amount of time. In this research acid solution is representing rain water, because rain water is containing a small amount of acid. The decrement of chemical content was characterized using XRF and XRD. The correlation between rock's porosity and chemical content with rock strength was determined using point load index.

3. Result and Discussion

3.1. X-Ray Diffraction (XRD) Analysis

X-Ray Diffraction (XRD) analysis was done to determine crystal phase of the rock sample. Sample from 4 different sampling location was characterized using XRD. To make the analysis process easier, 3 greatest peak were chosen. After being match with XRD data from ICDD (International Centre for Diffraction Data) it was known that the rock sample is consist of 3 minerals, Calcium Carbonate (calcite) with scale factor 0.874, Alumunium Calcium Silicon Oxide ($CaAl_2Si_2O_8$) with scale factor 0.063 and Calcium Iron Oxide ($Ca_4Fe_9O_{17}$) with scale factor 0.063. Because the scale factor of calcite is way higher than the other minerals, it can concludes that rock sample is mainly consist of calcite, while trace amount of the other minerals is impurities.

Table 1. XRD data obtained from rock samples

Sample	Chemical Formula	Mineral Name	Crystal	Content (%)
I	$CaCO_3$	Calcite	Rhombohedral	100%
A	$CaCO_3$	Calcite	Rhombohedral	100%
B	$CaCO_3$	Calcite	Rhombohedral	100%
C	$CaCO_3$	Calcite	Rhombohedral	100%

3.2. X-Ray Fluorosence (XRF) Analysis

X-Ray Fluorosence (XRF) analysis was done to determine the elemental composition of rock sample. XRF analyzer can not identify the sources of element detected, it can detect the element as it's oxide form [8]. Before testing using XRF, rock sample is pre-treated by soaking in an acid water (pH 6) with variation of soaking time. This treatment is carried out to find out the effect of rain water, represented as an acid water in this research, on the decrease of chemical content in rock sample.

XRF analysis result of the rock sample is shown in figure 1. From this image, it is known that the rock sample is mostly content of Ca and trace amount of Si and Fe that detected by the instrument as it's oxide form. After acid treatment, it is known that Ca content is decreased with increasing amount of acid soaking time. This result can be explain by a chemical reaction below :

$$CaCO_3 + 2H^+ \leftrightarrow Ca^{2+} + CO_2 + H_2O$$

From XRD data, it's already known that Ca as the biggest constituent of the rock sample is exist as calcite. Calcite will react with rain water (in this research is represented by acid water) and form a water soluble Calcium [9]. Before XRF test, all water is separated from the solid. This affect the amount of Ca content in the rock after acid soaking treatment.

IOP Conf. Series: Materials Science and Engineering **588** (2019) 012036 doi:10.1088/1757-899X/588/1/012036

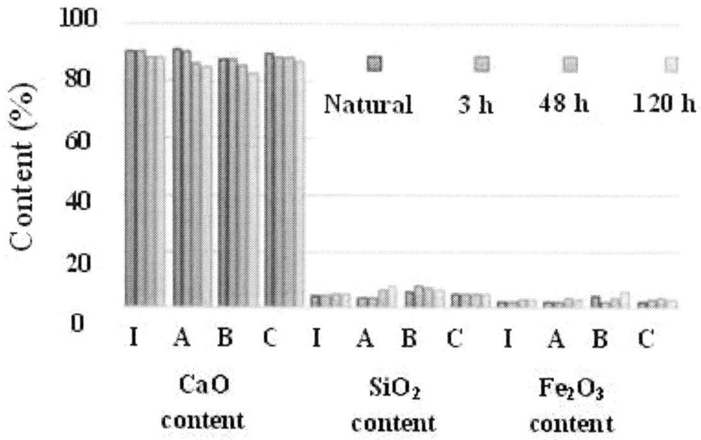

Figure 1. The decrement of rock samples' chemical content obtained by XRF analysis

3.3. Correlation between the decrement of CaO content, water adsorbed content, and porosity of rock sample

Figure 2 shows the correlation between the decrement of CaO content, water adsorbed content, and rock sample porosity. From this data it is known that the decrement of CaO content leads to the higher rock porosity, that makes water adsorbed content is increased.

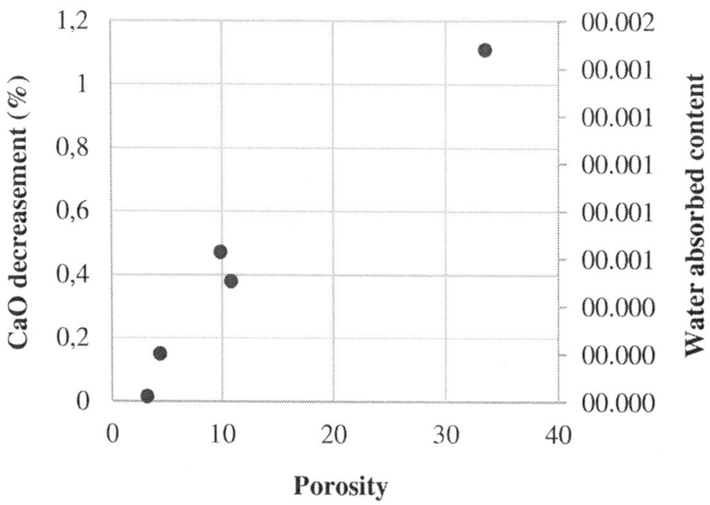

Figure 2. Correlation between CaO decrement, porosity and water abdorbed content

3.4. Correlation between Soaking Time, CaO content and Point Load Index

Figure 3 shows the correlation between acid soaking time, CaO content, and Point Load Index of the rock sample. Acid soaking time is varied to determine it's effect on rock strength. The variation made is 3 hours, 48 hours (2days), and 120 hours (5days). From this data, the longer acid soaking time is done, the higher CaO decreased from the rock sample, the lower CaO content remains on the rock sample, lead to a lower point load index. It means the strength of rock sample is low.

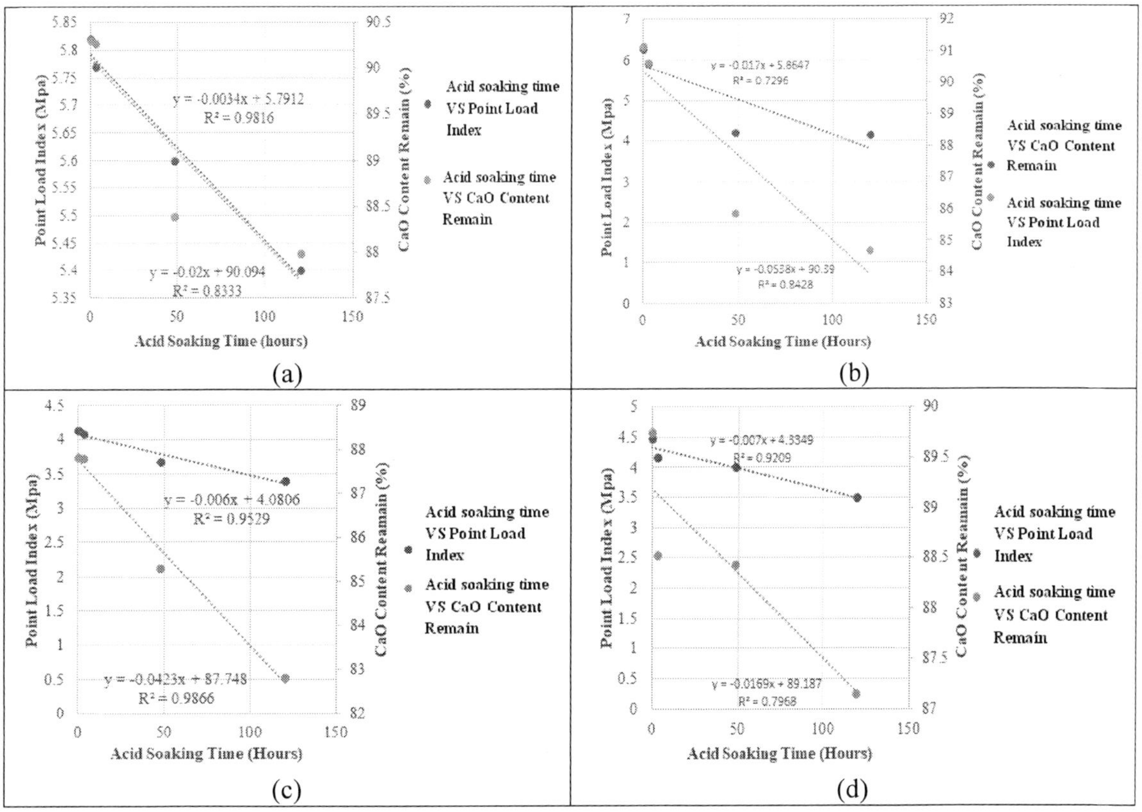

Figure 3. Correlation between acid soaking time, CaO content remain, and point load index in a) I point sampling location b) A point sampling location c) B point sampling location d) C point sampling location

The higher CaO content in the sample caused cementation force in the sample increased, so the more the strength of the rock caused by strong cement properties [6]. The strength of the rock in the sample is also affected by the iron content. Iron (Fe) as $FeCO_3$ is very reactive and stronger than $CaCO_3$. If limestone ($CaCO_3$) is substituted with iron (Fe), the properties will be more stable than the ordinary limestone. But for rock sample from point B, the amount of iron is very high while the strength of the rock is very weak. This happen due to a number of corroded iron, indicated by the presence of brown cracks which are rust of iron (Fe).

4. Conclusion

Kaliwadas Limestone Hills in Karangsambung, Kebumen, Central Java is mainly consist of calcite ($CaCO_3$). Calcite is mineral that has sementation characteristic. Sementation is characteristic that strengthen the rock. By contact with water rain that in this study is represented by acid water, this calcite mineral will form a water soluble chemical that makes the amount of calcite decreased over the time. During calcite decreasement process, bigger pore is formed in the rock. This lead to a higher aborbed water content that makes the rock has a lower load point index. It means that the rock that build the hill has a lower strength, that later can lead to landslide.

References

[1] Nurjaman F, Subandrio S, Ferdian D dan Suharno B 2018 Effect of Basicity on Beneficiated Chromite Sand Smelting Process Using Sub-merged Arc Furnace *AIP Conference Proceedings* **1964**

[2] Hartanto D, Iqbal, R M, Shahbihi, W E, Santoso, E, Fansuri, H dan Iryani, A 2017 Effect of H_2O/SiO_2 Molar Ratio on Direct Synthesis of ZSM-5 from Bangka's Kaolin without Pretreatment. *Malaysian Journal of Fundamental and Applied Sciences* **13** 4 817-820

[3] Iryani, A dan Hartanto, D 2018 Textile Dyes Removal by ZSM-5 from Bangka Kaolin *IOP Conf. Series: Journal of Physics* Conf. Series.

[4] Khusnun N F, Jalil A A, Triwahyono S, Hitam C N C, Hassan N S, Jamian F, Nabgan W, Abdullah T A T, Kamaruddin M J and Hartanto D 2018 Directing The Amount of CNTs in CuO–CNT Catalysts for Enhanced Adsorption-Oriented Visible-Light-Responsive Photodegradation of p –chloroaniline *Powder Technology* **327** 170–178

[5] Zi S C, Chandren S, Yuan L S, Razali R, Ho C S, Hartanto D, Mahlia T M I and Nur H 2016 New Method to Synthesize Mesoporous Titania by Photodegradation of Surfactant Template *Solid State Sciences* **52** 83–91

[6] Dutton S P, White C D, Wilis B J and Nuvakovic D 2002 Calcite Cement Distribution And Its Effect on Fluid Flow in a Deltaic Sandstone, Frontier Formation, Wyoming *AAPG Buletin* **86** 2017-21

[7] Ardhi H A, Azizi M A, Marwanza I, Hartami P N, Nugoro B and Saptono S 2017 Comparison Of Limit-Equilibrium Method Slope Stability Analysis With Finite Element Method Using Probabilistic Aproach *Proceeding seminar nasional geomekanika IV*

[8] Beckhoff B, Langhoff N, Kanngiefer B, Wedell R and Wolff H 2006 *Handbook Of Practical X-Ray Fluorescence Analysis* (Berlin : Springer)

[9] Lund K and Fogler H S 1975 Acidization-II The Disolution of Calcite in Hydrochloric Acid *Chemical Engineering Science* **30** 825-35

The effect of Titanium dioxide precursors variation to morphology of TiO_2/ ZSM-5 composite

A Iryani[1,2], R Kurniawati[1], S Jovita[1], F A Pramesti[1], H Nur[3,a] and M Santoso[1,b]

[1]Department of Chemistry, Institut Teknologi Sepuluh Nopember, Surabaya, Indonesia.
[2]Departement of Chemistry, Pakuan University, Bogor, Indonesia.
[3]Center for Sustainable Nanomaterials, Ibnu Sina Institute for Scientific and Industrial Research, Universiti Teknologi Malaysia, Johor Bahru, Malaysia

[a]hadi@kimia.fs.utm.my, [b]tsv09@chem.its.ac.id

Abstract. TiO_2 photocatalyst is one of the Advanced Oxidation Processes (AOPs) methods which is very potential to be applied to treat wastewater. Modification of TiO_2 with the addition of ZSM-5 as supporting material can increase the adsorption capacity and conductivity of water molecules. Those properties enable TiO_2/ ZSM5 composites to conduct photodegradation process within dye waste. TiO_2/ ZSM5 composites were prepared by sol gel method. The synthesized materials were characterized by X-ray diffraction (XRD), Energy Dispersive X-ray Scanning Electron Microscopy (SEM-EDX), and Brunauer-Emmett-Teller (BET). The effect of 2-Propanol to TiO_2/ZSM-5 Composite Morphology was studied. The results from X-Ray Diffraction analysis revealed that the peak indicating the presence of anatase TiO_2 on sample 2.5;5; and 10% TiO_2/ZSM-5 composite. The increasing effect of Titanium precursors to TiO_2/ ZSM-5 composite carried out agglomerates on external surface of ZSM-5 as seen in SEM which was supported by surface area data from BET.
Keyword : Photocatalyst, Sol- gel, TiO_2, ZSM-5

1. Introduction

At present, the development of Indonesia's textile industries is growing rapidly. The use of dyes in the textile industry causes problems, namely, waste produced is still coloured and difficult to degrade [1]. Dyes waste contains aromatic complex compounds which are difficult to decompose by microorganisms [2]. Due to these problems, it becomes very important to process dyes waste before it discharges into the water bodies. This treatment process is expected to reduce the negative impacts of dye waste.

Several methods have been developed to process and reduce dye waste. Those methods are adsorption, coagulation, and membrane filtration. However, adsorption and coagulation methods produce several secondary contaminants that require further processing, while the membrane filtration method is very expensive [3]. Currently, a new method is being developed to treat dyes waste, namely the AOPs method. This method employs several processes which involve hydrogen peroxide (H_2O_2), UV light, photocatalyst titanium dioxide, and several other processes to produce hydroxyl radical. This method produces far fever residues compared to conventional methods [4]. Due to this property, the AOPs method has attracted the attention of many researchers. The photocatalyst method using a TiO_2 semiconductor illuminated with UV light or visible light is one of the AOPs methods which is very potential to be applied to treat wastewater [5]. It is known that TiO_2 has high chemical stability, cheap, non-toxic and sufficient availability. The presence of those properties makes TiO_2 a very suitable material for the AOPs method. However, it is figured out that TiO_2 still has a deficiency, the band gap of TiO_2 is quite wide (3.2 eV). The wide band gap makes TiO_2 only active in the UV light region, thus limiting its usage and efficiency in its application process [6]. To overcome these shortcomings, such as hexagonal mesoporous silica (HMS) or zeolite are applied as additional porous support materials on TiO_2 [7] [8].

Zeolites as porous materials can increase surface area, hydrophilic or hydrophobic properties and expected to improve adsorption capacity and facilitating the process of substance diffusion [9]. One of materials which can be applied for high specification TiO_2 modification is ZSM-5. ZSM-5 has a pore size (5.3 Å), particle size, and it is easily synthesized according to research which has been conducted by Hartanto [10][11][12].

Modification of TiO_2 with the addition of ZSM-5 as supporting material can increase the adsorption capacity and conductivity of water molecules so that TiO_2/ZSM-5 composites possess high hydrophilic properties [13]. Those properties enable TiO_2/ZSM5 composites to conduct the photodegradation process within dye waste [14] [15].

When TiO_2 with particle size 30 Å is added to ZSM-5 with pore size 5.3 Å, TiO_2 will not be dispersed equivalently over the pores and will only affect the external surface, pore volume and mesopore volume becomes bigger [16][17]. This study focused on morphology and external surface area of TiO_2/ ZSM-5 composite materials after the addition of the various 2-Propanol.

2. Experimental

2.1. Materials

The materials used were commercial ZSM-5 with Si/Al = 50 from Acros Organic, titanium(IV) isopropoxide (TiIP) (97%, Aldrich), titanium dioxide (TiO_2) (98%, Acros Organic), aluminum oxide (Sigma-Aldrich), and 2-propanol (97%, Merck).

2.2. Methods

Synthesis TiO_2/ZSM-5 was carried out using sol-gel method. TiO_2 precursors, which is Titanium isopropoxide (TiIP) and 2-propanol was mixed under constant stirring for 5 minutes. Then ZSM-5 powder was added into the resulting solution. The amount of ZSM-5 was fixed while the amount of TiO_2 precursors was varied to analyze it's effect to composite morphology. Variation of TiO_2 precursors used is 2.5; 5; and 10 percent weight of ZSM-5.

Then the mixture of TiO_2 precursors and ZSM-5 was stirred for another minutes until a gel formed. The gel then dried for 12 hours in oven to form a white powder that further being calcined in 550 °C furnace for 3 hours. The resulting powder then characterized using XRD to identify phase and cristallinity of the resulting composite. SEM measurements was carried out to analyse morphological structure.

3. Result and Discussion

3.1. X-Ray Diffraction (XRD) Analysis

X-Ray Diffraction analysis was carried out to characterize the crystal structure of the material. Sample TiO_2/ZSM-5 composite with the variation of TiO_2 2.5; 5; and 10% weight was characterized by XRD, as shown in Figure 1.

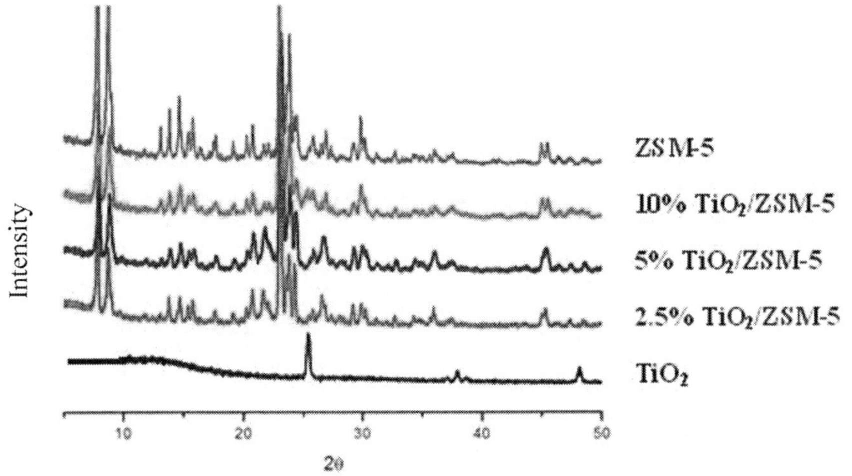

Figure 1. XRD patterns of TiO_2/ZSM-5 composite with variation of TiO_2 percent weight

From this diffractogram, all composite samples are indicating the presence of MFI structure (2θ =7-10° and 22-35°) and the anatase phase of TiO_2 ($2\theta = 25.3°$). The peak of anatase TiO_2 on sample 2.5;5; and 10% TiO_2/ZSM-5 composite is broader than the peak of TiO_2 pure because this peak merged with ZSM-5's peak around 25°. The addition of TiO_2 does not affect the crystal structure of ZSM-5. It caused by ZSM-5 suppresses the growth of TiO_2 during the synthesis process [8]

3.2. Scanning Electron Microscope – Electron Dispersion X Ray (SEM-EDX)

SEM-EDX is used to analyze the morphology and distribution of elements in a material. Commercial ZSM-5 was characterized by SEM, as shown in figure 2a. From this image, ZSM-5 has a hexagonal shape that indicates the sample has a high crystallinity [18]. Some sphere like shape around the hexagonal is impurities, likely to come from unreacted reactant during ZSM-5 synthesis. The addition of TiO_2 on the ZSM-5 surface retains the morphology of its forming material, namely hexagonal. The existence of this impurities makes it difficult to detect TiO_2 existence on the SEM image of the resulting composite, as shown in figure 2 c,d and e, so EDX analysis is needed.

Table 1. EDX results of TiO_2/ZSM-5 composite sample with variation of TiO_2 percent weight

Sample	% wt			
	Si	Al	O	Ti
2.5% TiO_2/ZSM-5	52.51	3.10	43.57	0.83
5% TiO_2/ZSM-5	45.31	2.69	49.95	2.04
10% TiO_2/ZSM-5	51.37	2.48	41.06	5.10

EDX analysis results are shown in table 1. Some amount of titanium elements in the composite sample is detected. When 2.5% weight of titanium precursor added, 0,83% weight of titanium element found in the composite sample. If the amount of titanium precursor doubled, the amount of titanium element in the composite sample also doubled. From table 1, it can conclude that the higher amount of titanium precursor added, the higher amount of titanium elements found in the composite sample.

Figure 2. SEM image of a) commercial ZSM-5 b) commercial TiO_2 c) 2.5% TiO_2/ZSM-5 composite sample d) 5% TiO_2/ZSM-5 composite sample and e)10% TiO_2/ZSM-5 composite sample

3.3. Nitrogen adsorption-desorption isotherms

Addition TiO_2 on the ZSM-5 surface increases the surface area of the composite than ZSM-5 pure. However, increasing in TiO_2 content resulting in a decrease in the specific surface area of TiO_2/ZSM-5 composites, as presented in Table 2. It is possible that the solids of TiO_2 formed outside the zeolite micropore allow some parts to cover the zeolite pore thereby reducing the surface area and zeolite porosity [8].

Table 2. Specific surface area of ZSM-5 and TiO_2/ZSM-5 composites (2.5; 5 and 10%)

Sampel	S_{BET} (m^2/g)
ZSM-5	324,76
2,5% TiO_2/ZSM-5	336,31
5% TiO_2/ZSM-5	332,76
10% TiO_2/ZSM-5	327,23

4. Conclusion

TiO_2/ ZSM-5 composite have been successfully synthesized from Titanium (IV) Isopropoxide and ZSM-5 precursors via the sol-gel method. The results from X-Ray Diffraction analysis revealed that the peak showed the presence of TiO_2 anatase in sample 2.5;5; and 10% TiO_2/ZSM-5 composite. The increasing effect of Titanium precursors to TiO_2/ ZSM-5 composite carried out agglomerates on external surface of ZSM-5 as seen in SEM which was supported by surface area data from BET.

Acknowledgments
This work was financially supported by the ITS Local Grand Program 2018.

References

[1] Naimah S, Silvie A A, Jati B N, Aidha N N, and Agustina A C 2014 Color Degradation in Textile Industrial Wastewater With Photocatalytic Method Using Nanocomposite TiO_2-Zeolite *J.Kimia Kemasan* **36** 225-236

[2] Jain R and Sikarwar S 2008 Removal of hazardous dye congo red from waste material *J. Hazard. Mater* **152** 942-8

[3] Hameed B, Mahmoud D and Ahmad A 2008 Sorption equilibrium and kinetics of basic dye from aqueous solution using banana stalk waste *J. Hazard. Mater* **158** 499-506.

[4] Rauf M A, Meetani M A, and Hisaindee S 2011 An overview on the photocatalytic degradation of azo dyes in the presence of TiO_2 doped with selective transition metals *Desalination* **276** 13-27

[5] N.F. Khusnun, A.A. Jalil, S. Triwahyono , C.N.C. Hitam, N.S. Hassan, F. Jamian , W. Nabgan,T.A.T. Abdullah, M.J. Kamaruddin, D. Hartanto 2018 Directing of the amount of CNTs in CuO-CNT catalysts for enhanced adsorption-oriented visible-light-responsive photodegradation of p-chloroaniline *Powder Technology* **327**170-178

[6] Fagan R, McCormack D E, Dionysiou D D, and Pillai S C 2015 A review of solar and visible light active TiO_2 photocatalysis for treating bacteria, cyanotoxins and contaminants of emerging concern *Materials Science in Semiconductor Processing* **42** 2-14

[7] Tawari A, Einicke and Glaeser R 2016 Photocatalytic oxidation of NO over composites of titanium dioxide and zeolit ZSM-5. *Catalysts* 316

[8] Zhuang Y, Song H, Li G and Xu Y 2010 Ti-HMS as a single-site photocatalyst for the gas phase degradation of benzene *Matterials Letters* **64** 2491-3

[9] Zhang J, Minagawa M, Yamashita H and Anpo M 2000 Photocatalytic decomposition of NO on Ti-HMS mesoporous zeolite catalysts *Catal Lett* **66** 241-3

[10] Hartanto D, Saputro, O, Utomo, W, Rosyidah, A, Sugiarso, D and Ersam, T 2016 Synthesis of ZSM-5 directly from kaolin without organic template: part-1 : effect of crystallization time *Asian Journal of Chemistry* **28(1)** 211-5

[11] Subandrio, Dahani W, and Purwiyono T T 2017 Optimizationof Gravity Chromite Processing with Shaking Table *Jurnal Petro* **6** 2

[12] Azizi M A, Marwanza I, Amala S A dan Hartanti N A 2018 Three dimensional slope stability analysis of open pit limestone mine in Rembang District, Central Java *IOP Conf. Series : Earth and Environmental Science* **212**

[13] Lakhane M, Khairnar R, Ryota K and Mahabole M 2016 Metal oxide blended ZSM-5 nanocomposites as ethanol sensors Bull. *Mater Sci* **39** 1438-92

[14] Takeuchi M, Kimura T, Hidaka M, Rakhmawaty D and Anpo M 2007 Photocatalytic oxidation of acetaldehyde with oxygen on TiO_2/ZSM-5 photocatalysts: effect of hydrophobicity of zeolites *Journal of Catalysis* **246** 235-40

[15] Zhou K, Hu Xin-Yan, Hsueh Chung-chung, Zhang Q, Wang J, Lin Yu-Jung and Chang Chang-Tang 2016 Synthesized TiO_2/ZSM-5 composites used for the photocatalytic degradation of azo dye: intermediates, reaction pathway, mechanism and bio-toxicity *Applied Surface Science* **383** 300-9

[16] Li X, Shen B and Xu C 2010 Interaction of titanium and iron oxide with ZSM-5 to tune catalytic cracking of hydrocarbons *Applied Catalysis A: General* **375** 222-9

[17] Kurniawati R, Iryani A and Hartanto D 2018 The effect of 2-propanol on the shifting band gap of ZSM-5/TiO_2 Composite Prepared Via Sol-Gel Method *AIP Proceeding* **2049** 020089

[18] Hartanto D, Iqbal R M, Shahbihi W E, Santoso E, Fansuri H, and Iryani A 2017 Effect of H_2O/SiO_2 molar ratio on direct synthesis of ZSM-5 from Bangka's kaolin without pretreatment *Malaysian Journal of Fundamental and Applied Sciences* **13(4)** 817-20

IOP Conf. Series: Materials Science and Engineering **588** (2019) 012038 doi:10.1088/1757-899X/588/1/012038

The Relevancy of Mineral Composition to Rock Shear Strength in Kaliwadas Hills, Karangsambung, Kebumen

W Dahani[1], I Marwanza[1,3], R Safitri[2], A B Pambudi[2], M I Rizki[1]
[1]Mining Engineering Department FTKE Trisakti Univesity Jakarta 11440
[2]Chemistry Department FS Institut Teknologi Sepuluh Nopember Surabaya 60111

Corresponding Author : irfanm@trisakti.ac.id

Abstract. The occurrence of landslides on the slopes due to a decrease in shear strength in the area of landslides in the limestone hills of Kaliwadas, Kebumen, Central Java, causes public facilities to be damaged and disrupting the activities of everyday citizens. Factors that influence rock shear strength are cohesion value and internal friction angle, while mineralogy and grain size is one of the factors derived from the environment which also affects the shear strength value. This study aims to determine the effect of mineral composition and grain size on rock shear strength. A number of rock samples were taken from the study site to be prepared, cylindrical samples for direct shear and petrographic strength tests with thin incision samples. Direct shear test results provide cohesion values and deep shear angles in limestone with peak and residual conditions of 0.22 Mpa and 39.73 ° respectively; 36.61 kPa and 29.28 ° while in claystone 67.86 kPa and 25.36 °; 13.41 kPa and 16.17 °. Petrographic analysis of rocks shows that limestone contains calcite (88%) and quartz minerals (5%), while claystone contains calcite (91%) and quartz minerals (2%). Based on the mineral content of limestone and claystone and their effect on rock shear strength, it can be concluded that limestones have a quartz mineral composition, cohesion values and deep shear angles that are better and larger when compared to rock shear strength in claystone and believed to be the cause of the slip field in the research location.
Keyword : slope, direct shear, mineralogy, thin section, petrography

1. Introduction

Karangsambung is a Sub-District in Kebumen Regency, which is an area that have a geological phenomena and various rare and unique types of rocks such as igneous, sedimentary and metamorphic rocks [1]. The rock has different types of minerals, for examples : calcite, chromite [2], quartz, kaolin and other minerals that can be used as natural resources as a raw material to synthesis inorganic material by various methods, for example hydrothermal method [3]. The inorganic material that has been synthesized can be used as textile dyes adsorbent [4], catalyst [5,6] or photocatalyst [7,8,9] for drying agents depend on their chemical and physical properties. On the other hand, the mechanical strength of rocks come from strength properties or resistance to external forces, the strength of rock depends on the composition of the mineral. Each mineral that compose rock has specific different for chemical and physical properties [10]. Among minerals that contained in rocks, quartz is a compact mineral with compressive strength reaching more than 500 MPa. Generally, the strength of rock in line with higher mineral content of quartz in rocks. Hardness and strength of rock are classified by the Fredrich van Mohs scale (1882). This study aims to determine the effect of mineral composition on rock shear strength. The mechanical strength of rock were measured directly using shear strength test to get cohesion value (c) and deep shear angle (°). Petrographic analysis was carried out to determine the mineralogical composition contained in rocks [11].

2. Research Methodology

2.1. Sampling of Rocks

Samples of rocks obtained were taken in various outcrops around the limestone hills of Kaliwadas, Karangsambung, Kebumen, Central Java. the location of research sample point can be seen in Figure 1.

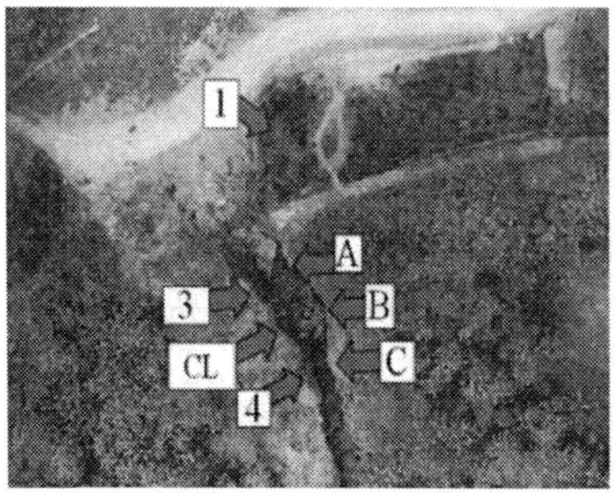

Figure 1. The sampling location of rocks

2.2. Rock Sample Preparation

Samples that have been taken from the research location then prepared according to the testing requirement. For direct shear strength test, the sample is prepared in a cylindrical form with a coring tool which is carried out in the Trisakti geomechanics laboratory, Jakarta which can be seen in Figure 2.

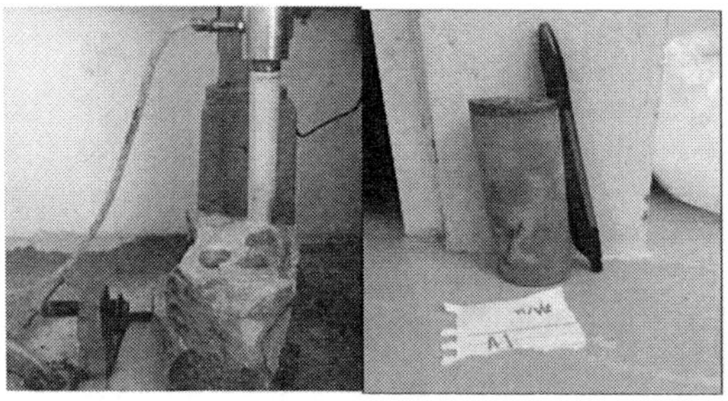

Figure 2. Sample in the cylindrical shape for testing shear strength

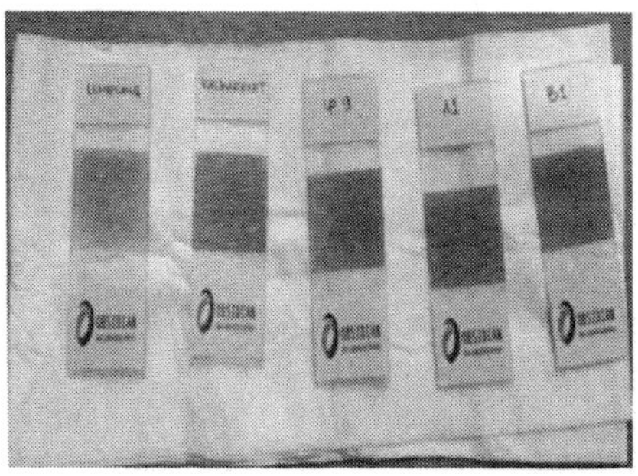

Figure 3. Samples with Thin Section

Petrographic analysis, the sample was sliced by sharpening the rock until thickness of sample reach ± 0.03 mm [12]. Sample prepared in the Obsidian laboratory, Bandung which can be seen in Figure 3.

2.3. Results of Rocks physical Properties Test

The results of rocks physical properties test in both lithology, namely limestone and claystone. The obtained physical properties of rocks such as natural content weight, dry content weight, saturated content weight, original moisture content, and porosity. These data is used as a parameter for the analysis of slope stability [13] and its relevancy to the mineral composition of rock in the research site which can be seen in Table 1.

Table. 1 Results of Physical Properties of Rock Test

No	Rocks	Real content weight (gram/cm^3)	Saturated content weight (gram/cm^3)	Dry content weight (gram/cm^3)	Real water content (%)	Porosity (%)
1	Limestone 1	2.02	2.06	1.85	9.61	21.77
2	Limestone A	2.13	2.21	1.89	12.55	31.65
3	Limestone B	2.07	2.15	1.84	12.16	30.6
4	Limestone C	2.03	2.08	1.76	15.47	32.27
5	Claystone	1.65	1.73	1.32	24.45	40.45

Based on the results of physical properties, there are differences in the weight value of the content of limestone and claystone due to lithological differences, in which claystone absorbs free water and expands in such a way that it becomes softer. The longer the clay is submerged, the condition becomes softer, thus decreasing the stability of the slope. This condition is critical for slopes which have a thick clay layer at the bottom of the slope.

2.4. Results of Rock Shear Strength Test

From the test results, the direct shear strength of the rock can be seen in Table 2. The cohesion and deep shear angle values for limestone at peak conditions ranged from 180 to 270 kPa for cohesion and 36 ° to 42 ° while in residual conditions 29 to 45 kPa for cohesion and 26 ° to 32 ° respectively. Claystone at peak conditions ranged from 58 to 77 kPa for cohesion and 24 ° to 27 ° while in residual conditions 12 to 15 kPa for cohesion and 15 ° to 17 °. From the results of the direct shear strength test obtained in two lithologies namely limestone and claystone there is a significant difference in value where in claystone is very low compared to limestone which is believed that claystone is the slip field on slopes that experience landslides [14].

Table. 2 Result of direct rock shear strength test

Unit	Sample Code	Peak C (kPa)	Peak Φ (°)	Residu C (kPa)	Residu Φ (°)
Limestone	BG-1	268.87	41.57	45.41	29.83
	BG-A	221.44	40.80	34.78	31.93
	BG-B	184.18	36.57	29.02	26.57
	BG-C	205.63	40.00	37.24	28.80
Claystone	CL-1	77.28	26.50	14.82	16.83
	CL-2	24.21	24.21	12.00	15.51

2.5. *Results of Petrographic Analysis*

From the results of petrographic analysis, observations were made with a special microscope, namely polarized light microscopy [15]. This method used polarized light at the bottom of the microscope that fired towards the objective lens.

The difference between polarized light microscopy with other is that there are 2 methods of observation in the form of plane polarized light and cross polarized light. The study used Nikon eclipse E200 Microscope. Observations can be seen in Figure 4 and Table 3 for mineral composition. Based on petrographic observations on limestone, the mineral composition of calcite ranges from 85% to 90% and quartz minerals 4% to 6% while claystone has 91% calcite minerals and 2% quartz minerals. Based on those result, there isn't clear difference between both lithologies, namely limestone and claystone, quartz mineral limestone is greater than + 3% from claystone, indicating that quartz mineral has the most role in rock shear strength compared to calcite minerals.

Table. 3 Mineral Composition of Rocks from Research Sites

No	Lithology	Mineral Composition		
		Quartz (SiO_2)	Calcite ($CaCO_3$)	Other Minerals
1	Claystone	2%	91%	7%
2	Limestone 1	6%	88%	6%
3	Limestone A	5%	87%	8%
4	Limestone B	4%	90%	6%
5	Limestone C	5%	85%	10%

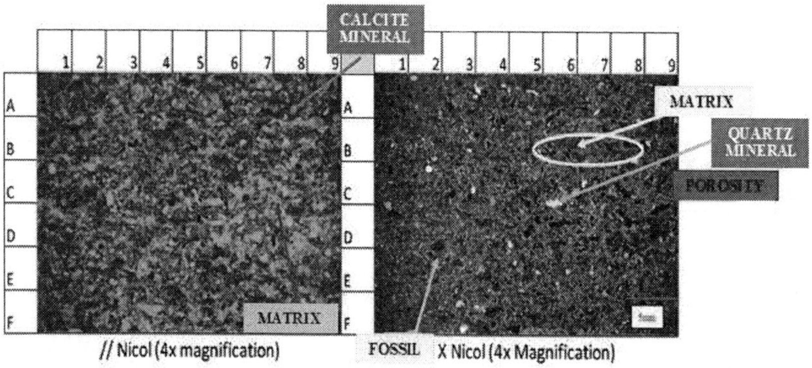

Figure 4. Petrographic Observation on Limestone Samples

3. Analysis and Discussion

Based on the results of direct rock shear strength test and petrographic observations on each rock sample, can be assembled into a graph to determine the effects of rock mineral composition on rock shear strength.

Based on the results of calcite mineral composition on limestone and claystone lithology, it was found that the difference of mineral composition percentage was not too great at 85% to 90% in limestone and claystone at 91%. From the value of limestone cohesion and claystone has a significant difference that limestone has cohesion value of 184 kPa up to 270 kPa and claystone 77 kPa for cohesion. Based on the influence of calcite mineral composition with rock cohesion in Figure 5, it was not very influential because calcite mineral has a hardness level of Mohs' scale of 3 including minerals that are soft enough.

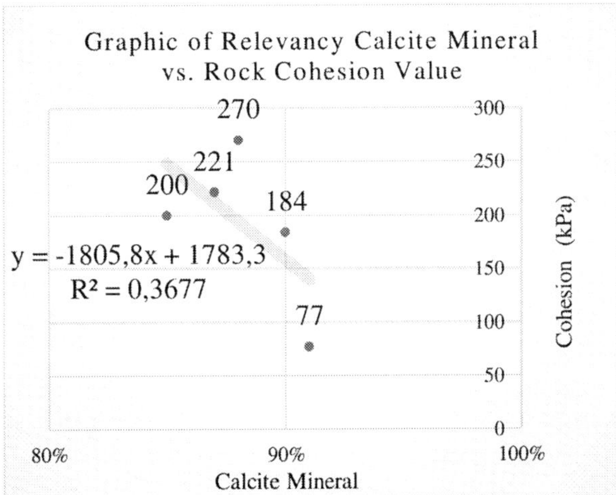

Figure 5. Graphic of Relevancy Calcite Mineral vs. Rock Cohesion Value

Based on the results of quartz mineral composition in limestone and claystone lithology, evidently it has a different mineral composition in limestone of 4% -6% and claystone is only 2%. From the value of limestone and claystone cohesion has a significant difference that limestone has cohesion value of 184 kPa up to 270 kPa and claystone 77 kPa.

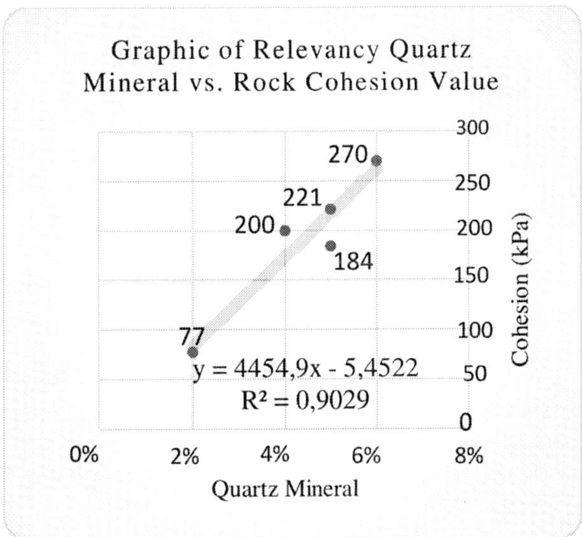

Figure 6. Graphic of Relevancy Quartz Mineral vs. Rock Cohesion Value

Based on the effect of quartz mineral composition with the value of rock cohesion in Figure 6, it caused a strong relevancy, an increase in quartz minerals causing an increase in the value of cohesion because quartz minerals are stable minerals and have a hardness level of 7 Mohs', including hard enough minerals.

4. Conclusion

Based on the results of the analysis, some conclusions can be drawn as follows:

1. Results of direct shear test provide cohesion and shear angle values in limestone at peak and residual conditions of 220 kPa and 39.73 ° ; and 37 kPa and 29.28° respectively,
2. Results of direct shear test provide cohesion and shear angle values in claystone at peak and residual conditions of 67.86 kPa and 25.36 °; 13.41 kPa and 16.17° respectively,
3. Based on the effect of calcite mineral composition ($CaCO_3$) with the value of rock cohesion it is not very influential because calcite minerals have a level of mohs' scale hardness of 3 including soft enough minerals..
4. The increase in quartz mineral (SiO_2) causes an increase in of the cohesion value because the quartz mineral is a stable mineral and has a hardness level of 7 mohs including minerals that are quite hard

References

[1] Rai M A dan Kramadibrata S *Mekanika Batuan* Bandung : Institut Teknologi Bandung.

[2] Nurjaman F, Subandrio S, Ferdian D dan Suharno B 2018 Effect of Basicity on beneficiated chromite sand smelting process using sub-merged arc furnace *AIP Conference Proceedings* **1964**

[3] Hartanto D, Saputro O, Utomo W P, Rosyidah A, Sugiarso D, Ersam T, Nur H dan Prasetyoko D 2016 Synthesis of ZSM-5 Directly from Kaolin without Organic Template Part-1 : Effect of Crystallization Time *Asian Journal of Chemistry* **28** 1 211-215

[4] Iryani, A dan Hartanto, D 2018 Textile Dyes Removal by ZSM-5 from Bangka Kaolin *IOP Conf. Series: Journal of Physics* Conf. Series.

[5] Hartanto D, Iqbal, R M, Shahbihi, W E, Santoso, E, Fansuri, H dan Iryani, A 2017 Effect of H_2O/SiO_2 molar ratio on direct synthesis of ZSM-5 from Bangka's kaolin without pretreatment. *Malaysian Journal of Fundamental and Applied Sciences* **13** 4 817-820

[6] Yulianti, C.H., Ediati, R., Hartanto, D., Purbaningtias, T.E., Chisaki, Y., Jalil, A.A., Hitam, C.K.N.L.C.K., Prasetyoko, D., (2014). Synthesis of CaOZnO Nanoparticles Catalyst and Its Application in Transesterification of Refined Palm Oil. *Bulletin of Chemical Reaction Engineering & Catalysis*, **9** (2): 100-110

[7] Khusnun N F, Jalil A A, Triwahyono S, Hitam C N C, Hassan N S, Jamian F, Nabgan W, Abdullah T A T, Kamaruddin M J, Hartanto D 2018 Directing the amount of CNTs in CuO-CNT catalysts for enhanced adsorption-oriented visible-light-responsive photodegradation of *p*-chloroaniline *Powder Technology* **327** 170-178

[8] Kurniawati R, Iryani A and Hartanto D 2018 The effect of 2-propanol on the shifting band gap of ZSM-5-TiO_2 composite prepared via sol-gel method *AIP Conference Proceedings* **2049** 020089

[9] Zi S C, Chandren S, Yuan L S, Razali R, Ho C S, Hartanto D, Mahlia T M I, and Nur H 2016 New Method to synthesize mesoporous titania by photodegradation of surfactant template *Solid State Sciences* 83-91

[10] Sulistya R, Hartanto D dan Murwani I K 2006 Characterization and Cobalt Distribution Study of Ceramic Body System : Bojonegoro Clay-Feldspar-Quartz *Indo. J. Chem.* **6** 2 138-143.

[11] Pettijohn F J 1957 *Sedimentary Rocks* Indian edition (New York : Harper & Row Publishers, Inc.)

[12] Hidayat M 2016 *Studi Mikrofasies dan Diagenesis Batuan Karbonat Sebagai Penentu Lingkungan Pengendapan di Daerah Nawungan dan Sekitarnya, Kecamatan Panggang, Kabupaten Gunungkidul, Daerah Istimewa Yogyakarta* (Yogyakarta : STTNAS)

[13] Hoek E, Bray, J W 1981 *Rock Slope Engineering Revised Third Edition* The Institution of Mining and Metallurgy London References

[14] Azizi M A, Kramadibrata S, Wattimena R K, Sidi I D, Adriansyah Y 2012 *Analisis Risiko Kestabilan Lereng Tambang Terbuka (Studi Kasus Tambang Mineral X)* 1st Symposium and National Seminar og Geomechanics (Yogyakarta : UPN Veteran Yogyakarta)

IOP Conf. Series: Materials Science and Engineering **588** (2019) 012038 doi:10.1088/1757-899X/588/1/012038

[15] Scholle P A, Dana S Ulmer-Scholle 2003 *A Color Guide to the Petrography of Carbonate Rocks: Grains, textures, porosity, diagenesis* (Oklahoma : American Association of Petroleum Geologists)

IOP Conf. Series: Materials Science and Engineering **588** (2019) 012039 doi:10.1088/1757-899X/588/1/012039

On The Synthesis of ZSM-5 Directly from Kaolin Bangka with Aging Time

D Hartanto[1,*], A B Pambudi[1], D N Cahyanti[1], and W P Utomo[1]

[1]Department of Chemistry, Institut Teknologi Sepuluh Nopember, Surabaya, Indonesia.

*Corresponding author : djokohar@ chem.its.ac.id

Abstract. ZSM-5 has been synthesized in this study using hydrothermal method with addition of ZSM-5 seed. Synthesis of ZSM-5 in this study using kaolin bangka as alumina source with aging time of 0, 12, 24 and 36 hours and the effect of aging time was also investigated. Samples were characterized by XRD, FT-IR, SEM and SEM EDX. ZSM-5 samples with aging 24 hours has the highest intensity at $2\theta = 23.03°$ with crystallinity of 69.60%. The vibration at wavenumbers 540 and 1220 cm^{-1} was shown in the sample with 24 hours aging, it was indicated that a large number of pentasil rings have been formed. These results are consistent with XRD data which shows that samples with aging 24 hours have the highest crystallinity. The size of ZSM-5 sample with an aging time 24 hours was about 4.5 μm. EDX results obtained Si / Al ratio in samples with aging time 0, 12 hours and 24 hours is 10.77; 11.23 and 8.34. All samples showed a Si / Al ratio above 5 which showed that the ZSM-5 had been formed. But morphologically, samples with aging 24 hours have a more uniform crystal shape.

1. Introduction

ZSM-5 is one of the microporous crystalline aluminosilicate zeolites with MFI structure which has a channel opening size of 5.2-5.6 Å [1,2]. Among zeolites family, ZSM-5 has an excellent heterogeneous acidity, great thermal and mechanical stability, well shape selectivity and interconnected two-dimensional micropore system which contribute to usage as catalyst in petroleum and petrochemical industry [3]. ZSM-5 are commonly synthetized by the hydrothemal method of the mother liquor containing presursor of silica, alumina and organic template as SDA [4,5]. However, the usage of Structure-directing agent (SDA), which is Tetrapropyl ammonium hydroxide (TPAOH) or tetrapropyl ammonium bromide (TPABr), in the synthesis of ZSM-5 cause many adverse problem such as coke deposition due incomplete combustion, harmful gas production from thermal decomposition, water contamination and high production cost [6,7]. Therefore, to overcome those adverse problem, many research has been conducted to synthesis ZSM-5 in the absence of organic template.

Bangka Belitung kaolin is promising natural resource as raw material to synthesis of ZSM-5. The element deposits can be estimated using geostatistics and geotechnical methods [8,9]. According to Hartanto et al. (2016) the use of sources of alumina and silica from natural materials such as kaolin, ash and rice flies can reduce the cost of basic ingredients for the synthesis processes. Kaolinite [$Al_2Si_2O_5(OH)_4$] is a two-layered aluminosilicate clay mineral, consisting of one alumina octahedron sheet and one silica tetrahedron sheet in a 1:1 stoichiometric ratio. Because its alumina content, kaolinite has been frequently used as the starting material for synthesis of zeolite such as SAPOs,

LTA, FAU and MOR [10,11,12]. Synthesis of ZSM-5 requires high Si/Al ratio, which means it necessary to further processing of the kaolinite. Some efforts have been devoted to synthesizing ZSM-5 zeolites from kaolinite via dealumination of the metakaolin or addition an alternative silica source [13,14].

Here, we report direct synthesis of ZSM-5 using kaolin bangka as the Si and Al sources without prior pre-treatment with seeding method. ZSM-5 seed was employed in this procedure as structure-directing agent (SDA) to obtain MFI-type framework, while LUDOX used as additional silica to control Si/Al ratio. The transformation of kaolin into zeolite was monitored by the change of functional group, phase and crystallinity at various aging condition.

2. Experimental

2.1. Materials

All materials used in this work were analytical grade. NaOH (Merck, Germany). LUDOX® HS-40 colloidal silica (30 % Si in water) was purchased from Aldrich, Germany. Kaolin (containing 57 % SiO_2 and 22 % Al_2O_3) was taken from Bangka Belitung, Indonesia without further purification. The self-synthesized ZSM-5 were used as the seed. Aqua DI was taken from a Millipore Milli-Q system and used within 1 week.

2.2. Methods

Synthesis of ZSM-5 without organic templates was used hydrothermal method and addition of ZSM-5 seed. ZSM-5 which was synthesized has a molar ratio : $10Na_2O : 120SiO_2 : 2Al_2O_3 : 1800H_2O$. The Synthesis was started by weighing an aquademineral which according to its molten composition of 16. 5927 grams. Aquademineral was divided into two parts. In the first half, 0.8 grams of NaOH were added and stirred until dissolved. After dissolved completely, 0.9266 grams kaolin bangka were added while stirring with a magnetic stirrer. Then, 22.5817 grams LUDOX was added slowly and the stirring rate was increased to 400 rpm. After the zeolite solution, half the second part of the aquademineral was added to the mixture. The stirring rate was increased to 550 rpm and stirred for 8 hours. After stirring was completed, aging has been done at room temperature with variations of 0, 12, 24 and 36 hours. The results of aging was added ZSM-5 seed by 3% by weight, which is 0.255 grams while stirring with magnetic stirrer for 30 minutes. Then the results of the mixture were put in a stainless steel autoclave for hydrothermal processes at a temperature of 175°C for 24 hours. The slurry of hydrothermal was neutralized with aquademineral until the pH of the solid neutral. After that, the samples was dried in oven at 110°C for 12 hours.

2.3. Characterization

Characterization of resulted product was measured by powder X-ray diffraction (XRD) patterns on PANalytical X'pert diffractometer with Cu Kα (λ=1.5418 Å) radiation generated at 40 kV, 40 mA. The crystallinity of synthetized product was calculated using reflection at 2θ = 23.01°. The functional group identification of product was carried out using Fourier Transform Infrared (FTIR) Shimadzu Instrument Spectrum One 8400S to analyse function group on finger print area. The morphology of the synthesized particles was examined by SEM (Model: Zeiss, EVO MA 10, Germany) operating with an accelerating voltage of 10 kV.

3. Result and Discussion

3.1. XRD Analysis

The XRD patterns from Bangka kaolin, ZSM-5 seed, ZSM-5 samples with aging time 0, 12, 24 and 36 hours are shown in Fig. 1. The XRD patterns from Bangka kaolin in Fig. 1.a. shows the peaks at 2θ = 12.32; 19.87; 20,34; 24.85; 26,61; 34.95; 35.40; 35.91; 38.37 and 39.22°. The characteristic diffraction peaks of kaolin can be observed at 2θ = 12.3; 24.8; and 38.3° [15]. Fig 1c, 1d, 1e and 1f shown peaks at 2θ = 7.9; 8.8; 8.9; 23.0 and 23.1 °. These peaks correspond to the characteristic peaks of the ZSM-5 with MFI structures [16]. The characteristic diffraction peaks of kaolin at 2θ = 12.3; 24.8; and 38.3 °

did not appear on the diffractogram of ZSM-5 with aging variation. This indicates that kaolin has dissolved and formed ZSM-5 crystals.

The XRD pattern from ZSM-5 samples with aging 0 hours (Fig. 1c) shows sharp peaks at 2θ = 7.91; 8.85 and 23.02 °. These sharp peaks indicated that the ZSM-5 crystal has been formed and allows the crystalline nucleus to have formed although aging time 0 hours. However, the crystallinity of ZSM-5 with aging time of 0 hours was still low due to crystal growth was not optimal.

XRD patterns from ZSM-5 samples with aging 12 and 24 hours are shown in Fig 1d and 1e. The results of the diffractogram from these samples have a pattern that is not different from the ZSM-5 sample with aging 0 hours. The highest peaks in aging samples 12 and 24 hours were at 2θ = 8.82; 8.85; 7.91; 7.94 and 23.03 °. Fig. 1e show the peak at 2θ = 23.03°, it was higher than the three other aging variation samples. The peak height indicated that the ZSM-5 sample with aging 24 hours has the highest intensity which means it also has the highest crystallinity.

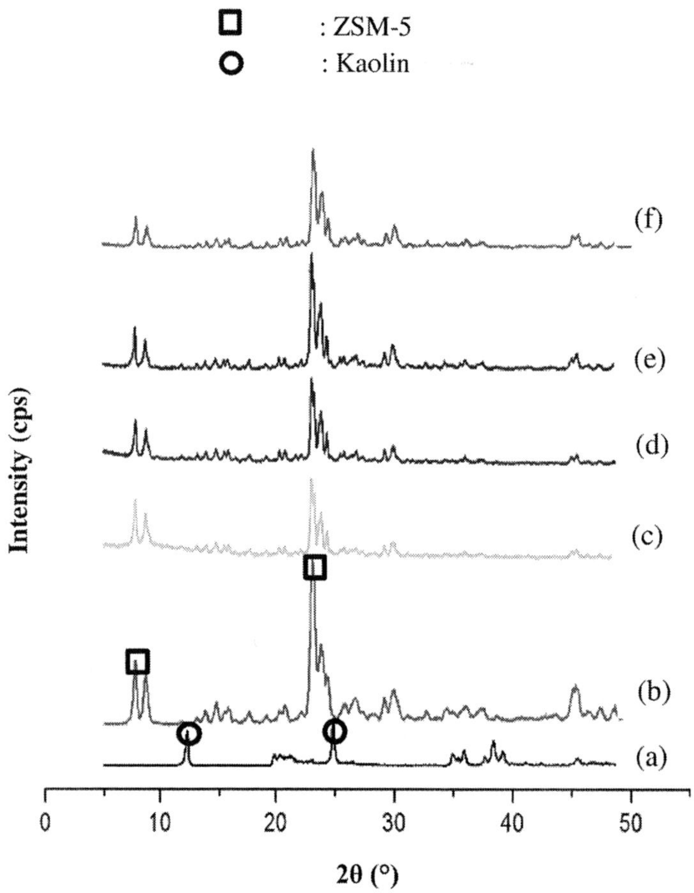

Figure 1. Diffractogram pattern of (a) Kaolin Bangka, (b) ZSM-5 seed, ZSM-5 sample with aging (c) 0 hours, (d) 12 hours, (e) 24 hours and (f) 36 hours.

ZSM-5 samples with an aging time of 36 hours had peaks that were identical to other aging time variation samples at 2θ = 8.01; 8.93 and 23.17 °. In the ZSM-5 sample with 36 hours aging variation, it have a decrease in crystallinity but not significant. In the ZSM-5 sample with aging 12 and 24 hours, crystal growth occurred, this was evidenced by the increase in crystallinity with increasing aging as shown in Table 1.

Aging time has a role in the crystallinity of the ZSM-5 samples formed. Table 1 shown that ZSM-5 samples with aging 0 hours have the lowest crystallinity. Crystallinity during aging 12 and 24 hours increased in a row but when aging 36 hours decreased crystallinity. The longer the aging time during synthesis, the percentage of crystallinity of ZSM-5 produced was greater but with prolonged aging time, the crystallinity of the ZSM-5 sample were reduced. The nucleation and growth of crystals from zeolite was occured during the aging process [17]. Therefore, each sample of ZSM-5 has a different amount of crystals. Increasing amount of crystal facet of the ZSM-5 sample, then the x-ray was reflected are increasing so that their intensities higher. Therefore, in this study aging time was very influential on the formation of ZSM-5 crystals where the optimum formation of ZSM-5 was aging 24 hours with crystallinity of 69.60%

Table 1. The intensity and crystallinity of ZSM-5 samples with aging time variation.

Sample	2θ (°)	Intensity (cps)	Crystallinity (%)
ZSM-5 seed	23,14	1813	100
Aging 0 hours	23,14	904	49,86
Aging 12 hours	23,02	971	53,55
Aging 24 hours	23,03	1262	69,60
Aging 36 hours	23,12	1088	60,01

3.2. *FT-IR Spectra*

The infrared spectra of kaolin samples, ZSM-5 seeds, ZSM-5 samples with aging time 0 hours, 12 hours, 24 hours and 36 hours are shown in Fig. 2. The infrared spectra of kaolin (fig. 2.a) show peaks at wave numbers 430, 470, 538, 700, 754, 788, 912, 1008, 1029 and 1110 cm^{-1}. This result was supported by Panda et al. (2010) reports that spesific peaks of kaolin are in the 912 cm^{-1} region indicating the bending vibration of the Al-Al-OH bond [18]. In the wave number regions 1032, 1100 and 1114 cm^{-1}, there is a Si-O bending bond. The wave number area of 755 cm^{-1} indicates the Si-O-Al bond. In the wave number area 693 and 541 cm^{-1} both show the vibration of the Si-O and Si-O-Al bending bonds.

The spectra of ZSM-5 seed (fig. 2b) show specific peaks at wave numbers 453, 563, 798, 1103, and 1226 cm^{-1}. Flanigen et al. (1971) reported that the peaks in the wave number are peaks that result from internal bond vibrations of SiO_4 or AlO_4 tetrahedral [19]. The peak at wave number 453 cm^{-1} shows the internal bond vibration of the Si-O-Si bond, the peak at 550 cm^{-1} shows the vibration of the pentasil ring frame which shows the formation of MFI structures [20]; the peak at 798 cm^{-1} shows the symmetrical stretching vibration of the Si-O-Si bond; and peaks at 1103 cm^{-1} and 1226 cm^{-1} show asymmetric stretch vibrations of Si-O-Si bonds. Similar results have also been reported by Prasetyoko (2006) [21].

The infrared spectra of the ZSM-5 sample in Fig. 2(c),(d),(e) and (f) show the five wave number peaks of about 452, 545, 792, 1090 and 1222 cm^{-1}. While the peaks in the regions of 430, 470, 700, 754, 912, 1110 cm^{-1} which are typical of kaolin have been lost, this indicates that the kaolin bonds are broken and begin to form new bonds. This is in accordance with the report of Mohammed et al. (2005) that the absorption band that appears around 1200-1000 cm^{-1} shows the internal bond of the tetrahedral asymmetric stretch vibration of the Si-O-T where T is Si or Al [22]. Vibration of external asymmetry stretching is indicated by the absorption band at 1200 cm^{-1} while the internal vibration of asymmetry is indicated by the absorption band of 1100 cm^{-1}. Absorption band around 788 cm^{-1} shows internal stretch vibration. Goncalves et al (2008) reported that the fifth ring in the ZSM-5 pentasil zeolite structure can be shown from the absorption band around 546 cm^{-1} [23]. The band around 455 cm^{-1} shows the tetrahedral bending vibration of TO_4. The existence of the ZSM-5 can be determined by the appearance of peaks in these areas. Wave numbers that appear around 1225 and 545 cm^{-1} are sensitive peaks of the MFI structure [24].

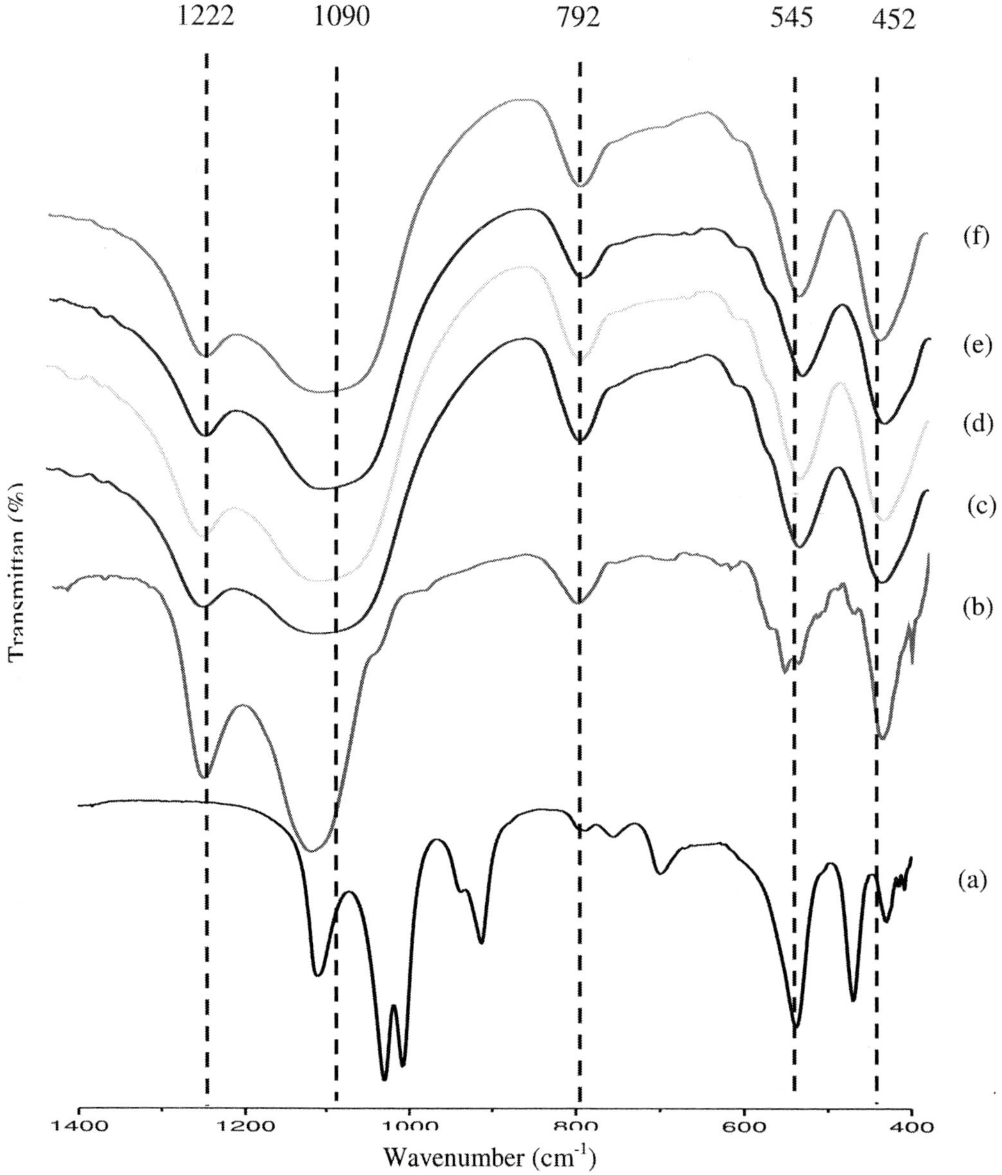

Figure 2. IR spectra of (a) Bangka Kaolin, (b) ZSM-5 seeds, ZSM-5 samples with aging (c) 0 hours, (d) 12 hours, (e) 24 hours and (f) 36 hours

The areas in wave numbers around 1220 and 540 cm^{-1} did not have a sharp peak in the ZSM-5 sample without aging. This shows that there are still few MFI structures formed in the sample without aging. These results support the results of the XRD pattern of ZSM-5 samples without aging which indicates that the sample without aging still has low crystallinity. Meanwhile, samples with an aging time of 12 hours showed a fairly sharp peak at wave numbers 540 and 1220 cm^{-1}, but not too significant when compared with aging samples 0 hours. The peak is also found in other aging variations. These results indicate that the structure of ZSM-5 has been well formed in the sample with variations in aging time 12, 24 and 36 hours. The sharp peak intensity in wave numbers 540 and 1220 cm^{-1} is show in the sample with 24 hours aging indicating that a large number of pentasil rings have been formed. These results are consistent with XRD data which shows that samples with aging 24 hours have the highest crystallinity. However, it appears that the peak intensity in the sample with aging 36 hours in the wave number area 540 and 1220 cm^{-1} decreases. This indicates that the number of pentasil rings that mark the ZSM-5 structure has been reduced. This result is also consistent with the results obtained from XRD data which showed that the crystallinity of the sample decreased at 36 hours of aging. The peaks from the results of infrared spectroscopy characterization on samples with aging times 0, 12, 24 and 36 hours are shown in Table 2.

Table 2. The wavenumbers of infrared spectra from ZSM-5 samples with variations in aging time.

Sample	Stretching vibration asymmetry (cm^{-1})		Stretching vibration symmetry (cm^{-1})	Pentacile structure (cm^{-1})	Bending vibration T-O (cm^{-1})
0 hours	1222	1091	792	545	453
12 hours	1222	1091	792	545	449
24 hours	1222	1089	792	545	451
36 hours	1222	1089	792	545	453

3.3. SEM images and SEM EDX

Morphology and size of crystals from solids were observed using Scanning Electron Microscopy (SEM) equipped with Energy Dispersive X-ray to determine the abundance of the elements present in the sample. Fig. 3, Fig. 4 and Fig. 5 are the morphology of ZSM-5 samples with aging times 0, 12 and 24 hours. The crystal formed is like a rectangle (hexagonal) with a sloping elbow.

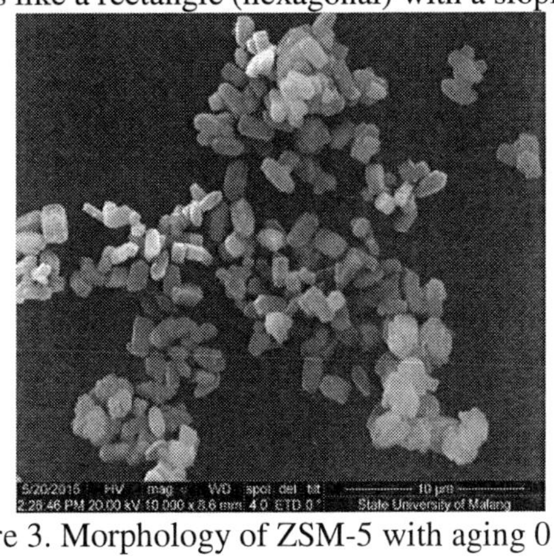

Figure 3. Morphology of ZSM-5 with aging 0 hours

Figure 4. Morphology of ZSM-5 with aging 12 hours

Figure 5. Morphology of ZSM-5 with aging 24 hours on enlargement

The Crystal Morphology from ZSM-5 samples with aging 0 hours in Fig.3. looks uniform. This crystal uniformity shows that ZSM-5 crystals have formed even without aging. This is consistent with the XRD results of aging 0-hours ZSM-5 samples that appear typical of ZSM-5 peaks. The size of formed ZSM-5 in aging 0 hours samples was around 1.7 μm. In the ZSM-5 sample with an aging time 12-hours, ZSM-5 crystal hexagonal formed with a sloping elbow but clustered. The size of the ZSM-5 formed is about 2.6 μm. Morphology from the ZSM-5 sample with aging 24 hours was not different from the aging sample 12 hours but more uniform, not clustered and the size of the sample was larger which was about 4.5 μm. Crystal sample uniformity with 24-hours aging was supported by XRD results with the highest crystallinity of 69.60%

From the results of the SEM analysis above shows that the bulk size increases as the aging time increases. At the time of aging the crystal growth process occurs, the longer the aging process is given, the more time for the crystal growth process so that the size of the crystal formed increases. The elements contained in the ZSM-5 sample synthesized with aging variations 0, 12 and 24 hours can be seen from the EDX results. Table 3. shows the EDX results from each sample having the same content, namely Si, Al, Na, and O. The distribution of these elements is quite evenly distributed from the three samples and has almost the same percentage levels. Based on EDX results obtained Si / Al ratio in samples with aging time 0, 12 hours and 24 hours is 10.77; 11.23 and 8.34. All samples showed a Si / Al ratio above 5 which showed that the ZSM-5 had been formed. But morphologically, samples with aging 24 hours (Fig. 5) have a more uniform crystal shape.

Table 3. EDX results of ZSM-5 sample with aging time variation

Sample	Atom percent (%)				
	Si	Al	Na	O	Si/Al
0 hours	24,57	2,28	2,36	66,07	10,77
12 hours	27,63	2,46	2,55	60,50	11,23
24 hours	25,23	3,06	2,67	63,58	8,34

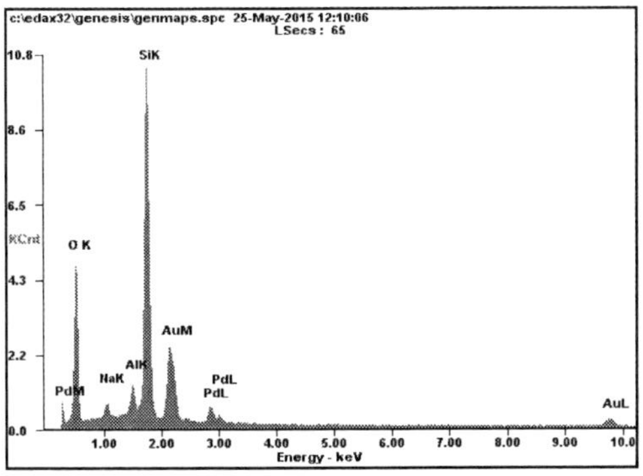

Figure 6. EDX results of ZSM-5 samples with an aging time of 24 hours

4. Conclusion

ZSM-5 has been synthesized using hydrothermal method with addition of ZSM-5 seed and kaolin bangka as alumina source. Synthesis ZSM-5 was used variations aging time of 0, 12, 24 and 36 hours to determine which ZSM-5 sample has a good crystallinity. ZSM-5 sample with 24-hours aging has the highest crystallinity of 69.60% was supported by XRD results with the larger particle size was about 4.5 μm. In XRD analysis, the ZSM-5 sample with aging 12 and 24 hours, crystal growth occurred, this was evidenced by the increase in crystallinity with increasing aging but with aging 36 hours it have a decrease in crystallinity but not significant. In FT-IR analysis, the areas in wave numbers around 1220 and 540 cm^{-1} did not have a sharp peak in the ZSM-5 sample without aging. Meanwhile, samples with an aging time of 12, 24 and 36 hours showed a fairly sharp peak at wave numbers 540 and 1220 cm^{-1}. These results indicate that the structure of ZSM-5 has been well formed. From the results of the SEM analysis shows that the bulk size increases as the aging time increases. At the time of aging the crystal growth process occurs, the longer the aging process is given, the more time for the crystal growth process so that the size of the crystal formed increases.

References

[1] Shi L, Li N, Wang L, Liu R, Lin S, Liu J, Wang R and Li Y 2018 Creating intraparticle mesopores inside ZSM-5 nanocrystals under OSDA-free conditions and achievement of high activity in LDPE degradation *Microporous and Mesoporous Materials* **258** 178

[2] Hartanto D, Yuan L S, Sari S M, Sugiarso D, Murwani I K, Ersam T, Prasetyoko D and Nur H 2016 Can kaolin function as source of alumina in the synthesis of ZSM-5 without an organic template using a seeding technique *J. Malaysian of Fundamental and Applied Sciences* **12**

[3] Xue T, Chen L, Wang Y M and He M Y 2012 Seed-induced synthesis of mesoporous ZSM-5 aggregates using tetrapropylammonium hydroxide as single template *Microporous and Mesoporous Materials* **156** 97

[4] Subandrio, Dahani W, and Purwiyono T T 2017 Optimization of Gravity Chromite Processing with Shaking Table *Journal Petro* **6** 2.

[5] Wang P, Shen B, Shen D, Peng T and Gao J 2007 Synthesis of ZSM-5 Zeolite from Expanded Perlite/Kaolin and its Catalytic Performance for FCC Naphtha Aromatization *Catalysis Communications* **8** 1452

[6] Yue Y, Kang Y, Bai Y, Gu L, Liu H, Bao J, Wang T, Yuan P, Zhu H, Bai Z and Bao X 2018 Seed-assisted, template-free synthesis of ZSM-5 zeolite from natural aluminosilicate minerals *Applied Clay Science* **158** 177

[7] Dey K P, Ghosh S and Naskar M K. 2013 Organic Template-Free Synthesis of ZSM-5 Zeolite Particles using Rice Husk Ash as Silica Source *Ceramics International* **39** 2153

[8] Marwanza I, Badaruddin S P and Azizi M A 2018 Copper Resource Estimation in PT X Batu Hijau, Regency of West Sumbawa, West Nusa Tenggara Province using Geostatistical Method *IOP Conference Series: Earth Environmental Science* **212**.

[9] Azizi M A, Marwanza I, Amala S A and Hartanti N A 2018 Three Dimensional Slope Stability Analysis of Open Pit Limestone Mine in Rembang District, Central Java *IOP Conference Series: Earth Environmental Science* **212**.

[10] Wang T, Lu X, and Yan Y 2010 Synthesis, characterization and crystallization mechanism of SAPOs from natural kaolinite *Microporous and Mesoporous Materials* **136** 138

[11] Novembre D, Sabatino B, Gimeno D and Pace C 2011 Synthesis and characterization of Na-X, Na-A and Na-P zeolites and hydroxysodalite from metakaolinite *Clay Minerals* **46** 339

[12] Mignoni, L.M., Petkowicz, I. D., Machado, F.R.C.N., Pergher, B.C.S 2008 Synthesis of Mordenite Using Kaolin as Si and Al Source *Applied Clay Science* **41** 99

[13] Holmes S M, Khoo S H, and Kovo A S 2011 The direct conversion of impure natural kaolin into pure zeolite catalysts Green Chemistry **13** 1152

[14] Pan F, Lu X, Wang Y, Chen S, Wang T and Yan T 2014 Organic template-free synthesis of ZSM-5 zeolite from coal-series kaolinite *Materials Letters* **115** 5-8

[15] Sibel K, Önal S, Sarıkaya Y and Bozdoğan I 2005 Characterization of Silica Polymorphs in Kaolins by X-ray Diffraction Before and After Phosphoric Acid Digestion and Thermal Treatment *Analytica Chimica Acta* **552 201-206**

[16] Treacy M, M J and Higgins J B 2007 *Collection of Simulated XRD Powder Patterns for Zeolites Fifth (5th) Revised Edition Elsevier*

[17] Zhdanov S P and Samulevich N N 1981 *Proc 5th Conference on Zeolites* (London: Heyden) p 75

[18] Panda A K., Mishra B G, Mishra D K and Singh R K 2010 Effect of sulphuric acid treatment on the physico-chemical characteristics of kaolin clay *Colloids and Surfaces A Physicochemical and Engineering Aspects* **363** 98-104

[19] Flanigen E M, Szymanski H A and Khatami H 1971 Infrared Structural Studies of Zeolites Framework in Molecular Sieve Zeolites I *Advances in Chemistry* **101**.

[20] Fang Y M and Hu H Q 2006 An Ordered Mesoporous Aluminosilicate with Completely Crystalline Zeolite Wall Structure *J. Am. Chem. Soc* **128** 10636-10637

[21] Prasetyoko D, Ramli Z, Endud S, Hamdan H, and Sulikowski B 2006 Synthesis Of Zeolite Beta Directly From Rice Husk Ash : Effect Of Reaction Composition On Crystallinity Of Zeolite Beta J. Indo Chem **6(1)** 11–15

[22] Mohamed M M, Ali I O, Eissa N A, 2005 Effect of thermal treatment on surface and bulk properties of Fe/ZSM-5 zeolites prepared by different methods *Microporous and Mesoporous Material* **87** 93

IOP Conf. Series: Materials Science and Engineering **588** (2019) 012039 doi:10.1088/1757-899X/588/1/012039

[23] Goncalves M L, Dimitrov L D, Jorda M H, Wallau M, Ernesto A and Urquieta-Gonzalez 2008 Synthesis of mesopori ZSM-5 by crystallisation of aged gels in the presence of cetyltrimethylammonium cations *Catalysis Today* **133–135** 69–79

[24] Shukla D B and Pandya V P 1989 Estimation of Crystalline Phase in ZSM-5 Zeolites by Infrared spectroscopy *J. Chemical Technology Biotechnol* **44** 147- 154

IOP Conf. Series: Materials Science and Engineering **588** (2019) 012040 doi:10.1088/1757-899X/588/1/012040

Preparation and Morphological Property of $Co_3O_4/Ba_xSr_{1-x}Co_{0.8}Fe_{0.2}O_{3-\delta}$ (x=0.5-0.7) Membranes using Starch as Binder Agent

W.P. Utomo[1], A.S. Wijayanti[1], S.D. Nurherdiana[1], R.M. Iqbal[1,2], D. Hartanto[1] and H. Fansuri[1,a]

[1]Department of Chemistry, Faculty of Science, Institut Teknologi Sepuluh Nopember, Surabaya, Indonesia
[2]Department of Chemistry, Universitas Palangka Raya, Palangka Raya, Indonesia

[a]Corresponding author: h.fansuri@chem.its.ac.id

Abstract. $Ba_{0.5}Sr_{0.5}Co_{0.8}Fe_{0.2}O_{3-\delta}$ (x=0.5-0.7) or BSCF was known as promising material to prepare MIEC membrane for oxygen separation. Several modifications have been performed to improve catalytic activity of BSCF membranes. One of the promising methods was coating of the membrane with catalyst layer such as Co_3O_4. This work investigates the simple and effective method to coat Co_3O_4 layer on the surface of $Ba_xSr_{1-x}Co_{0.8}Fe_{0.2}O_{3-\delta}$ (x=0.5-0.7) flat sheet membranes resulting Co_3O_4/BSCF xx82 membranes. BSCF were prepared from its initial powders by dry pressing method followed by sintering at 1150 °C for 4 hours. The resulted BSCF dense membranes were coated with Co_3O_4 using starch as binder agent in the form of suspension. The coating suspension consist of 62.5% w/w of starch and 37.5% w/w of Co_3O_4. The coating suspension was directly coated on the surface of the BSCF dense membranes followed by calcination at 800 °C for 3 hours. The surface morphology and cross section of membranes were observed using Scanning Electron Microscope (SEM) equipped with Energy Dispersive X-ray (EDX). SEM images show that the method succeeds to coat Co_3O_4 layer on the surface of BSCF membranes. However, the distribution of the Co_3O_4 was not uniform and some particles form aggregate. The particle size of the membranes was about 0.5-2.0 μm whereas the aggregate size is around 2-5 μm. The presence of Co_3O_4 layer was also indicated by the increasing of Co intensity in the surface of membranes as analyzed by EDX.

1. Introduction

Methane is the main gas that composed natural gas. Its abundance reaches approximately 70-90% of the natural gas composition [1]. This gas is also produce as the by product of the oil refinery [2]. Due to its abundance, methane is being potential to be developed as alternative fuel to substitute the using of oil. To increase the affectivity of the methane usage, as well as the increasing the product value, methane could be converted to be more added value product such us syngas (CO+ H_2). The conversion methane to be syngas could be carried out by partial oxidation of methane (POM)[3], [4]. However, POM required controllable oxygen supply.

The control of oxygen supply is crucial factor in POM. The lack of oxygen could only resulted in low yield, on the other hand, the excess of oxygen could convert methane to be CO_2 and H_2O as product of total oxidation. This oxygen controlled supply in POM could be performed using mixed ionic and electronic conducting membranes (MIEC) [5]. This membrane could permeate oxygen in the form of

ion resulting high selectivity, up to100% theoretically [6]. Among MIEC membrane materials, perovskite based materials, composed of strontium-cobalt-iron (SCF) are promising material due to their high oxygen permeability [7]. Recently, the $Ba_{1-x}Sr_xCo1\text{-}_yFe_yO_{3-\delta}$ (BSCF)'s are getting more interest from many researchers due to their high oxygen permeability and their excellent stability toward reducing atmosphere [8], [9]. Ghadimi et.al. reported that this increasing of oxygen permeability is caused by the decreasing of activation energy of ion diffusion through crystal lattice [10]. Moreover, Faraji et.al. reported that this material could release its oxygen ion in relatively low temperature compare to the conventional material in POM operation [11]. The selectivity of this material for CO was also excellent reaching 91.5% after addition of Ni/ZrO. The 50-60% amount of Ba in BSCF could also stabilize oxidation state on the B cation resulting higher stability in the structure[12]t. BSCF membranes also exhibited good mechanical and thermal properties [13].
Despite of BSCF's high oxygen permeability, the oxygen permeability of these membranes should still be improved to reach the commercial target. In MIEC membrane, the oxygen permeability is determined by two factors: the ionic bulk diffusion along the crystal lattice and the surface exchange reaction in the membrane surface [7]. The faster surface exchange reaction, the higher oxygen permeability. Thus, one pathway to increase the oxygen reaction is increasing the surface exchange reaction by coating the BSCF surface with catalyst. Previously, Leo at.al. used Ag for coating layer on BSCF membranes. The presence of Ag layers could indeed increase membrane's permeability. However, the Ag particles are concentrated on the BSCF grain boundaries [14]. Leo at.al. coat the BSCF membrane using Pd nanoparticle and revealed that this coating could enhance the oxygen permeability up to 10 times at 700 °C [15]. The using of these noble metals clearly enhanced the oxygen permeability of BSCF membranes, however, this material is expensive thus make it less reliable for industrial scale.
Recently, Co_3O_4 is getting interest to be used as catalyst layer on BSCF surface as its good activity toward reaction involving oxygen, both in oxygen reduction reaction (ORR) or oxygen evolution reaction (OER). Li et.al. reported that the coating of BSCF membrane using Co_3O_4 nanorod could enhance the oxygen permeability up to 11.5 times higher than unmodified BSCF membrane at 850 °C [16]. Wang et.al. reported that the coating of Co_3O_4 nanoparticle on BSCF membrane's surface could enhance the oxygen permeability up to 4 times higher that unmodified membranes [17]. However, the methods to coat the catalyst layer on the BSCF membrane's surface were often complex involving several steps and releasing of dangerous compounds during calcination. The precursor of Co_3O_4 was usually come from nitrate ion that release NO_x during calcination that harmful for the environment. Moreover, the using of relatively complex technique such as hydrothermal requires complex procedure that could be resistant for the application on industrial scale. Due to these reasons, in this recent work, we try to propose the simple and effective way to coat Co_3O_4 layer on the BSCF xx82 membrane's surface. The coating process was performed using Co_3O_4 particles, and starch as binder agent to bind Co_3O_4 particles in membranes surface prior to calcination. The calcination after coating process leads to formation of Co_3O_4/BSCF xx82 membranes. The morphological properties of the resulted membranes were discussed.

2. Eexperimental

2.1. Material

Material used for the synthesis of BSCF are barium carbonate ($BaCO_3$) p.a. 99.5%, strontium carbonate ($SrCO_3$) p.a. 99.9%, cobalt(II,III) oxide (Co_3O_4) p.a. 99.5% and iron(III) oxide (Fe_2O_3) p.a. 97%. All the reactants were purchased form Merck and were directly used without further purification. The preparation of composite membrane was carried out using starch (technical grade) purchased from Brataco chemical.

2.2. Procedure

2.2.1 Synthesis of BSCF xx82

BSCF oxides were synthesized using solid state method. All reactants were dried in the oven at 105 °C for 1 our prior to the weighing. After drying, the reactants were placed in desiccator for cooling. The cooled reactants were then weighted according their stoichiometric amounts. The stoichiometric amount of reactants were then mixed and were ground together in porcelain pastel and mortar for 2 hours to form homogeneous mixture of reactants, indicated by the blending of the reactants color. After grinding, the reactants were calcined at 890 °C for 2 hours. This calcination is repeated once more time. The intermediate grinding for 30 minutes was applied between two calcination process. The obtained powders were calcined again at 1000 °C for 3 hours twice by applying intermediate grinding also. The heating rate of this calcination process is 20 °C /minute.
The resulted powders were characterized using X-ray diffraction (XRD) using CuKα (λ = 1.54060 A) irradiation at 2θ of 20-90°. The obtained XRD pattern were then matched with $Ba_{0,5}Sr_{0.5}Co_{0.8}Fe_{0.2}O_{3-\delta}$ standard obtained form Powder Diffraction File (PDF) card number 01-075-6980 published by JCPDS-International Center for Diffraction Standard 2012.

2.2.2 Preparation of Co_3O_4/BSCF xx82
The Co_3O_4/BSCF xx82 membranes were prepared by coating the starch-Co_3O_4 suspension to the surface of BSCF xx82 dense flat sheet membranes. For this purpose, the procedure is consisted of two steps. The first step is preparation of BSCF xx83 dense membrane. In this step, BSCF xx82 powders were sieved using 400 mesh sieve. The BSCF particles passing the sieve (particle size -400 mesh) were used for flat sheet membranes. Each BSCF xx82 with particle size -400 mesh were weighed for 1 g. The powders were then placed into 13 mm mold. The powders were pressed into pellets with pressure of 4 tons. The pressing was handled for 15 minutes. The obtained pellets were then sintered at 1150 °C for 4 hours. This sintering process is carried out using heating rate of 3 °C /minute. The shrinkage of flat sheet membrane was calculated using Equation 1.

$$\text{Shrinkage (\%)} = \frac{\text{Vi-Vf}}{\text{Vi}} \text{x } 100\% \qquad 1$$

were v_i is initial volume before and v_f is final volume after sintering. The density of flat sheet membrane was examined physically using water droplet test. The surface of membrane was observed using Scanning Electron Microscope (SEM) with 20kv accelerating voltage and Secondary Electron (SE) imaging.
The second step is coating the flat sheet membrane using starch-Co_3O_4 suspension. Prior to the coating process, the suspension is prepared by mixing 0.3 g starch with 2 mL water. The mixture was stirred to form starch suspension. After forming suspension, 3 mL of hot water was added and the stirring is continued to form starch gel. After obtaining gel, 0.3 g Co_3O_4 was added to the starch gel, and the stirring is continued on the hotplate to form homogeneous starch-Co_3O_4 suspension. The starch-Co_3O_4 suspension was then coated to the one surface of the each of BSCF xx82 membranes. The coated membranes were let dried overnight. After getting dried, the membranes were calcined at 800 °C for 3 hours with heating rate of 3 °C /minute to remove the starch. This process yielded Co_3O_4/BSCF xx82 composite membranes, i.e. Co_3O_4/BSCF 5582, Co_3O_4/BSCF 6482 and Co_3O_4/BSCF 7382 respectively. The membrane's surface and cross section is examined using SEM equipped with Energy Dispersive X Ray (EDX).

3. Result and Discusion

3.1 Synthesis of BSCF xx82
BSCF 5582, BSCF 6482 and BSCF 7382 were synthesized using solid state method. This method could produce material with high yield as well as high crystallinity. The reactants used in this work are in the form of carbonate and oxide with high purity. The selection of this reactant type is based on the property of the material. Previously, Aliyatulmuna et.al. had synthesized $La_{1-x}Sr_xCo_{0.8}Fe_{0.2}O_{3-\delta}$ (LSCF) by solid state method using nitrate compound, i.e. strontium nitrate ($Sr(NO_3)_2$. This compound is hygroscopic thus during the weighing, the compound may adsorp water vapor from the air resulting

the inappropriate amount of strontium. As the result, the resulted LSCF contained impurity [18]. For making sure the amount of reactants, all of the reactants were dried in oven prior to weighing. After weighing, all of reactants were mixed, ground together and then calcined at 890 °C. This calcination step is aimed to form perovskite as reported by Maulidah [19] who state that the cobalt based perovskites were formed in the calcination temperature of 880 °C. The next calcination step at 1000 °C is the optimation of the perovskite formation process as well as increasing BSCF crystallinity. Each calcination step is repeated twice, with intermediate grinding between calcination. This intermediate grinding is aimed to reduce the particles sizes and to form new contact area between the particles that can enhance the reaction [20].

The yielded powders from synthesis process were then characterized using XRD to confirm the formation of perovskite oxide [18], [21]. The XRD patterns of the three samples (BSCF 5582, BSCF 6482 and BSCF 7382) were then compared to the BSCF 5582 standard pattern from PDF Card 01-075-6980. The comparison of XRD patterns form the three samples with the standard is shown in Figure 1. Figure 1 shows that the three samples possess good accordance with the standard pattern as shown in the presence of BSCF 5582 characteristic peaks around 31, 45 and 55° in all samples. This confirms the formation of perovskite structure in all samples. The three samples were also have good intensity indicating that the samples possess high crystallinity. This high crystallinity is quite important in application of this material as oxygen membrane separation because the high crystallinity could reduce the resistance of oxygen ion diffusion due to grain boundary of the particles during permeation process. This XRD result also suggest that the three BSCF's could be proceed further for preparing Co_3O_4/BSCF xx82 membranes.

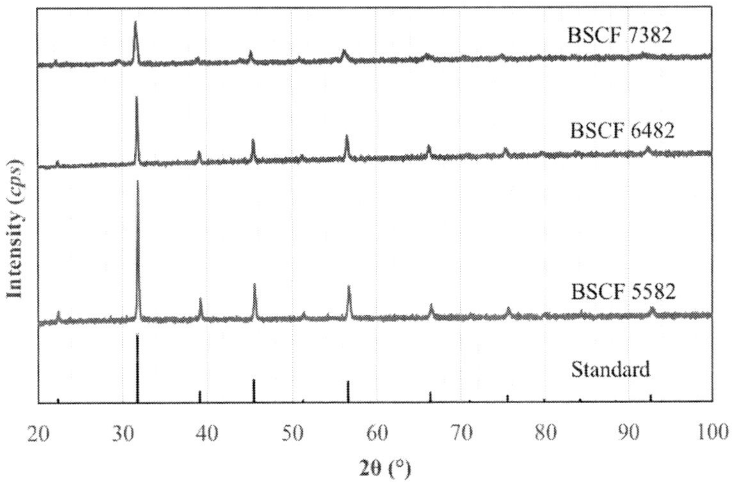

Figure 1. XRD patterns of three samples a) BSCF 5582, b) BSCF 6482 and c) BSCF 7382

3.2 Preparation BSCF Dense Membrane

Preparation of Co_3O_4/BSCF membrane was carried out by dry pressing method from the initial BSCF powders to form dense membrane followed by coating of one side of the dense flat membrane using starch-Co_3O_4 suspension. The calcination is then carried out to the coated flat BSCF membrane to remove the starch binder.

This process began with the preparation of green membranes from BSCF xx82 powders. This step was carried out by pressing 1 g of each BSCF xx82 powders with pressure of 5 tons followed by sintering process at 1150 °C for 4 hours at heating rate of 3 °C/minute [13]. The temperature of 1150 °C is the suitable temperature for the BSCF type membrane as in higher temperature, the BSCF will be melted due to the presence of Ba^{2+} that could be role as flux in solid solution. This flux species could decrease the melting point of the solid materials. On the other hand, the lower temperature could not produce

the dense membranes because the sintering of the BSCF particles is not optimally yet. This could lead to the presence of pores in the membrane. The pores in BSCF membrane lead to the reducing of membrane's ability to separate oxygen gas. In BSCF membrane separation system, the oxygen should be separated in the form of oxygen ions, not oxygen molecules, by diffusion of these ions via oxygen vacancies that exist in the BSCF crystal lattice [9]. The presence of pores lead to the diffusion of oxygen in the form of oxygen molecules and will be risky also to pass the other gas molecule via the membranes pore. This process could reduce the BSCF membrane's selectivity toward oxygen.
During the sintering process, the particles of BSCF xx82 membranes were sintered. This process made the physical change on the membrane before and after sintering. Before sintering, in general, the physical appearance of BSCF xx82 is brittle black disc solid. After sintering, the color of the membrane changed to lighter color similar to shiny silver as shown in Figure 2. This changing indicates that the BSCF xx82 particles were sintered producing the dense membrane [18].

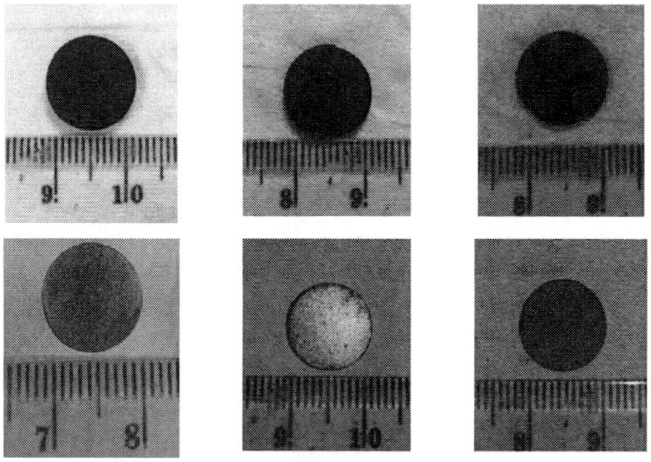

(a) (b) (c)

Figure 2. Membranes of a) BSCF 5582, b) BSCF 6482 and c) BSCF 7382. Upper side: before sintering, down side: after sintering at 1150 °C

The densification of the membranes could also be indicated by the reduction of membrane's dimension before and after sintering. Before sintering, the BSCF xx82 membranes posses similar dimension i.e.12.7 mm in diameter and 2.2-2.4 mm in thickness. These dimensions are varied after sintering. Based on this dimensional changing, the shrinkage of membranes could be calculated as shown in Table 1. Based on data in Table 1, the shrinkage of the membranes reaches 30-40% from its initial volume. This indicates that sintering of the particles is successfully occurred, where the BSCF xx82 particles were melted and bound each other removing the space between the particles. The highest shrinkage is exhibited by BSCF 6482 membrane, followed by BSCF 5582 and BSCF 7382. This result is in good accordance with the previous report [13]. This shrinkage could also be influenced by the membrane's morphology.

Table 1. Shrinkage of BSCF xx82 membrane after sintering

BSCF membrane	Diameter (mm)		Thickness (mm)		Volume (mm^3)		Shrinkage (%)
	Initial	Final	Initial	Final	Initial	Final	
5582	12.9	11.6	2.2	1.7	287.4	181.1	37.0
6482	12.7	11.6	2.4	1.7	383.9	278.5	40.9
7382	12.7	11.7	2.2	1.7	181.1	182.7	34.4

IOP Conf. Series: Materials Science and Engineering **588** (2019) 012040 doi:10.1088/1757-899X/588/1/012040

Previously, Utomo et.al. reported that Ba^{2+} could be acted as flux, a species that lower the melting temperature of the solid material. The increasing of Ba^{2+} amount lead to the higher decreasing of melting point of the materials thus the materials would be easier to be softened in lower temperature. Also, the particles will be easer to join. As the result, the whole materials would become denser as the particles join together and the pores were removed [22]. The dimension (volume) of the material will be reduced tremendously. This makes the BSCF 6482 posses higher shrinkage than BSCF 5582 because BSCF 6482 posses higher amount of Ba^{2+} in its A-site compare to BSCF 5582. However, the membrane sintering also causes the reduction of B-site cation that release oxygen gas [13]. As mentioned before, the higher amount of Ba^{2+}, the softer membrane material during sintering process. In this case, BSCF 7382 posses highest amount of Ba^{2+}. Due to the softening of the membrane particles, the oxygen gas released by the reduction of B-site cation is easier to suppress the BSCF 7382 particles resulting larger pore volume in the membrane's body. The formation of larger pore in the membrane's body would expand the membranes volume in BSCF 7382 thus the shrinkage would be reduced as can be seen from shrinkage data in Table 1.

To physically test the density of membranes, the water is dropped on the top surface of the membranes. If the membrane could resist the water droplet on its surface without absorbing it, thus the membrane was dense. This method is usually called water droplet test. The result of this water droplet test is shown in Figure 3. Figure 3 clearly shows that the water is resisted on the membrane's surface. This indicates that the membrane is dense as no pore in the membrane that could absorp the water due to capillarity. However, in dense membrane, the pores may still exist, as indicated by the different in shrinkage value. However, in the dense membrane, the pores that generally exist are the isolated pores, meaning that the pores are not connected between each others thus could not provide the pathways to water to be absorp in the body of membrane via pores.

(a) (b) (c)

Figure 3. Water droplet test for BSCF xx82 membranes
a) BSCF 5582, b) BSCF 6482 and c) BSCF 7382

SEM analysis was then carried out to the surface of the membrane to confirm the density in the membrane's surface. The SEM images of the three membrane's surface were shown in Figure 4. Figure 4 shows that all membrane's surfaces were dense with clear grain boundary. However, the there are several points in the surface of BSCF 6482 as indicated by the rectangular line. EDX analysis of this point revealed that these points contain calcium that is not used in this experiment. The presence of calcium is predicted as impurity coming from ruins of firebrick in the furnace during sintering process. The calcium is one of the element that compose firebrick on furnace [23]. The result of the shrinkage calculation, water droplet test and SEM images show that the dense membrane was ready to be processed for coating with starch-Co_3O_4 suspension.

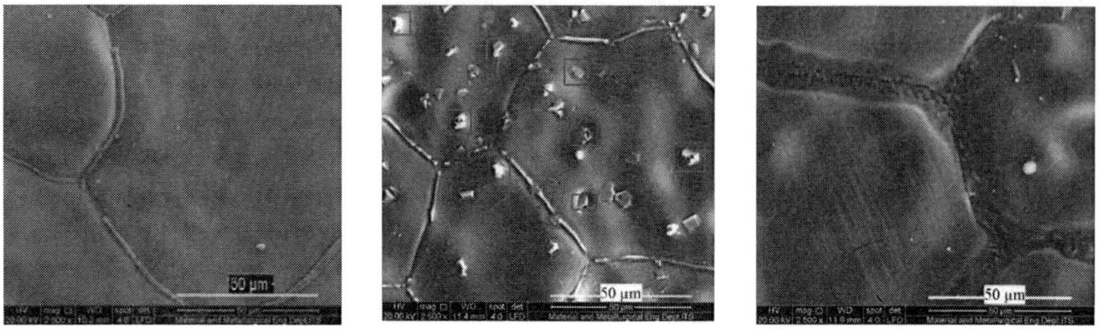

(a) (b) (c)

Figure 4. SEM images of flat sheet membrane's surface
a) BSCF 5582, b) BSCF 6482 and c) BSCF 7382

3.3 Preparation and characterization of Co_3O_4/BSCF xx82

The resulted BSCF membranes were then coated with starch-Co_3O_4 suspension. The starch- Co_3O_4 suspension was coated in one surface of the membrane and then aged for 24 hours. In this work, after coating the BSCF xx82 membrane's surface with starch-Co_3O_4 suspension, the coated membranes were then calcined at 800 °C. In this process, the starch would begin to decompose at 700 °C left the Co_3O_4 particles in the membrane's surface. The BSCF xx82 and Co_3O_4 particles were then partially sintered forming deposit of Co_3O_4 in the surface of BSCF xx82 membrane forming Co_3O_4/BSCF xx82 composite membranes.

Scanning Electron Microscope (SEM) was then used to investigate the surface of BSCF membranes after calcination. The surface morphology of the three composite membranes was shown in Figure 5. Figure 5 shows that the Co_3O_4 particles were successfully coated to the surface of BSCF xx82 membranes. This confirms that the coating method using starch as binder agent succeeds to coat Co_3O_4 layer in the BSCF xx82 surface. However, the dispersion of Co_3O_4 particles was not perfectly uniform indicated by the presence of BSCF surface that does not coated by the Co_3O_4 particles as pointed by the arrow sign. Also, it seems that the Co_3O_4 particles form aggregate [24]. To further analyze the Co_3O_4 particle's shape and size, as well as the aggregate size, the analysis was conducted to the BSCF 5582 as representative sample. The SEM image of Co_3O_4 coating in BSCF 5582 is shown in Figure 6.

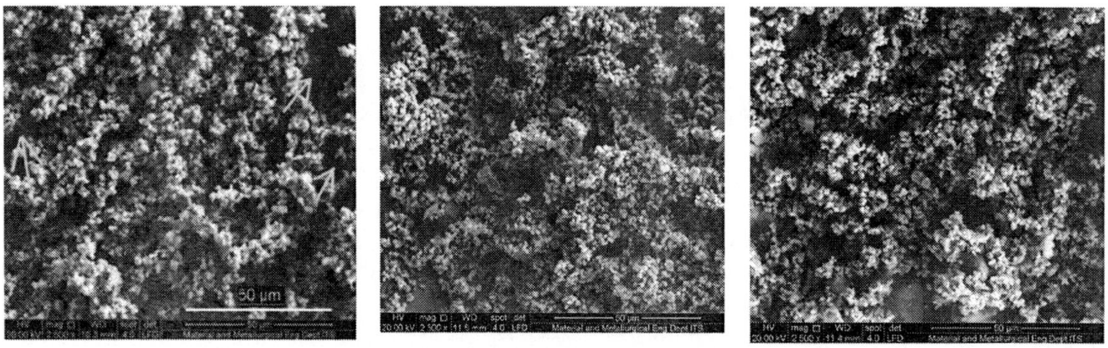

(a) (b) (c)

Figure 5. SEM images of Co_3O_4/BSCF membrane's surface
a) Co_3O_4/BSCF 5582, b) Co_3O_4/BSCF 6482 and c) Co_3O_4/BSCF 7382

IOP Conf. Series: Materials Science and Engineering **588** (2019) 012040 doi:10.1088/1757-899X/588/1/012040

Figure 6. Co_3O_4 layer in Co_3O_4/BSCF 5582 membrane

Figure 6 shows that the Co_3O_4 particles have firm angles [25]. These particles are bound each other forming aggregate. The measurement of this particles size reveled that the size of the particle is around 0.5-2.0 μm. The aggregate size (as shown in dash square sign as example) is varied from 2-5 μm. This shows that the particles size as well as the aggregate is larger than the previous result as reported by Wang et.al [17] and Chen et.al [16]. Wang et.al. succeeded to coat Co_3O_4 nanoparticle on the surface of BSCF membrane. The mean particle size was about 100 nm in diameter and 20 μm in thickness. Whereas, Li et.al. succeeded to coat Co_3O_4 nanorod on the surface of BSCF. The particle size of this nanorod was about 100 nm in diameter and 2 μm in length. Both reports succeded to coat a nano-sized Co_3O_4 particles in the BSCF membrane which can not be achieved in this work. This phenomenon is mainly caused by the initial reactant of Co_3O_4 layer. Wang et.al. and Li et.al. used $Co(NO_3)_2{\cdot}6H_2O$ as starting material to form Co_3O_4. Despite of difference coating method (using hydrothermal and dip coating method), the starting material was dissolved in the certain solvent to form solution. The Co_3O_4 particles were formed during calcination of BSCF surface coated with solution containing cobalt ion. The heating under air atmosphere creating the Co_3O_4 that coated on the BSCF surface. This process allow to the formation of Co_3O_4 particles from the smallest size to the larger one when the cobalt ion in the solution contact with oxygen in the air to form Co_3O_4. The Co_3O_4 particles then grew along with time and temperature (bottom-up process). This process is different with this recent work. In our works, the Co_3O_4 layer is composed form the initial Co_3O_4 powders. The Co_3O_4 particles had been existed prior to coating process. Thus, the smallest particle in the Co_3O_4 layer would be as big as the initial Co_3O_4 powder/particles size. Furthermore, the calcination process could also initiate the sintering process between the Co_3O_4 particles forming larger particles and/or aggregate as can be seen in Figure 6. This phenomenon occurs in the three samples.

The SEM analysis was then continued to the cross section of the three Co_3O_4/BSCF xx82 membranes to convince the coating of Co_3O_4 particles on the surface of BSCF xx82 membranes. The SEM images completed with membrane's mapping using EDX were shown in Figure 7. It can bee seen form Figure 6 that there is strict border between BSFF xx82 surface with Co_3O_4 layer. The BSCF membranes exhibit excellent density form its cross section. Several pores appear on the membrane's body, however, the amount of pore is minor compare to the whole membranes body. Beside that, the pores are isolated which means that the pore can not become the pathway for oxygen molecules to diffuse. The Co_3O_4 particles were indicated by the irregular particle coated on the surface of the membranes. The thickness of this layer is about 2-10 μm. The mapping using EDX in the cross section exhibits the increasing intensity of Co element confirming the presence of Co_3O_4.

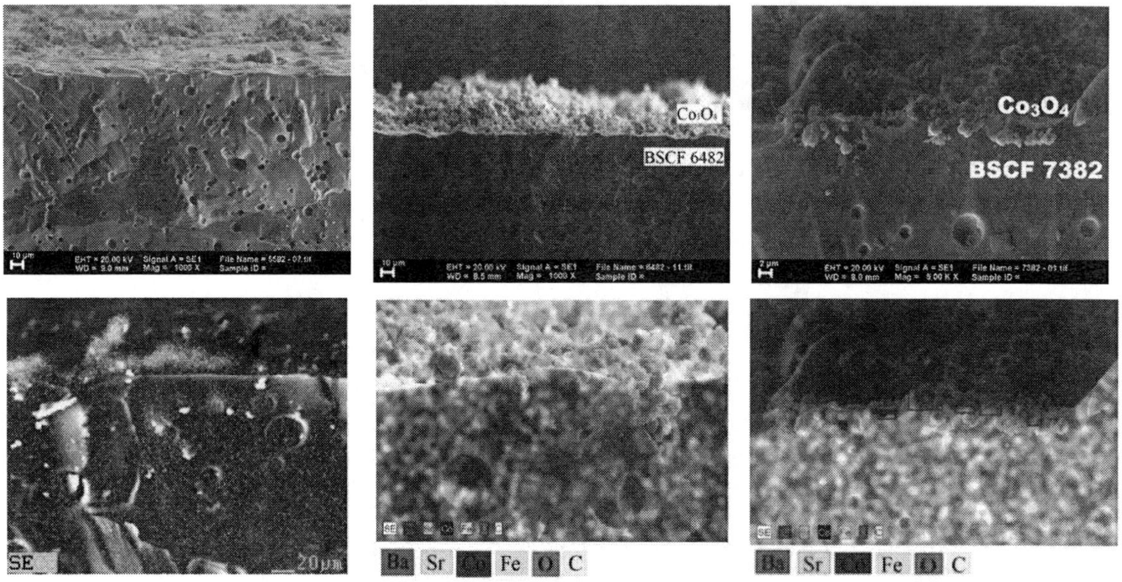

(a) (b) (c)

Figure 7. Mapping images of membrane's cross section of a) Co_3O_4/BSCF 5582, b) Co_3O_4/BSCF 6482 and c) Co_3O_4/BSCF 7382

4. Conclusion

The Co_3O_4 particles are successfully coated to the surface of BSCF xx82 using starch as binder agent. This coating process forms composite membranes of Co_3O_4/BSCF 5582, Co_3O_4/BSCF 6482 and Co_3O_4/BSCF 7382. The different composition of membranes body leads to the difference of the membranes shrinkage. BSFF 5582 exhibits the highest shrinkage. However, the observation using SEM revealed that all membranes are dense. The Co_3O_4 particles are distributed irregularly and the particles form aggregates. The particle size of the Co_3O_4 varied from 0.5-2.0 μm whereas the aggregate size is around 5-10 μm. The Co_3O_4 layers were thin enough with thickness of 2-10 μm. This result indicates that the coating of Co3O4 layers using starch as binder gent could be potential and effective method to fabricate Co_3O_4/BSCF xx82 composite membranes.

5. Acknowledgement

The authors greatly acknowledge the LPPM Institut Teknologi Sepuluh Nopember (ITS) for funding this research under Penelitian Pemula grant 2018 with contract number of 1359/PKS/ITS/2018.

6. References

[1] V J Ferreira 2013 *Development of Nanostructured Catalysts for The Oxidative Coupling of Methane* Doctoral Thesis, Porto, Portugal

[2] C F Zuhra 2003 *Penyulingan, Pemrosesan dan Penggunaan Minyak Bumi*

[3] Z Shao, G Xiong, H Dong, W Yang, and L Lin 2001 *Synthesis, oxygen permeation study and membrane performance of a Ba0.5Sr0.5Co0.8Fe0.2O3−l oxygen-permeable dense ceramic reactor for partial oxidation of methane to syngas* p. 20

[4] H Luo *et al* 2010 Performance of a ceramic membrane reactor with high oxygen flux Ta-containing perovskite for the partial oxidation of methane to syngas *Journal of Membrane Science* **vol. 350** no. 1–2, pp. 154–160

[5] Z Shao, H Dong, G Xiong, Y Cong, and W Yang 2001 Performance of a mixed-conducting ceramic membrane reactor with high oxygen permeability for methane conversion *Journal of Membrane Science* **vol. 183** no. 2, pp. 181–192

[6] W Chen, N Nauels, H J M Bouwmeester, A Nijmeijer, and L Winnubst 2015 An accurate way to determine the ionic conductivity of mixed ionic–electronic conducting (MIEC) ceramics," *Journal of the European Ceramic Society* **vol. 35** no. 11, pp. 3075–3083

[7] J Sunarso *et al* 2008 Mixed ionic–electronic conducting (MIEC) ceramic-based membranes for oxygen separation *Journal of Membrane Science* **vol. 320** no. 1–2, pp. 13–41

[8] S Baumann *et al* 2010 Influence of sintering conditions on microstructure and oxygen permeation of Ba0.5Sr0.5Co0.8Fe0.2O3−δ (BSCF) oxygen transport membranes *Journal of Membrane Science* **vol. 359** no. 1–2, pp. 102–109

[9] A Behrouzifar, A A Asadi, T Mohammadi, and A Pak 2012 Experimental investigation and mathematical modeling of oxygen permeation through dense Ba0.5Sr0.5Co0.8Fe0.2O3−δ (BSCF) perovskite-type ceramic membranes *Ceramics International* **vol. 38** no. 6, pp. 4797–4811

[10] A Ghadimi, M A Alaee, A Behrouzifar, A A Asadi, and T Mohammadi 2011 Oxygen permeation of BaxSr1−xCo0.8Fe0.2O3−δ perovskite-type membrane: Experimental and modeling *Desalination* **vol. 270** no. 1–3, pp. 64–75

[11] S Faraji, K J Nordheden, and S M Stagg-Williams 2010 A comparative study of Ba0.5Sr0.5Co0.8Fe0.2Ox (BSCF) and SrFeCo0.5Ox (SFC) ceramic membranes used for syngas production *Applied Catalysis B: Environmental* **vol. 99** no. 1, pp. 118–126

[12] S Tanasescu *et al* 2013 Effects of A-site composition and oxygen nonstoichiometry on the thermodynamic stability of compounds in the Ba–Sr–Co–Fe–O system *Journal of Solid State Chemistry* **vol. 200**, pp. 354–362

[13] H Fansuri *et al* 2017 Preparation of dense BaxSr1-xCo0.8Fe0.2O3 membranes: Effect of Ba2+ substituents and sintering method to the density, hardness and thermal expansion coefficient of the membranes *Advanced Materials Letters* **vol. 8** no. 7, pp. 799–806

[14] A Leo, S Liu, and J C Diniz da Costa 2009 The enhancement of oxygen flux on Ba0.5Sr0.5Co0.8Fe0.2O3−δ (BSCF) hollow fibers using silver surface modification *Journal of Membrane Science* **vol. 340** no. 1, pp. 148–153

[15] A Leo, S Smart, S Liu, and J C Diniz da Costa 2011 High performance perovskite hollow fibres for oxygen separation *Journal of Membrane Science* **vol. 368** no. 1, pp. 64–68

[16] S Li, J Cheng, Y Gan, P Li, X Zhang and Y Wang 2015 Enhancing the oxygen permeation rate of Ba0.5Sr0.5Co0.8Fe0.2O3−δ membranes by surface loading Co3O4 nanorod catalysts *Surface and Coatings Technology* **vol. 276** pp. 47–54

[17] Y Wang, J Cheng, M Huang, M Liu, M Li, and C Xu 2017 Effects of surface modification with Co 3 O 4 nanoparticles on the oxygen permeability of Ba 0.5 Sr 0.5 Co 0.8 Fe 0.2 O 3-δ membranes *Applied Surface Science* **vol. 416** pp. 574–580

[18] A Aliyatulmuna, W P Utomo, H Fansuri, and I K Murwani 2017 Thermal Expansion, Microhardness and Oxygen Permeation of La1-xSrxCo0.8Fe0.2O3-δ Membranes *Asian Journal of Chemistry* **vol. 29** no. 10, pp. 2191–2196

[19] N Maulidah 2010 *Sintesis dan Karakterisasi Oksida Perovskit La1-xSrxCo1-yFeyO3-δ (0,0≤x,y≤0,5) Dengan Metode Solid-State* Surabaya, Undergraduate Thesis, Institut Teknologi Sepuluh Nopember

[20] A R West, 1987 *Solid State Chemistry and Its Applications* John Wiley & Sons

[21] A B Pambudi, R Kurniawati, A Iryani, and D Hartanto 2018 Effect of calcination temperature in the synthesis of carbon doped TiO2 without external carbon source *AIP Conference Proceedings* **vol. 2049** no. 1, p. 020074

[22] W P Utomo 2014 *Kajian Permeabilitas Ion Oksigen Pada Membran Rapat Asimetris La0,7Sr0,3Co0,8Fe0,2O3-δ (LSCF 7382), SrCo0,8Fe0,2O3-δ (SCF 182) dan Ba0,5Sr0,5Co0,8Fe0,2O3-δ (BSCF 5582)* Surabaya, Thesis, Institut Teknologi Sepuluh Nopember

[23] S Lukin and G A Serova *Certain Properties Of Calcium-Oxide Refractories* p. 14

[24] C H Yulianti *et al* 2014 Synthesis of CaOZnO Nanoparticles Catalyst and Its Application in Transesterification of Refined Palm Oil," *Bulletin of Chemical Reaction Engineering & Catalysis* **vol. 9** no. 2, pp. 100–110

[25] S C Zi *et al* 2016 New method to synthesize mesoporous titania by photodegradation of surfactant template *Solid State Sciences* **vol. 52**, pp. 83–91

IOP Conf. Series: Materials Science and Engineering **588** (2019) 012041 doi:10.1088/1757-899X/588/1/012041

Experimental study of influence injection moulding on shrinkage material sisal-glass fiber biocomposite using response surface methodology (RSM)

M A Al-amin [a] , D Harnany [b] and I M L Batan[c]

Mechanical Engineering, Institut Teknologi Sepuluh Nopember, Sukolilo,Surabaya, Indonesia, 60111

[a]muhammadakbar.018@gmail.com , [b]dinnyharnany@gmail.com
[c]londbatan@me.its.ac.id

Abstract. Shrinkage is one of the properties of plastic materials, which the greater the shrinkage percentage value can result poor product quality and possible product defects. Therefore, it is important to know the shrinkage value of a material so that the geometry design of a product can be achieved properly. In this final project, shrinkage control is carried out by adjusting injection process parameters, until minimum shrinkage is obtained. The parameters set to control shrinkage values are melting temperature, holding pressure, and injection pressure. According to ASTM D955 to determine the shrinkage value of a material the specimen is made through the injection process. To obtain the minimum shrinkage value, an experimental design was carried out by using response surface method to reach the optimum injection process parameters with Box-Behnken parameter design. As the next step, the injection specimens which made from sisal fiber biocomposite material with composition of 80% Polypropelene, 10% MAPP, and 10% sisal fiber with injection process parameters according to the design. The injection result specimens were measured by calipers, Minitab 18 software was used to process the injection result. This result is used to determine the optimum value of the injection process parameters. Furthermore, with the optimum injection process parameters, the injection process is carried out again with the same material. The specimens obtained were measured to carry out a confirmation test, so that the minimum shrinkage value was obtained. From the statistical response surface analysis using Minitab 18 software, the optimum injection process parameters were obtained at 185^0 C, holding pressure 35 bar and injection pressure 60 bar with shrinkage values of 1.05%.

1. Introduction

Shrinkage of products written in units of percent of products where the value of shrinkage of each plastic material varies. In the process of injection molding in the manufacture of plastic biocomposite materials many components with complex designs and require a high degree of accuracy. The shrinkage properties of a material can cause a decrease in the precision of a product. In 2017 Deni Gumelar in his assignment finally designed injection molding of shrinkage specimens that were adjusted to the ASTM D955 standard and could be used on HAITIAN MA 900 / 260e injection machines. Then Tubagus Bima [5] conducted a study and analysis of shrinkage of biocomposite material with a mixture of sisal fiber and rice husk, in which studies and analyzes of biocomposite materials were printed into shrinkage specimens that were in accordance with ASTM D955 standards.

The material used in the simulation and injection molding process is a mixture of 10% biocomposite rice husk, 5% MAPP, 85% PP and mixed biocomposite 10% sisal fiber, 5% MAPP, 85% PP. The results showed varied shrinkage values, this was caused by problems with the mold. Kazmer[2] defines parameters that have the most significant effect on shrinkage, including pressure and temperature. The holding time parameter, injection time (time) has a non-significant impact on shrinkage. This research will be carried out using a HAITIAN MA 900/260e injection machine, there are several parameters that can be used including Melt Temperature, Pressure Holding, and Injection Pressure. Determination of process parameters using the Response Surface method which is one of the concepts of statistical approaches using simple and strong experimental designs[3]. In the injection molding process, several previous studies adopted the method to determine good results in determining the parameters that affect shrinkage material. The objectives of this research include knowing the effect of melt temperature, Holding Pressure, Injection Pressure, on shrinkage that occurs in sisal fiber biocomposite material. Knowing the optimal injection process parameters to minimize shrinkage values in ASTM D955 specimens using Response Surface Methodology (RSM). Injection machines used in data collection to determine parameters that affect shrinkage on standard specimens ASTM D955 is an injection machine made in HAITIAN China with the type MA 900/260E. The experimental design based on the *box-behnken* design[3] can be shown in Table 1.

Table 1. Box-Behnken Experiment Design

StdOrder	RunOrder	Blocks	Parameter Proses		
			MT (^{0}C)	HP (Bar)	IP (Bar)
7	1	1	185	30	60
4	2	1	205	35	55
5	3	1	185	30	50
14	4	1	195	30	55
15	5	1	195	30	55
10	6	1	195	35	50
11	7	1	195	25	60
8	8	1	205	30	60
2	9	1	205	25	55
3	10	1	185	35	55
1	11	1	185	25	55
13	12	1	195	30	55
6	13	1	205	30	50
9	14	1	195	25	50
12	15	1	195	35	60

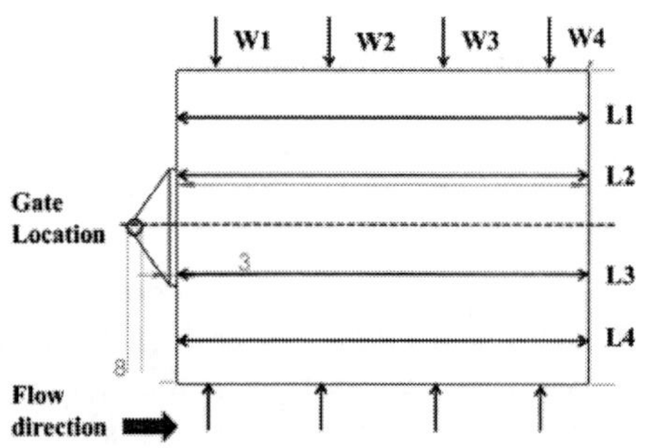

Figure 1. Measurement location of the length and width of the specimen

2. Specimen Measurement

The results of data retrieval experiments based on the *box-behnken* design are shown in Table 2 with Melt Temperature (MT), Holding Pressure (HP), and Injection Pressure (IP) shrinkage (Sh) response.

Table 2. Measurements in the direction of flow (L) and x-flow (W)

Percobaan Ke-	Parameter Proses			Sh L (%)	Sh W (%)
	MT (^{0}C)	HP (Bar)	IP (Bar)		
1	185	30	60	1.17	1.48
2	205	35	55	1.26	1.50
3	185	30	50	1.18	1.49
4	195	30	55	1.26	1.53
5	195	30	55	1.26	1.53
6	195	35	50	1.20	1.46
7	195	25	60	1.23	1.50
8	205	30	60	1.25	1.52
9	205	25	55	1.27	1.53
10	185	35	55	1.09	1.40
11	185	25	55	1.24	1.58
12	195	30	55	1.23	1.51
13	205	30	50	1.28	1.58
14	195	25	50	1.18	1.55
15	195	35	60	1.09	1.38

The process of retrieving data where the shrinkage L results (in the direction of flow) in table 2, get the lowest shrinkage value on the 10th and 15th experiments at 1.09%. The injection parameters in the 10th trial included melt 185^{0}C temperature, holding pressure 35 bar and injection pressure 55 bar. The 15th experiment used injection parameters melt temperature 195^{0}C, holding pressure 35 bar and injection pressure 60 bar. The extraction process for shrinkage W results (x-flow) in Table 2, the lowest shrinkage value in the 15th experiment was 1.38%. The injection parameters in the 15th experiment included melt temperature of 195^{0}C, holding pressure 35 bar and injection pressure 60 bar. Figure 2 [left] is one example of measuring specimens in sisal-glass fiber biocomposite material in the direction of the flow of plastic coming while Figure 2 [right] sisal-glass fiber biocomposite material is x-flow to the flow.

Figure 2. Examples of Measurement in ASTM D955 shrinkage test specimens [left] sisal-glass fiber biocomposite in the direction of flow, [right] sisal-glass fiber biocomposite x-flow

2.1. Comparison of Sisal-Glass Biocomposite Value

Comparison of graphs of shrinkage material values is shown in Figure 3. The flow rate of sisal-glass fiber biocomposite material (L) has a lower percentage than the x-flow (W) direction of flow of plastic arrival. Plastic and biocomposite materials have anisotropic properties which have different properties depending on the direction of flow. Anisotropic material where the molecule or fiber becomes parallel in the flow field during the filling process and packing process as shown in Figure 4. Shear flow during the filling and packing process will tend to direct the plastic melt towards the flow due to a large part of the thickness of a specimen.

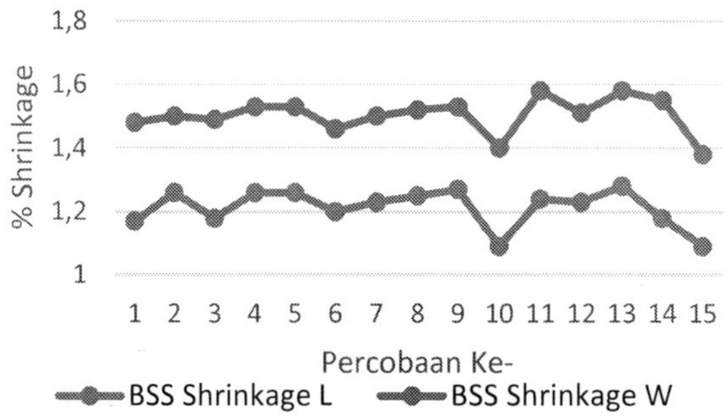

Figure 3. Comparison of percentage of shrinkage in terms of plastic flow

Phenomenon can be influenced by the biocomposite material properties including density and Coefficient of Thermal Expansion. The percentage of shrinkage produced is smaller than the direction of flow of plastic, this can be caused by sisal fiber material distributed in the direction of plastic flow so that it can withstand shrinkage that occurs in the direction of the flow of plastic. This can also be caused by the lower value of the coefficient of thermal expansion material which also tends to have a lower shrinkage value.

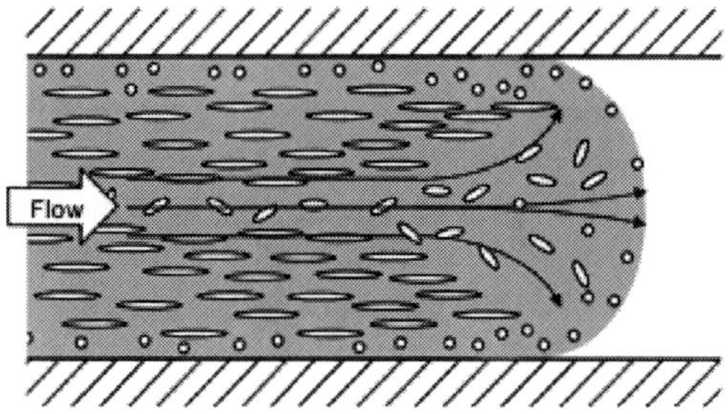

Figure 4. Orientation of plastic flow leads to Anisotropic[2]

2.2. Paired Sample-T Analysis

Figure 5. shows a graph of the average percentage of shrinkage values in each specimen point. The 1st and 4th points have a lower shrinkage percentage value than the 2nd and 3rd point. This is because the lowest plastic liquid pressure is at the center and increases from the center to the mold wall. The cooling process in plastic fluids tends to begin in the contact area with the cavity side toward the center of the specimen. This has a significant effect on the orientation of the flow of polymer molecules and a large percentage of the value of shrinkage. The percentage of shrrinkage at point 1 is lower than at point 4, because the direction of the flow of plastic starts at the 1st point and ends at the 4th point table 3.

Based on data Table 3. shows T_{hitung} = 33.02, while T_{table} for *two-tail* $\alpha = 0.05$ with v= 15-1= 14 regions received in distribution *t* is < 2.145. Thitung value > T_{table}, then H_0 is rejected. The statement that can be concluded is that the measurement of x-flow is different with the direction of flow in the same parameter.

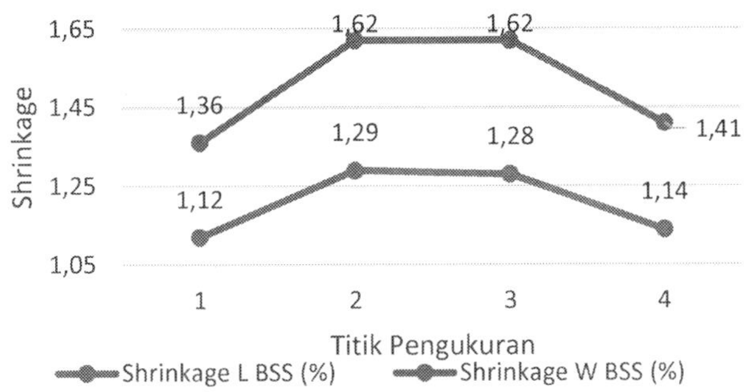

Figure 5. Percentage of Each Point Shrinkage

Table 3. Paired sample-T test results

Descriptive Statistics

Sample	N	Mean	StDev	SE Mean
tegak lurus	15	1.5027	0.0566	0.0146
searah	15	1.2127	0.0606	0.0157

Estimation for Paired Difference

Mean	StDev	SE Mean	95% CI for μ_difference
0.29000	0.03402	0.00878	(0.27116, 0.30884)

μ_difference: mean of (in flow and x-flow)

Test

Null hypothesis H_0: μ_difference = 0

Alternative hypothesis H_1: μ_difference ≠ 0

T-Value	P-Value
33.02	0.000

2.3. Response Surface Methodology Analysis

Analysis of estimation of second-order models of continuous Sisal-glass fiber biocomposite material the flow of plastic to shrinkage values using Minitab 18 software can be seen in Table 4. and Table 5.

Table 4. Second-order BSS Shrinkage in the direction of flow

Analysis of Variance

Source	DF	Adj SS (10^{-4})	Adj MS (10^{-3})	F-Value	P-Value
Model	9	50.59	5.621	31.23	0.001
Linear	3	29.10	9.700	53.89	0.000
Melt Temperature	1	18.05	18.05	100.2	0.000
Holding Pressure	1	9.800	9.800	54.44	0.001
Injection Pressure	1	1.250	1.250	6.94	0.046
Square	3	10.09	3.364	18.69	0.004
Melt Temperature*Melt Temperature	1	0.092	0.092	0.51	0.506
Holding Pressure*Holding Pressure	1	5.908	5.908	32.82	0.002
Injection Pressure*Injection Pressure	1	4.523	4.523	25.13	0.004
2-Way Interaction	3	11.40	3.800	21.11	0.003
Melt Temperature*Holding Pressure	1	4.900	4.900	27.22	0.003
Melt Temperature*Injection Pressure	1	0.100	0.100	0.56	0.490

Holding Pressure*Injection Pressure	1	6.400	6.400	35.56	0.002
Error	5	0.900	0.180		
Lack-of-Fit	3	0.300	0.100	0.33	0.808
Pure Error	2	0.600	0.300		
Total	14	51.493			

Model Summary

S	R-sq	R-sq(adj)
0.0134164	98.25%	95.11%

In Table 4 shows the output from Minitab for the regression analysis of variance in the shrinkage response. In determining the regression significance test there is a linear relationship of the shrinkage response variable to the process parameters using the hypothesis:

$$H_0: \beta_1 = \beta_2 = \cdots = \beta_k = 0$$

$$H_1: \beta_1 \neq 0 \text{ at least one } j$$

The Analysis of Variance (ANOVA) table uses F-value because it looks for variance where several samples are owned from several populations with the same sign. Variations of all the independent variables, namely the parameters of the melt temperature injection process, holding pressure, and injection pressure both linear, square and 2-way interaction on dependent shrinkage variables using the *distribution F* table. Estimation of the second order model of shrinkage response can be known for parameters the process with the value that can be explained R^2 = 98.25%. The output data from ANOVA can be concluded that the second order estimation model (square) is appropriate. To examine whether or not a second-order model is significant, it can be seen the p-value of regression in table 5. The p-value of 0.001 is smaller (<) than the value of α of 5%, this means that the process parameters have a significant impact on the second-order model.

Table 5. Estimation of Model Equation Coefficient

Coded Coefficients

Term	Coef	SE Coef	T-Value	P-Value
Constant	1.25000	0.00775	161.37	0.000
Melt Temperature	0.04750	0.00474	10.01	0.000
Holding Pressure	-0.03500	0.00474	-7.38	0.001
Injection Pressure	-0.01250	0.00474	-2.64	0.046
Melt Temperature*Melt Temperature	0.00500	0.00698	0.72	0.506
Holding Pressure*Holding Pressure	-0.04000	0.00698	-5.73	0.002
Injection Pressure*Injection Pressure	-0.03500	0.00698	-5.01	0.004
Melt Temperature*Holding Pressure	0.03500	0.00671	5.22	0.003
Melt Temperature*Injection Pressure	-0.00500	0.00671	-0.75	0.490
Holding Pressure*Injection Pressure	-0.04000	0.00671	-5.96	0.002

In Table 5. shows the testing of each regression coefficient on the resulting regression model, it can occur in adding variables or deleting one or more existing variables. Adding variables to the second-order regression model can cause the sum of square for regression to increase and the number of squared errors (error sum of square) to decrease. Adding variables that do not affect the estimation of the model can increase the mean square error, thereby reducing the usefulness of estimating the model. Hypothesis to test the significance of each individual regression coefficient:

$$H_0: \beta_j = 0$$

$$H_1: \beta_j \neq 0$$

If $H_0: \beta_j = 0$ is failed to be rejected, then it can be indicated that x_j can be removed from the estimated model.

In estimating the second-order regression model in Table 5 the factors that did not have a significant impact included quadratic (square) *melt temperature*melt temperature* and interaction factor (2-way interaction) *melt temperature*injection pressure*. The p-value of the factor is α> 0.05, so that these factors can be eliminated. Parameters that do not have a significant impact can be removed and then re-analyzed using Minitab 18, these results are shown in Table 6.

Table 6. Results Of ANOVA Eliminate the significant process parameters

Analysis of Variance

Source	DF	Adj SS (10^{-4})	Adj MS (10^{-3})	F-Value	P-Value
Model	7	50.40	7.200	46.14	0.000
Linear	3	29.10	9.700	62.16	0.000
Melt Temperature	1	18.05	18.05	115.6	0.000
Holding Pressure	1	9.800	9.800	62.80	0.000
Injection Pressure	1	1.250	1.250	8.01	0.025
Square	2	10.00	5.001	32.05	0.000
Holding Pressure*Holding Pressure	1	6.058	6.058	38.82	0.000
Injection Pressure*Injection Pressure	1	4.651	4.651	29.80	0.001
2-Way Interaction	2	11.30	5.650	36.21	0.000
Melt Temperature*Holding Pressure	1	4.900	4.900	31.40	0.001
Holding Pressure*Injection Pressure	1	6.400	6.400	41.01	0.000
Error	7	1.092	0.156		
Lack-of-Fit	5	0.492	0.098	0.33	0.864
Pure Error	2	0.600	0.300		
Total	14	51.493			

Model Summary

S	R-sq	R-sq(adj)
0.0124918	97.88%	95.76%

Coded Coefficients

Term	Coef	SE Coef	T-Value	P-Value
Constant	1.25308	0.00600	208.82	0.000
Melt Temperature	0.04750	0.00442	10.76	0.000
Holding Pressure	-0.03500	0.00442	-7.92	0.000
Injection Pressure	-0.01250	0.00442	-2.83	0.025
Holding Pressure*Holding Pressure	-0.04038	0.00648	-6.23	0.000
Injection Pressure*Injection Pressure	-0.03538	0.00648	-5.46	0.001
Melt Temperature*Holding Pressure	0.03500	0.00625	5.60	0.001
Holding Pressure*Injection Pressure	-0.04000	0.00625	-6.40	0.000

To prove the results in Table 6 the tests are carried out as follows:

1. Model suitability testing

To find out the suitability of the model above, then testing includes:

a. R^2 Test

Table 3.5 shows that as many as 97.88% of the response variations can be explained by this estimation.

b. Lack Of Fit Test

The lack of fit test for the model obtained p-value of 86.4% or greater than the degree of significance α = 5%. This value indicates that the regression model is appropriate.

2. Testing Of Regression Coefficients

a. Individually

In testing with α <0.05, injection process parameters in table 3.5 have p-value <0.05. this represents a process parameter that has a significant effect on shrinkage response.

b. Simultaneously

Based on data Table 3.5 shows Fcount = 46.14, while Ftable = F (7; 14; 0.05) = 3.53. The value of Fcount> Ftable can be concluded that the process parameters have a significant impact on the second-order regression model.

3. Testing Residual Assumptions

To check the adequacy of the model, not only do testing lack of fit, there is one test, namely residual analysis. Residual analysis is iudentik, independent, and normally distributed with mean zero on certain variations. Some tests carried out using residual analysis:

a. Identity

In Figure 6 below, it can be seen that the plots between residuals and fitted values appear randomly around the gold price and do not form a certain pattern. This shows that identical (homogeneous) residual variance is fulfilled.

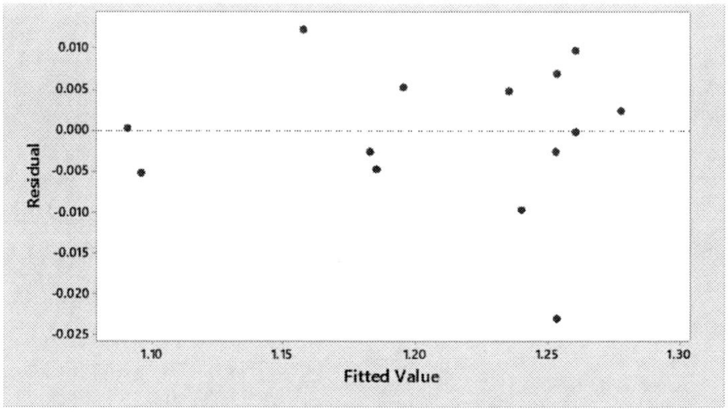

Figure 6. Residual graph vs. fitted value shrinkage BSS in the direction of flow

b. Normality

Testing of residual normal assumptions was carried out by the Kolmogorov-smirnnov test. The hypothesis used for normal testing:

H_0 : Residual regression models are normally distributed.

H_1 : Residual regression models are not normally distributed.

Testing of residual normal assumptions was carried out by the Kolmogorov-smirnnov test. The test results with a degree of significance $\alpha = 5\%$ are shown in Figure 7. The Kolmogorov-Smirnov (KS_{hitung}) statistical value is 0.212, while the Kolmogorov-Smirnov value of the table (KS_{tabel}) is $\alpha = 0.05$ and the number of observations is 0.34. Because $KS_{hitung} < KS_{tabel}$ then the residual of the estimation model has normal distribution.

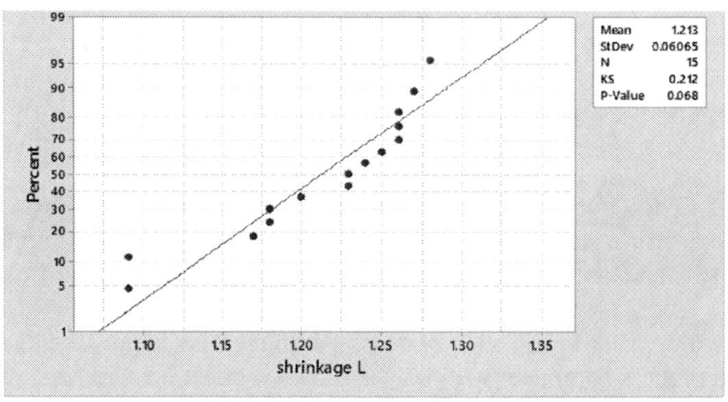

Figure 7. Graph Normal Probabbility Shrinkage

c. Independence
Observations on this research have been carried out independently and there was no correlation between observations. This is evidenced by the AutoCorrelation Function (ACF) plot graph in Figure 8. the graph shows that all correlations are still in the interval.

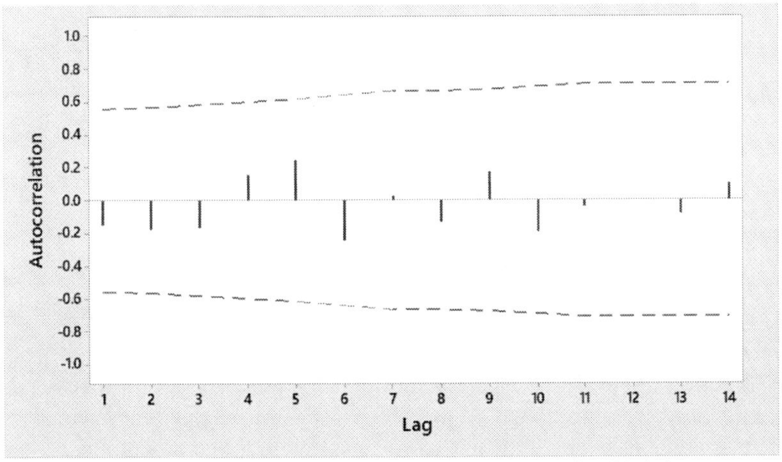

Figure 8. Graph of AutoCorrelation Function for Shrinkage in the direction flow

With several tests that have been done above, it is known that the model that satisfies the relationship between the injection molding process parameters, namely melt temperature, holding pressure, injection pressure to the shrinkage response in sisal fiber biocomposite material can be formulated with the following empirical equation:

$$\hat{Y}_{shrinkage\ BSS} = 1.25308 + 0.0475x_1 - 0.035x_2 - 0.0125x_3 - 0.04038{x_2}^2 - 0.03538{x_3}^2 + 0.035x_1x_2 - 0.04\,x_2x_3.$$

with $\hat{Y}$: Estimated model value for shrinkage response
x_1 : Parameters of the process of *melt temperature*
x_2 : Parameter of the *holding pressure*
x_3 : Parameter of the *injection pressure*

2.4. Surface Plot Graph

The graph of the surface plot shown in Figure 9 [left] with the x axis of the injection pressure process parameter and the y axis of the holding pressure process against the shrinkage response. Figure 9 shows the injection injection and holding pressure injection process parameters for the shrinkage response in sisal-glass fiber biocomposite material.

The surface plot is in the form of a hill with a process melt temperature parameter of 195^{0}C, while a contour plot graph is shown in Figure 9 [right]. The percentage point that shows the smallest shrinkage value is in the injection pressure process parameter of 60 bar and holding pressure of 35 bar

IOP Conf. Series: Materials Science and Engineering **588** (2019) 012041 doi:10.1088/1757-899X/588/1/012041

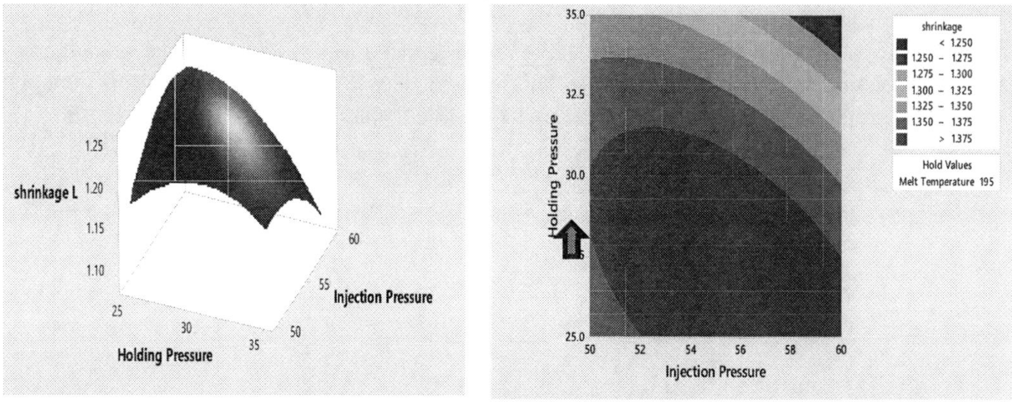

Figure 9. Graph of surface [left] and contour plot [right] shrinkage vs Injection pressure, Holding pressure in the direction of flow

2.5. Parameter Optimazion

The desirability function approach is used for the melt temperature (0C) process parameter, holding pressure (bar), injection pressure (bar) to produce an Ŷshrinkage (%) response. The desirability function uses The smaller is better with the lower sisal-glass fiber biocomposite material = 1.09%, upper = 1.28%, weight = 1. The mathematical model is processed using Minitab 18 software, so that a graph of combination parameters can produce optimal shrinkage response.

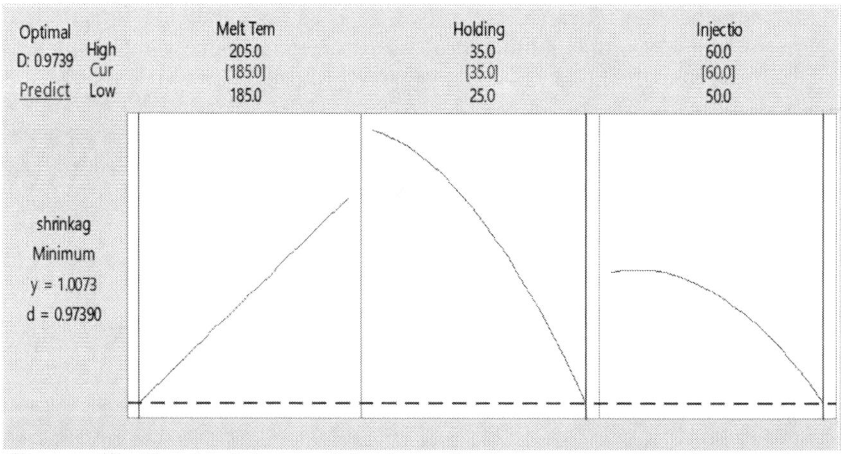

Figure 10. Graph of the combination of process parameters that produces the optimal response of the flow direction Sisal Biocomposite Fiber material

The analysis in Figure 10 shows the combination of process parameter settings for the flow of sisal biocomposite material. Process parameters to get the optimum percentage value, namely to melt 185^{0}C temperature, hold pressure 35 bar, and injection pressure 60 bar. This combination is predicted to get a shrinkage response of 1.0073% with a desirability value of 1.00.

The graph in Figure 10, where the higher the pressure is given, the smaller the percentage shrinkage value will be. The holding pressure graph has a trendline that is more influential than the parameters of the melting temperature and injection pressure process. It is caused by the fact that during the freezing process in the cavity, the plastic compression properties result in high shrinkage values. This phenomenon can be prevented by providing high holding pressure so as to minimize the compressibility of plastic. Uniform and high pressure can control and produce a low shrinkage percentage value. It can be useful to determine the geometry tolerance of the specimen to be produced.

2.6. Confirmation Trial

Confirmation experiments were carried out by recovering data from the HAITIAN MA 900 / 260e injection molding machine. Process parameters used are based on optimization charts in BSS material Figure 11. The results of the specimens were then re-measured to determine the shrinkage percentage value of each material

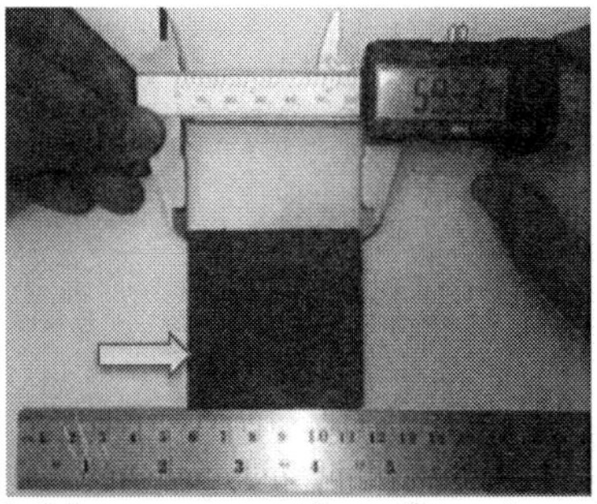

Figure 11. Measurement of confirmation of the flow direction of Sisal Fiber Biocomposite material

In Figure 11 is one of the confirmation tests for measuring the shrinkage value percentage in the 1st experiment of sisal-glass fiber biocomposite material in the direction of the 1st replication plastic flow at the 4th point. Confirmation experiments at optimum values were carried out 3 times and replicated 3 times. The results of the confirmation experiment are shown in Table 7.

Table 7. Confirmation Experiment Results for optimum values

Material	Percobaan Ke-	Parameter Proses *MT* (⁰C)	*HP* (bar)	*IP (bar)*	Respon Shrinkage (%)
Biocomposite	1	185	35	60	1.03
	2	185	35	60	1.06
	3	185	35	60	1.06

In Table 8 the value of the shrinkage percentage for sisal fiber biocomposite material in the optimization conditions is 1.05% smaller than the shrinkage percentage resulting in the initial condition of 1.09%. The reduction in the percentage of shrinkage value is very significant at 3.67% of the initial price. The percentage value is still greater than the value of optimization based on statistical analysis that is equal to 1.0073%. The difference between statistical analysis and optimization conditions can be caused by interference with the injection process parameters in addition to controlled process parameters but not too significant.

Table 8. Comparison of optimization results with initial conditions

Material		Parameter proses TM(⁰C)	HP(bar)	IP (bar)	Respon Shrinkage (%)
Biocomposite	Kondisi Awal	195	35	60	1.09
	Analisa Statistik	185	35	60	1.0073
	Kondisi Oprimasi	185	35	60	1.05

3. Conclusion

From the results of the experimental studies and analyzes that have been carried out, there are some conclusions, namely:

1. The parameters of the injection molding process melt temperature, holding pressure, and injection pressure have a very significant effect on the response of sisal-glass fiber biocomposite material due to the p-value of 0.001 <α.
2. The effect of melt temperature process parameters on shrinkage response can be caused by density and Coefficient of Thermal Expansion, while the parameters of holding pressure and injection pressure can be attributed to the compressibility of sisal fiber biocomposite material.
3. The parameters of the injection molding process melt temperature (x_1), holding pressure (x_2), and injection pressure (x_3) have the regression equation based on the following response surface:
$$\hat{Y}_{shrinkage\ BSS} = 1.25308 + 0.0475x_1 - 0.035x_2 - 0.0125x_3 - 0.04038{x_2}^2 - 0.03538{x_3}^2 + 0.035x_1x_2 - 0.04\ x_2x_3.$$
4. Based on the help of MINITAB 18 process parameters to obtain a minimum shrinkage value in sisal-glass fiber biocomposite material melt temperature 185^{0}C, holding pressure 35 bar, injection pressure 60 bar. Confirmation experiments produced a shrinkage value of 1.05%

References

[1] ASTM International 2001 ASTM-D955. Standard Test Method of Measuring Shrinkage form Mold Dimension Of Thermoplastiks.
[2] D O Kazmer 2007 *Injection Mold Design Engineering* (New York: Hanser Publisher)
[3] D C Montgomery 2001 *Design and Analysis of Experiments Fifth Edition* (New York: John Wiley & Sons, Inc)
[4] J Shoemaker 2006 *Moldflow Design Guide* (Massachusetts: Moldflow Corporation)
[5] Tubagus 2018 Simulasi Dan Studi Eksperimental Untuk Mengetahui Sifat Penyusutan (Shrinkage) Material Biokomposit Sesuai Standard ASTM D955 – 08. Surabaya.
[6] H Zhou 2013 *Computer Modeling For Injection Molding Simulation, Optimization, and Control* (New Jersey: John Wiley Sons, Inc.)

Antioxidant Evaluation of *Ganoderma lucidum* Extracts

Fahmi Hidayat[1] and Sri Fatmawati[1,a]

[1] Department of Chemistry, Faculty of Science, Institut Teknologi Sepuluh Nopember (ITS), Kampus ITS Sukolilo, Surabaya 60111, Indonesia.

[a]fatma@chem.its.ac.id

Abstract. *Ganoderma lucidum* is one of the many mushrooms utilized by the people as a medicine for cancer, asthma, diabetes and hearth disease. The purpose of this experiment is to determine the antioxidant activity of the extracts from *G. lucidum* using different solvents. The simplicial powder of *G. lucidum* are subjected to maceration using water, methanol, ethyl acetate, methylene chloride, and *n*-hexane; and evaporated the solvent until a viscous extract was obtained. The methanol extract shows highest antioxidant activity with IC_{50} value of 251.95 µg/mL.

1. INTRODUCTION

The use of plants as source of medicine in treating disease is an ancient practice. In fact, plants produce a diverse range of bioactive molecules as an antioxidant. These are chemical substances present in plants and include alkaloids, saponins, steroids, carbohydrates, glycosides, tannins, flavonoids and steroids, making them rich sources of different types of medicines. The presence of these substances is an indicator of the pharmacological property as well as the nutritive value of the plant [1], [2].

Free radicals can be produced from outside the body or from metabolic processes in the body when producing energy. Exposure to free radicals from outside can come from environmental pollution such as: contamination of pesticides, heavy metals and the use of preservatives and dyes [3]–[6]. Excess free radicals In living organisms result in chain reactions of free radicals that cause damage lipids, proteins and DNA, and cause damage to cells so that cell function decreases [7]. Damage caused by free radicals can lead to various diseases such as cancer, heart disease, cataracts, decreased immune system and diabetes [8]–[10].

In recent times, attention has been reverted back to plants as sources of therapeutic agents due to their higher properties. These include among others reduced cost, relative lower incidence of adverse reactions compared to modern conventional pharmaceuticals. In recent years much attention has been devoted to natural antioxidant and their association with health benefits [11]. Natural resources are potential sources of natural antioxidants. It produces various antioxidative compounds to counteract reactive oxygen species (ROS) in order to survive [12].

The genus *Ganoderma* has more than 400 species that live in tropical and subtropical regions of the world. *Ganoderma lucidum* generally grows throughout the year on dead wood or trees, such as oak, maple, elm, willow, and magnolia. Some species of the genus *Ganoderma* are generally used as medicine, including: *G. lucidum, G. luteum Steyaert, G. atrum Zhao, G. tsugae Murrill, G. applanatum Pat., G. australe Pat., G. capense Teng, G. tropicum Bres., G. tenue and G. sinense*. Even the species *G. lucidum* has been known in China as an effective medicinal mushroom thousands of years ago [13]–[15].

G. lucidum commonly called lingzhi is a medicinal mushroom of the Polyporaceae family. It is a group of wood degrading mushrooms with hard fruiting bodies [13]. Traditionally, for generations *G. lucidum* have been used by people in China, Japan, Taiwan and other Asian countries to treat various diseases, such as: asthma, diabetes, heart disease, cancer and other chronic diseases [16].
Secondary metabolites of *G. lucidum* mushrooms are derived from groups of terpenoids, polysaccharides, proteins and glycoproteins. These compounds are known to have activities as anti-inflammatory, antioxidant, antidiabetic, antihypertensive, antimicrobial, immunostimulant, antitumor, anti-inflammatory, neuroprotective, AChE inhibitors, cardioprotective, antiherpetic, antimutagenic, antifungal and anti-HIV [17]–[25].
The present study is designed to determine antiradical activity fo *G. lucidum* extracts using DPPH assay. *G. lucidum* is one of the medicinal mushrooms with high antioxidant activity. Extraction of *G. lucidum* using a water solvent by ultrasonic method can produce polysaccharide extract with a inhibition value of DPPH of 53.62 % at a concentration of 47.87 mg/mL [26]. In addition, Joseph *et al.* extracted *G. lucidum* using chloroform extract was produced with a inhibition value of 58.81 % at a concentration of 300 mg/mL [27].

2. EXPERIMENTAL

2.1. Chemicals
Organic solvents including aqua DM, nethanol, ethyl acetate, methylene chloride and, *n*-hexane. For the assay are used methanol from Merck in high grade, 2,2-Diphenyl-1-picrylhydrazyl (DPPH) was purcased from Tokyo Chemical Industries (Tokyo, Japan), and gallic acid as positive control.

2.2. Preparation of G. lucidum Extract
G. lucidum were collected in Bantul, Central Java, Indonesia. The dried fruiting body of *G. lucidum* is cut into small pieces then grounded into powder using an electric milling machine. The powdered sample were extracted using maseration with water, methanol, ethyl acatate, methylene Chloride, and *n*-hexane for 24 hours at room temperature. After filtered through filter paper, the extracts were concentrated under reduced pressure using Rotary evaporator to give crude extract.

2.3. Antioxidant Activity
The antioxidant activities of *G. lucidum* were determined on the basis of the free radicals scavenging effect of DPPH free radical model as described Hidayati *et al.* (2017) with slight modification [4]. Each extract (10 mg) dissolved in 1 mL methanol to get extract solution. 33 μl each extract solution was added 1 ml DPPH radicals solution $6x10^{-5}$ M, then the mixture was shaken vigorously and left to stand for 20 min at room temperature in the dark. The absorbance of blank was then measured at 515 nm and The scavenging activity was expressed as IC_{50} (μg/mL). Gallic acid was used as the positive control. The percentage scavenging effect was calculated as:

$$\text{Inhibition } (\%) = ((A_1 - A_2)/A_1) \text{ x } 100\% \qquad (1)$$

where A_1 was the absorbance of blank and A_2 was the absorbance in the presence of the extract.

3. RESULT AND DISCUSSION

3.1. Extraction of G. lucidum
The extraction process using 20 g *G. lucidum* powder with 300 mL various solvents provide varying extraction yield 0.60 g (3.02%) water extract; 0.64 g (3.19%) methanol extract; 0.34 g (1.71%) ethyl acetate extract; 0.28 g (1.40%) dichloromethane extract; and 0.13 g (0.67%) *n*-hexane extract.

3.2. *Antioxidant Activity*
The hydrogen atom or electron donation abilities of the corresponding extracts and standards were measured from the bleaching of the purple color methanol solution of DPPH. The antioxidant activity of various solution extracts of *G. lucidum* are given in Figure 1.

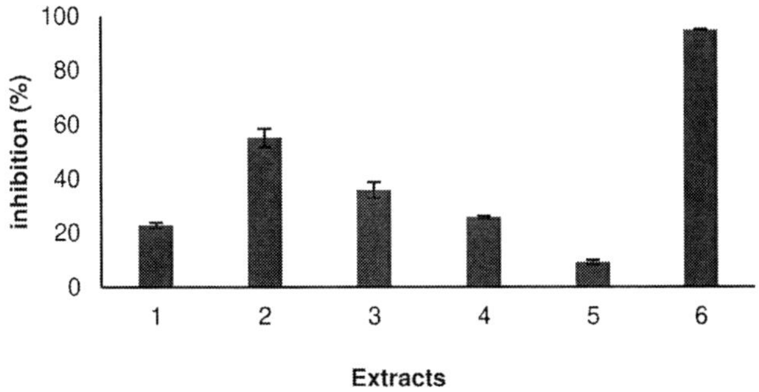

Figure 1 DPPH Scavenging Activity at concentration 319.46 µg/mL such as 1 (water); 2 (methanol); 3 (ethyl acetate); 4 (methylene chloride); 5 (*n*-hexane) and; 6 (gallic acid).

The *G. lucidum n*-hexane extracts has 8.77 % DPPH radical scavenging properties. While excellent activities were shown by the methanol 57.98 %, ethyl acetate 37.19 %, methylene chloride 27.10 % and water 21.58 %. The highest antioxidant activity of the methanol extract among the other extracts was suspected because methanol extract contained compounds that play active role as antioxidants. The methanol has more polarity and solubility. So that, in addition to get high yields of extracts, extracts of active antioxidants were also obtained [28]. Extraction with methanol is known to be obtained polysaccharides from *G. lucidum* where this component has good activity as an antioxidant [26][29]. Because methanol extract give highest antioxidant activity and more then 50 % of inhibition value, furthermore this extract was measured for IC_{50}. The presented curves were plotted between concentration and percentage of inhibition methanol extract was showed in Figure 2. Methanol extract gives IC_{50} value of 251.95 µg/mL and gallic acid as postive control had IC_{50} value of 0.4 µg/mL.

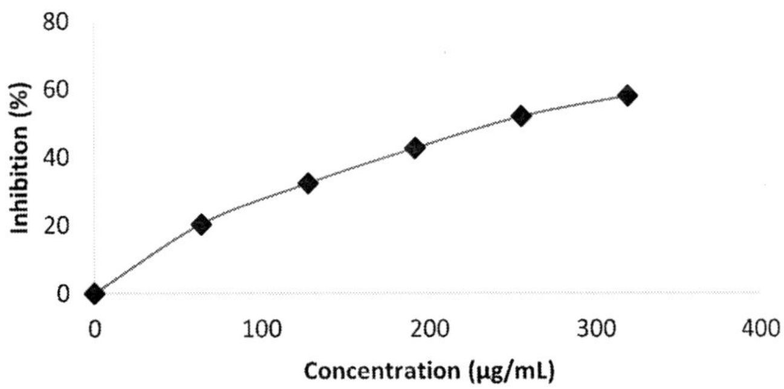

Figure 2 DPPH Scavenging Activity of Methanol Extract

4. CONCLUSION
Methanol extract from *G. lucidum* showed highest activity compared with other extracts as antioxidant. Methanol extract showed IC_{50} value of 251.95 µg/mL, that include moderate activity as antioxidant.

References

[1] Jack I R and Okorosaye-Orubite K 2008 *J. Appl. Sci. Environ. Manag.* **12** 63–65.
[2] Partap S and Pandey S 2011 *J. Pharmacogn. Phytochem* **1** 26–37.
[3] Purnomo A S, Ashari K and Hermansyah F 2017 *J. Microbiol. Biotechnol.* **27** 1306–1315.
[4] Hidayati M D, Ersam T, Shimizu K and Fatmawati S *Indones. J. Chem.* **17** 49–53.
[5] Purnomo A S, Nawfa R, Martak F, Shimizu K and Kamei I 2017 *Curr. Microbiol.* **74** 320–324.
[6] Sariwati A, Purnomo A S and Kamei I 2017 *Curr. Microbiol.* **74** 1068–1075.
[7] Laili K, Fatati N, Inneke P F, Purnomo A S, Mardi S, Taslim E and Fatmawati S 2018 *Res. J. Chem. Environ.* **22** 146–150.
[8] Devasagayam T P A, Tilak J C, Boloor K K, Sane K K, Ghaskadbi S S and Lele R D 2004 *Japi* **52** 794–804.
[9] Percival M 1998 *Clin. Nutr. Insights* **31** 1–4.
[10] Rahimi-Madiseh M, Malekpour-Tehrani A, Bahmani M and Rafieian-Kopaei M 2016 *Asian Pac. J. Trop. Med.* **9** 825–831.
[11] Ayofe A M, Oluwasegun A, Adesina B A and Ogochukwu U 2017 *Am. J. Essent. Oils Nat. Prod.* **5** 33–36.
[12] Fitriana W D, Ersam T, Shimizu K and Fatmawati S 2016 *Indones. J. Chem.* **16** 297–301.
[13] Baby S, Johnson A J and Govindan B2015 *Phytochemistry* **114** 66–101.
[14] Sanodiya B S, Thakur G S, Baghel R K, Prasad G B K S and Bisen P S 2009 *Curr. Pharm. Biotechnol.* **10** 717–742.
[15] Wasser S P 2005 Reishi or Ling Zhi (*Ganoderma lucidum*) *Encycl. Diet. Suppl*. 603–622.
[16] Fu Y, Liu W, Zu Y, Shi X, Liu Z, Schwarz G and Efferth T 2009 *Food Chem.* **112** 71–76.
[17] Chu T T W, Benzie I F F, Lam C W K, Fok B S P, Lee K K C and Tomlinson B 2012 *Br. J. Nutr.* **107** 1017–1027.
[18] El-Mekkawy S, Meselhy M R, Nakamura N, Tezuka Y, Hattori M, Kakiuchi N, Shimotohno K, Kawahata T and Otake T 1998 *Phytochemistry* **49** 1651–1657.
[19] Jiang Y, Gao H and Turdu H 2017 *Bioorg. Chem.* **75** 50–61.
[20] Kaur R, Singh V and Shri R 2017 *Biomed. Pharmacother.* **92** 1055–1061.
[21] Lakshmi B, Ajith T A, Jose N and Janardhanan K K 2006 *J. Ethnopharmacol.* **107** 297–303.
[22] Liu J, Yang F, Ye L, Yang X, Timani K A, Zheng Y and Wang Y 2004 *J. Ethnopharmacol.* **95** 265–272.
[23] Poucheret P, Fons F and Rapior S 2006 *Cryptogamie,Mycologie* **27** 311–333.
[24] Wang H and Ng T B 2006 *Peptides* 27 27–30.
[25] Zhou Y, Qu Z, Zeng Y, Lin Y, Li Y, Chung P, Wong R and Hägg U 2012 *Exp. Toxicol. Pathol.* **64** 673–680.
[26] Chen T, Wu Y, Wu J, Dong Z and Wu J 2014 *J. Taiwan Inst. Chem. Eng.* **45** 57–62.
[27] Joseph S, Sabulal B, George V, Shima T P and Janardhanan K K 2009 *Sci. Pharm.* **77** 111–121.
[28] Askin R, Sasaki M and Goto M 2010 *Food Bioprod. Process.* **88** 291–297.
[29] Li J, Zhang X and Liu Y 2016 *LWT - Food Sci. Technol.* **70** 16–23.

IOP Conf. Series: Materials Science and Engineering **588** (2019) 012043 doi:10.1088/1757-899X/588/1/012043

Usage of power plant waste for airport pavement surface

N Widayanti and E Ahyudanari

Department of Civil Engineering, Faculty of Civil, Environmental, and Geological Engineering, Institut Teknologi Sepuluh Nopember (ITS), Kampus ITS Sukolilo, Surabaya, 60111, Indonesia.

ervina@ce.its.ac.id

Abstract. A research has been conducted to analyze usage of power plant waste and characteristics of asphalt concrete mixture for airport pavement using artificial aggregate. The artificial aggregate was made of power plant wasted related to waste energy that has not been fully utilized. This testing method used to determine the marshall stability and flow values of asphalt concrete, and to measure how much fly ash from power plant waste can be reduce to utilized as airport pavement surface. The result of marshall testing provides the new material that made of power plant wasted was eligible to used as a replacement material for airport pavement surface and the mixed compositions can be applied for airport runway pavement which is using artificial aggregate was expected to reduce power plant wasted.

1. Introduction

As infrastructure development in Indonesia, the need for natural aggregate as a base material for pavement is increasing. Previously, the fulfillment of aggregate needs has been obtained from quary spread in Indonesia. The need is so great that the resources obtained from nature are depleting. If excessive aggregate exploitation is carried out, it is possible that Indonesia will experience a shortage of natural aggregate as a base material for pavement. On the other hand, this behavior can also threaten environmental sustainability. For this reason, there needs to be an effort to find alternatives so that the use of natural aggregates can be reduced, one of which is by utilizing waste.

Based on PLN data, the trend of FABA (fly ash and bottom ash) production in Java Island until 2027 will reach 11.18 million tons per month, which consists of PLN 6.5 million tons and Independent Power Producers (IPP) of around 4.6 million tons. In 2016, ash production has reached 4.6 million tons per month, consisting of ash from PLN of 3.26 million tons and IPP of 1.33 million tons [1]. Related to the data, there is waste that has not been fully utilized. Many studies have been conducted to find alternative uses of the waste. One of them is the processing/recycling of fly ash into a useful product. According to previous research, fly ash-based geopolymer is potentially used as an artificial aggregate material for as a replacement aggregate in a concrete asphalt mixture [2].

The main differences in design considerations for highway and airport pavements arise from the characteristics of traffic using them. An airport pavement may have to be designed to withstand equivalent single wheel loads (ESWL) of the order of 50 t (approximately 50 tons), whereas the maximum single wheel load allowed on the road pavement by most highway authorities is about 10 t (approximately 10 tons). Furthermore, the wheel tire pressure of an aircraft of about 1200 kPa (175 psi) is nearly twice the value of a normal truck tire. These differences greatly influence the material requirements for the pavements [3].

Along with the development of technology, the increasing mobility of the Indonesian people in the use of air transportation is one of the important considerations in determining the type of aircraft based on its capacity. The weight of an aircraft or the wheel load carried by the airport runway pavement is also greater based on the type of aircraft used. The weight of the load carried by the pavement that continues to increase will have an impact on the quality of a pavement. Changes in aircraft characteristics, especially aircraft dimension shown as follows [4]:

- Gear geometry:
  B777 – Tridem; Combo gears A380-800 tandem/tridem
- A/C Loadings:
  Boeing 747 (1995) – 833,000 lb vs. A380-800 (2017) – 1.24×10^6 lb
- Wheel Load:
  B777-300 (1995) - 57,000 lb versus A350-900 (2017) – 70,000 lb
- Higher tire pressures:
  215 psi (1995) vs. 241 psi (2017)

Tire Pressures may be restricted on asphaltic concrete, depending upon the quality of the asphalt mixture and climatic conditions. Tire pressure effects on an asphalt layer relate to the stability of the mix in resisting shearing or densification. A poorly constructed asphalt pavement can be subject to rutting due to consolidation under load. The principal concern in resisting tire pressure effects is with stability or shear resistance of lower quality mixtures [5]. An airport has a flexible (asphalt-surfaced) pavement runway having the operating gross weights and tire pressure shown in Table 1.

Table 1. Using Airplane and Tire pressure for a Flexible Pavement [5]

Airplane	Operating Weight, lbs	Tire Pressures (psi)
B727-200	185,000	148
B737-300	130,000	195
A319-100	145,000	196
B747-400	820,000	200
B767-300ER	370,000	190
DC8-63	330,000	194
A300-B4	370,000	205
B777-200	600,000	215

The performance of asphalt mixtures can be tested using a Marshall tool. The testing is intended to determine the resistance (stability) of asphalt concrete mixtures to the flow of the mixture of asphalt and aggregate. Flow is a state of changing the shape of an asphalt concrete mixture that occurs as a result of a load to a failure limit in mm or 0.01". The correlation of the stability value with the tire pressure can determining the loading that can bear by asphalt mixture from the aircraft wheel.

In this study, the test results can be used as a consideration for the potential utilization of waste as an alternative to asphalt mixtures for airport runway and it can be seen the pavement capacity in resisting the load. Furthermore, this research is part of author' thesis. To meet the element need to be considered for pavement (fatigue resistance), therefore, the results of Marshall testing on this research are needed for fatigue testing as material for subsequent authors' research [6].

IOP Conf. Series: Materials Science and Engineering **588** (2019) 012043 doi:10.1088/1757-899X/588/1/012043

2. Materials and Methods

2.1. Materials

In this study, the gradation of the mixture of asphalt met the requirements for Federal Aviation Administration 2014 (Table 2) as illustrated in Figure 1. The median gradation was selected as the target gradation.

Table 2. Grading Limits [7]

Sieve Size (mm)		Percentage by Weight Passing Sieves
		Gradation 2
1"	25	-
3/4"	19	100
1/2"	12	79-99
3/8"	9	68-88
No. 4	4.75	48-68
No. 8	2.36	33-53
No.16	1.18	20-40
No. 30	0.600	14-30
No. 50	0.300	9-21
No. 100	0.150	6-16
No. 200	0.075	3-6

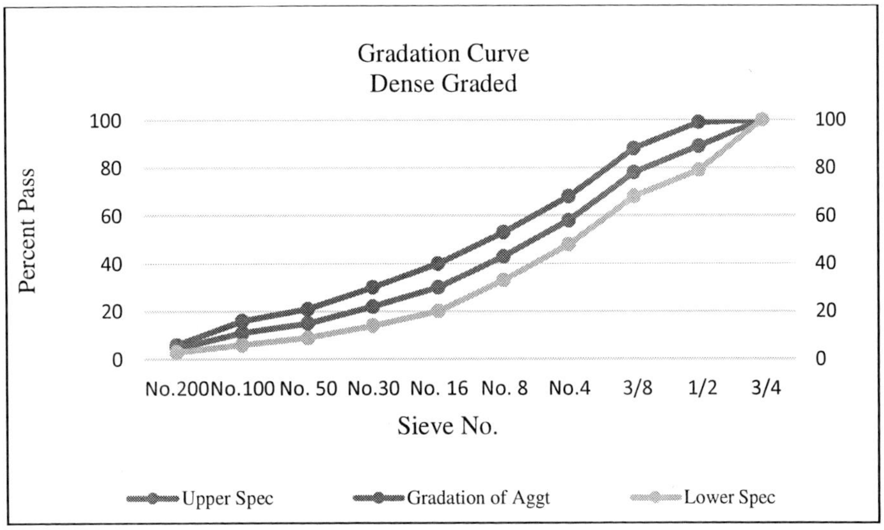

Figure 1. Gradation Curve

The materials used for this research include coarse fraction of natural and artificial aggregates, fine aggregates, filler, and bitumen. The coarse and fine aggregates were collected from Laboratorium

Transportasi dan Perhubungan ITS. The mineral fillers used are the crushed stone dust (CSD). The 60/70 penetration grade bitumen was chosen for the study.

Fly ash which is used for the artificial aggregates were collected from Paiton Power station, Probolinggo, East Java. Procedure for making artificial aggregates follows the procedure in the previous study [8]. The artificial aggregate is shown in Figure 2.

Figure 2. Artificial Aggregates

2.2. Methods

The specification of the asphalt mixture met the requirements for Federal Aviation Administration (FAA) 2014. The specification is shown in Table 3.

Table 3. Specification Requirements [7]

Test Property	Pavement Design for Aircraft Gross Weights of 60.000 Lbs (27216 kg) or more or Tire pressures of 100 psi or more
Number of blows	75
Stability, pounds (Newtons) minimum	2150 (9560)
Flow, 0.01 inch. (0.25 mm)	10-16
Target air voids (percent)	3.5
Percent Voids in Mineral Aggregate (minimum)	15%

This study evaluates the characteristics on the Marshall stability and flow of airport surface course. This was achieved by preparing asphalt mixtures with natural aggragate and artificial aggregate to Marshall test. Based on previous research, optimum stability of ratio for artificial and natural aggregate is 25% : 75% [9]. The mixture composition of the airport pavement surface is shown in Table 4.

Table 4. The Mixture Composition

Sieve Size		Lower Spec (%)	Median Spec (%)	Upper Spec (%)	Percentage by Weight Passing Sieves (%)		Rasio Aggregate 25% : 75%	
							AA	NA
in	mm				(%)	(gr)	%	%
3/4	19	100	100	100	0.0	0	-	-
1/2	12	79	89	99	11.0	132	33	99
3/8	9	68	78	88	11.0	132	33	99
No.4	4.75	48	58	68	20.0	240	60	180
No.8	2.36	33	43	53	15.0	180	45	135
No.16	1.18	20	30	40	13.0	156		156
No. 30	0.6	14	22	30	8.0	96		96
No.50	0.3	9	15	21	7.0	84		84
No.100	0.15	6	11	16	4.0	48		48
No.200	0.075	3	4.5	6	6.5	78		78
Pan					4.5	54		54
Jumlah					100	1200	171	1,029

2.2.1. Marshall Test. Marshall Test aims to determine the optimum bitumen content, then determine the stability and flow of asphalt mixture. The method used to make airport pavement surface will be follow Marshall procedure of AASTHO T 245-74, or ASTM D 1550-E27.

The specimen is asphalt mixture with a diameter of 101.6 mm (4 inches) and a height of 76.2 mm (3 inches). Compaction for the Marshall test is done by blows 75 times per field using a compactor. After the test object is compacted, then stored at room temperature for 24 hours, then the objects are weighed in air, in the water and in saturated surface dry (SSD) conditions to obtain bulk specific gravity. Then soaked at a temperature of 60±1°C for 30-40 minutes and ready to conduct stability and flow testing. The Marshall stability machine is shown in Figure 3.

Figure 3. Marshall Testing Machine

3. Results and Discussions

3.1. Marshall Test

The optimum bitumen content was found to be 5.10%. The result of marshall testing can be seen in Table 5. The results show that the stability increases with increasing bitumen content up to the optimum bitumen content and thereafter decreases.

Table 5. Result of Specification Requirements

Properties	Specification	Bitumen Content (%)				
		4.50	5.00	5.50	6.00	6.50
Stability	2150 lbs	3128	3797	3917	3969	3932
Flow	10-16 inch	12	11.6	11.2	13.6	16.8
Target Air Voids (%)	3.5	5.80	3.82	2.89	1.06	0.77
Voids In Mineral Aggr. (%)	15%	18.50	17.74	17.89	17.30	18.01

The stability value met the FAA specifications requirements. The increase in stability can be attributed to improved adhesion between the aggregate and bitumen. However, more tests like the indirect tensile fatigue test would be carried out in future studies to confirm this. The Marshall test result are shown in Figure 4.

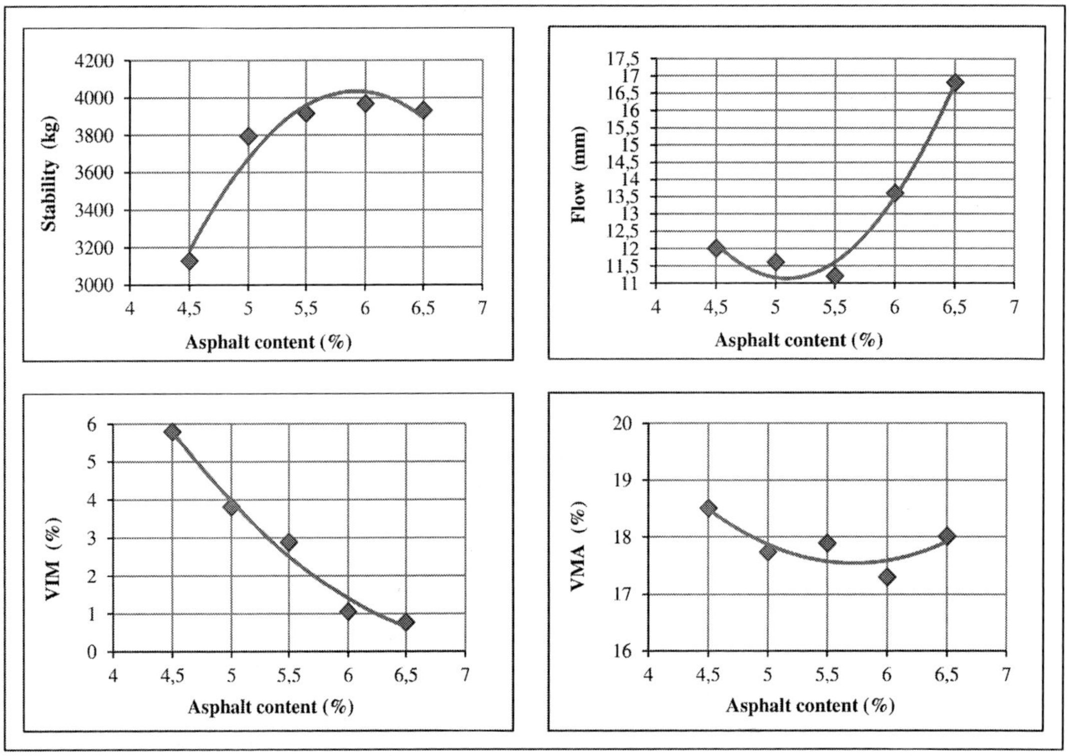

Figure 4. Marshall Characteristics

The flow values of the asphalt mixtures against bitumen content are shown in Figure 5. It can be seen from the results that the flow value inversely proportional with stability. All of variation of bitumen content meet the requirements except in 6.5%. This shows that the bitumen content is a state of asphalt which has flow value in resisting changes in the shape of a mixture when receiving a load.

From Table 5 it can be seen that voids in mineral aggregate (VMA) meet the specification requirements, but the values of target air voids (VIM) only meets the requirements in bitumen content of 5.0%. This indicates there is a part of the asphalt mixture which is not filled by aggregate or by bitumen. The higher the VIM value, the easier the asphalt to be oxidized so that it causes reduced adhesion between the aggregate granules so as to enable revelling and stripping. The VIM value also affects the durability of the pavement layer, the higher the VIM value shows the greater the cavity in the mixture so that the mixture is porous. To improve the VIM value, modifications can be made to the penetration of bitumen or compaction temperature.

3.2. Utilization of Power Plant Waste

To measure how much fly ash from power plant waste can be reduce to utilized as airport pavement surface, the method is calculated the volume asphalt mixture of airport runway pavement such as the following:
The needs calculation of artificial aggregate is 171 gram per sample [6].
Geometric standard design of airport pavement surface [10] :

Thickness : 4 inch = 0.102 m (HMA Surface Course)
Runway Width : 45 m
Runway length : 2000 m

- Pavement volume (*Vp*)
  Vp = width × length × thinkness
  Vp = 45 m × 2000 m × 0.102 m
  Vp = 9180 m^3

- Sample volume (*Vs*)
  $Vs = \pi \times r^2 \times t$
  $Vs = 3.14 \times 0.05 \text{ m} \times 0.0694 \text{ m}$
  $Vs = 0.0108958 \text{ m}^3$

- Multiplier factor (Fp)
  Fp = $Vp \div Vs$
  Fp = 9180 m^3 ÷ 0.0108958 m^3
  Fp = 842,526.5

Then, to find out how much power plant wasted (fly ash) needed per standard design of airport pavement surface can be seen such as the following:

- Fly ash reduction
  Fly ash = Total weight of fly ash per sample × Fp
  = 171 gr × 842,526.5
  = 144,072,027.8 gr
  = 144.072 ton

So, the amount of fly ash that can be reduce from power plant wasted for runway airport pavement is equal to 144.072 ton per standard design of airport pavement surface.

4. Conclusions

From research conducted can be concluded as follows:

- The Asphalt mixture using artificial aggregate-fly ash geopolymer is potential to used as an alternative mixture pavement for airport pavement surface.
- The asphalt mixture of artificial aggregates has a fairly high value of stability.
- The test result indicate that the specimen of asphalt mixture with artificial aggregate for airport pavement surface met the value of the specification requirements in stability, flow (except in 6.5% bitumen content) and voids in mineral aggregates. While the values of target air voids (VIM) only meets the requirements in bitumen content of 5.0%.
- The amount of fly ash can be used in runway airport pavement is equal to 144.072 ton per standard design of runway.

References

[1] Adi 2016 PLN Minta Dipermudah Regulasi Pemanfaatan Limbah FABA PLTU *http://petroenergy.id/article/pln-minta-dipermudah-regulasi-pemanfaatan-limbah-faba-pltu* [Online]
[2] Karyawan I D M A, Ahyudanari E and Ekaputri J J 2017 Potential Use of Fly Ash Base-Geopolymers Aggregate Substitution in Asphalt Concrete Mixtures *Int. J. Eng. Technol.* **9** issue 5 pp 3744–52
[3] Fwa T F 2003 *Highway and Airport Pavement Design* (Singapore: CRC Press LLC) chapter 62
[4] Roesler J 2017 Airport Pavements: Reflection on 20 Years of Design Material and Construction Innovations
[5] Federal A A 2005 Advisory Circular in *AC 150/5335-5A* **1** pp 2–4.
[6] Widayanti N 2019 Fatigue Analysis of Flexible Pavement of Asphalt Concrete Mixture With Artificial Aggregate Made Using Fly Ash Geopolymer *Thesis Teknik Sipil* Institut Teknologi Sepuluh Nopember
[7] Federal A A 2014 Advisory Circular in *AC 150/5370-10G* pp. 217–256.
[8] Yuliana H A, Karyawan I D M A, Murtiadi S, Ekaputri J J and Ahyudanari E 2019 The Effect of Slope Granulator on The Characteristic of Artificial Geopolymer Aggregate Used in Pavement *J. Eng. Sci. Technol.* **14** issue 3
[9] F Teknik 2018 Thesis Teknik Sipil Universitasi Mataram [Unpublished]
[10] Federal A A 2016 Airport Pavement Design and Evaluation *AC 150/5320-6E* pp 1

IOP Conf. Series: Materials Science and Engineering **588** (2019) 012044 doi:10.1088/1757-899X/588/1/012044

Absorption of indigosol dye waste from batik home industry at ex redlight district, Dolly, Surabaya using activated carbon

Yatim Lailun Ni'mah[1*], Ita Sulistyani[1], Suprapto Suprapto[1], Kartika Sari Dewi[1]

[1] Department of Chemistry, Faculty of Natural Sciences, Institut Teknologi Sepuluh Nopember (ITS), Kampus ITS Sukolilo, Surabaya 60111, Indonesia

* Corresponding author: yatimnikmah@gmail.com

Abstract. Research on adsorption of indigosol batik dye waste that derived from batik home industry from ex redlight district, Dolly, Surabaya using activated carbon has been done. The variation of contact time and adsorbent dosage was optimized. This research was conducted by a batch method. The contact time was varied from 10-70 minutes. The adsorbents dosage was varied from 2-14 mg/25 mL. The dilution factor was varied from 160-240 for initial concentration. This research was aimed to obtain the optimum conditions of indigosol adsorption onto activated carbon. Spectrophotometer UV-Vis shows that the equilibrium of indigosol adsorption by activated carbon was achieved at 50 minutes contact time. In addition, the highest indigosol removal by the addition of 10 mg activated carbon in 25 mL indigosol solution with the dilution factor of 200 for initial concentration. The result, 72,73% was optimum indigosol removal. From the analysis results, it can be concluded that the optimum conditions for indigosol adsorption using activated carbon were achieved at 50 minutes contact time with an adsorbent dosage of 10 mg/25 mL and the dilution factor was 200 for initial concentration

1. Introduction

Surabaya is the capital of East Java province where local government, politics, commerce, industry, education, and culture were located. Surabaya is also a city that has the 6th largest redlight district in Southeast Asia of which include Dolly, Bangunsari, Kermil, Klakah Rejo, and Moroseneng [1]. The development process of the prostitution activity is supported by market mechanisms that make prostitution more profitable [2]. Finally, on June 19, 2014, Dolly redlight district was closed by Dr. Ir. Tri Rismaharini, Mayor of Surabaya. To cope with the employment problem for the citizens affected by the closing, some NGO provides capital assistance that worth Rp 3 million to create new business and developing Dolly area as tourist attractions[1]. Many people in the area set up and make business in the field of textiles craft. Rumah Batik is one of a new home industry located at the West Street VIII B No. 31 Putat Jaya, Dolly. The common coloring agents used in Rumah Batik were Remasol, Indigosol, and Naphthol. The dye compounds were dangerous if directly discharge to the environment. However, the communities surrounding Rumah Batik Dolly were not aware of the risk as well as treatment to handle wastewater that produced.

The textile dye is a compound which can provide color on the surface of the fabric. The majority of textile dyes have complex organic structures. Indigosol is one of the most commonly used dyes in textile [3.4]. Indigosol waste that exists in the environment requires special attention because of its

IOP Conf. Series: Materials Science and Engineering **588** (2019) 012044 doi:10.1088/1757-899X/588/1/012044

toxicity to the environment. Indigosol can cause skin disease and damage the eye cornea [5-7]. For years, the methods to reduce indigosol waste was studied i.e. precipitation, adsorption, and photocatalysis [8-9]. Adsorption is one of the most effective methods for wastewater treatment that used by industry to reduce the inorganic and organic pollutants [10]. Many of the textile industry was used activated carbon to absorb commercial dye wastewater. Activated carbon was adsorbent which is economical and has high efficiency [11]. Adsorption by batch adsorption was carried out by the addition of adsorbent to a solution that contains substances to be absorbed [12]. The absorption efficiency was determined by UV-Vis Spectrophotometry technique. The difference of the solution absorbance before and after adsorption was compiled to obtain optimum adsorption condition[13].

2. Experimental

Materials

The material used in this research i.e. Indigosol Dye that was obtained from a local store and used without further purification, academic for dye solvent, activated carbon (Merck KGaA) that was used as absorbent and indigosol waste that obtained from Rumah Batik, Dolly, Surabaya. Dye removal was monitored by UV-Vis Spectrophotometer (Genesy 10S)

2.1 Standard Calibration Curve

A stock solution of 1000 ppm was prepared by dissolution of 0.25 grams of indigosol yellow powder in 250 mL of aquademin. A stock solution of 100 mL was diluted into 1000 mL to obtain a solution of 100 ppm. Furthermore, the maximum wavelength was determined by wavelength scanning of 100 ppm solution using UV-Vis spectrophotometer. After that, out of 100 ppm solution was further diluted to 10 ppm 20 ppm, 30 ppm, 40 ppm, 50 ppm, and 60 ppm in 10 mL measuring flask. The absorbance of each solution was plotted as a function of its concentration to make a calibration curve.

2.2 Preparation of Indigosol Waste

Indigosol waste from Rumah Batik of 5 ml was diluted to 1000 ml. Further dilution by a dilution factor of 200 was carried out so that the sample absorbance was in the range of calibration curve. Any particulate that presents in the sample was removed by mean of centrifugation at 250 rpm for 30 minutes followed by decantation. The filtrate obtained was analyzed using spectrophotometer UV-Vis at wavelength ranges of 350-800 nm.

2.3 Contact Time Optimization for Indigosol Natural Waste

Activated carbon of 10 mg was added into 25 mL of Indigosol waste solution in a 250 mL beaker. The mixture was stirred at varied contact time i.e. 10, 20, 30, 40, 50, 60, and 70 minutes. From each contact time variation, the filtrate was tested for absorbance using a UV-Vis spectrophotometer at the maximum wavelength.

2.4 Adsorbent Dosage on Indigosol Natural Waste

Indigosol waste of 25 mL was put into a 250 mL beaker. Activated carbon of 2, 4, 6, 8, 10, 12, and 14 mg was added to the indigosol solution. The mixture was stirrer at optimum contact time. Every time the variation of the dose of the adsorbent is done 3 times the test to get triple data. The remaining filtrates were then tested for their absorbances using a UV-Vis spectrophotometer at the maximum wavelength.

2.5 Initial Concentration on Indigosol Natural Waste

Indigosol waste of 25 mL was put into a 250 mL beaker. The variation of initial concentration were 160, 180, 200, 220, and 240 of the dilution factor. Every initial concentration was analyzed using a spectrophotometer UV-Vis to get the initial absorbances. Activated carbon in optimum adsorbent dosage was added to the indigosol solution. The mixture was stirrer at optimum contact time. Every time the variation of the initial concentration is done 3 times the test to get triple data. The remaining filtrates were then tested for their absorbances using a UV-Vis spectrophotometer at the maximum wavelength.

3. Results and Discussion

3.1. The Calibration Curves of Yellow Indigosol Dyes

The maximum wavelength of yellow indigosol dye solution can be seen in Figure 1

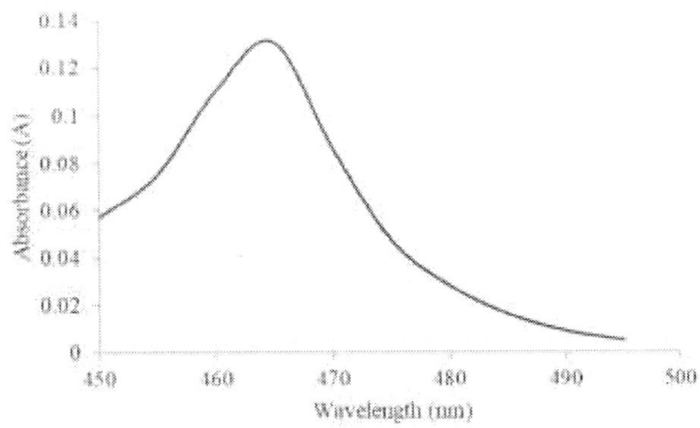

Figure 1. The maximum wavelength Indigosol Yellow solution

Figure 1 shows that the maximum wavelength of the indigosol yellow batik dye solution was at 465 nm. The wavelength was in accordance with the yellow complement color wavelength that lies between 435-480 nm [14]. The absorbances of indigosol yellow dye solution with concentration from 10 - 60 ppm were measured using UV-Vis spectrophotometer at maximum wavelength, 465 nm. The absorbances obtained were presented in Table 1.

Table 1. The absorbance of Indigosol Yellow solution 10 ppm – 60 ppm

Solution concentration (ppm)	Absorbance (A)
10	0.137
20	0.25
30	0.365
40	0.464
50	0.576
60	0.673

The plot of absorbance as Indigosol Yellow concentration function was presented in Figure 2 below:

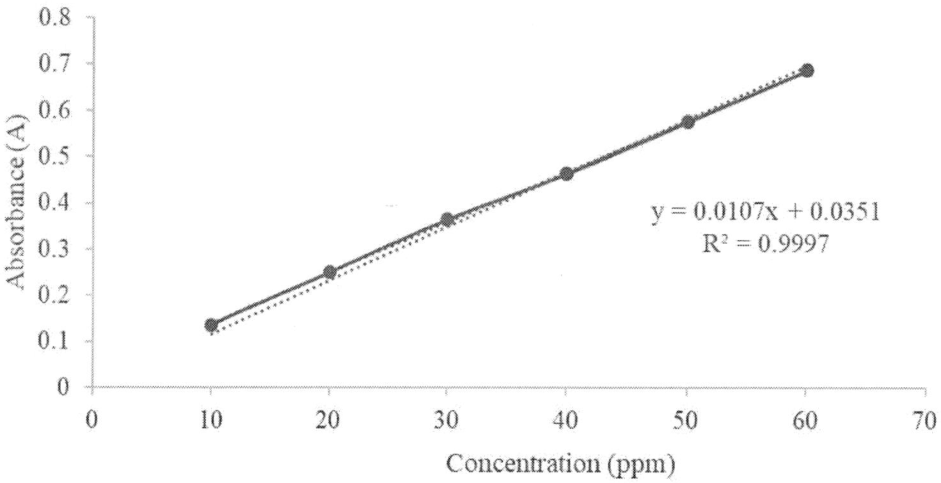

Figure 2. Calibration curve of Indigosol Yellow solution

The regression equation that obtained from Figure 2 was used to determine the concentration of indigosol waste sample as shown in Equation 1.

$$y = 0{,}0107\,x + 0{,}0351 \quad (1)$$

Information :
y = absorbance
x = Indigosol Yellow concentration

3.2 The Maximum Wavelength of Indigosol Waste Solution
Initial Indigosol waste solution was diluted 200 times and centrifuged to separate any particulates in the solution. The maximum wavelengths of indigosol waste solution were shown in Figure 3.

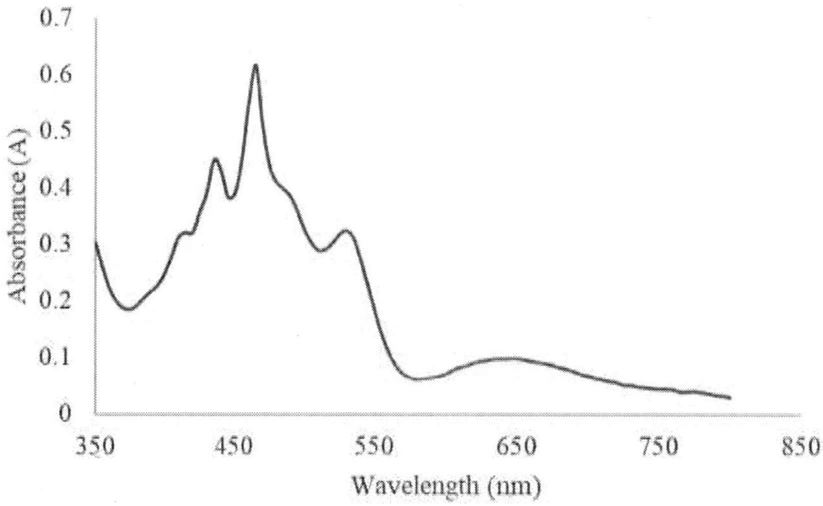

Figure 3. Maximum wavelength of Indigosol waste solution

Based on the graph, the maximum wavelength of Indigosol waste above was at 465 nm with an absorbance of 0.616. The absorbance data obtained, then calculated to obtain the initial concentration of indigosol waste sample solution through Equation 1. From the linear regression equation of the calibration curve, it was obtained that the initial concentration of the Indigosol waste solution was 54.29 ppm

3.3 Contact Time Optimation of Indigosol Waste Solution
The contact time optimation that obtained from triple measurement showed that the standard deviation was ≤ 2%. Thus, the measurement precision statistically satisfied [15]. The % Removal was calculated following Equation 2 below:

$$\%Removal = \frac{C_{init} - C_{final}}{C_{init}} \times 100\% \quad (2)$$

Information :
C_{init} = Initial concentration of Indigosol Yellow
C_{final} = Concentration of Indigosol Yellow after adsorption

The averages of % Removal of Indigosol Yellow at varied contact time were presented in Table 2.

Table 2. The average of % Removal at varied contact time

Contact time (minutes)	% Removal average (%)
10	47.34 ± 1.70
20	57.50 ± 0.30
30	65.53 ± 0.50
40	70.81 ± 0.10
50	72.53 ± 0.10
60	72.59 ± 0.20
70	72.93 ± 0.26

The correlation between contact time and the average of % Removal was plotted in Figure 4 below:

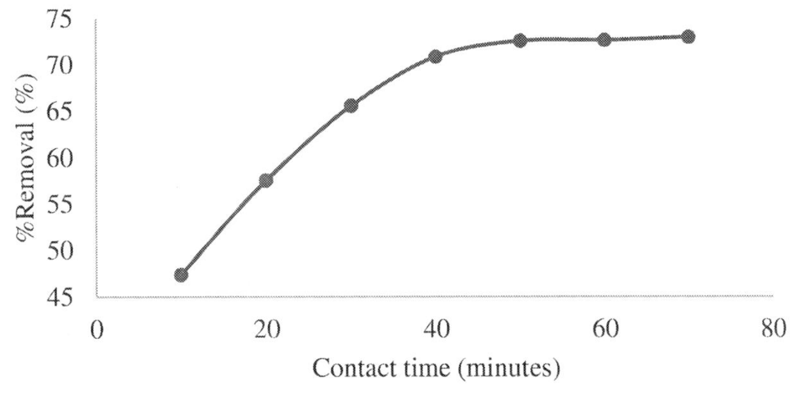

Figure 4. Correlation between contact time and average % removal of Indigosol waste

From Figure 4, it was obtained that the optimum contact time for adsorption of indigosol waste using activated carbon was 50 minutes. Thus, the optimum contact time will be used for the next optimation.

3.4 The Optimation of Adsorbent Dosage on Indigosol Waste Adsorption
The averages % Removal at varied adsorbent dosage were tabulated in Table 3. The average of % removal was plotted as the dosage of adsorbent function. The graph of the % Indigosol removal at varied adsorbent dosage can be seen in Figure 5 below:

IOP Conf. Series: Materials Science and Engineering **588** (2019) 012044 doi:10.1088/1757-899X/588/1/012044

From Figure 5, it can be seen that the optimum adsorbent dosage for indigosol waste adsorption using activated carbon was 10 mg/25 mL of Indigosol solution.

Table 3. The average of % Removal at a varied adsorbent dosage

Adsorbent dosage mg/25 mL	*% Removal average (%)*
2	55.43 ± 0.30
4	56.69 ± 0.10
6	62.60 ± 1.05
8	69.60 ± 0.80
10	72.65 ± 0.34
12	72.24 ± 0.70

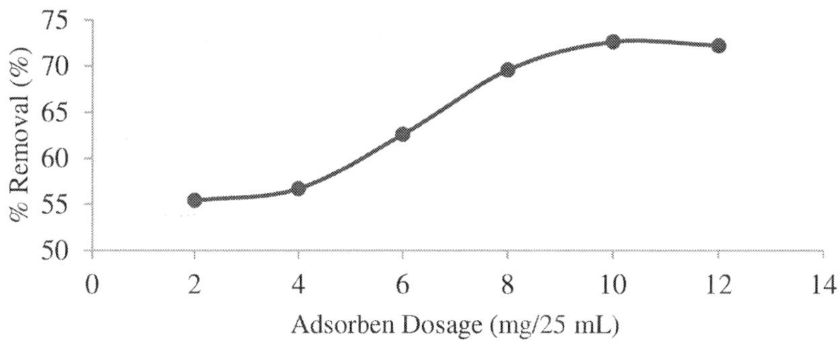

Figure 5. The activated carbon dosage for Indigosol Yellow waste adsorption

3.5 The Optimation of Initial Concentration on Indigosol Waste Adsorption

The averages % Removal at varied initial concentration were tabulated in Table 4.

Table 4. The average of % Removal at an initial concentration

Dilution Factor	*% Removal average (%)*
160	65.34 ± 0.17
180	65.92 ± 1.22
200	72.73 ± 0.31
220	72.99 ± 1.76
240	72.94 ± 0.99

The average of % removal was plotted as the dilution factor function. The graph of the % Indigosol removal at varied initial concentration can be seen in Figure 6 below:

From Figure 6, it can be seen that the optimum initial concentration for indigosol waste adsorption using activated carbon was 200 of Indigosol solution.

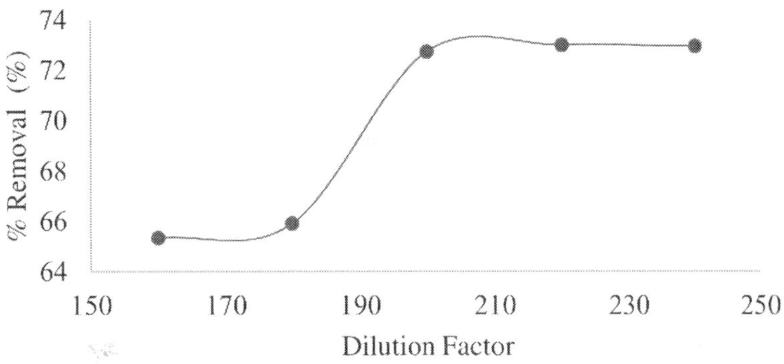

Figure 6. The initial concentration for Indigosol Yellow waste adsorption

4. Conclusion

Based on the optimization results, it can be concluded that the optimum adsorption contact time for Indigosol Yellow waste solution onto activated carbon was 50 minutes with optimum removal of 72.53%. In addition, the optimum adsorbent dosage of activated carbon was 10 mg/25 mL of solution with optimum removal of 72.65%. And the optimum initial concentration of Indigosol Yellow natural waste solution is dilution factor of 200 with optimum removal of 72.73%.

Acknowledgments

The author would like to thank the Instrumentation and Analytical Science Laboratory, Fundamental Laboratory, Chemistry Department FS ITS, and LPPM ITS for their support until the completion of this research, as well as all parties involved in research and drafting of this research paper.

References

[1] Amirah, A. 2015. Cultural, Social, And Economic Perspectives in Making a Criminal Policy (Dolly Prostitution). The Journal of Macrotrends in Social Science, 1(1). 45 – 53

[2] Moefad, A.M. 2015. Komunikasi Masyarakat Eks Lokalisasi Pasca Penutupan Dolly. Jurnal Komunikasi Islam. 5, 145 -184.

[3] F.D. Snell, L.S. Ettre, Encyclopedia of Industrial Chemical Analysis, fifth ed., John Wiley and Sons Inc., New York, 1973.

[4] C.J. Donaldson, D. Khan, The safety of intravenous indigo carmine to assess ureteric patency during transvaginal uterosacral suspension of the vaginal vault, J. Pelvic Med. Surg. 15 (2009) 11–15.

[5] N. Barka, A. Assabbane, A.Y. Nounah, A. Ichou, Photocatalytic degradation of indigo carmine in aqueous solution by TiO2-coated non-woven fibres, J. Hazard. Mater. 152 (2008) 1054–1059.

[6] D.A. Fungaro, L.C. Grosche, A.S. Pinheiro, J.C. Izidoro, S.I. Borrely, Adsorption of methylene blue from aqueous solution on zeolitic material and the improvement as toxicity removal to living organisms, Orbital 2 (2010) 235–247

[7] I. Othman, R.M. Mohamed, F.M. Ibrahem, Study of photocatalytic oxidation of indigo carmine dye on Mnsupported TiO2, J. Photochem. Photobiol. A 189 (2007) 80–85.

[8] T.M. Elmorsi, Y.M. Riyad, Z.H. Mohamed, H. Abd, Mh. El Bary, Decolorization of Mordant red 73 azo dye in water using H2O2/UV and photo-Fenton treatment, J. Hazard. Mater. 174 (2010) 352–358.

[9] M.N. Chong, B. Jina, C.W.K. Chow, C. Saint, Recent developments in photocatalytic water treatment technology: a review, Water Res. 44 (2010) 2997–3027.

[10] Kant R. Adsorption of dye eosin from an aqueous solution on two different samples of activated carbon by static batch method. J Water Resour Prot 2012;4(2):93–8.

[11] Rehman MSU, Kim I, Han JI. Adsorption of methylene blue dye from aqueous solution by sugar extracted spent rice biomass. Carbohydr Polym 2012;90(3):1314–22

[12] Sulistiyana. 2011. Studi Adsorpsi Kation Ca dan Mg (Penyebab Kesadahan) Menggunakan Selulosa Bakteri Nata De coco dengan Metode *Batch*. Prosiding Seminar Nasional Waste Management I. ISBN 978-602-95595-4-5, pp. 432-438.

[13] Skoog, D. A., Donald, M. West, F. James Holler, Stanley R. Crouch, 2000. *Fundamentals of Analytical Chemistry* .Hardcover: 992 pages, Publisher: Brooks Cole

[14] Underwood. 2002. Analisis Kimia Kuantitatif. Edisi. Keenam. Jakarta. Penerbit Erlangga. Hal 394, 396-404

[15] Harmita. (2004). Petunjuk Pelaksanaan Validasi Metode dan Cara Perhitungannya. Majalah Ilmu Kefarmasian. 1. 117-135. 10.7454/psr.v1i3.3375.

IOP Conf. Series: Materials Science and Engineering **588** (2019) 012045 doi:10.1088/1757-899X/588/1/012045

Alternative of stone ash as a sand replacement in fc '40 mpa quality concrete mixture on pressure stress

R Hidayata, Iswinarti, and Asnun Parwanti

Department of Civil Engineering, Universitas Darul 'Ulum, Jombang, Indonesia

ruslanh.1964@gmail.com,

Abstract. Stone ash is the result of breaking stone by using a stone crusher. Stone ash is currently a by-product in the industry the amount of stone is not small. At present stone ash is not very salable for sale because the use in the construction industry is very little considering the use of sand as fine aggregate is still used for concrete mixtures. The method used in this study is a reduction in the amount of sand used in the concrete composition replaced with stone ash. Where the percentage of the use of stone ash is determined at 5%, 10%, 15% and 20% of the amount of sand, using concrete quality Fc '40 Mpa. From this study, it is known that the more mixed the ash, the lower the compressive strength. Each 5% increase in the proportion of rock ash mixture to the weight of sand results in a decrease in strength. For 0% of stone ashes produce Fc '= 40.17 MPa; 5% of stone ash produces Fc '= 39.37 MPa; 10% of stone ash produces Fc '= 38.32 MPa; 15% of stone ash produces Fc '= 37.38 MPa; 20% of rock ash produced Fc '= 36.8 Mpa. From the research carried out the addition of rock ash more than 5% decreased the concrete compressive strength required because by using a mixture of 5% stone ash, the compressive strength of concrete only reached 39.37 Mpa.

1. Introduction

In the field of construction, the material that is often used is concrete. The use of concrete is the main choice because concrete is a basic material that is easily formed at a relatively cheap price compared to other construction. Concrete is a mixture of Portland cement, coarse aggregate, fine aggregate and water. Concrete hardening will occur immediately because of the bond between water and cement, where the mass of concrete will grow stronger with increasing age of concrete. In making concrete, the selection of materials used is very important, especially for obtaining concrete quality with special properties, which desired for certain purposes in the most economical way.

Along with the scarcity of sand raw materials, nowadays in some concrete construction works, sand is replaced with stone ash, stone ash is the result of processing broken stones using stone crusher. Stone ash is currently a by-product in the industry, the amount of which is not small. At present stone ash is not very salable for sale because the use in the construction industry is very little considering the use of sand as fine aggregate is still used for concrete mixes. For this reason, a normal concrete composition is tested by testing the compressive strength of concrete by using stone ash as a substitute for sand for concrete mixture.

According to [4] that the more mixture of rock ash, the lower the compressive strength. Every 20% increase in the proportion of rock ash mixture to the weight of sand results in a decrease in strength. For 0%

of stone ashes, Fc '= 20.67 MPa; 20% of stone ash produced Fc '= 19.44 MPa; 40% of stone ash produced Fc '= 18.14 MPa; 60% of stone ash produces Fc '= 17.03 Mpa; 80% of stone ash produces Fc '= 15.94 MPa; 100% rock ash produces Fc '= 15.01 Mpa. From the research carried out the addition of rock ash more than 20% did not reach the concrete compressive strength required because by using a mixture of 20% stone ash, the compressive strength of the concrete only reached 19.44 Mpa. For this reason, it is necessary to test the compressive strength of the concrete with the composition of adding rock ash below 20% to the weight of sand as a substitute for sand, in order to make a better change in the compressive strength of the concrete mixture.

2. Literature Review

2.1 Cement

Cement is a solid, hard object with a strong compressive strength but weak to tensile, with its base material being: cement, fine aggregate (sand), coarse aggregate (gravel), water and other additives as admixed material (stone ash). Cement is an important bonding material and is widely used in physical construction around civil construction, compounds contained in mothballs cement (CaO), sandililate (SiO2), soilLi (Al2O3), seedsBesi (Fe2O3), magnesium (MgO), sulfur (SO3) , soda or potash (Na2O + K2O). If added, cement water will become a cement paste. If added with fine aggregate, cement paste will become mortar and if combined with coarse aggregate into fresh concrete mixture which after hardening will become hard concrete. The function of cement is to bind aggregate granules until they form a period (solid and fill air cavities between aggregate grains. Cement is a very complex industrial product, with different mixtures and structures, cement can Concrete regulation 1989 (SKBI. 4,53,1989) division of cement porland into five types [9] those are:

a) Type 1
   Porcelain cement which in its use does not require special requirements such as other types, is used for public buildings that do not need to use special requirements.
b) Type II
   Porcelain cement which is in use requires resistance to sulfate and moderate hydration heat. Used for construction of buildings and concrete which are continuously in contact with dirty water or ground water or for foundations that are retained in soil containing aggressive water (sulfate salts) and sewerage or buildings that are directly related to swamps.
c) Type III
   Portland cement in its use requires high initial strength in the initial phase after binding occurs, this type of cement is used in areas with low temperatures, especially areas that have winter (winter season)
d) Type IV
   Porland cement which in its use requires low hydration heat, is used for large jobs, such as weir work, large foundation or other large work
e) Type V
   Porcelain cement which in its use requires high resistance to sulfate, is used for buildings related to seawater, industrial waste water, buildings affected by aggressive chemical gases or vapors and for buildings associated with high percentage of sulfuric groundwater.

2.2 Fine Agregate

Fine aggregate (sand) is a natural mineral that functions as a filler in a concrete mixture that has a grain size of less than 5 mm or passes no. 4 filter and is stuck on filter no.200. fine aggregate (sand) comes from the natural disintegration of rock breakers. Types of sand:

a) Excavated sand
   This class of sand is obtained directly from the ground surface or by digging it first. This sand is usually free of salt content
b) River Sand
   This sand is obtained directly from the river, which is generally fine-grained, rounded due to friction, less adhesion between granules due to strong grain. Because the grain size is small, it is good to use for plastering the walls as well as other necessities.
c) Sea Sand
   Sea sand is sand taken from the beach, the grain is smooth and round due to friction, this sand is the ugliest sand because it contains lots of salt. These salts absorb water from the air and this results in the sand always getting wet. Therefore, it is better for sea sand (beach) not to be used in concrete mixes.

2.3 Coarse Agregate

Coarse aggregates come from natural disintegration of natural rocks or in the form of broken stones produced by stone breakers with a grain size of more than 5 mm or retained by no. 4 filter.

2.4 Water

Water is an important basic material for making concrete. Water is needed to react with cement and as a lubricant between aggregate grains to be easily worked and compacted. Water containing dangerous compounds contaminated with salt, oil, sugar, or other chemicals when used in concrete mixtures will reduce the quality of the concrete and can even change the properties of concrete.

Because cement paste is the result of a chemical reaction between cement and water, it is not the ratio of the amount of water to the total weight of the mixture that is important, but rather the ratio of water to cement or commonly called cement water ratio.Low water content causes hard to work concrete (not easy to flow), and high water content causes low concrete strength. Besides that the excess water will be together with cement moving the surface of the freshly poured fresh concrete and then become foam and form a thin layer known as laitance. This thin membrane will reduce adhesion between the concrete layer and is a weak connection area. If there is a mold leak, water with cement can also come out so there are small nests. Apart from the amount of water, the quality of water must also be maintained, the impurities contained in the water can cause the strength of the concrete and its durability to decrease, the effect on the concrete including the length of the initial bonding time.

2.5 Stone Ash

Stone ash is a building material which is the result of the process of breaking stone slabs that are used for concrete combinations. This aggregate is a mineral filler / filler that passes filter No. 200 size <0.075 mm , obtained from the by-product of a stone-breaking plant. Abu-stone stone can be said to have a lot of volume and is still in the development stage to reduce the use of sand in concrete mixes. smooth. Stone Ash is often used as a side material as a combination of mortar or concrete. Stone ash is easy to obtain and can be considered cheap in terms of price.

High quality concrete testing with a strong pressure plan of 25 MPa, with cylindrical test material 15 x 30 cm, with a total sample of 10 samples, each variation using a mixture of Ash Bat as cement substitute material. Concrete testing was carried out at the age of 7 and 28 days with the results of the optimum pressure strength of 40 Mpa, namely the addition of Aboriginal Ash as much as 12% and at the age of 28 days[5]

2.6 Work Abilitty

This property is a measure of the level of ease of stirring to be complained, transported, poured and compacted. Comparison of the composition of the concrete constituent materials together affects the nature of workability (workability). Elements that affect the nature of ease of work include: addition of cement, gradation of sand and gravel mixtures, use of rounded relief items, and how to compact concrete

2.7 Concrete Slump

Concrete slump experiments are a way to measure concrete stirredness, ie stir / liquid thickness which is useful in concrete work. The lower the slump value indicates that the mixture is getting thicker. Concrete slump checking is intended to determine the consistency of concrete and workability in accordance with the predetermined requirements shown in Figure 1, while the Test Slump values are in Table 1

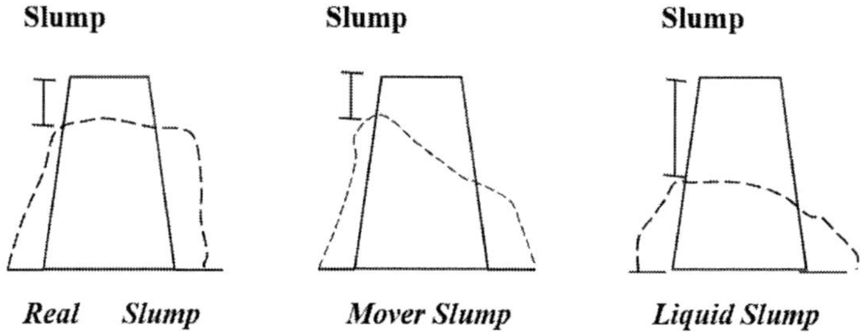

Figure 1. Possibly Slump

Table 1. Slump score

Types	Slump (mm) Max	Slump (mm) Min
Walls, Plates, Foundations and Bones for Bones	125	50
The foundation of the palm is not boned, kaison and underground construction	90	25
Plate, beam, column and wall	150	75
Pavement	75	50
Mass concretion	75	25

Source : PBI 1971

3. Methodology

The research was carried out in the concrete laboratory of the Faculty of Engineering, University of Darul Ulum Jombang. Before making the test specimens for research on the "Alternative of Ash Stone as a Substitute of Sand in Fc '40 Mpa Concrete Mixture Against Workability and Compressive Strength", it is necessary to check the constituent material. The stages for conducting research in the laboratory are carried out in 7 stages, Figure 2

1. Provision of concrete constituents
2. Material inspection

IOP Conf. Series: Materials Science and Engineering **588** (2019) 012045 doi:10.1088/1757-899X/588/1/012045

3. Mix Design
4. Making test specimens
5. Curing
6. Testing the compressive strength of concrete aged 7 days, 14 days, 21 days and 28 days
7. Analysis of Test Results

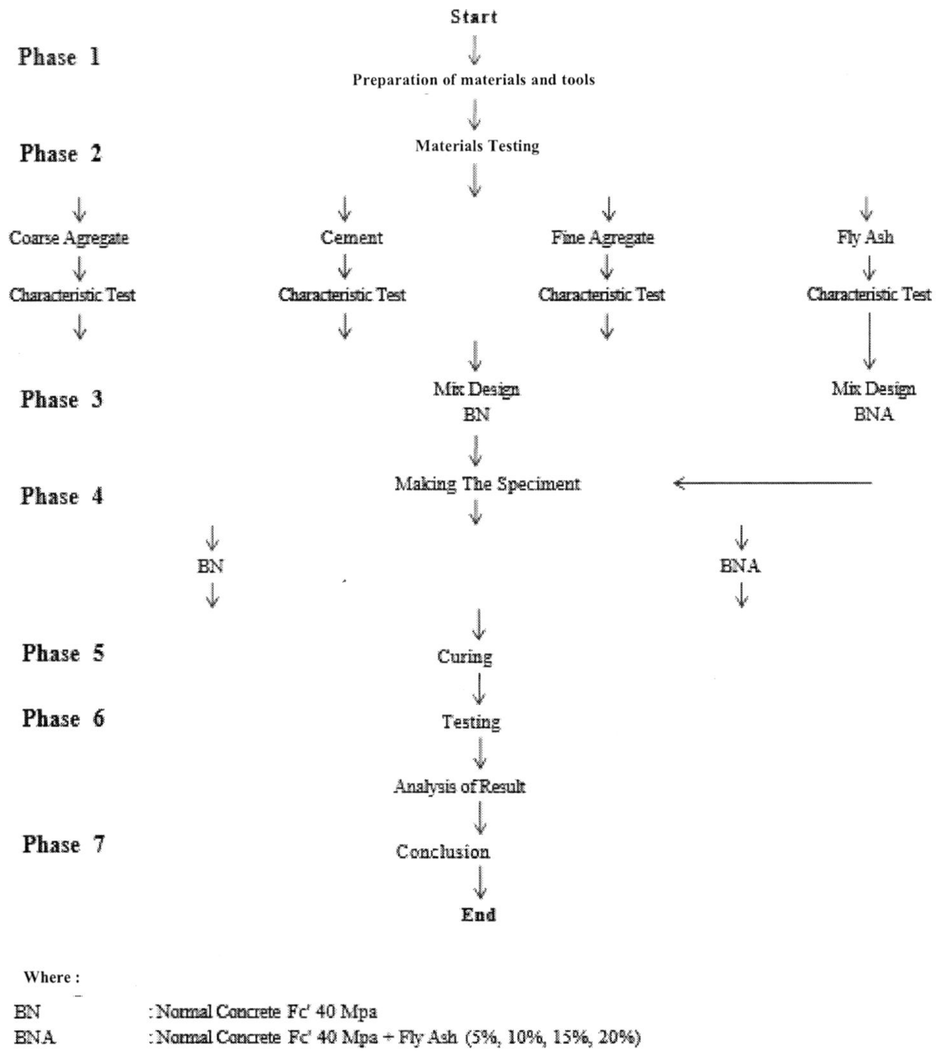

Figure 2. Research Flow Chart

4. Result

4.1 Concrete Mixture

IOP Conf. Series: Materials Science and Engineering **588** (2019) 012045 doi:10.1088/1757-899X/588/1/012045

The results of the proportion of concrete mix design Fc '= 40 MPa in 1.0 m^3 concrete are presented in Table 2 and Table 3.

Table 2. Design of Normal Concrete Mixtures

Parameter	Value
Standard deviation (s)	Mpa
Value-added (m)	10 Mpa
The concrete compressive strength required, at 28 days (from RKS) and fourth sheet) Fc'	40 Mpa
Average compressive strength (Fcr = Fc'+ m)	50 Mpa
Cement type (Choices : normal or quickly hardens)	I
Agregate Type **A type of fine aggregate (choose: natural / fractional)** **B type of coarse aggregate (choose: natural / fraction)**	 **Natural** **Pieces**
Cement water factor (from attachment III and attachment IV)	0,39
Slump value	6 Cm
Maximum size of aggregate grain	2 Cm
Water requirement per cubic meter	189,9 Liter
Cement requirements per cubic meter	487 Kg
Type of fine aggregate (write: 1,2,3 or 4)	Zona 3
The proportion of the weight of the fine aggregate to the mixture	35 %
Mixed aggregate specific gravity	2,675
Estimated weight of concrete per cubic metr	2420 Kg
Mixed aggregate needs per cubic meter of concrete	1743,11 Kg
Need for fine aggregate per cubic meter of concrete	610,09 Kg
The need for coarse aggregate per cubic meter of concrete	1133,02 Kg

Source : SK SNI 03-2847-2002

Table 3. Result of proportion of concrete mixing design

Plan of Making Concrete		**The need for concrete base material**			
Volume	**Weight**	**Water**	**Cement**	**Coarse**	**fine aggregate**
1 m^3	2420 kg	189,9 Ltr	487 kg	610,09 kg	113,02 kg
Stir	12,83 kg	1,01 Ltr	2,58 kg	3,24 kg	6 kg

4.2 Test Press Normal Concrete

The compressive strength of concrete is the amount of load per unit area, which causes the concrete test object to be destroyed when loaded with a certain compressive force, which is produced by the press machine. Testing of normal concrete compressive strength is carried out on cylindrical test objects using Compression Machine compressive strength testing machine. The test was carried out after the concrete reached the ages of 7, 14, 21 and 28 days. As for the results of testing the normal concrete compressive strength can be seen in table 4 and figure 3.

Table 4. Normal Compressive Strength Results

No	(Mpa)	Days
1	26,11	7
2	35,35	14
3	38,16	21
4	40,17	28

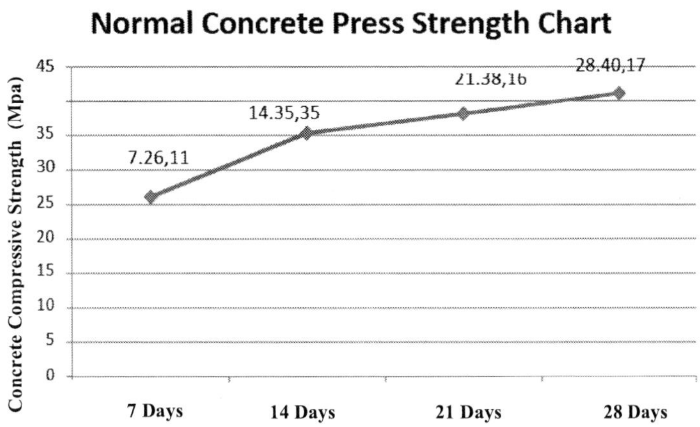

Figure 3. Normal Strength Compressive Graph

4.3 Concrete Press Test With Substitute Ash Stone

The compressive strength of concrete is the amount of load per unit area, which causes the concrete test object to be destroyed when loaded with a certain compressive force, which is produced by the press machine. Testing of concrete compressive strength with rock ash replacement material is carried out on the cylinder test object by using Compression Machine compressive strength testing machine. The test was carried out after the concrete reached the ages of 7, 14, 21 and 28 days. The results of testing the compressive strength of concrete with a substitute for rock ash can be seen in table 5 and figure 4.

Table 5. Normal Compressive Strength Results

No	Stone Ash %	Mpa 7 Days	Mpa 14 Days	Mpa 21 Days	Mpa 28 Days
1	5	25,59	34,64	37,40	38,37
2	10	24,91	33,73	36,41	38,32
3	15	24,30	32,89	35,51	37,38
4	20	23,58	31,92	34,46	36,28

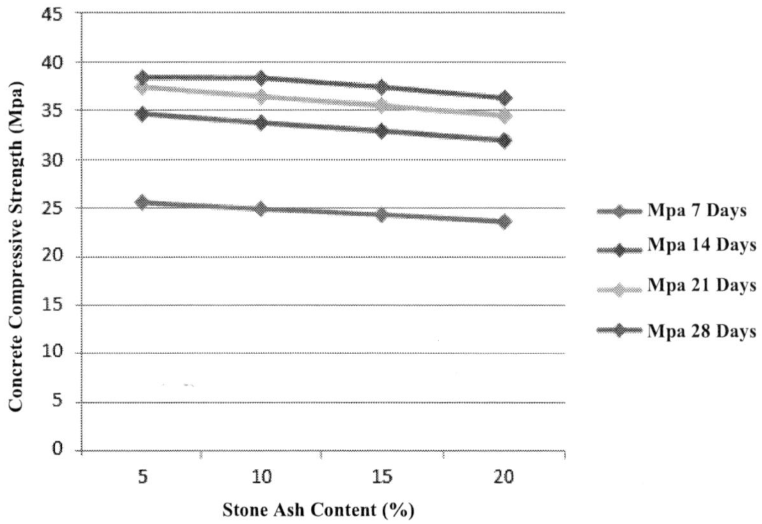

Figure 4. Result Graph of Concrete Press Strength with Substitute Material of Stone Ash

From the graphic above (Figure 4) it can be seen that the results of the replacement of sand with stone ash for concrete compressive strength at 28 days are as follows:

- Replacement of 5% stone ash to the concrete compressive strength decreased by 39.37 MPa this strength was reduced by 0.63% from the normal concrete compressive strength
- Replacement of 10% of rock ash to the compressive strength of the concrete decreased by 38.32 MPa this power was reduced by 1.68% from the normal concrete compressive strength
- Replacement of 15% of rock ash against concrete compressive strength decreased by 37.38 MPa this strength was reduced by 2.62% from normal concrete compressive strength
- Replacement of 20% rock ash to the compressive strength of the concrete decreased by 36.28 MPa this strength was reduced by 3.72% from the normal compressive strength of the concrete

5. Conclusion

From the results of research conducted in the laboratory, the following conclusions can be drawn:

1. From the test results, the normal concrete compressive strength is 40.17 Mpa
2. Replacement of 5% stone ash to concrete compressive strength decreased by 39.37 Mpa this strength was reduced by 0.63% from the normal compressive strength of concrete, replacement of 10% stone ash to the concrete compressive strength decreased by 38.32 Mpa, this power was reduced by 1.68% from normal concrete compressive strength, Replacement of 15% stone ash to concrete compressive strength decreased by 37.38 Mpa this strength was reduced 2.62% from normal concrete compressive strength, 3). The replacement of 20% of rock ash against the compressive strength of the concrete decreased by 36.28 MPa this strength was reduced by 3.72% from the normal compressive strength of the concrete
3. Testing Results Slump Value On Normal Concrete Of 9.5 cm, On Replacement Of 5% stone ash of 8 cm, On Replacement Of 10% stone ash of 7 cm, On Replacement 15% stone ash of 6.5 cm, On Replacement 20% ash stone as big as 6 cm.

References

[1] Antono A 1995 *Concrete Technology* (Yogyakarta : Universitas Atma Jaya)
[2] Antono A 1995 *Civil Engineering Construction Materials* (Yogyakarta : Universitas Atma Jaya)
[3] Dipohusodo I 1994 *Reinforced Concrete Structure* (Jakarta : PT. Gramedia Pustaka Utama)
[4] Kurnyawan D 2014 *The Effect of Stone Ash as a Substitute for Sand for Concrete Making* (Jember : Universitas Jember)

[5] Sukana I 2014 *Effect of Use of Fly Ash and Ash Stone as Partial Substitutes for Cement in High Quality Concrete Press Strength* (Yogyakarta : Universitas Kristen Immanuel)
[6] Arif and Anton 2014 *The Effect of the Use of Fly Ash and Ash Stone as a Substitute for Some Cement in 40 Mpa Quality Concrete Press Strength* (Yogyakarta : UII)
[7] Murdock L J and Brook K M 1986 *Concrete Materials and Practices* (Jakarta : Erlangga)
[8] ASTM C 33-97 Standard specifications for fine aggregate and coarse aggregate
[9] SNI T-15-1990-03 Procedures for Making a Normal Concrete Mix Plan

IOP Conf. Series: Materials Science and Engineering **588** (2019) 012046 doi:10.1088/1757-899X/588/1/012046

Application of savonius turbine behind the propeller as energy source of fishing vessel in indonesia

Madi[1], M E N Sasono[2], Y S Hadiwidodo[1] and S H Sujiatanti[3]

[1] Department of Ocean Engineering, Institut Teknologi Sepuluh Nopember
[1] Department of Technology Management, Institut Teknologi Sepuluh Nopember
[1] Department of Naval Architecture Engineering, Institut Teknologi Sepuluh Nopember

madi.arranim@gmail.com

Abstract. The energy wasted by the propeller of vessel can be gained as benefit by converting it using turbine as the source of electrical energy. The electrical energy can be used as the main energy source of fisherman vessel. The method in this research is redesigning the fisherman vessel combined with savonius turbine to obtain the correct dimension. Vessel dimension is built within Loa = 30.71 m; B = 7.67 m; H= 3.5 m; T = 2.2 m. From those dimensions can be obtained the coefficient between the propeller into the pitch 0.8 m, with the turbine diameter 0.5 m and the position of the turbine 0.1 m from the propeller. Savonius turbine captures electrical energy that produced by the low fluid flow that comes out from the propeller rotation, and will be converted by the generator inside the turbine to become energy. The power produced by the rotation of 1.31 rpm of the turbine is 25.689 kilowatt. It is enough to supply the engine generator with the power of 20.1 kilowatt from the power that produced by the turbine as big as 25.689 kilowatt to lighting the fisherman vessel. This technology can be applied massively to the fishermen vessel as renewable energy.

1. Introduction

Indonesia is a maritime country which still need to be developed especially in infrastructure aspect. Sea area of Indonesia reaches 5.8 million km^2 and the length of the longest beaches in the world after Canada, approximately 81,000 kilometers in length [1]. In addition, Indonesia has abundant natural resources in the sea and a high esthetic value. Ironically, Indonesian coastal communities are still in poverty. One contributing factor is the price of fuel oil which are expensive. The price of fuel without subsidies lead to sea for fishing capital increase, thus reducing expenditures fishing effort, especially in the season when conditions are hostile and famine.

In recent years, oil fields in Indonesia are gradually becoming scarce. With the system of regional autonomy that is running now, it is difficult for foreign oil companies to operate for dealing with small kings in the area. Meanwhile, domestic demand has exceeded production capacity. As a result, scarcity of fossil energy could lead in an energy crisis. Based on data from the Directorate General of Oil and Gas at the Ministry of Energy and Mineral Resources of Republic Indonesia of proven reserves reached 3.741 billion barrels, or 0.3% of world reserves, while potential reserves of 3.666 billion barrels. In other word,

IOP Conf. Series: Materials Science and Engineering **588** (2019) 012046 doi:10.1088/1757-899X/588/1/012046

the fuel consumption rate of the national average reached 40 million kiloliters per year, Indonesia's oil reserves will be exhausted within 10-12 years. Not surprisingly, the title of the country's largest net importer of oil in Southeast Asia had girded Indonesia. Indeed, today is a bleak future, knowing the country is used to provide a large income from oil exports for the community. Based on that data, the government uses a strategic move to raise fuel prices in overcoming the threat of imbalance budget as a result of energy subsidies is too great. However, these measures will make the fishing communities complain due to higher capital to look for fish in the sea. Rising fuel just before the fishing community have spent money amounting to Rp 200,000 for capital ship fuel, especially when fuel prices increased, the issued capital by fishermen will be higher. In fact, revenues generated by the uncertain benefit. As for the impact of the increasing high fuel prices in Indonesia, the coastal communities who are mostly fishermen are getting poorer. This is evident from the condition of the vessel used by fishermen to catch fish at night in conditions of total darkness, without the illumination of the power source. It thus makes the fishermen trouble to find fish in the sea, so the fish gets a little search results and impact on the economic needs of coastal fishing communities.

In addition, based on the facts of a conventional propeller used on fishing vessels until now is still the primary means of ship propulsion. In the workings of the system, the propeller will accelerate the flow from the front to the back part of the ship, so that it will produce two streams are axial and rotational. The flow velocity will determine the size of the loss of energy for ship propulsion systems. Classifies energy that occurs in the system ship propulsion of propeller to 60% of energy used, 20% of energy is lost in the form of axial jet flow, 13% is lost in the form of friction and 7% is lost in the form of stream rotation [2]. Energy lost 40% of the lap behind the propeller in a fishing vessel. Minimizing of axial energy loss by using vane turbine in the propeller slipstream [3]. Based on the above problems, it is expected that the use of renewable energy more efficient, innovative, inexpensive and environmentally friendly, with a touch of technology that is used by the fishermen. Therefore, they invented fishing vessel with the latest designs utilizing renewable energy as a source of energy, with application of savonius turbin behind the propoller. Savonius turbine chosen because have the twisted rotor has very good starting characteristics [4].

2. Problem Formulation and Objectives

Issues to be resolved in this study are:

1. How to get a source of electrical energy is more economical, efficient, effective, and environmentally friendly, for fishing vessels illumination sources ?
2. How to design a fishing vessel as an innovative power generation technologies to harness the energy lost in the rear propeller for lighting a fishing vessel ?

The objectives of this study are:

1. Obtain a source of electric energy more economical, efficient, effective, and environmentally friendly, for the lighting source of fishing vessels.
2. Designing a fishing vessel as an innovative power generation technologies to harness the energy lost in the rear propeller for lighting a fishing vessel.

3. Research Method

In this research,the following steps is :

1. Literature Review and Collecting Data. The study of literature contains a series of search activities and assessment of the sources of relevant and reliable in gathering the material and become a reference in this study.
2. Identification and modelling system. Identifying the existing case, then be re-examined, and produce an assessment conclusions for the initial design process. For the modeling of this fishing vessel, the rear propeller mounted savonius turbine to capture the energy is wasted as shown figure 1:

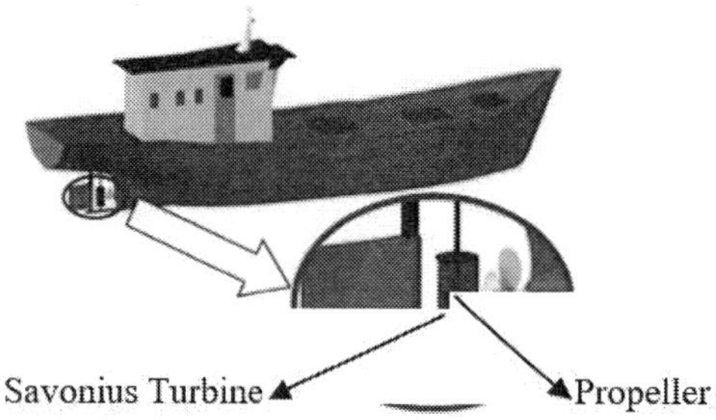

Figure 1. Design of fishing vessel

3. Identification and modelling system. Identifying the existing case, then be re-examined, and produce an assessment conclusions for the initial design process. For the modeling of this fishing vessel, the rear propeller mounted savonius turbine to capture the energy is wasted as shown figure 2 :

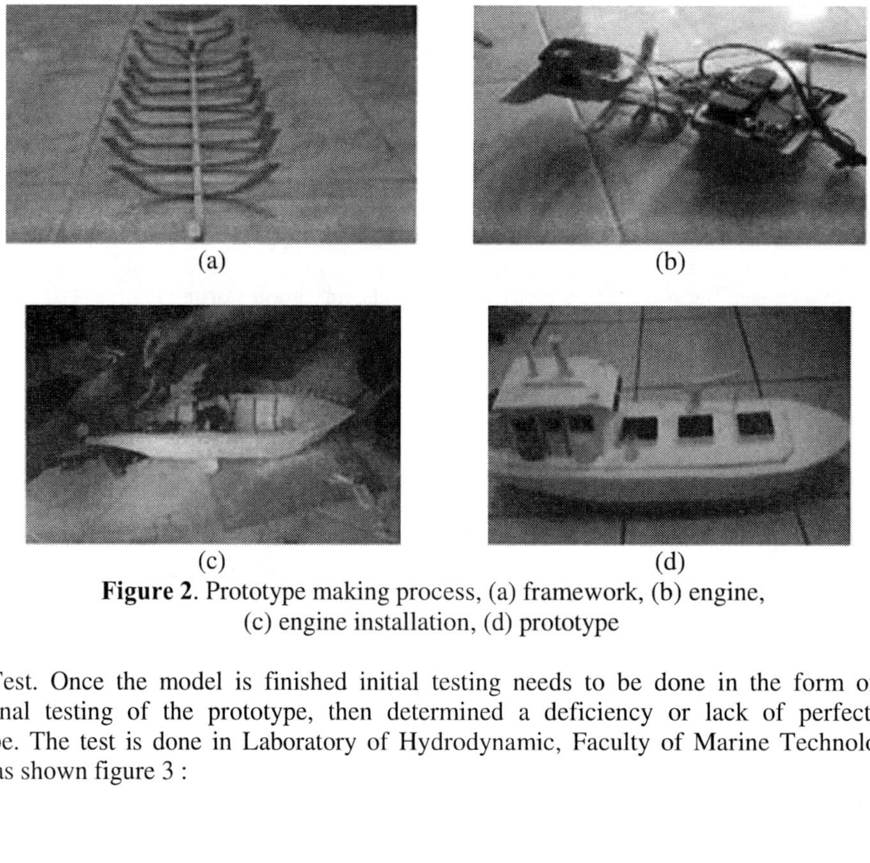

Figure 2. Prototype making process, (a) framework, (b) engine, (c) engine installation, (d) prototype

4. Initial Test. Once the model is finished initial testing needs to be done in the form of an initial operational testing of the prototype, then determined a deficiency or lack of perfection of the prototype. The test is done in Laboratory of Hydrodynamic, Faculty of Marine Technology ITS, is wasted as shown figure 3 :

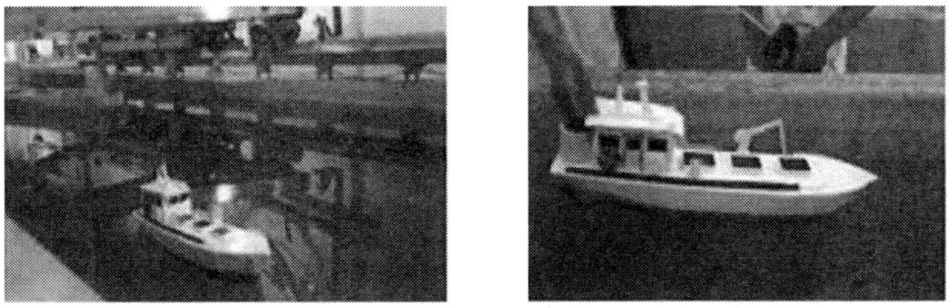

(a) (b)

Figure 3. Fishing vessel trial test in hydrodynamics laboratory ITS, (a) stability test, (b) testing electrical energy result

5. Evaluation and model improvement. Prototype will be refined shape and the system works. Data evaluation results on the initial tests are used as a reference for improvement. The execution-execution minor form of painting and installation of trinkets models will also be done at this stage.
6. Further Testing. In this section, collect data on the effectiveness of the prototype after experiencing improvement. In addition, we tested the feasibility of the system overall work created on the model. Once the model is completed perfected in the previous stage has now arrived for the actual test.
7. Analysis and Discussions. Data from the advanced testing above were used for analysis and discussion on the prototype system as well as the feasibility of using the system. Analysis and discussion carried out to assess the results of design and suitability of the problems encountered.
8. Conclusions and Recommendations. The last process is, the convergence of ideas with the data collection process analysis and calculation, so it can be taken conclusion that aimed to answer the purposes mentioned earlier based on the analysis results. As well as recommendation for various parties in developing a fishing vessel and applications in Indonesia.

4. Literature Review

In this research,the following steps is :

1. Fishing Vessels. Fishing vessels is a ship specially designed to catch fish, which the operation for his arrest is a bit far from the berths and in its operation may take days, so it is equipped with a refrigerated fish boxes that catch is not perishable. Even on vessels equipped with a modern fish canned fish factory. All vessels must have made based on their respective functions as well as the fishing vessel that has a function to catch fish that comes by fishing gear.
2. Propeller. Propeller is a tool for generating thrust is now the most widely used. Propeller rotated by a shaft-driven prime mover in the engine room. There are at least three types of propeller, fixed pitch propeller, propeller with leaves fixed to the casing (nozzle), controllable pitch propeller is wasted as shown figure 4 :

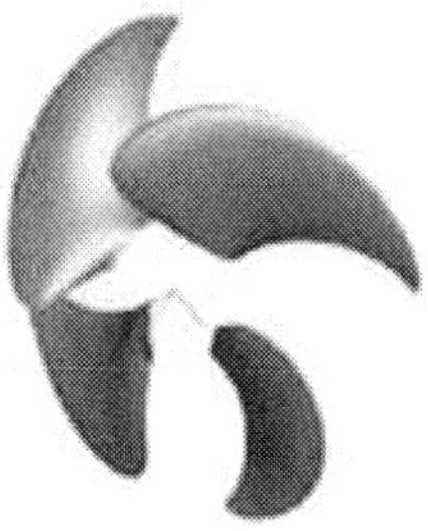

Figure 4 . Vessel propeller [5]

The principle is using a hydraulic system that is, by flowing fluid oil towards a housing located on the propeller shaft, the said housing there is a rotor which is connected to the leaf blades (blade), so if the drained fluid in the forward direction then the oil will push fin dividing the rotor and pushed so that the feather with a certain angle, if the flow direction is reversed, the leaf blades will spin the opposite direction. Operation can be done with two systems, namely, pull-push rod piston system and hub system. In the push-pull rod systems use long rods connected to the hub of the shaft ship propeller. While on the hub system piston, the piston rod is placed in the propeller hub.

3. Propeller Slipstream and Losses Energy. Conventional Propeller, until now it is still the primary means of ship propulsion. In the application, there are still large enough wasted energy in propeller is 40%. Waste stream from the propeller slipstream is referred to as flow. Slipstream propeller flow is very complex and very little has been understood. Leathard, showing the measurement point area by using the model propeller speed KCD 19. The area around the propeller flow is very complicated and is influenced by the following factors: unsteadiness, three dimensionality and high levels of turbulence. Slipstream flow axial velocity can be calculated by using the formula Maynord [6].

$$V(xp)_{max} = 1.45 V_2 \left(\frac{X_p}{D_p} \right) \quad (1)$$

Hydrocom, a software to design the CRP (Contra Rotating Propeller) provides empirical formula to calculate the characteristics of the flow in and out.

4. Sea Currents in Stern of Ship. Sea currents involved is the difference between the speed of the ship with the speed of water flow to the propeller. By the time the ship moves, the friction of water on the surface of the hull causing water layer follows the direction of motion of the vessel. Dividing this difference by the ship's speed or flow rate of water produced two currents join coefficient (wake fraction). This opinion was expressed by Taylor (Taylor wake fraction) and by Froude (Froude wake fraction).
5. Savonius Turbine. Savonius turbine is a tool to convert potential energy into a fluid into mechanical energy. This mechanical energy is then converted into electricity by a generator. Water turbine was developed in the 19th century and is widely used for electric power generation. In a Hydro Electric Power Plant (HEPP) water turbine is the main equipment in addition to the generator.

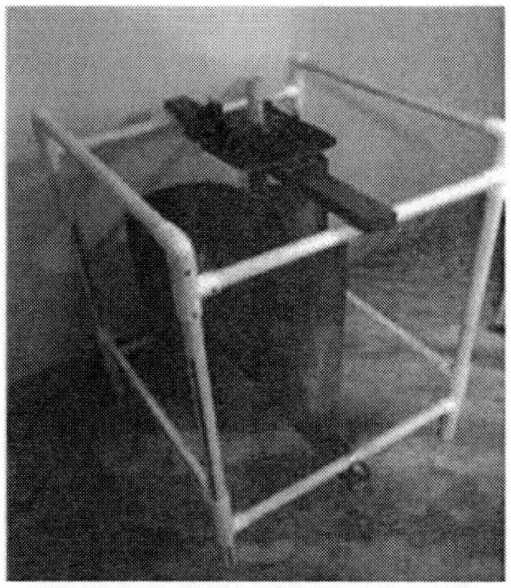

Figure 5 . Savonius turbine [7]

The principle of the water turbine which converts the potential energy of water into mechanical energy. Mechanical energy is converted to electric generator into electrical power. The type of water turbine which is rarely used, namely, the vertical axis Savonius because it has the characteristics of a good starting torsi, easy in making and can work from all directions.

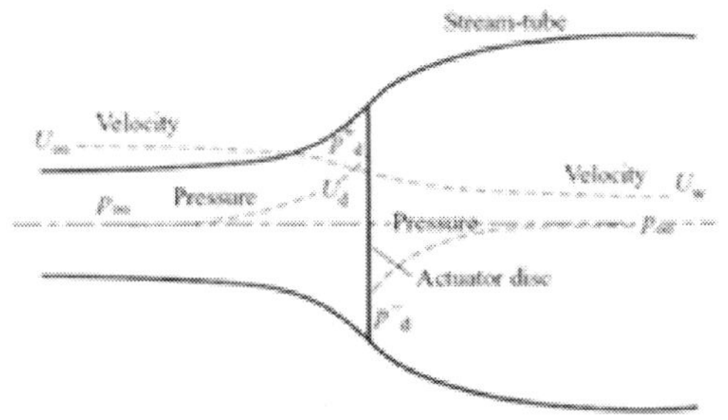

Figure 6 . Turbine analysis by Model Actuator Disc [3]

5. Results

In this study, we will analyze based on laboratory experiments and mathematical analysis. The initial data required vessel in the calculation is as follows:

Table 1. Fishing vessel main dimensiony.

Scale	Quantity and Unit
Loa	30,72 m
Lpp	20,416 m
B (breadth)	7,67 m
T (Draught)	2,2 m
H (high)	3,5 m
Displacement	84,1 m^3
Vs (Speed)	10 knot
CT (coefficient of resistance)	0,0038

1. Ship Resistance. Ship resistance (RT) is the amount of force that must be overcome to move the ship at a speed V_s and ρ is specific mass of sea water.

$$R_T = C_T x0.5x\rho xV_s^2 xS$$
$$R_T = 0.0038\, x0.5x1.025x100x158$$
$$R_T = 30.77kN \quad (2)$$

2. Calculation of Effective Power Vessels. Effective power or EHP is the power required to transform and move the ship in the water or to pull the vessel to the speed (Vs). Known 1 HP = 0,746 kW.

IOP Conf. Series: Materials Science and Engineering **588** (2019) 012046 doi:10.1088/1757-899X/588/1/012046

$$EHP = \frac{R_T}{1000} xV$$

$$= 30.77$$

$$EHPx10\ \sout{1000}$$

$$EHP = 0.308HP$$
$$EHP = 0.230kW \tag{3}$$

3. Ship Resistance. Ship resistance (RT) is the amount of force that must be overcome to move the ship at a speed V_s and ρ is specific mass of sea water.

$$DHP = \frac{EHP}{Pc}$$
$$DHP = \frac{0.308}{0.675}$$
$$DHP = 0.465HP$$
$$DHP = 0.340kW \tag{4}$$

4. Thrust Horse Power Calculation. THP is a big thrust generated by the rotation of the propeller. Where the thrust is being used as a force to move the ship. Null efficiency or $\eta H = 1.072$.

$$THP = \frac{EHP}{\eta H}$$
$$THP = \frac{0.308}{1.072}$$
$$THP = 0.287HP$$
$$THP = 0.214kW \tag{5}$$

5. Speed Calculation Axial Flow Propeller Slipstream Generated. Based on the equation (1) obtained large axial velocity is generated by the propeller slipstream of 16.1 m/s.
6. Calculation of Turbine Savonius. There are some that should be analyzed mathematically in the manufacture of turbines that will be installed in the rear propeller, that is:
   a. Angular Velocity. It involves angle of rotation with $\pi = 3,14$.

$$\omega = \frac{2n\pi}{}$$

$$\omega = 1.31rpm \tag{6}$$

   b. Torque. It needs force and radius of turbin to calculate.

$$T=FR$$
$$T = 19.61Nm \tag{7}$$

c. Turbine Power. The power of Savonius turbine depends on torque and angular velocity.

$$P_T = T\omega$$
$$P_T = 25.689kW$$
$$P_T = 25689W \quad (8)$$

Power generated by turbine of 25.689 kW is enough to supply genset with a power of 20.1 kW

d. Hydraulic Power. Hydraulic power is the power that enters the turbine in the form of the potential power of water.

$$P_T = \rho HD\left(\frac{U^3}{2}\right)$$
$$P_T = 1025\,x\,1.03x0.5x\left(\frac{3.5^3}{2}\right)$$
$$P_T = 12281W \quad (9)$$

e. Turbine Efficiency. Turbine efficiency obtained from the amount of power that comes out (PT) is proportional to the input power (Ph).

$$\eta_T = \frac{Ph}{PT}x100\%$$
$$\eta_T = \frac{12281}{25689}x100\%$$
$$\eta_T = 47.8\% \quad (10)$$

7. Fishing Vessel Cost Comparation. Fishing vessel with savonius turbine compared to conventional fishing vessel in order to know difference in economy aspect. The price used is based on regional price standard of goods and services in Surabaya, Indonesia.

Table 2. Production Cost

Parts	Conventional Fishing Vessel (Rp)	Fishing Vessel with Savonius Turbine (Rp)
Manufacture Hull	1,359,457,000	1.359,457,000
Vessel Equipment	329,825,000	329,825,000
Job Engine	228,750,000	228,750,000
Fishing Equipment	55,500,000	55,500.000
General Cost	149,196,800	149,196,800
Turbine (titanium)	0	2,460,000
Generator	0	23,400,000
Battery (4 units)	0	4,800,000
Total	**2,122,728,800**	**2,153,388,800**
Difference		**30,660,000**

Table 3. Operation Cost

Parts	Conventional Fishing Vessel (Rp)	Fishing Vessel with Savonius Turbine (Rp)
Operation Cost (One Day)	680,700	548,100
Operation Cost (OneMonth)	20,421,000	17,523,000
Operation Cost (One Year)	245,052,000	210,276,000
Save		34,776,000

Based on the table above, in the construction process fishing vesels as new innovative vessels with the addition of Savonius turbines, only requires a fairly low cost. And based on that, we can be able to save operational cost of Rp 34,776,000.

6. Conclusions and Recommendations

Fishing vessel by utilizing the energy lost from round the propeller of fishing vessels were converted by the turbine into electrical energy. The electrical energy produced will be used as a power source for lighting the fishing vessels. With the implementation of fishing vessel that produces energy by 25.689 kW electrical energy source then obtained a more economical, efficient, innovative, and environmentally friendly fishing communities. Fishing vessel applied in the coastal areas of Indonesia, is expected to facilitate the fishing communities for fishing at night to fulfill their work activity. Moreover, it can help the economy of the fishing community to use a fishing vessel that can save Rp 34,776,000.

Water flows from propeller can be an environmentally friendly alternative energy solution for diesel fuel substitute on a fishing vessel. In order to produce this technology massively, it must be supported by every stakeholders which consist of government, private sector, and communities. Concept development and marketing strategy is also the key to sustainable commercialization and good management of technology. It refers to several action such as invesment analysis, product development, market testing, etc.

7. Acknowledgements

This research done by assisstance of several parties. Author thanks to the team in Laboratory of Hydrodynamic, Faculty of Marine Technology, Institut Teknologi Sepuluh Nopember (ITS) for support in making process of vessel design. Another appreciation is also given to all parties that cannot be cannot be mentioned.

References

[1] R Dauhari 2006 Pre-and Post-Tsunami Coastal Planning and Land-Use Policies and Issues in Indonesia *Proceedings of The Workshop on Coastal Area Planning and Management in Asian Tsunami-affected Countries* Thailand

[2] Glauert and Glover 1987 Minimizing axial energy loss by using vane turbine in the propeller slipstrea *Applied mechanics and Material Journal*

[3] S Leksono, I K A P Utama, I M A Djoni and W D Aryawan 2014 Minimizing axial energy loss by using vane turbine in the propeller slipstrea *Applied mechanics and Material Journal*

[4] Nahidul Khan, Masek V, Tariq I 2009 Performance of Savonius rotor as a water current turbine. *Journal of Ocean Technology Maritime and Port Security* **4** (2)

[5] Thomas S 2011 Propeller Design and Propulsion Concepts for Ship Operation in Off-Design Conditions *Second International Symposium on Marine Propulsors smp 11* Germany

[6] Maynord 2000 Slipstream Propeller Jurnal (ISSN: 1660-9333).

[7] A Pradesh and Visakhapatnam 2014 Design and Fabrication of Savonious Wind Mill. P. L. N. V. Aashrith Int *Journal of Engineering Research and Applications* **4** (6) (Version 1) pp 70-79. Dept. of Mechanical Engineering Andhra University College of Engineerin : India.

IOP Conf. Series: Materials Science and Engineering **588** (2019) 012047 doi:10.1088/1757-899X/588/1/012047

Energy harvesting and vibration reduction analysis on cantilever piezoelectric double vibration absorber (CPDVA) mechanism

W Hendrowati, H L Guntur and A A A Daman

Dynamic System and Vibration Laboratory, Department of Mechanical Engineering, Institut Teknologi Sepuluh Nopember (ITS), Surabaya, Indonesia

wiwiek@me.its.ac.id

Abstract. The CPDVA (Cantilever Piezoelectric Double Vibration Absorber) is a mechanism that combines the working principles of DVA and energy harvesting to reduce vibration while generating electrical energy. The working principle of CPDVA uses two masses as an absorber of the vibration system ie main absorber mass and piezoelectric mass. The CPDVA mechamism is designed to get the vibration reduction value at the first natural frequency (36.544rad/s) of system. The simulation results show the optimum value can be achieved if the design of CPDVA mechanism using piezoelctric as much as 4000 pieces. Because, by using 4000 piezoelectric on the CPDVA mechanism, the highest reduction value and electrical power can be achieved.

1. Introduction

Dynamic vibration absorber (DVA) is widely used passive vibration control device. When a mass-spring system, referred to as primary system, is subjected to a harmonic excitation at a constant frequency, its steady state response can be suppressed by attaching a secondary mass-spring system or DVA. However, a DVA consisting of only mass and spring has a narrow operation region and its performance deteriorates significantly when exciting frequency varies. The key design parameters of a damped DVA are its tuning parameter and damping ratio [1]. Pachpute [2], Kefu [3] and Daman [4] have found the response of the system without DVA and with damped DVA. From the experimental results, the system that produced the maximum reduction was Dynamic Vibration Absorber (DVA) with the optimum ratio of mass ratio (μ) = 0.2, the damping ratio (ζ) = 0.125 and the tuning ratio (f) = 0.7 to 0.8.

Research on energy harvesting and simultaneous vibration control device are less reported. The device is used to transfer the primary structural vibrations to the secondary structure, which is DVA. A properly designed DVA absorb all the vibration energy that was input to the primary structure and itself undergoes large vibrations. The two main vibration based energy harvesting technologies are electromagnetic and piezoelectric. The electromagnetic harvester generates power from changes in magnetic field induced due to motion of the host structure [5, 6]. Piezoelectric energy harvester generates power from strain in piezoelectric materials attached to the host, which vibrates in response to external excitations [7, 8, 9]. Patel's research [10] investigated the use of cantilever piezoelectric to determine the effect of piezoelectric length and mass on natural frequency and its deflection, as the deflection affects the generated electrical energy. Liao [11] investigated the effects of placement on damping of various vibration modes, and the effects of patch size on optimal placement. For small piezoelectric patches, the optimal location is very close to the location of the overall maximum bending moment.

G.A. Hassaan [12] conducted a DVA system research to reduce vibration in the main system and serve as a harvester of electrical energy. It is found that the optimum natural frequency of absorber-harvester with mass ratio between 0.05 to 0.45 and damping ratio between 0.1 to 0.4. W. Similarly, Mineto's research [13], de Silva [14] which states that the greatest electrical energy occurs at a natural frequency.

Rahma Efendy [15] examined the application of the mechanism of CPVA (Cantilever Piezoelectric Vibration Absorber) using a simulation method. The results of the simulation show that the generated power and the highest CPVA reduction percentage is 3.52E-7 watt and 20,36% at natural frequency. Besides Wiwiek Hendrowati [16] also examined piezoelectric applications in the form of Multilayer Piezoelectric to harvest the kinetic energy of shock absorber. By pairing the Multilayer Piezoelectric Vibration Energy Harvesting (ML PZT VEH) mechanism to shock absorber, the shock absorber performance is undisturbed and the wasted energy can be utilized into electrical energy, i.e. 6.23 volts and 1.6 m Watt.

This paper discusses the analysis of vibration reduction and the electrical energy generated by CPDVA mechanism. The CPDVA (Cantilever Piezoelectric Double Vibration Absorber) is a mechanism that combines the working principles of DVA and energy harvesting to reduce vibration while generating electrical energy. The design of this mechanism consists of mass, spring and piezoelectric. It is said that the double vibration absorber is because the working principle of DVA uses two masses as an absorber of the vibration system.

2. Methodology

In this paper, a vibration simulator (a system without CPDVA) is modeled with the main mass M_1, springs K_1 and dampers C_1, which is driven by a motor rotating at certain frequency ω and amplitude X_0, as shown in figure 1. Then the input in the form of displacement is forwarded to the main mass M_1 by spring exciter K_0, so that the main mass obtains a harmonic force and will oscillate with amplitude X_1.

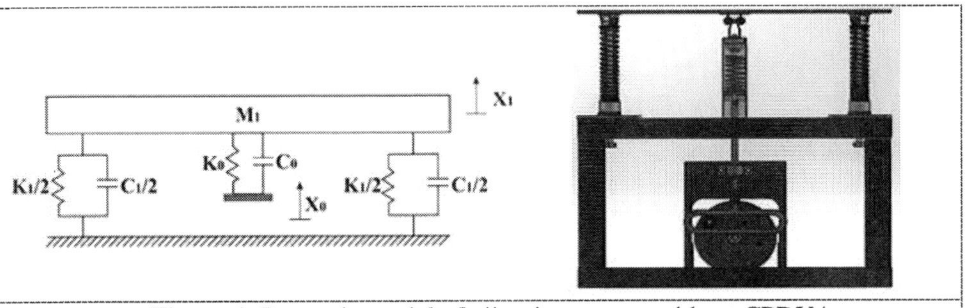

Figure 1. Dynamic model of vibration system without CPDVA

From the free body diagram of the system without CPDVA, the mathematical equations are obtained based on Newton's law as follows :

$$\Sigma F = m_1\ddot{x}_1$$
$$m_1\ddot{x}_1 + Fc_1 + Fk_1 - Fc_0 - Fk_0 = 0$$
$$m_1\ddot{x}_1 + (c_1 + c_0)\dot{x}_1 + (k_1 + k_0)x_1 = c_0\dot{x}_0 + k_0x_0 \tag{1}$$

In figure 2, CPDVA is added to the main system, consisting of main absorber (absorber mass M_3, spring K_3 and damper C_3) and piezoelectric cantilever : piezoelectric mass M_{2eq}, spring K_{2eq} and

IOP Conf. Series: Materials Science and Engineering **588** (2019) 012047 doi:10.1088/1757-899X/588/1/012047

damper C_{2eq}. This CPDVA serves as a vibration damper and an electric generator. Where, Piezoelectric is mounted on the absorber frame and the main absorber.

When the main mass oscillates with an amplitude of X_1, the absorber frame and the main absorber move with an amplitude of X_2. The relative movement between the absober frame and the main absorber will maintain a maximum piezoelectric deflection of 6mm. While the excessive kinetic energy will be absorbed by the main absorber that moves with an amplitude of X_3. The dynamic model of systems with CPDVA can be seen in figure 2.

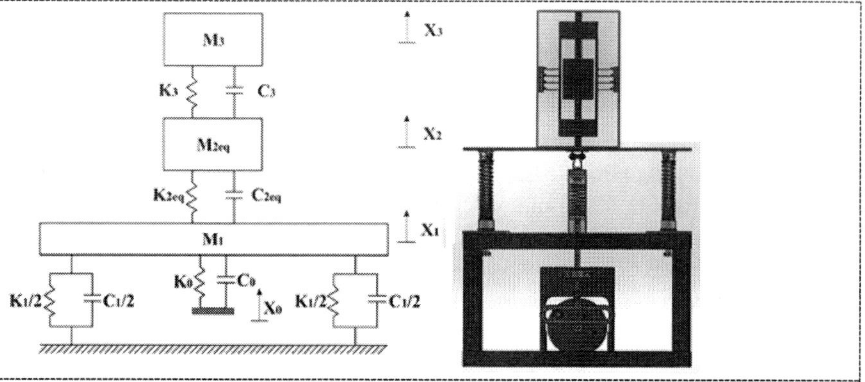

Figure 2. Dynamic model of vibration systems with CPDVA

From the free body diagram of the main system with CPDVA, the mathematical equations are obtained based on newton law as follows:

$$\Sigma F = m_1\ddot{x}_1$$
$$m_1\ddot{x}_1 + Fc_1 + Fk_1 + Fc_{2eq} + Fk_{2eq} + Fe = Fc_0 + Fk_0$$
$$m_1\ddot{x}_1 + (c_1 + c_0 + c_{2eq})\dot{x}_1 + (k_1 + k_0 + k_{2eq})x_1 - c_{2eq}\dot{x}_2 - k_{2eq}x_2 + Fe = c_0\dot{x}_0 + k_0x_0 \quad (2)$$

where
$Fe \quad = n\Gamma k_v(x_1 - x_2)$
$k_{2eq} = nk_p + k_2$
$c_{2eq} \quad = nc_p + c_2$
$m_{2eq} = nm_p + m_2$
$n \quad$ = amount of piezoectric cantilever

The equation of translational motion of the piezoelectric mass and absorber mass are seen in equation

$$\Sigma F = m_{2eq}\ddot{x}_2$$
$$m_{2eq}\ddot{x}_2 + Fc_3 + Fk_3 - Fc_{2eq} - Fk_{2eq} - Fe = 0$$
$$m_{2eq}\ddot{x}_2 + (c_3 + c_{2eq})\dot{x}_2 + (k_3 + k_{2eq})x_2 - c_3\dot{x}_3 - k_3x_3 - c_{2eq}\dot{x}_1 - k_{2eq}x_1 - Fe = 0 \quad (3)$$

$$\Sigma F = m_3\ddot{x}_3$$
$$m_3\ddot{x}_3 - Fc_3 - Fk_3 = 0$$
$$m_3\ddot{x}_3 + c_3\dot{x}_3 + k_3x_3 - c_3\dot{x}_2 - k_3x_2 = 0 \quad (4)$$

IOP Conf. Series: Materials Science and Engineering **588** (2019) 012047 doi:10.1088/1757-899X/588/1/012047

Natural Frequency Analysis

To obtain natural frequency of system, a fundamental analysis is done. In the analysis , the damping constant and the external force are assumed to be zero (extremely small). The value is changed by substituting $\ddot{x} = -\lambda x$ and $\lambda = \omega^2$, so the above equation becomes:

$$\begin{bmatrix} -m_1\omega^2 + (k_0 + k_1 + k_{2eq} + n\Gamma k_v) & -(k_{2eq} + n\Gamma k_v) & 0 \\ -(k_{2eq} + n\Gamma k_v) & -m_{2eq}\omega^2 + (k_{2eq} + k_3 + n\Gamma k_v) & -k_3 \\ 0 & -k_3 & -m_3\omega^2 + k_3 \end{bmatrix} \begin{Bmatrix} x_1 \\ x_2 \\ x_3 \end{Bmatrix} = \begin{Bmatrix} 0 \\ 0 \\ 0 \end{Bmatrix} \quad (5)$$

The deflection that occurs in piezoelectric will produce electrical voltage which is proportional to the deflection, according to equation 7.

$$V_p = \frac{3d_{31}E_p w_p t_p}{4C_p} x_p = k_v x_p \quad (6)$$

$$\Gamma = \sqrt{k_{31}k_p C_p} \quad (7)$$

Table 1. Parameters of Cantilever Piezoelectric

Parameter	Symbol	Value	unit
Piezoelectric mass	m_p	6×10^{-4}	kg
Piezoelectric Spring Coefficient	k_p	0.575	N/m
Piezoelectric capacitance	C_p	244×10^{-10}	F
Piezoelectric thickness	t_p	1×10^{-4}	m
Piezoelectric Width	w_p	6×10^{-3}	m
Piezoelectric length	l_p	12×10^{-3}	m
Piezoelectric strain constant	d_{31}	1.1×10^{-10}	C/N
Electromechanical Coupling Factor	k_{31}	12	%
Modulus Young	E	3×10^9	N/m^2

3. Analysis And Discussion

A vibration system that is given an excitation force with a certain amplitude and frequency will vibrate. Systems without CPDVA has a natural frequency of 36.544rad/s. And after adding CPDVA to the vibration system, the system has 3 natural frequencies of 31,056rad/s, 46,735rad/s, and 92.12rad/s. The additional CPDVA to the vibration system aims to reduce vibration at the first natural frequency of the system. Figure 3 shows the main mass bode diagram without CPDVA and with CPDVA.

Figure 4 presents, the comparison of the displacement respon of each operating frequency of the system without and with CPDVA. On system without CPDVA, at a frequency of 36,544 rad /s, the main mass has the highest amplitude compared to the amplitude at other frequencies, i.e 31.056rad/s and 46.735rad/s, because the first natural frequency of the system is 36,544 rad / s. While on a system with CPDVA, at this frequency, the main mass has the lowest amplitude compared to the amplitude at other frequencies. This shows that the additional CPDVA can reduce the main system vibration at its first natural frequency.

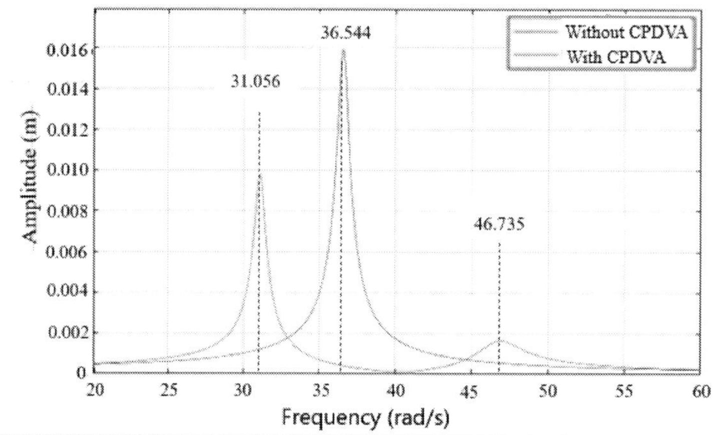

Figure 3. Bode diagram of the system without and with CPDVA

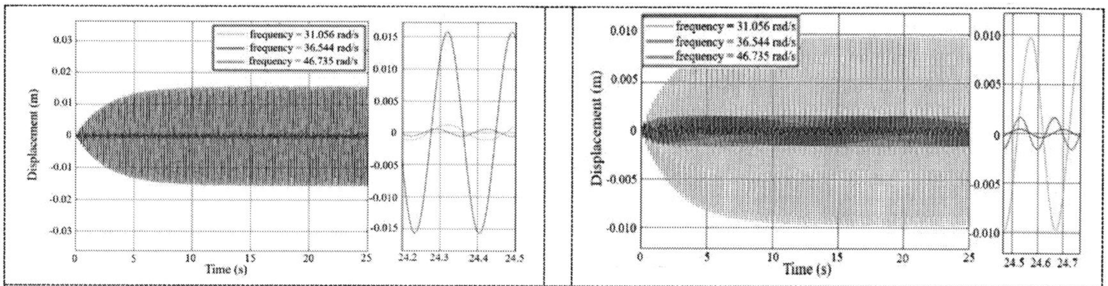

(a) Without CPDVA (b) With CPDVA

Figure 4. Comparison of the displacement response from the main mass

Figure 5. shows the displacements response of the main mass, with and without CPDVA at various operating frequencies. The biggest reduction occurs in the first natural frequency of the system, i.e. at 36.44rad/s.

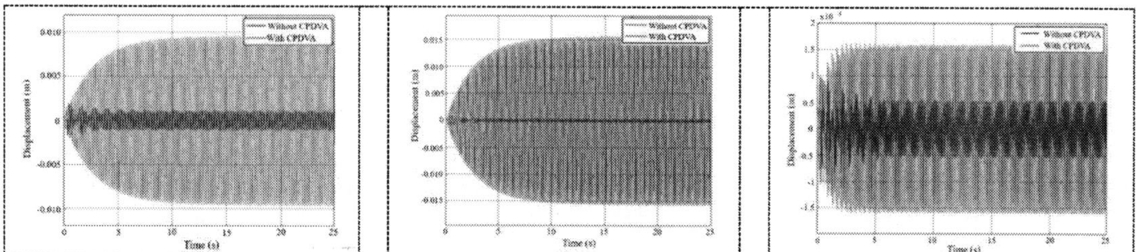

(a) 31.056rad/s (b) 36.544rad/s (c) 46.735rad/s.

Figure 5. The displacements response of the main mass, with and without CPDVA at various operating frequencies

The percentage of reduction in displacement response from the main mass can be calculated as follows

$$\%red = \frac{RMS\ without\ CPDVA - RMS\ with\ CPDVA}{RMS\ without\ CPDVA} \tag{8}$$

The relation of the frequency variation to the reduction is simulated and the result is presented in figure 6.

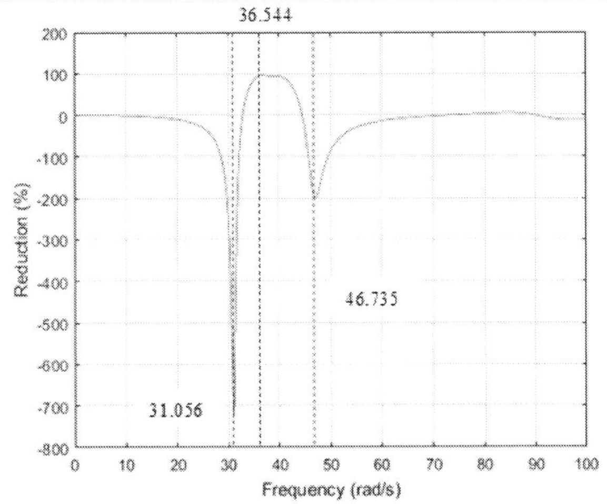

Figure 6. Percentage of reduction at operating frequency variations

From the simulation results, it can be seen that CPDVA is able to reduce the main system vibration with the percentage of reduction value above 0%. The percentage of reduction value below 0% indicates that, the additional CPDVA increases the vibration amplitude of the main mass. Thus, the main mass vibration amplitude that can be reduced is at the frequency interval between 32.91 rad/s to 44.59 rad/s. While the operating frequency is avoided around the frequency of 31,056 rad/s and frequency of 46,735 rad/s because it is in the lowest valley of the graph.

The mechanism of CPDVA which reduces vibration from the main mass will use its energy to drive piezoelectric in the CPDVA mechanism. The response of Piezoelectric is the deflection of the piezoelectric which will produce electrical energy, as in Figure 7.

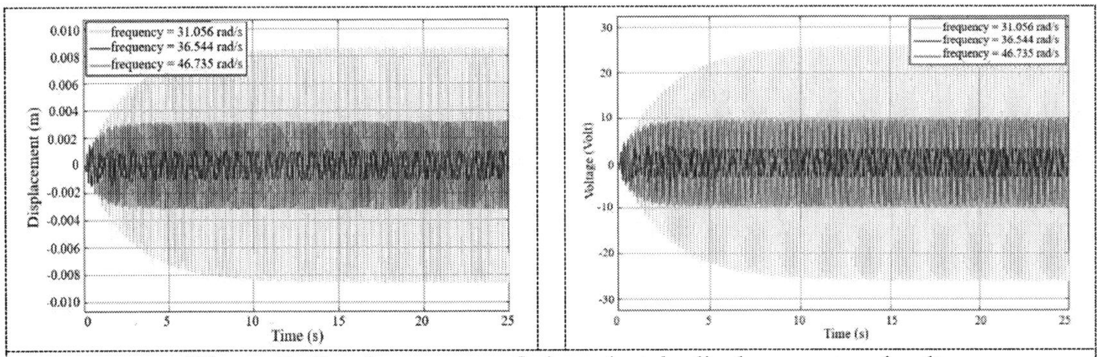

Figure 7. Graph response of piezoelectric displacement and voltage

Figure 8 shows the electrical power that can be produced by the CPDVA mechanism with variations in operating frequency and simulated with number of piezolectric 500 pieces and amplitude of 0.02 m.

IOP Conf. Series: Materials Science and Engineering **588** (2019) 012047 doi:10.1088/1757-899X/588/1/012047

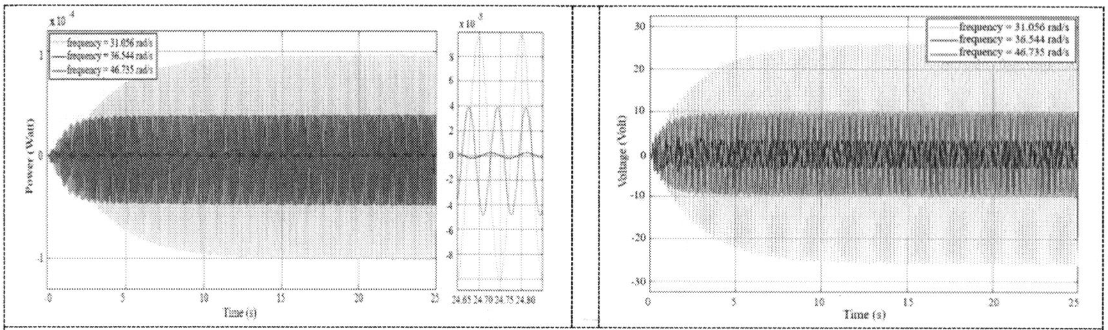

Figure 8. The generated Electrical power from piezoelectric against operating frequency

Figure 9 shows that increasing in the value of reduction will decrease in the generated power. So, the system with CPDVA does not have the optimum frequency to get the largest reduction and power generation value. However, from this graph, it can be concluded that the biggest reduction value is occur at operating frequency is 36,544 rad/s.

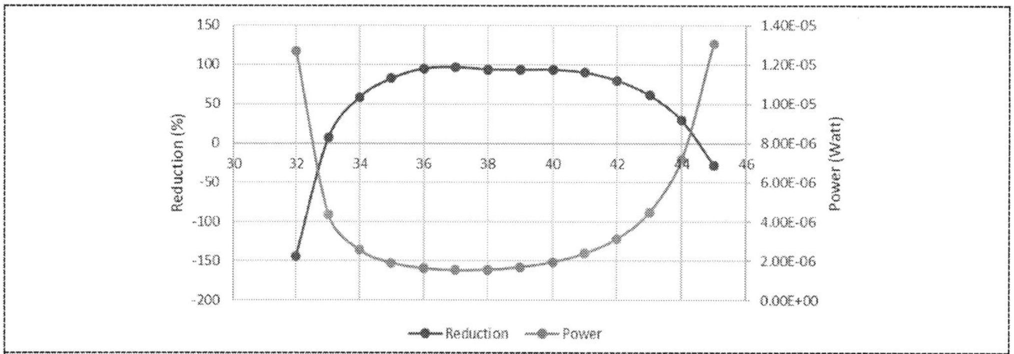

Figure 9. Percentage of reduction and power generation against the operating frequency

The increase in the amplitude of the excitation force to the system only cause an increase in response of the main mass without CPDVA or with CPDVA. Likewise, increasing the excitation force amplitude will not shift the natural frequency value without or with CPDVA. So that the percentage value of the resulting reduction is also relatively the same. This can be seen in figures 10 and 11

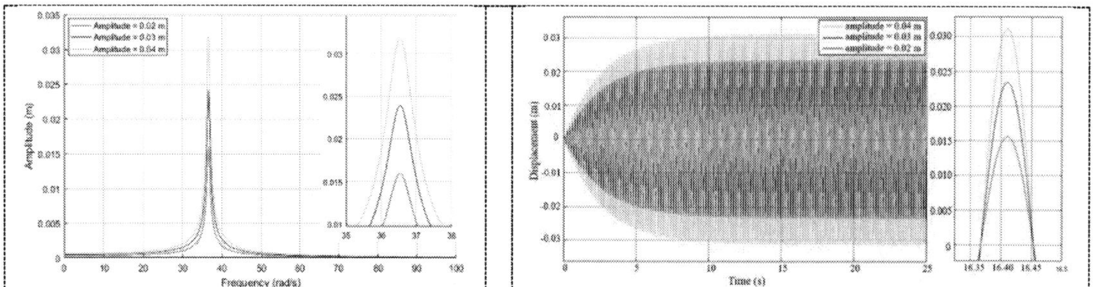

Figure 10. Bode diagram of the main mass with amplitude variations without CPDVA

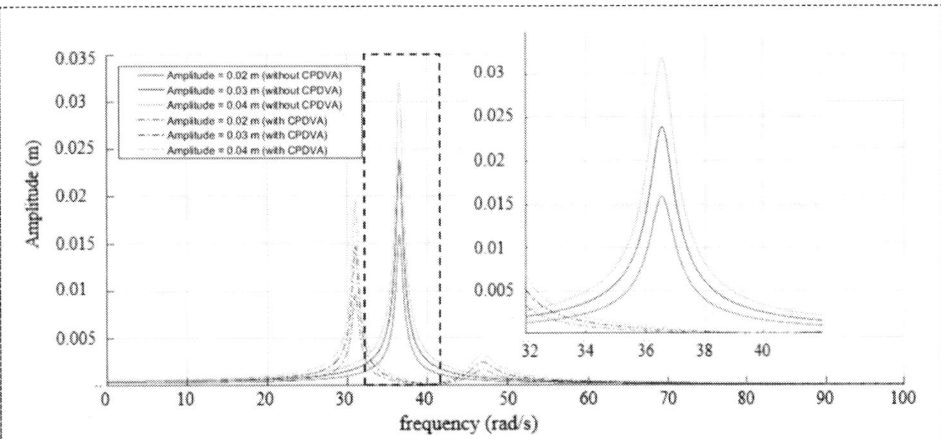

Figure 11. Bode diagram of the main mass with CPDVA

However, increasing the value of amplitude will increase the deflection of piezoelectric and as a results the generated power will also increase at each working frequency, as shown in figure 12.

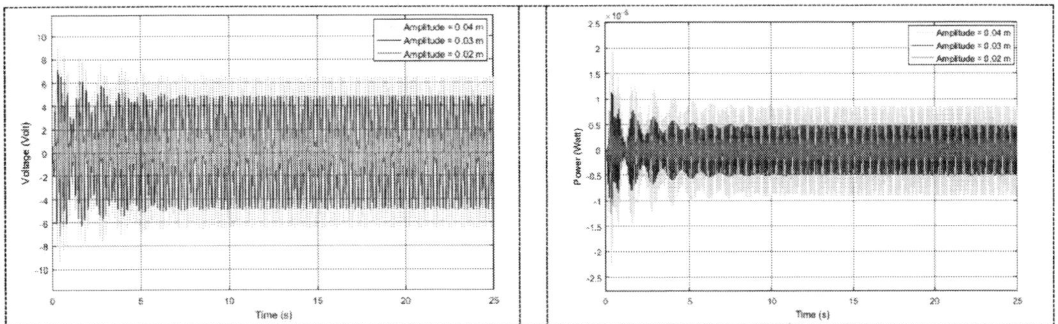

Figure 12. Voltage and power generation of piezoelectric to amplitude variations

The amount of piezoelectric in the CPDVA mechanism will affect the natural frequency. Increasing the amount of piezoelectric will increase the piezoelectric stiffness so that the natural frequency will be achieved faster. This will also affect the percentage of the reduction value. In addition, if more piezoelectric is used, it will increase the generated voltage by the piezoelectric, as shown in Figures 13 and 14.

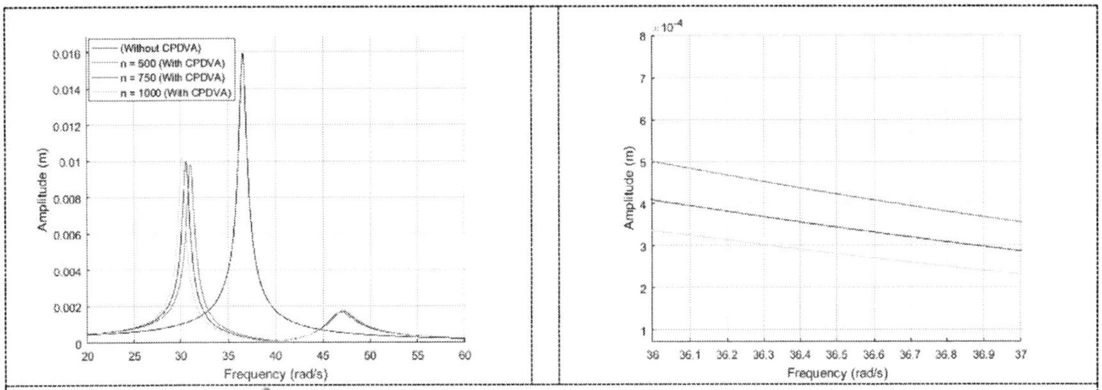

Figure 13. Bode diagram of the main mass with variations in piezoelectric quantities

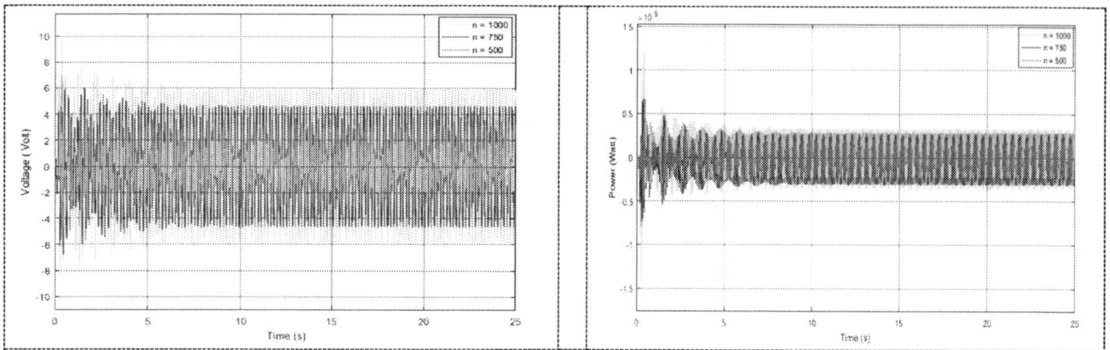

Figure 14. Graph of piezoelectric voltage responses to variations in piezoelectric quantities

From Figure 15. it is shown in the graph that the optimum value is between the reduction value and the electrical power generated by the CPDVA mechanism. The optimum value of the CPDVA mechanism is obtained when using piezoelectric as much as 4000 pieces. The simulation results show that by using 4000 piezoelectric, the highest reduction value and the electrical power produced is also greater.

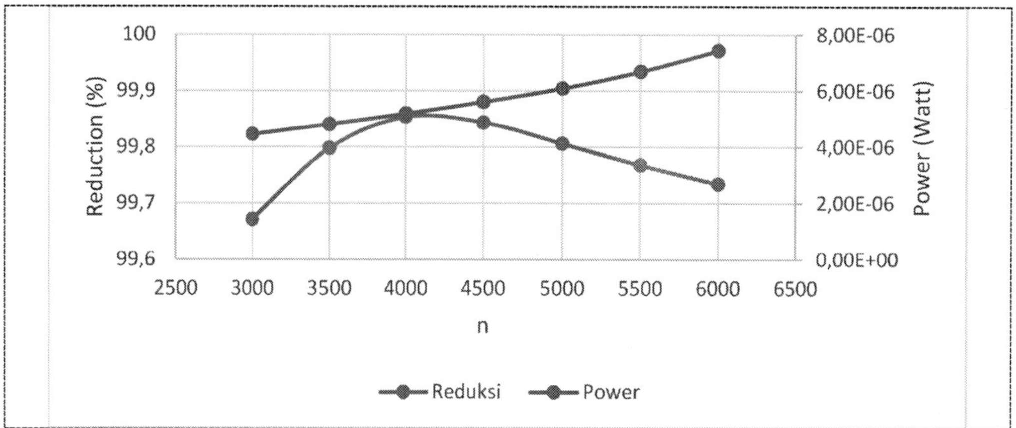

Figure 15. The optimum value of reduction and generated power of the CPDVA mechanism

4. Conclusion

In this investigation, the design of CPDVA mechanism was presented. The CPDVA mechanism has the biggest reduction value at operating frequency is 36,544 rad/s, but increasing the excitation force amplitude does not change the natural frequency value without or with CPDVA. The optimum value of the CPDVA mechanism design is obtained when using piezoelectric as much as 4000 pieces. Because, by using 4000 piezoelectric on the CPDVA mechanism, the highest reduction value and electrical power can be achieved.

5. References

[1] Ormondroyd J and Hartog JD 1928 *Theory of the dynamic vibration absorber* Transactions of the American Society of Mechanical Engineers vol **50** pp 9–22
[2] Pachpute AZ, Bawa PB 2016 *Optimum Design of Damped Dynamic Vibration Absorber* Imperial Journal of Interdisciplinary Research (IJIR)

[3] Kefu L, Gianmarc C 2010 *Optimal Design Of Damped Dynamic Vibration Absorber For Damped Primary Systems* Transactions of the Canadian Society for Mechanical Engineering vol **34** no 1

[4] Daman A A A, Harus L G, Susatro 2016 *The influence of dynamic vibration absorber to reduce the vibration of main system with 2-DoF* Proceedings Of The International Mechanical Engineering And Engineering Education Conferences (IMEEEC 2016) DOI: 10.1063/1.4965762

[5] Williams C and Yates R 1996 *Analysis of a micro-electric generator for micro systems* Sensors and Actuators A-Physical vol **52** pp 8–11

[6] Kulkarni S, Koukharenko E, Torah R, Tudor J, Beeby S, O'Donnell T and Roy S 2008 *Design, fabrication and test of integrated micro-scale vibration-based electromagnetic generator* Sensors and Actuators A-Physical vol **145** pp 336–342

[7] Sodano H, Park G and Inman D 2005 *Comparison of piezoelectric energy harvesting devices for recharging batteries* J.of Intelligent Material Systems and Structures vol **16** pp 799–807

[8] Halvorsen E 2008 *Energy harvesters driven by broad-band random vibrations* J.of Microelectromechanical Systems vol **17** pp 1061–1071

[9] Ali S and Adhikari S 2013 *Energy harvesting dynamic vibration absorbers* Journal of Applied Mechanics ASME 80(4) 041104

[10] Patel R, McWilliam S and Popov A A 2014 *Optimization of piezoelectric cantilever energy harvesters including non-linear effects* Smart Mater. Struct. **23** 085002 pp 17

[11] Liao Y and Sodano H A *Optimal placement of piezoelectric material on a cantilever beam for maximum piezoelectric damping and power harvesting efficiency* Smart Materials and Structures Vol **21** Number 10

[12] Galal A H 2014 *Optimal Design of A Vibration Absorber-Harvester Dynamic System* International Journal of Research in Engineering and Technology (IJRET)

[13] Mineto A T 2010 Modeling of a Cantilever Beam for Piezoelectric Energy Harvesting (Brazil: Departamento de Engenharia Mecânica, Escola de Engenharia de São Carlos, USP)

[14] Silva S D 2012 *Vibration Energy Harvesting Using Piezoelectric Transducer and Non Controlled Rectifiers Circuits* J. of the Braz. Soc. of Mech. Sci. & Eng., vol. XXXIV pp 378-385

[15] Efendy W R and Hendrowati W 2018 *Analysis And Modellingof Reducing Translational Vibration Responseinmain System And Electrical Energy Generated By Dynamic Vibration Absorber Mechanism With Cantilever Piezoelectric (Cpva) Method* Arpn Journal Of Engineering And Applied Sciences Vol **13** No 2

[16] Hendrowati W, Guntur H L, Sutantra I N 2012 *Design, Modeling and Analysis of Implementing a Multilayer Piezoelectric Vibration Energy Harvesting Mechanism in the Vehicle Suspension* Engineering vol **4** pp 728-738 http://dx.doi.org/10.4236/ eng.2012.411094 Published Online (http://www.SciRP.org/journal/eng)

Acknowledgments

Authors wishing to acknowledge support from Penelitian Terapan Unggulan Perguruan Tinggi (PTUPT) that has been funding this study.